20th International Symposium on Aviation Psychology (ISAP 2019)

Dayton, Ohio, USA
7-10 May 2019

ISBN: 978-1-7138-0617-2

Printed from e-media with permission by:

Curran Associates, Inc.
57 Morehouse Lane
Red Hook, NY 12571

Some format issues inherent in the e-media version may also appear in this print version.

Printed with permission by Curran Associates, Inc. (2020)

For permission requests, please contact Wright State University – Conferences & Events at the address below.

Wright State University – Conferences & Events
3640 Colonel Glenn Highway
Dayton, Ohio 45435
USA

Phone: (937) 775-1000

http://www.wright.edu

Additional copies of this publication are available from:

Curran Associates, Inc.
57 Morehouse Lane
Red Hook, NY 12571 USA
Phone: 845-758-0400
Fax: 845-758-2633
Email: curran@proceedings.com
Web: www.proceedings.com

20

International Symposium on Aviation Psychology

For your convenience, the table of contents is linked to the body of the document and is keyword searchable

Accident Investigation

Pilot and ATC Training

Mental Health Issues in Pilots

Managing Air Traffic Challenges

Cognition and Decision Making

A Decade of Lessons From the MTSU NASA FOCUS (Flight Operations Center - Unified Simulation) Lab

Human-Machine Teaming

Wednesday Posters

Designing for Display Support

Safety and Effective Aviation Operations

Physiological Measures in Aviation

Communication and Coordination

Displays for Orientation

Pilot Fatigue and Physiological Stress

Single-Pilot Operations

Visual Displays for Pilots

Real-World Perspectives

Fatigue

Recovery from Unexpected Events

Thursday Posters

Aviation Selection

Helicopter Human Factors

Weather Hazards in General Aviation

Human-Autonomy Teaming: An Evolving Interaction Paradigm

Collegiate Aviation Training

Weather Information for General Aviation

Approaching Machine Common Sense

UNDERSTANDING CURRENT WAYS OF REPORTING RUNWAY INCURSION INCIDENTS AT TOWERED AIRPORTS

Divya Bhargava
Purdue University
West Lafayette, IN.
Karen Marais
Purdue University
West Lafayette, IN.

Runway incursions occur when an unauthorized aircraft, vehicle or pedestrian operates on a runway. While most incursions are near-misses, they have the potential of turning into fatal accidents such as the Tenerife accident. Despite various efforts to reduce runway incursions, the number of incursions has been increasing. Learning from past incidents can help us develop effective preventive strategies but lack of in-depth investigations limits our understanding of the causes of incursions. At towered airports, the controller on duty reports the incident to the FAA using a form asking them to describe the incident in their own words. Our research question is whether the current form contributes to the lack of detail in incident reports. To answer this question, we interviewed controllers and asked how they view incident reporting and the factors they consider while doing so. In this paper, we report the results of interviews with air traffic controllers.

Nomenclature

FAA – Federal Aviation Administration
MOR – Mandatory Occurrence Report
ATSAP – Air Traffic Safety Action Program
ATO – Air Traffic Organization
ATM – Air Traffic Manager
CIC – Controller in charge
OM – Operations Manager
QAQC – Quality Assurance and Quality Control

Runway incursions are a significant threat to runway safety. The FAA defines a runway incursion as an incorrect presence of an aircraft, vehicle, or person on the protected area of the airport surface designated for the landing and takeoff of aircraft (FAA, 2016). The 1977 runway incursion involving two Boeing 747s at Tenerife that resulted in 583 fatalities (Stroeve et al., 2013) remains the deadliest accident in aviation history. An aviation accident is "an occurrence associated with the operation of an aircraft which takes place between the time any person boards the aircraft with the intention of flight until such a time as all such persons have disembarked in which: (a) a person is fatally or seriously injured; or (b) the aircraft sustains damage or sustainable failure; or (c) the aircraft is missing or completely inaccessible" (ICAO, 2001). Considering the potential that runway incursions have of resulting in an accident, the FAA's estimate of about three runway incursions occurring each day in the U.S. is concerning (FAA, 2012a). With 1264 incidents in 2014 to 1832 incidents in 2018, runway incursion incidents have been increasing each year over the last four years (FAA, 2019). Fortunately, most of these runway incursions are near-misses or incidents, that is, an occurrence, other than an accident, associated with the operation of an aircraft that affects or could affect the safety of the operations (ICAO, 2001).

One way to reduce runway incursions is to learn from historic accident and incident data. Runway incursion accident reports by the NTSB often help us understand the causes of accidents in detail before recommending corrective actions. Runway incursion incidents, which occur more frequently than accidents are not always investigated by the NTSB. At towered airports, they are reported by the controller on duty at the time of the incident.

When any surface event, including a runway incursion, occurs at a towered airport, the controller on duty is required to fill out a Mandatory Occurrence Report (MOR). The concerned FAA departments, such as the Air Traffic Organization (ATO), then review these reports and may contact the controller if they feel the incident needs further investigation. These reports help identify patterns in incident causation and provide a focal point for safety discussions. The controllers who fill this form out may not be trained investigators and hence may not always know how to report incidents such that we can learn from them. Some reports are extensive, stating possible underlying causes, while some reports just mention how the incursion occurred, not necessarily specifying why it occurred. Our research question is whether the current incident reporting form discourages controllers from looking further into the incident and reporting all the underlying causes. Additionally, we want to understand how controllers view incident reporting and the factors they consider while reporting the incidents. These insights can help us identify potential ways to help controllers create better reports.To answer these questions, we interviewed air traffic controllers to gain insight into the reporting process and incident reporting in general. We asked them to identify what forms they use for reporting incursions, their experiences using the form, and their opinions on it. In this paper, we first describe the interview method we used to conduct our study. We then present the results of the interviews with controllers and conclude the paper with thoughts on future work.

Interview Method

We developed a semi-structured interview with a list of 18 questions. The questions focused on three areas: (1) what kinds of forms do controllers use; (2) what happens after an incident is reported, that is, who reviews or investigates the incident; and (3) the controllers' experience using the form, and their opinions on it. The semi-structured format allowed the controllers to be more descriptive in their responses and allowed us to ask follow-up questions. The controllers' participation in the study was voluntary. We obtained approval for the research from Purdue's Institutional Review Board.

We requested permission to talk to controllers at three airports in Indiana. We received written confirmation from one of these airports and are waiting on a response from the other two. We have thus far interviewed four controllers in person. The controllers we interviewed were in positions where they were either in charge of reporting events or in managerial positions where they reviewed the reporting of events or assisted in the investigation of events. In the next section, we present a summary of the discussions we had with the controllers.

Results

We categorize the controllers' responses into four broad categories. First, we discuss their views on incident reporting, then give an overview of the reporting process, details of the reporting forms controllers use to report incidents, and finally their views on the forms.

Controllers' Views on Incident Reporting

The controllers all agreed that incident reporting is important. They believe that for smaller airports without equipment to track and record surface movements, reporting of incidents is the only way to become aware of unsafe events occurring at that airport. In their opinion, the change in aviation's safety culture with an emphasis on fixing systemic issues rather than punishing the culprit has motivated them to report incidents. They feel that reporting incidents will help reduce the likelihood of future similar incidents.

The Reporting Process

In this paper, we focus on reporting incidents. The controller in charge (CIC) and the Air Traffic Manager (ATM) at one of our airports gave us an insight into the incident reporting process. Overall the details that the controllers provided on the reporting process conform to FAA guidelines. Controllers fill out a Mandatory Occurrence Report (MOR) using an online tool called CEDAR to provide details of the incident. The FAA's guidelines mention that if controllers do not have access to CEDAR, they can fill out and submit Form 7210-13 instead (FAA, 2012b).

When an incident occurs, the first step is to determine whether it was a significant event or not. The FAA gives a list of potentially significant events, such as security incidents, or when the separation between aircraft is less than 33% of the FAA separation standard. This list of events is not all-inclusive and the FAA guidelines suggest that each situation should be considered based on individual circumstances. In case of a significant event, the CIC indicated that they must report the incident using an MOR within an hour of the event. The person filling out the form must indicate that it is a significant event by checking 'Yes' on Question 4 shown in Figure 1.

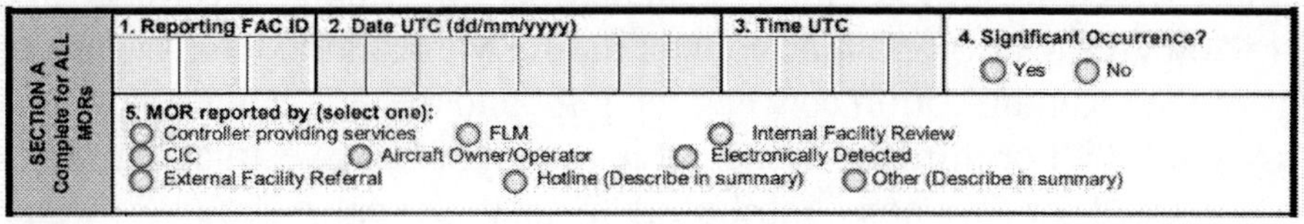
SECTION A Complete for ALL MORs

1. Reporting FAC ID | 2. Date UTC (dd/mm/yyyy) | 3. Time UTC | 4. Significant Occurrence? ○ Yes ○ No

5. MOR reported by (select one):
○ Controller providing services ○ FLM ○ Internal Facility Review
○ CIC ○ Aircraft Owner/Operator ○ Electronically Detected
○ External Facility Referral ○ Hotline (Describe in summary) ○ Other (Describe in summary)

Figure 1. An excerpt of Form 7210-13 that controllers may use to report an incident. The online form has a similar question pertaining to the significance of the event.

The controller on duty must notify at least the supervisor or the CIC. The CIC will get in touch with the ATM, who informs the Regional Operations Center (ROC). The CIC told us that in case of a significant event, they would be calling a lot of people in managerial positions to notify them of the incident, and answering a lot of questions related to the incident.

If the controller selects non-significant on the form, they must complete the MOR before the end of their current shift. This timeline conforms to the FAA guidelines on MORs. One of the controllers provided us with a flow chart that shows the steps involved in submitting an MOR. Figure 2 shows a modified version of the flow chart. The type of MOR controllers fill out depends on the occurrence. As Figure 2 shows, MORs include surface separation, airport environment, and terrain/obstruction. In the case of runway incursions, controllers need to consider whether it was a possible pilot or vehicle deviation. If that were the case, the controller must specify whether a Brasher Warning was issued to the pilot. A Brasher warning is issued to the flight crew instructing them to make a note of the occurrence and collect their thoughts for future coordination with Flight Standards regarding enforcement actions or operator training (FAA, 2012b). If the warning was not issued, the controller needs to explain why not.

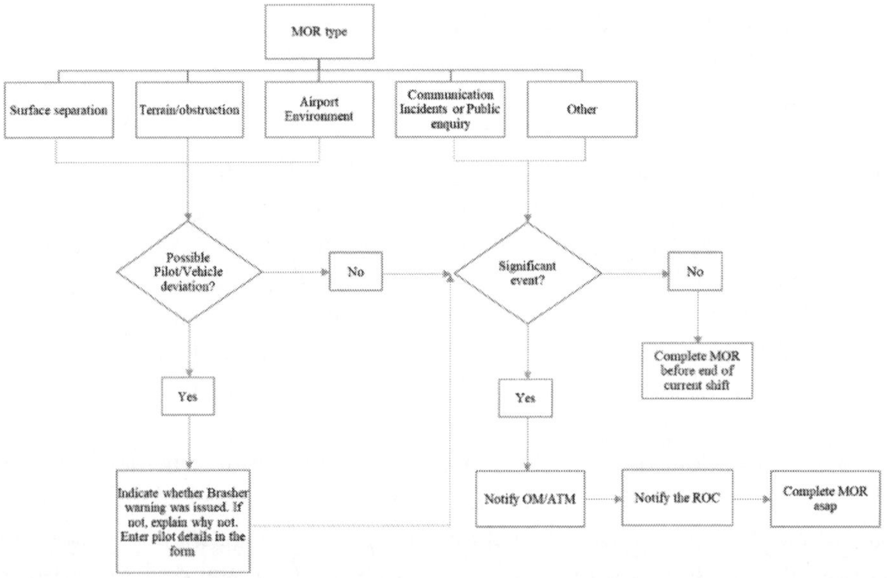

Figure 2. A modified version of the Event Reporting Flow Chart. One of the controllers printed out the original chart for us.

The ATM explained that controllers submit the MOR to the service center (East, West, or Central) under which the airport falls. The department of Quality Assurance and Quality Control (QAQC) reviews the MORs and may contact the ATM for more details. They may ask for recordings of the communication between the pilot and the ATC, interview the CIC, speak to the pilot, or ask the NTSB or third parties to get involved in the investigation. QAQC analyzes the reports to find trends or common contributing factors to events. They may issue an Internal Compliance Validation (ICV) to the ATM, recommending steps to reduce the frequency of unsafe events. The department of QAQC also does random checks to ensure that events are reported. For example, they may review ATSAP reports or pilot-submitted reports to check if a specific event is missing an MOR. In such cases, the controller may lose their job for not reporting the event.

The Reporting Form

The online form has a drop-down menu with types of MOR. The controllers said that the list is quite extensive in terms of types of occurrences, and that the terminology used in the form is easy to understand. The time taken to fill out the MOR depends on how busy they are and the event's significance. Controllers must immediately reports significant events to their manager (OM or ATM). If the event is not significant, the controller may decide to wait till the airport is not too busy. In case they are busy, they may make a note of the event and fill out the form later or call another controller to take over while they fill out the form. One of the controllers stated that non significant events are quick and easy to report. Sometimes, the controller may talk to the pilots involved to find out their side of the story. In case of a pilot deviation, the controllers must give pilot details such as name and licence number for further investigation.

Controllers' Views on The Form

One of the controllers stated that the MORs often help identify *what* happened, while the Air Traffic Safety Action Program (ATSAP) reports help to identify *why* it happened. ATSAP is a non-punitive program that encourages controllers to report incidents. The controllers felt that ATSAP has helped develop a strong safety culture among controllers. One of them said that ATSAP reports are often more detailed than MORs—one of the reasons being that the ATSAP form probes the controllers by asking additional related questions. By referring to the ATSAP reports, the personnel investigating the incident not only understand the incident better but can also identify any cases of under-reporting. For example, if the report submitted by the pilots is vague or the pilot's account contradicts the controller's, the investigators can contact the involved personnel to get more clarity on the event and raise awareness of such issues. The controllers mentioned that they receive training on how to report incidents.

The controllers had varied opinions on the online MOR form. One said that the online form was better than the previously used paper form because it asks only those questions that pertain to the specific type of MOR selected. Another controller said that the form was user friendly and easy to fill out. While these two controllers said that they do not dislike anything about the form in particular, another controller said that the questions the form asks are too basic. They pointed out that the form lacks objectivity as it asks the controller to describe the incident in their own words. The controller said that with an open-ended question, the person filling out the form uses their discretion in reporting the details of the incident. This person may choose to not report certain facts if they think they are not significant enough. Additionally, the controllers filling out the form may be under time pressure, or busy at work, and hence may only report the bare minimum.

Conclusion and Future Work

Our first reseach question was whether the current reporting form discourages controllers from looking further into the incident and reporting all the underlying causes. Two controllers, including a supervisor, explained that they mostly viewed the MOR as a means of reporting *what* happened and not necessarily *why*. The FAA guidelines simply mention that the MOR must be complete enough to describe *what* happened. They said that they tend to provide more details in

ATSAP reports than MORs because the ATSAP questions encourage them to look deeper into the incidents' causes. The simple format of the MOR form may not necessarily encourage controllers to look deeper into the causes–something they are willing to do when asked more detailed questions. Our next research question was to understand how controllers view incident reporting and the factors they consider while reporting incidents. All four controllers we interviewed advocated incident reporting and believe that effective reporting can improve safety. The controllers in managerial positions always supported the FAA in further investigations of an incident and followed the procedures. The biggest factor they consider in reporting is the significance of the event. Not surprisingly, they are more invested if the incident is significant and will look deeper into the underlying causes. The controllers we interviewed did not have conflicting opinions on any of the questions we asked. The reporting process they described agrees with the FAA guidelines. In current work, we are creating an alternative reporting tool that is still quick and easy to fill out, but will encourage deeper consideration of the causes.

Acknowledgments

We thank the controllers for answering our questions and providing us a deeper insight into the reporting process. We also acknowledge the controllers who showed interest in our study and reached out to their colleagues to help us gain varied perspectives.

References

ICAO. (2010). Annex 13, Convention on International Civil Aviation Aircraft Accident and Incident Investigation. International Civil Aviation Organization, Montreal, Quebec, Canada.

ICAO. (2013). Safety Management Manual (SMM). Doc 9859 AN/474. International Civil Aviation Organization, Montreal, Quebec, Canada.

FAA. (2012a). Appendix 1 - Runway Incursion Avoidance. Retrieved from https://www.faa.gov/airports/runway_safety/media/pdf/PHAK%20-%20Appendix%201%20-%20April%202012.pdf

FAA. (2012b). Air Traffic Organization Occurrence Reporting. ORDER JO 7210.632. Effective January 30, 2012.

FAA. (2015). Runway Safety, Runway Incursions. Retrieved from https://www.faa.gov/airports/runway_safety/news/runway_incursions/

FAA. (2019). Runway Safety Statistics. Retreived from https://www.faa.gov/airports/runway_safety/statistics/ [Retreived March 2019].

Stroeve, S. H., Blom, H.A.P., Bakker, G.J. (2013). Contrasting Safety Assessments of a Runway Incursion Scenario: Event Sequence Analysis versus Multi-Agent Dynamic Risk Modelling. Reliability Engineering and System Safety 109, 2013, p. 133- 149.

A CASE STUDY OF TAXIWAY LANDING (1982-2016)

Linfeng Jin
Chien-tsung Lu
Purdue University
West Lafayette, Indiana

The paper reviews the 26 "landing on the taxiway" cases happened between 1982 and 2016 recorded by the National Transportation Safety Board (NTSB) aviation accident/incident database, it evaluates causal and contributing factors such as visibility, navigation, preparation, fatigue, experiences, age and more affecting pilots' operations. Also, personnel injury/fatality and severity of the aircraft damage are extracted from the NTSB's accident/incident databases to conduct the inductive research. Some interesting findings in the paper includes the experienced pilots landing on the taxiway, and different trends of mistakes between general aviation (GA) and commercial operation. Based on these findings, the authors have given several recommendations to mitigate the possibility of landing on the taxiway.

On February 17, 2017, Harrison Ford told the Federal Aviation Administration (FAA) tower controller, "I'm the schmuck landed on the taxiway (BBC, 2017)." The word appeared when Harrison Ford was making a post-incident report to the tower after Mr. Ford landed his Aviat Husky—N89HU on Taxiway C at John Wayne airport instead of the Runway 20L as cleared by the tower controller (Thurber, 2017). "Landing on the taxiway" is rare and peculiar in aviation industry; however, the consequence could be catastrophic due to causality and economic loss, which needs to be studied for the possible prevention programs. This paper will review the 26 "landing on the taxiway" cases happened between 1982 and 2016 recorded by the National Transportation Safety Board (NTSB) aviation accident/incident database, it evaluates causal and contributing factors such as visibility, navigation, preparation, fatigue, experiences, age and more affecting pilots' operations. Also, personnel injury/fatality and severity of the aircraft damage will be extracted from the NTSB's accident/incident databases to conduct the inductive research.

Literature Review

In the 21st Century, air transportation is the safest way to travel; in 2004, the absolute number of 430 fatalities with the respect to1.8 billion passenger-kilometers (Stoop & Kahan, 2005). And in 2013, the World Health Organization (WHO) listed the European region as the lowest fatality rate on road, at 9.3 per 100,000 population (World Health Organization, 2018). In the sky, aviation is safer. According to the Federal Aviation Administration (FAA), the number of life loss in general aviation (GA) has dropped from 471 to 347from fiscal year 2010 to 2016, respectively the GA fatal accidents per 100,000 Hours has dropped from 1.10 to 0.89 per 100,000 hours (FAA, 2018).

There are few research projects concerning landing on the wrong runway or taxiway. In a recent study of landing on the wrong runway, it showed that most of pilots were low time pilots and wrong runway landings often took place in good visibility conditions (Jin & Lo, 2017). It has been shown that way-finding and situation and the sensation of being lost in aviation are the common reasons behind landings on the wrong runways, a comparison of airports in the vicinity of a destination airport and the use of Global Positioning System (GPS) to assist in an identification procedure as a navigation strategies are recommended for the prevention of wrong runway landings (De Voogt & Van Doorn, 2007).

In the aviation history, we have seen several major innovations to avoid aviation accidents, prevent aviation accidents from happening, and improve aviation safety in return. The Swiss Cheese model proposed by Dr. James Reason (1997) attributed aviation accidents into four levels: organizational influences, unsafe supervision, preconditions for unsafe acts, and the unsafe acts themselves, and there are defenses for these four levels, when the defenses were broken up, and the accidents will appear (Reason, 1997). Drawing on the Reason's concepts of active and latent failures, Dr. Shappell and Dr. Wiegmann developed the Human Factors Analysis and Classification System (HFACS) to investigate and analyze human aspects of aviation, and later it was used in training and accident prevention, the HFACS describes four levels of failure: 1. Unsafe Acts, 2. Preconditions for Unsafe Acts, 3. Unsafe Supervision, and 4. Organizational Influences. Under the four levels, there are more finite detailed reasons (Shappell & Wiegmann, 2000).

In this research, the authors chose the SHELL model as the fundamental theory to conduct a case study of taxiway landing accidents, because it does not only look at the human elements (liveware) of accidents, but also other contributory categories like hardware, software, environment and their iterative relationship in a holistic way (Hawkins & Orlady, 1993).

Methodology and Data Analysis

To illustrate the methodology and find answers for research questions, the authors reviewed 26 events and generated a list of contributory factors. Case study was chosen to conduct the inductive study. By definition, a case study is "a method used to study an individual or an institution in a unique setting or situation in as intense and as detailed a manner as possible. (Salkind, 2012) (Leenders & Erskine, 1978)" The unique situation in all the cases are "landing on the taxiway." All the cases were recorded in the NTSB Aviation Accident Database & Synopses by the NTSB investigation professionals. And the events were investigated, and reports were written in a generally standardized way, which guaranteed the inter-rater reliability of the research data including not only factual data like pilot flight hours, local airport weather information etc., but also the probable causes and findings (Salkind, 2012). The authors also read the reports and made the conclusions of probable causes and findings which matched the NTSB ones. The subjects investigated entailed most civil aviation operations in the United States, so the result shall be predictively valid for civil aviation cases (Salkind, 2012).

NTSB Aviation Accident Database & Synopses Searching

The authors searched the key word of "landing on the taxiway" in the NTSB aviation database, it returns totally 47 event records dated between 04/01/1965 and 03/10/2016 involving landing on the taxiway. After the careful reading of each report several times, the authors concluded there are only 26 reports of the events dated between 7/4/1982 and 3/10/2016. From the report, the authors extracted the following variables: event data, number of injuries, degrees of injury (none, minor, serious), death toll, aircraft damage (none, substantial, destroyed), pilot/copilot flight hour, pilot/copilot rating, pilot/copilot age, visibility, light, wind speed, aircraft make, aircraft model, landing gear configuration, runway heading, taxiway heading, operation in terms of aviation regulation, and event factors provided by the NTSB accident investigation personnel in the probable causes and findings. By the interpretation of each report, the authors also came up with additional factors contributed to these events, and divided them into these categories: pilot factors, aircraft factors, and software. Using all the data gathered above, the authors are able to answer the following questions under the framework of SHELL model.

Findings

Using the SHELL model (Hawkins & Orlady, 1993), the authors would like to answer the following questions:

Consequence: What were the aircraft damages, human injuries or fatalities associated with the events?

In terms of aircraft damage, the authors see about 73% (19/26) of aircraft involved with "landing on the taxiway" have experienced the substantial damage, 8% (5/26) of them have experienced no damage, and 7% (2/26) of them have been destroyed. And it is noteworthy that the ones with no damage are all aircraft operated under commercial operations including two Part 121, one part 135, and one foreign air carrier. There were 21 events with no injury or death occurred, three events with one minor injury, one event with one serious injury and one minor injury, and one event with one serious injury and one death.

Liveware: What was the pilots' background related to the events? (Pilot flight hours and ratings etc.)

Table 1.
Count of Pilot Ratings

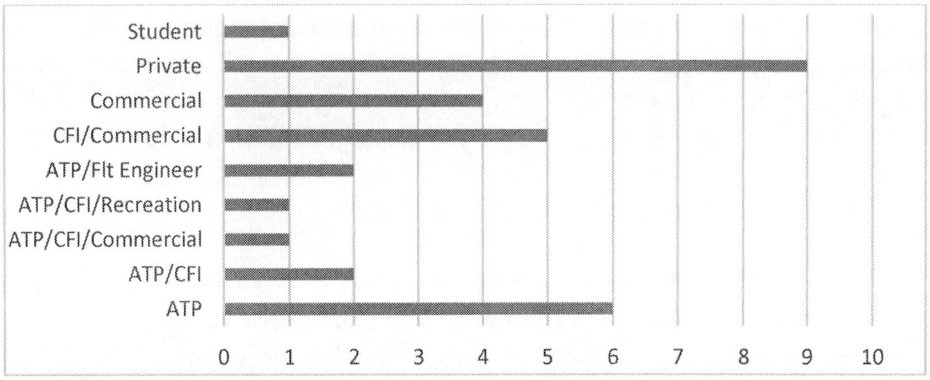

As we can see, there were totally 31 pilots involved with the 26 events of landing on the taxiway. And it is a surprising fact that two thirds of the pilots (21/31) hold ratings commercial or higher, which means they are professional pilot by the FAA standards (FAA, 2019). The average and median flight hour of the pilots are 4917.52 and 2207, and the average age of them is 46.25 years old.

Environment: What were the weather like during the events? (Visibility, Wind Speed, Light Condition)

The average visibility of the events is seventeen statute miles, and there're a couple of recordings of high visibilities, so the authors also counted the mode of the visibility of them: 10 statute miles. And all the five aircraft operated under the commercial operation rules, the airport authority were responsible for not providing clear runway indications for the pilots to land: taxiway lights and runway lights mixing, runways covered by the snow, and water. The average wind speed of the events is eight knots per hour. The light condition distribution was the following: 21 days, one dusk, one night, and one night/bright.

Liveware: What are the pilots' factors in the events?

18 out of 26 events could be identified with the piloting factors involved, and five of them are operated under the commercial operation categories: Part 121, Part 135, and foreign air carrier operation. Surprisingly, all the pilots in the five events had committed the same mistake: selected the wrong runway and landed on the parallel taxiway. The rest of the pilots were operating aircraft under Part 91 and one unknown condition. The authors used the HFACS model to analyze the pilots' factors and divided them into these five categories: skill-based errors, decision-errors, perceptual errors, routine violations, and exceptional violations. The authors found that the primary culprits (16 counts) are skill-based errors which were mainly controlling errors: failing to keep the speed, direction, and attitude. The secondary culprits (7 counts) were perceptual errors which were similar to the transport pilots' mistakes: wrong identification of runway. There were three decision-errors, one routine violation, and no exceptional violation.

Hardware: What were the aircraft conditions in the events? (Aircraft make/model, landing gear types, w/ or w/o mechanic problems)

By their respective aircraft manufacturers and models, there were five transport category aircraft, one helicopter, and twenty GA airplanes. In terms of landing gear configuration, there were one helicopter, two tailwheels and twenty-three tricycles. There were eleven aircraft with mechanic problems that could contributed to the taxiway landing events. And seven of them were related to the aircraft engines; causing the partial or full power loss.

Software: What were the software issue in these events?

There were fifteen events could be identified with the software issues: the operators did not have the knowledge and did not follow the procedure. For the Part 135 or Part 121 or the foreign carrier, the pilots were unfamiliar with the landing procedures and landed good aircraft on the taxiways could be counted as the Controlled Flight into Terrain (CFIT). And we also see one case from the Delta Air Lines that two tired crew landed on the taxiway by mistake after a long international flight due to the lack of fatigue management and one crew incapacitated due to food poison (NTSB, 2010). On the other hand, the GA group in these cases have a variety of software problems: misreading of the Exhaust Gas Temperature (EGT) gauge, poor fuel management, negligence of fuel-sampling in the pre-flight check, misunderstanding of trim controls, lack of situational awareness for the cold weather operation, ignorance of airport landing restriction, lack of situational awareness for the high density/high temperature takeoff, and deviation of landing gear extension procedure. There was also a case that the wrong installation of mixture cable by the mechanic caused the emergency landing on the taxiway after takeoff.

What were other causes of the events?

There was one event of illegal transportation of 250 kilograms of cocaine by air, and the suspect landed aircraft on the taxiway, and the person ran away with the injuries.

Conclusion

From this study, we found that two thirds of taxiway landing pilots had commercial pilot rating or above, and the average and median flight hour of the pilots are 4917.52 and 2207which were different from the previous similar study result that low-time/inexperienced pilots made most of the landings on the wrong runway (Jin & Lo, 2017). And the visibilities were generally good during the events, and eleven GA aircraft with mechanic problem were contributing to the taxiway landings. Finally, we recommend that the commercial pilots should be familiar with the landing procedures like a good memory of key reference points on the destination airports, added simulator training of landing on the new airports, and the application of fatigue risk management to the pilots in the airlines. For GA community, we recommend that the mechanics should take care of aircraft mechanic condition before each flight, and the pilots should maintain situational awareness in the extreme weather environments like hot, humid, cold weathers, and geographic locations like long cross-country flight, plateau operations.

References

BBC. (2017, March 25). *Harrison Ford: 'I'm the Schmuck That Landed on the Taxiway'.* Retrieved January 19, 2019, from The British Broadcasting Corporation Web site: https://www.bbc.com/news/av/world-us-canada-39391431/harrison-ford-i-m-the-schmuck-that-landed-on-the-taxiway

De Voogt, A. J., & Van Doorn, R. R. (2007). Approaches and Landings at Wrong Airports: Analysis of 54 Incidents and 11 Accidents, 1981-2004. *Aviation, Space, and Environmental Medicine*, 117-120.

FAA. (2018, November 26). *Fact Sheet – General Aviation Safety.* Retrieved January 19, 2019, from The Federal Aviation Administration Web site: https://www.faa.gov/news/fact_sheets/news_story.cfm?newsId=21274

FAA. (2019, January 19). *61.133 Commercial Pilot Privileges and Limitations.* Retrieved February 21, 2019, from Legal Information Institute Web site: https://www.law.cornell.edu/cfr/text/14/61.133

Hawkins, F. H., & Orlady, H. W. (1993). *Human Factors in Flight.* Ashgate.

Jin, L., & Lo, E. (2017). A Study of Accidents and Incidents of Landing on Wrong Runways and Wrong Airports. *International Symposium on Aviation Pyschology* (pp. 107-112). Dayton: Wright State University.

Leenders, M. R., & Erskine, J. A. (1978). *Case Research: The Case Writing Process.* London: University of Western Ontario.

NTSB. (2010). *The National Transportation Safety Board Aviation Final Accident Report.* Washington, DC: The National Transportation Safety Board.

NTSB. (2016, October 06). *National Transportation Safety Board Accident Report.* Retrieved from the National Transportation Safety Board Web site: https://app.ntsb.gov/pdfgenerator/ReportGeneratorFile.ashx?EventID=20160304X22545&AKey=1&RType=Final&IType=LA

NTSB. (2019, February 4). *History of The National Transportation Safety Board.* Retrieved February 15, 2019, from The NTSB Web site: https://web.archive.org/web/*/https://www.ntsb.gov/about/history/Pages/default.aspx

Reason, J. T. (1997). *Managing the Risks of Organizational Accidents.* Brookfield: Ashgate Publishing.

Salkind, N. J. (2012). *Exploring Research.* Boston: Pearson.

Shappell, S. A., & Wiegmann, D. A. (2000). *The Human Factors Analysis and Classification System-HFACS.* Springfield: the National Technical Information Service.

Stoop, J. A., & Kahan, J. P. (2005). Flying is the Safest Way to travel: How Aviation was a Pioneer in Independent Accident Inverstigation. *European Journal of Transport and Infrastructure Research*, 115-128.

Thurber, M. (2017, May 16). *FAA Opts Out of Punishment for Harrison Ford Taxiway Landing.* Retrieved from Aviation International News Web site: https://www.ainonline.com/aviation-news/general-aviation/2017-05-16/faa-opts-out-punishment-harrison-ford-taxiway-landing

World Health Organization. (2018, November 26). *Global Health Observatory (GHO) Data-Road Safety.* Retrieved January 19, 2019, from The World Health Organization Website: https://www.who.int/gho/road_safety/en/

Lost of Situation Awareness During Commercial Flight Over Colombian Mountains

OSCAR SARMIENTO MEJIA
CIVIL AVIATION AUTHORITY OF COLOMBIA
BOGOTA COLOMBIA SOUTH AMERICA

facts about the flight:

According to the report received by the commander of the aircraft, the 48 hours prior to the event were instructional flights with high workloads, as they were aspirants who expected to be linked to the company, the assignment of the flight began in Medellin(Colombia) under the condition of additional crew member (tripadi) to the city of Cali, there he made his first flight covering the Cali-Bogota route, without any novelty with the student, they had a stopover in Bogota of approximately three (3) hours and departed to Cali in order to perform an initial operational training (IOT) for the first officer.

This is how the instructor served as pilot monitor (PM) in the left chair of the A-318 and the first officer, co-pilot, was as a pilot flying (PF) in the right chair, it was a night flight taking off at 21:25 Local time, with an approximate arrival itinerary at 22:33 LT it was a dark night, they flew under IFR conditions, during the descent phase performing a standardized arrival procedure called "STAR MANGA8" at 22:33 LT when the aircraft was 40.5 miles from Cali VOR and 13,980 feet above In the area of the central mountain range, in an area where the minimum height for IFR flights was 17,000 feet, the instructor noticed that the FMS (Flight Management System) of the aircraft was badly programmed and it was there when he began to give directions to the first official, to correct the error, which led to the aircraft crossing the MANGA position below the flight level (FL) 170, it was there when the instructor neglected the vertical profile of the aircraft and automatically activated the alarm of the GPWS (Ground Proximity Warning System) when they were approximately 13,980 feet above the mountain terrain, the device turned on indicated in imperative and loud tone "PULL UP" "PULL UP" "TERRAIN" "TERRAIN", in that moment the instructor takes control of the aircraft and declares "the plane is mine" and carried out the evasive maneuver ascending to a flight level of 20,000 feet, there proceeded to level the aircraft and when all the parameters were normalized, he return the controls The student after inquiring about whether or not he was able to continue with the control of the aircraft, upon obtaining an affirmative response, the co-pilot finished the descent and the normal approach to Alfonso Bonilla Aragón Airport that serves the city of Cali Colombia South America .

The instructor said that he was very well trained to face the situation; since the evasive maneuver was executed according to the flight technique and recommendations of the manufacturer of the aircraft, taking into account the confusion factor that the student presented and the low situational awareness that they had at the precise moment when the GPWS He gave the warning notice, The commander is fully aware that no one is exempt from making mistakes, but is committed to further develop competencies that allow him to mitigate the chain of error by being more aware of the situations that may arise, managing better the time in which the flight instruction is given under permanent monitoring of the environment and the lateral and vertical profile of the aircraft by the crew under initial operational training, adhering strictly to the established procedures. " It is striking that when inquiring about the performance of the Air traffic Controller (ATC), he indicated that they was under radar surveillance and the control authorized them to approach; However, if they had radar contact, why did not they warn them about the dangerous descent below the minimum allowed?

According to the statement made by this experienced instructor commander, it is important to clarify that after the event that took place on March 27 of last year year, the assignment of this captain ended with a flight the next day on the CALI-MEDELLIN-JFK route. the city of NY USA; Now according to the Colombian Aeronautical Regulations "114,500 Actions post - accident / serious incident or incident (a) Preventive cessation of aeronautical functions : (1) While assessing the incidence or not of their operational or technical performance in an accident or serious incident; the members of the crew involved will be suspended from any aeronautical activity of land or flight as appropriate, automatically and without any requirement, before which this staff will refrain from exercising the privileges of their license ... ". According to the above, why did the airline not suspend this crew member immediately after this serious incident? once his last assignment was completed, he was unable to make flights because he was on free days. After this, he rejoined his work, but was found to have him on land (GRND) to allow the investigation of this particular event.

On the other hand, analyzing the event itself, carrying out the evasive maneuver did not take much time, just a matter of seconds, because according to its version the procedure that was implemented was to disconnect the autopilot and start an immediate ascent with maximum angle of pitch and maximum power that for the reaction time would take around 30 Seconds, to subsequently level again with 20,000 feet approximately without any conflict with other aircraft to be alone in the area that was flying, if eventually had other aircraft in At the top level, the TICAS of the aircraft would have been activated without any doubt and its version is totally trustworthy.

This flight was under radar surveillance and the control authorized them to approach; However, if they had radar contact, why did the air traffic controller not warn them about the dangerous descent too low of the mínimum flight level in this area allowed?.

According to the statement made by this experienced instructor commander, it is important to clarify that after the event that took place on March 27 last year , the assignment of this captain ended with a flight the next day on the CALI-MEDELLIN-JFK route. the city of NY USA; Now according to the Colombian Aeronautical Regulations "114,500 Actions post - accident or serious incident (a) the crew involved can't return to his flight duties due to Preventive cessation of aeronautical functions : (1) While assessing the incidence or not of their operational or technical performance of the crew in an accident or serious incident ; the members of the crew involved will be suspended from any aeronautical activity of land or flight as appropriate, automatically and without any requirement, before which this staff will refrain from exercising the privileges of their license ... ". According to the above, why did the airline not suspend the crew member immediately after this serious incident? once his last assignment was completed, he was unable to make flights because he was on free days. After this, he rejoined his work, but was found to have him on land (GRND) to allow the investigation of this particular event.

On the other hand, at looking closely the event itself, carrying out the evasive maneuver did not take much time, just a matter of few seconds, because according to its version the procedure that was implemented was to disconnect the autopilot and start an immediate ascent with maximum angle of pitch and maximum power that for the reaction time would take around 30 Seconds, to subsequently level again with 20,000 feet approximately without any conflict with other aircraft to be alone in the area that was flying.

In my opinion there is not any doubt about his version is totally trustworthy

Situation Awareness In Aviation System

Situation awareness (SA) is the perception of environmental elements and events with respect to time or space, the comprehension of their meaning, and the projection of their future status.

"Level 1 SA_ Perception of the elements in the Environment.

A pilot needs to perceive important issues during his flight such as other aircraft ,terrain,system status and warning signals."

"Level 2 SA Comprehension of the current situation.

Its based on a synthesis of level 1 elements, include an understanding of the significance of those elements in light of one's goals .
A novice pilot may be capable of achieving the same level 1 SA as more experienced pilots , but may fall far short of being able to integrate various data elements along with pertinent goals in order to comprehend thje situation as well."

Level 3 SA_ Projection of the future status.

"Amalberti and Deblon(1992) found that a significant portion of experienced pilots time was spent in anticipating possible future ocurrences . This gives them the k nowledge (and time) necessary to decide on the mosty favorable course of action to meet their objetives."

Actions of the airline after facts :

the airline acknowledged that the instructor commander had taken an active part in the reported incident and proceeded with the execution of immediate actions in accordance with Colombian regulations RAC 114, article 114.500.
the pilot was subjected to two periods of flight simulator emphasizing the following aspects: training oriented to the flight line, use and modes of the EGPWS, LOFT with a scenario of correct application of procedures in the vertical handling of approach modes and descent calculations , as well as the handling of workloads with respect to the terrain MEA, MORA, MOCA, workloads with respect to the position of the aircraft.

After satisfactory training you must complete at least 50 hours of route check, once completed you must submit again to training for the validation of the results and obtain the approval to return to the line of flight .

During the leveling of this crew, it was established that their resilience structure was optimal, which facilitated his rapid internalization and corrections of their own errors, reflecting their adequate understanding and risk management.

At the moment, this commander its classified as a safe pilot for the airline's flight line and is currently in operational monitoring although he no longer performs instructor duties.

RECOMMENDATIONS

Commander: follow-up of the process of reincorporation to the flight activity will be carried out, through technical operational reports by the company quarterly during six months of its effective entry to the flight line.

Aeronautical Authority: according to the version of the commander when they were reported with air traffic from the city of Cali they were informed that they were under radar contact and they authorized the approach; However, if they were monitored by radar, why did not the control warn them that they had a flight level lower than that allowed in that area? It would be convenient to investigate if this aircraft, which was the only one in the airspace at that time, was actually being monitored by radar, in which case, because they were not warned that they had descended below the limits allowed for the mountainous area who were flying over at that time?

It would be advisable that the air traffic controller be cited for evaluation by Aeronautical Medicine for the reasons already explained in the aeronautical recommendations.

For the Airline: taking into account that this case was less than 30 seconds from becoming a CFIT accident, the airline is required to provide timely information on some aspects of it, to document the study carried out in order to improve the As far as possible, operational safety by a human factor: the airline is required to review, among other aspects: the training process of its flight crew members, bearing in mind that in this case the student pilot had hardly any operational experience in the flight team. only 35 hours. The rules on the training processes on commercial flights must be improved, although it is written which are the critical phases and the respect that must be had to the sterile cabin, there is still no clarity in this respect in reference to when it should stop imparting the instruction to attend fortuitous situations that may compromise the safety of the flight as in this case.

It would be advisable to cite the air traffic controller to evaluation with Aeronautical Medicine, bearing in mind that he was monitoring the plane under radar surveillance and he not warning about the fact that the plane was not fliging under safe level

Request a copy of the recording of the communications of the controller with the aircraft in its phase of approach to the Alfonso Bonilla Aragón airport in the city of Cali Colombia

Acknowledgements

to my wife and my daughter for their patience and support during the days of preparation of this presentation, to Dra Patricia Barrientos who with her encouragement whom i starting together our first investigations of air accidents and always trust in my professionalism in this field and finally the support that i received of Civil Aviation Authority of Colombia and his goverment that gave me the opportunity to present in this magnanimous event my appreciations on these air safety problems related with human factor , that with the support of the international community, we have to face it in a determined way so that we do not have more funerary technology as I call innovations in the sector because of the serious accidents that take human lives.

References

Mica R. Endsley SA Technologies
Situation Awareness In aviation System
(2007)

Toward a Theory of Situation Awareness in Dynamic Systems
Mica R. Endsley

First Published March 1, 1995 Research Article
https://doi.org/10.1518/001872095779049543

INDIVIDUALIZED LANDING FLARE TRAINING USING BOTH FLIGHT PERFORMANCE AND PSYCHOPHYSIOLOGICAL MEASURES

Jorg Onno Entzinger
The University of Tokyo
Tokyo, Japan

In this paper, we propose the analysis of various measures of eye and heart-rate data in addition to the flown trajectories and landing result, to get a better understanding of a trainee's learning phase and optimize the time spent on and exercises used for the flare training. A problem often experienced is that the trainee knows what to do and tries to a level that (s)he even believes to be doing just that, without actually doing it. Objective, visual feedback can be used to provide the trainee with tangible points to focus on, instead of the often heard comments like "a little too early" or "a little stronger". We present the outcomes of a series of experiments we carried out with students as well as experienced pilots in our fixed base flight simulator.

The landing flare is arguably the most difficult routine maneuver for pilots to learn. As a visuo-motor skill, it is generally learned through practice rather than teaching, while its timing--- just seconds before landing--- is critical and the consequences of errors can be large. A too early or too strong flare will result in floating or ballooning, with the risk of runway overrun or a go-around, while a too late or too weak flare will result in a hard landing or even a crash. To worsen things, the increasing stress (early) trainees experience just before the landing may render flare training ineffective, in particular if they are unable to focus on the necessary cues (which might vary depending on the flare method being taught).

The research presented in this paper is part of the larger "Pilot's Individualized Learning using Objective Data" (PILOD) project. In this project, subjects receive introductory flight simulator training in 6 sessions of ca. 1 hour each. During the first, fourth, and last sessions we record simulator data, eye data, and heartrate data while the subject flies a fixed set of scenarios. We provide the participants personalized feedback with an overview of their progress and suggest some specific training exercises based on the results of the data analysis. In this paper we analyze which factors are related with flare performance, and how we could use these to support trainees to master the flare.

Definitions & Background

Definitions of Glide, Flare, Floating, and Ballooning

In the final approach to landing, the pilot tries to make the airplane descend along a straight line, called the glide slope or glide path (diagonal dotted line in the bottom-left graph of Figure 1). We call this phase the '*Glide*'. If the pilot would maintain this path until hitting the runway, the sinkrate at touchdown would be too high and the pitch may be too low, causing a hard landing, nose wheel landing, or crash. Therefore, shortly before touchdown, the pilot pulls the control column (indicated with the blue triangles in Figure 1) to initiate the '*Flare*', resulting in an increased pitch angle and reduced sinkrate, followed by a smooth but not too soft touchdown.

Ideally, the flare is a smooth transition from the glide to touchdown. In practice however, and particularly in turbulent weather, we generally see one or more of the following phenomena:

- The flare is made in multiple steps (the column is not pulled once, but multiple times). This is mostly not a problem, and may even be desirable in some cases. Only if the flight becomes unstabilized (e.g., large pitch or sinkrate fluctuations due to overcompensated over-control) this is undesirable.
- The flare is somewhat strong, causing the aircraft to '*Float*' over the runway (the thick orange line segments in Figure 1). Floating increases the risk of runway overrun. In this research we define floating as a sinkrate below 100ft/min.
- The flare is too strong, causing '*Ballooning*' (the thick red line segments in Figure 1). This means the sinkrate becomes negative, and the aircraft starts climbing again. Apart from the runway overrun risk, there is a risk of stall, and a risk of hard landing or crash due to over-controlling after the pilot notices the aircraft is ballooning.

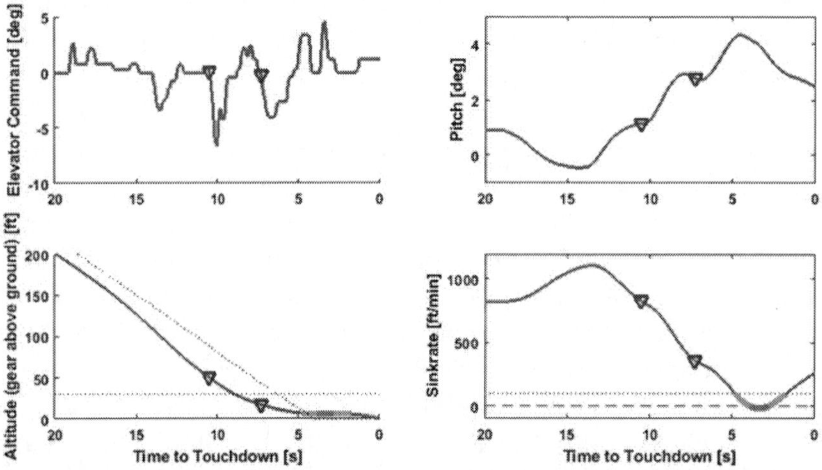

Figure 1. Example of the flare maneuver. In this case, the flare is carried out in 2 steps (the initiation points are indicated by the blue triangles). The floating and balooning regions are indicated by the thick orange and red line segments, respectively.

Cues for Flare Initiation and Guidance

For an acceptable touchdown sinkrate, the flight path should be changed from 3° to 1° or even a little bit less. The flight path can be identified from the out the window view as the center of expansion of the visual scene (Figure 2). If the aircraft is stabilized, it will be the only point in the visual scene which seems motionless. We will call the process of identifying this point and controlling the aircraft to make it coincide with the desired flight path direction `*Aiming*'.

Although the aiming method tells us what to do, it does not tell us when to start doing it, nor how fast to do it. Training manuals generally mention a certain altitude (10~20ft for general aviation, or ±30ft for jet airliners) to start the flare initiation, often with a remark that this should be adapted based on several factors, most importantly the sinkrate. This brings us to the question how to determine altitude (and sinkrate) during the final seconds before touchdown.

A study by Benbassat (2005) showed that the "shape of the runway or runway markings", the "end of the runway or the horizon" and "peripheral vision" are the most commonly used visual cues to determine their altitude above the ground level. Even though the questionnaire covered a relatively homogeneous population at 2 schools, these results seem to be in line with other literature (for an overview, see Entzinger (2010)). Interestingly, the importance of "peripheral vision" is mainly noticed by expert pilots, and almost never mentioned by novices.

(a) 3° path (b) 1° path (c) 0° path (level flight)

Figure 2. The center of expansion of the optical flow field proides a visual cue of the flight path. In the glide phase the pilot should control the aircraft such that the center of expansion coincides with the aiming point markers. In the flare, the pilot should control the aircraft such that center of expansion shifts to a position near the far end of the runway.

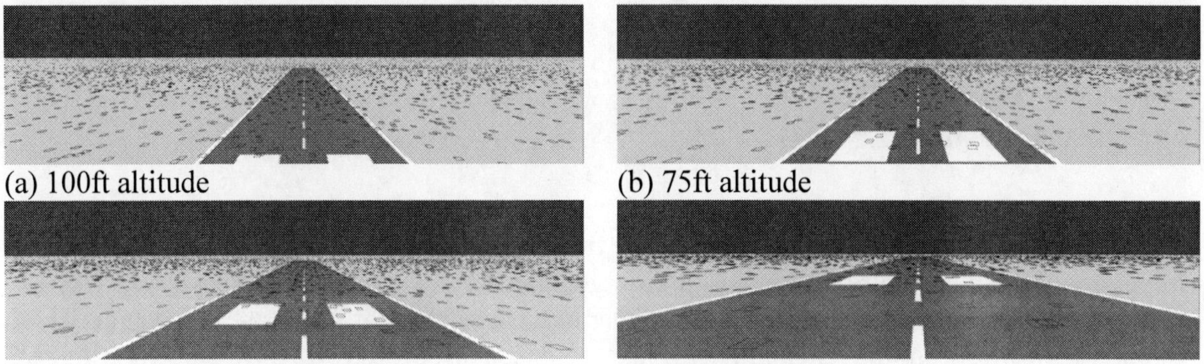

(a) 100ft altitude (b) 75ft altitude

(c) 50ft altitude (on glide slope) (d) 25ft altitude

Figure 3. Visual cues to altitude are provided by the runway shape (apparent angle between the sidelines), the apparent vertical distance between the far runway end and the horizon, and the relative size of nearby texture. (All frames are taken at the same longitudinal position.)

Experiments

Overview & Hypotheses

We consider the following 4 flight performance parameters for the evaluation of the flare: sinkrate at touchdown, amount of floating, amount of ballooning, and longitudinal position of the tochdown point. The last one is largely affected by glide slope deviations, floating, and balloning, and therefore only an indirect objective for training.

There are several reasons to measure not only flight performance, but also a pilot's psychophysiological factors. These factors can help explain *why* the trainee could not achieve a good performance. On the other hand, if a good performance was obtained, they can indicate whether it was due to skill or luck; important additial knowledge considering the limited number of trials we usually get to analyze.

The heart rate (HR) is related to stress and arousal (Roscoe, 1992), while the heart rate variability's power spectrum density (PSD) in the frequency band from 0.06 to 0.14 Hz is said to be suppressed in cases of high mental effort (ME) (Aasman, Mulder, & Mulder, 1987; Vicente, Thornton, & Moray, 1987). Low ME during the flare can indicate panic (if accompanied by a high HR) or giving up (if accompanied by a decreasing HR). We hypothesize that high ME in the flare phase will lead to better flare performance.

The pupil diameter can be used as a measure for cognitive and memory workload (Beatty, 1982; Simpson & Hale, 1969). Recently it has also been shown that the pupil dilates when looking at a peripheral drift motion illusion (Beukema, Olson, Jennings, & Kingdom, 2017). We have seen pupil dilation in pilots and some of the trainees. We hypothesize that a significant increase in pupil diameter during the flare phase will lead to better flare performance. We also hypothesize pupil dilations will be larger when the subject uses peripheral vision.

Subjects

For the main analysis, we use the data of 22 university students, of which 21 male and 1 female. Some of them had flight experience in a glider or hanglider, or played flight simulator games. None had significant experience with the large jet simulator used in the experments.

Five subjects took part in an additional experiment to clarify the effect of peripheral vision use on pupil dilation. One was a retired B747 captain pilot, another one was a student with extensive experience in our simulator, and the other 3 subjects just finished the introductory course of the main experiment.

The experiment protocol was approved in advance by the ethics committee of The University of Tokyo's School of Engineering. Each subject provided written informed consent before participating.

Materials & Methods

The experiments were carried out in the fixed-base Boeing 747 flight simulator at The University of Tokyo (Figure 4). The visuals are generated by Microsoft Flight Simulator 2004, but custom software is used to simulate the dynamics. The simulator states are logged at 20Hz. All approaches were simulated landings to Tokyo Haneda airport runway 34R, starting trimmed and on glide slope. We recorded eye-data at 30 Hz using the Takei TalkEyeLite eye-mark recorder and removed artifacts and outliers. The increase of pupil diameter was calculated as the slope of a straight line fit of the pupil diameter vs. time for the altitude interval 50~10ft.

Electrocardiograms (ECG) were recorded at 250Hz using the ParamaTech EP-301. We then calculated the instantaneous heart rate (HR) and the mental effort (ME) was calculated over a 16s sliding window.

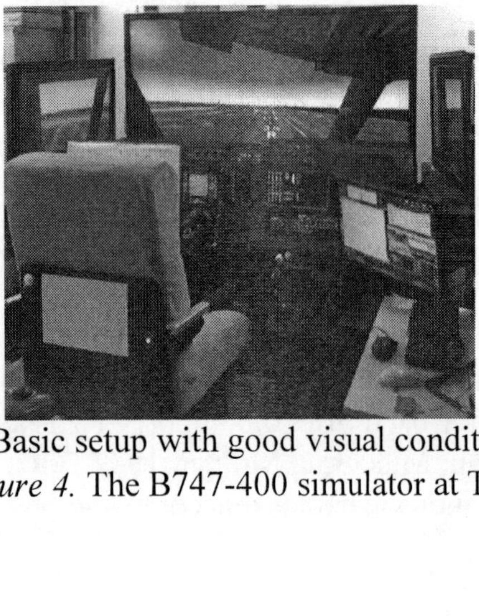

(b) Only peripheral vision

(c) Only foveal vision

(a) Basic setup with good visual conditions

Figure 4. The B747-400 simulator at The University of Tokyo and specific test conditions.

Results

As expected, we saw strong training effects on the flare performance parameters. The sinkrate at touchdown, amount of floating, and touchdown location improved significantly at 1% level and the amount of balooning at 5% level. Figure 5(a) shows that high mental effort results in significantly lower sinkrates at touchdown, as we expected. For the effect of pupil size increase on tochdown sinkrate, we could not find general trend. However, interestingly, there seems to be an oppisite effect before and after training. Subjects who were able to touchdown at consistently low sinkrates before training actually showed a decrease in pupil diameter (Figure 5 (b)). After training, trials with in which a strong pupil size increase was observed had a significantly lower sinkrate at touchdown (Figure 5 (c)).

The additional experiment with different fields of view showed an stronger increase of pupil size during the flare when using only peripheral vision then when everything was visible, but only for the experienced subjects (Figure 6). Subject SRE showed an opposite effect. More experiments will be needed to verify the hypothesis.

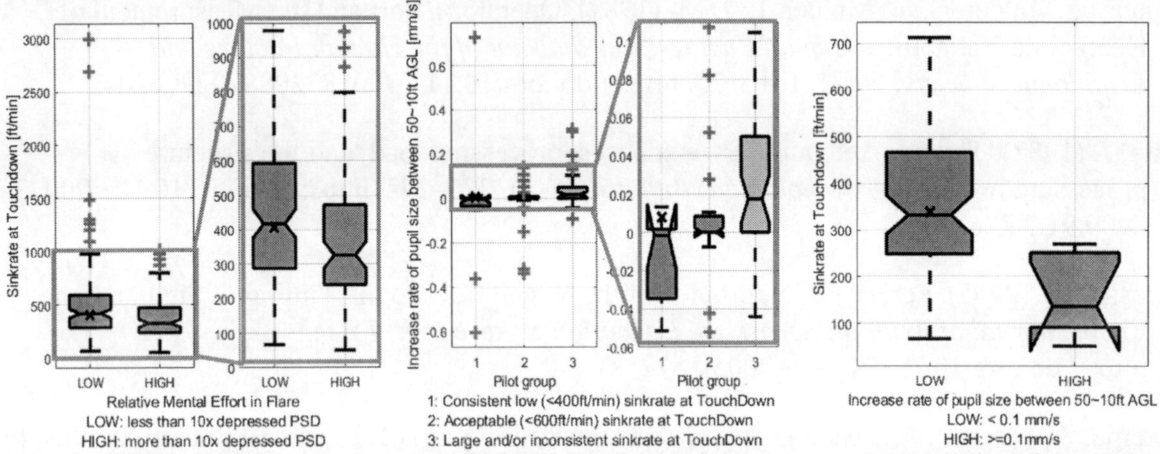

(a) Mental effort. (b) Pupil size before training (c) Pupil size after training

Figure 5. Main experiment analysis results. Boxes span from the 25th to the 75th percentile. Box plots whose notches do not overlap have different medians at the 5% significance level. Red + marks indicate outlier values (outside 1.5 times the interquartile range).

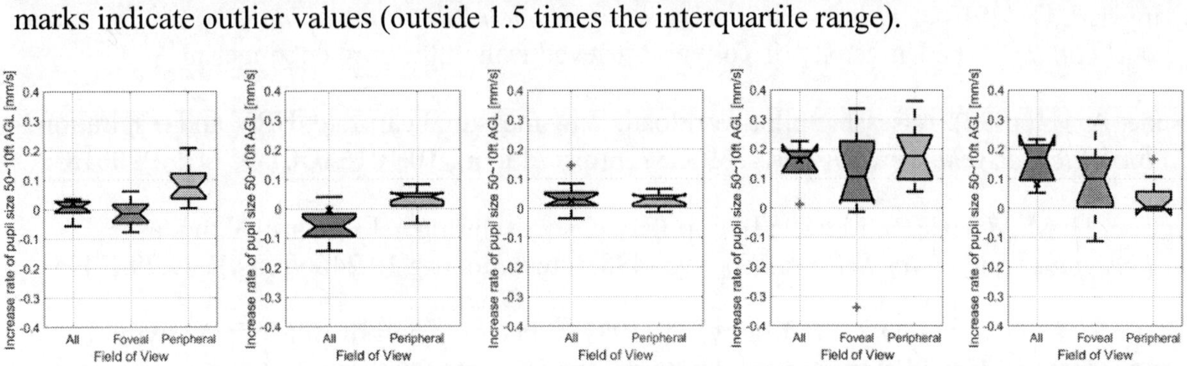

(a) Retired airline pilot (b) Intensively trained subject (c) Subject SSY finished basic training (d) Subject SRD finished basic training (e) Subject SRE finished basic training

Figure 6. Experienced subjects show significantly larger pupil size increases when using only peripheral vision, while this effect is not observed for less experienced subjects.

Discussion

Psychophisiological parameters such as heart rate variability based mental effort and pupil dilation short before touchdown seem to be related to flare performance. These could be used to identify a trainee's needs, such as starting with a simple rule of thumb if mental effort is low, or practicing the use of peripheral vision if pupil dilation is weak. In addition, the results suggest aiming practice may be more useful for beginners, while using peripheral vision to estimate altitude and sink rate may help more experienced subjects to further hone their skills.

Acknowledgements

This research is financially supported by JSPS KAKENHI Grant Number JP16K21002.

References

Aasman, J., Mulder, G., & Mulder, L. J. M. (1987). Operator Effort and the Measurement of Heart-Rate Variability. *Human Factors: The Journal of the Human Factors and Ergonomics Society*, *29*(2), 161–170. https://doi.org/10.1177/001872088702900204

Beatty, J. (1982). Task-evoked pupillary responses, processing load, and the structure of processing resources. *Psychological Bulletin*, *91*(2), 276–292. https://doi.org/10.1037/0033-2909.91.2.276

Benbassat, D. (2005). Perceived Importance of the Visual Cues to the Timing of the Landing Flare. *Journal of Aviation/Aerospace Education & Research*, *15*(1). https://doi.org/10.15394/jaaer.2005.1512

Beukema, S., Olson, J. A., Jennings, B. J., & Kingdom, F. A. A. (2017). Pupil dilation to illusory motion in peripheral drift images: Perception versus reality. *Journal of Vision*, *17*(8), 1. https://doi.org/10.1167/17.8.1

Entzinger, J. O. (2010). *Analysis of Visual Cues for Human Pilot Control in the Final Approach and Landing*. The University of Tokyo. Retrieved from http://jorg.entzinger.nl

Roscoe, A. H. (1992). Assessing pilot workload. Why measure heart rate, HRV and respiration? *Biological Psychology*, *34*(2–3), 259–287. https://doi.org/10.1016/0301-0511(92)90018-P

Simpson, H. M., & Hale, S. M. (1969). Pupillary changes during a decision-making task. *Perceptual and Motor Skills*, *29*(2), 495–498. https://doi.org/10.2466/pms.1969.29.2.495

Vicente, K. J., Thornton, D. C., & Moray, N. (1987). Spectral Analysis of Sinus Arrhythmia: A Measure of Mental Effort. *Human Factors: The Journal of the Human Factors and Ergonomics Society*, *29*(2), 171–182. https://doi.org/10.1177/001872088702900205

DESIGN OF AIR TRAFFIC CONTROL WEATHER RELATED TRAINING PROGRAM

Emma Pierson, Chen Ling, Ali Alshaqah, Joseph James, Shengyong Wang
The University of Akron
Akron, Ohio

Essential components of a new scenario-based air traffic control (ATC) training platform whose effectiveness is being analyzed are outlined with respect to its use in the decision-making skills of trainees when confronted with emergency situations. The custom designed platform allows the trainee to interact with the program such that the 10-minute ramification of a proposed aircraft redirection can be explored visually before a decision is made. Actual previous extreme weather incidences are used. Testing of the platform is ongoing with ATC students from Kent State University. Data from subjective pre- and post-questionnaires as well as objective decision parameters are currently being taken.

The "En-Route Air Traffic Control for Weather-Related Training" program was developed using MATLAB's Graphical User Interface Development Environment (GUIDE). Fundamentally, the purpose of the program is to provide a platform for experimentation on factors that could potentially enhance the weather-related training for ATC personnel. The training program utilizes past radar data from which a weather-related aircraft incident or accident occurred. The research hypothesis states that this method of scenario-based training for ATC personnel will help reduce similar, avoidable mistakes from happening in the future. Further, if an air traffic controller is trained through these types of scenarios, they should be able to extrapolate that knowledge to apply to different, but similar, emergency situations. The National Transportation Safety Board (NTSB) has reported that the recurrent training of air traffic controllers in emergency situations (such as extreme weather) is often short and not specific (NTSB, 2015). Further, the NTSB noted that it "is critical to provide controllers with specific examples to help them identify unstated emergencies and handle aircraft in the safest manner possible" (NTSB, 2015). Using NEXRAD (Next-Generation Radar) data files from previous occurrences of aircraft incidents/accidents to create radar images for the training program has allowed the research team to explore this type of training and plan for ongoing and future experiments.

Design of Training Program

The "En-Route Air Traffic Control for Weather-Related Training" program consists of two main components: the radar image where data blocks representing aircrafts move along their route towards their destinations and the control section where changes to each aircraft's altitude, ground speed, and heading can be made. To create the radar images, weather data files were retrieved from NEXRAD data files – Level II provided by Amazon Web Services Platform. The NEXRAD (Next-Generation Radar) network consists of 160 high-resolution Doppler radar sites. These sites detect precipitation and atmospheric movement and disseminates data in approximately five-minute intervals from each site (Ansari S. et al, 2018). The NEXRAD Level II Doppler data was transformed to imitate the radar images available through the En Route Automation Modernization (ERAM) system. This transformation was completed using the

Python ARM Radar Toolkit (Py-ART). The toolkit is an open source Python module that contains weather radar algorithms and utilities (Helmus & Collis, 2016).

First, the NEXRAD station location and event time must retrieve the data files from the Amazon Web Services platform. Then using Py-ART capability, the weather data is transformed to remove any reflectivity data below 5 dBZ and above 75 dBZ. The data is then classified using three reflectivity level ranges. Weather level 1 (light precipitation) consists of reflectivity levels ranging from 5 dBZ to 29 dBZ and is represented on the radar image using midnight blue (RGB of 25, 25, 112). Weather level 2 (moderate precipitation) consists of reflectivity levels ranging from 30 dBZ to 49 dBZ and is represented on the radar image using green (RGB of 0, 100, 0). Weather level 3 (heavy precipitation) consists of reflectivity levels ranging from 50 dBZ to 75 dBZ and is represented on the radar image using cyan (RGB of 0, 200, 200). If no precipitation is present (reflectivity outside of 5-75 dBZ), the radar screen is black. The colors used for weather level 1, weather level 2, and weather level 3 all correspond to the colors used for ERAM (Federal Aviation Administration, 2013). The radar intensity ranges (dBZ levels) were structured to follow closely with the NEXRAD radar intensity ranges for light, moderate, and heavy precipitation (Boyette, 2006).

Data blocks which represent the position and movement of each aircraft are then added to the radar screen. They consist of an aircraft marker (square symbol) and alphanumeric data that includes the flight identification, altitude, and ground speed. They are designed to provide enough information about the aircraft to make realistic decisions with respect to separation and weather. In addition to the aircraft marker and alphanumeric data, a gray line projects from the top left corner of the aircraft marker as a visual indicator of both the aircraft heading and ground speed. Figure 1 shows the radar image with eight aircrafts present.

Figure 1. Radar image with eight data blocks representing aircrafts.

The program code uses the aircraft heading and ground speed to calculate the horizontal, Equation 1, and vertical, Equation 2, on screen movement of the aircraft marker and alphanumeric data at one second intervals.

$$x_i = x_{i-1} + \Delta x \tag{1}$$

$$y_i = y_{i-1} + \Delta y \tag{2}$$

Where x_i is the new current x-position of the aircraft, x_{i-1} is the x-position of the aircraft one second interval prior, and $\Delta x = GS\sin(H)$, is the change in the x-position of the aircraft during the one second time interval. For the Δx calculation, H is the aircraft heading at that interval and GS is the aircraft ground speed at that interval. All variables for y_i, the new current y-position of the aircraft, are similarly defined with $\Delta y = GS\cos(H)$. In addition, the end-point of the gray line projecting from the top left corner of the aircraft marker shows where the aircraft will be two minutes later if it continues at its current heading and ground speed. The needed position end-point to plot this line is calculated using Equation 3 and Equation 4:

$$x_{ind\_end} = (x_i) + GS\sin(H) \cdot t_{2\,min} \tag{3}$$

$$y_{ind\_end} = (y_i) + GS\cos(H) \cdot t_{2\,min} \tag{4}$$

Where x_{ind_end} and y_{ind_end} are the x- and y-position of the end-point of the plotted direction indicator, respectively, and $t_{2\,min}$ = 120s is the time needed to find the coordinates of this point 2 minutes later. The background code of the computer program then plots the line from (x_i, y_i) to $(x_{ind\_end}, y_{ind\_end})$.

The aircraft marker and alphanumeric data are colored yellow (RGB of 238, 243, 174) to signify that the user is responsible for the safety of that aircraft. This follows the coloring used by the ERAM system to show that a data block is under the responsibility of an air traffic controller (Federal Aviation Administration, 2013). For the purpose of weather-related training experimentation, the data block was designed to be a simplified version of what is available to controllers in the field.

Control Section

During the test scenario the program allows the participant to have control over each aircraft's altitude, ground speed, and heading. Also available to the user of the program is a predictive tool that visualizes the next ten minutes of an aircraft's route. Each aircraft that is included in the test scenario is shown in the control section of the program. To make changes to an aircraft, the participant must select the radio button next to the flight identification of that aircraft. To reduce the number of mistakes made during a test scenario, when an aircraft is selected in the control section the aircraft marker and the alphanumeric data of the corresponding data block changes from yellow to red. This color change allows the participant to confirm the changes are being made to the correct aircraft on the radar image. Once an aircraft is selected, the edit boxes below the radio buttons fill with the information of that aircraft. The user can then make changes to any parameter that will continue to ensure the safety of the aircraft. These changes are applied to the equation that controls the movement of the data block on the radar image. The control section of the program is located to the right of the radar image. Figures 2(a) and 2(b) show the components of the control section display that allow for the decision-making of the participant and the corresponding changes to the data block on the radar image, respectively.

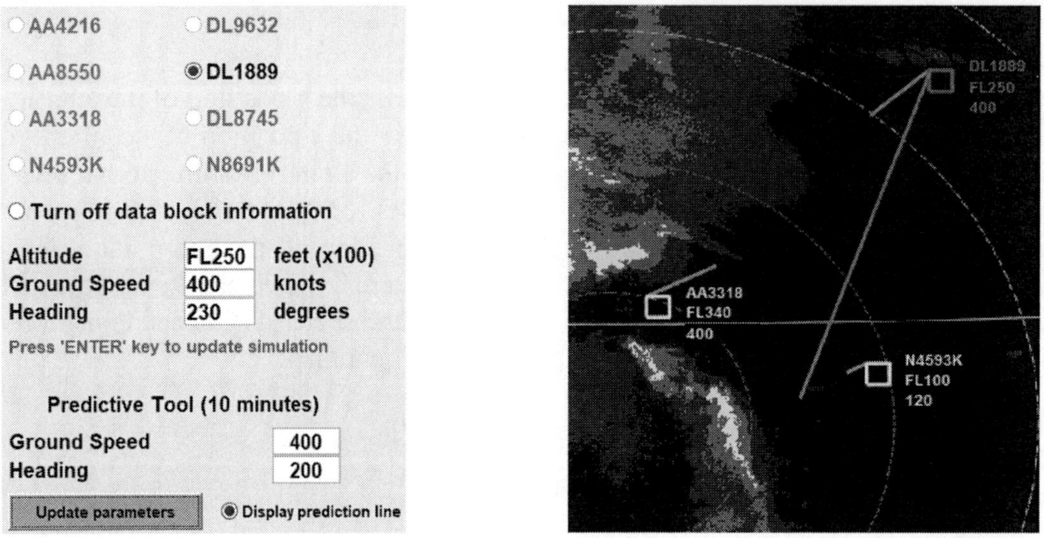

(a) (b)

Figure 2. Aircraft control and predictive capabilities: (a) Control Block and (b) Radar Image

The predictive tool makes it possible for the participant to see what a proposed new route of a selected aircraft would be. The predictive tool uses the edit box inputs for ground speed and heading to calculate what the x-position and y-position of the aircraft will be in ten minutes. Equation 5 and Equation 6 show the how the predicted x- and y-positions of the aircraft marker in ten minutes,x_{10} and y_{10} respectively, are calculated:

$$x_{10} = (x_i) + GS_{10}\sin(H_{10}) \cdot t_{10\,min} \tag{5}$$

$$y_{10} = (y_i) + GS_{10}\cos(H_{10}) \cdot t_{10\,min} \tag{6}$$

Where x_i and y_i are the current x- and y-positions, respectively, of the aircraft marker, H_{10} is the edit box input heading, GS_{10} is the edit box input ground speed, and $t_{10\,min} = 600$s is the time needed to find the predicted position of the aircraft marker ten minutes later. The background program code then plots the predictive line from (x_i, y_i) to (x_{10}, y_{10}). The user can either use the tool with the aircraft's current ground speed and heading or the participant can view how a different ground speed and/or heading will change the aircraft's route.

In the example shown in Figure 2(a), flight DL1889 was selected. The current ground speed of the aircraft was 400 knots with a heading of 230°. The user wanted to see a projected ten-minute route of the aircraft corresponding to the ground speed remaining at 400 knots but changing the heading to 200°. Notice that in Figure 2(b), the data block for the selected flight changed to red, the 2 minute current route indicator remained grey and the 10 minute new route indicator was added using an orange line.

Weather Scenario Selection

The weather scenario chosen for the design verification was based on an aircraft incident that occurred on August 8, 2015. According to the National Transportation Safety Board (NTSB) report, Delta Airlines flight DL1889 was heading to Salt Lake City, Utah from Boston, Massachusetts when around 02:00 coordinated universal time (UTC), while traveling near the Nebraska/Colorado border, the aircraft encountered hail which resulted in an emergency landing of the aircraft. As a result of the hail the aircraft had damage to the windshield, airframe, the leading edges of the wings, the aircraft skin, and the nose cone (NTSB, 2015).

Discussion/Future Work

Currently the research team is conducting experiments with air traffic control students at Kent State University. The experimental procedure is registered and approved through the University of Akron Institutional Review Board (IRB) since it involves participation from human subjects. The experiments are designed to record subjective results from each participant based on their ratings of workload, confidence, and benefit. Objective results such as final aircraft routes and time of decision-making are also recorded. Figure 3 shows an example of how these results can be viewed with the full routes of the eight aircrafts on the radar image. In this case, the route remained uninterrupted during the occurrence of severe weather to show how vectoring is needed to avoid the occurrence of severe weather. The color of each route corresponds to the color of the destination marker. The analysis and conclusions made based on these results will be released at a later date.

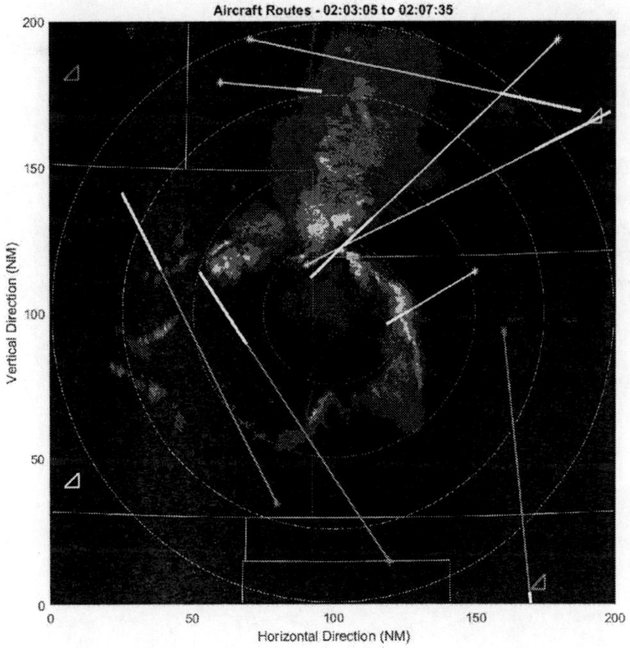

Figure 3. Aircraft routes displayed on the radar image.

Acknowledgements

This study is funded by the Federal Aviation Administration (FAA) Center of Excellence (COE) for Technical Training and Human Performance (TTHP)

References

Ansari, S., Del Greco, S., Kearns, E., Brown, O., Wilkins, S., Ramamurthy, M., ... & Gold, A. (2018). Unlocking the potential of NEXRAD data through NOAA's Big Data Partnership. *Bulletin of the American Meteorological Society*, *99*(1), 189-204.

Boyette, D. (2006, October). Radar talk turned into Plain Speaking. *The Front: NOAA's National Weather Service, 6(3),* Retrieved from https://www.weather.gov/media/publications/front/06nov_Front.pdf

Federal Aviation Administration. (2013). *Development, Validation, And Development of a Revised Air Traffic Control Vision Test: Incorporating Advanced Technologies and Oceanic Procedures and En Route Automation Modernization Systems.* Retrieved from https://www.faa.gov/data_research/research/med_humanfacs/oamtechreports/2010s/media/201318.pdf

Helmus, J.J. & Collis, S.M., (2016). The Python ARM Radar Toolkit (Py-ART), a Library for Working with Weather Radar Data in the Python Programming Language. Journal of Open Research Software. 4(1), p.e25. DOI: http://doi.org/10.5334/jors.119

National Transportation Safety Board. (2015). *NTSB Identification: OPS15IA020.* Retrieved from https://www.ntsb.gov/_layouts/ntsb.aviation/brief2.aspx?ev_id=20150811X22104&ntsbno=OPS15IA020&akey=1

National Transportation Safety Board. (2015). Safety Recommendation Report: Emergency Training for Air Traffic Controllers. *Report No: ASR1604.* Retrieved from https://www.ntsb.gov/investigations/AccidentReports/Reports/ASR1604.pd

INVESTIGATING THE EFFECT OF MICRO-QUADCOPTER FLIGHT ON UAS INSTRUCTION

William Weldon
Purdue University
West Lafayette, Indiana
Kristoffer Borgen
Purdue University
West Lafayette, Indiana
Dr. Brian Kozak
Purdue University
West Lafayette, Indiana

Abstract

This paper examines student perceptions of micro-quadcopter flights in a higher education setting through a case study format. The purpose of the flight activity discussed is to allow students to familiarize themselves with quadcopter flight characteristics, as early as possible, at their own pace in a low stress environment. Through a series of interviews with students who had taken a course where this activity was performed it was found that students enjoyed the activity and found it engaging. Some students felt that the activity provided them confidence for later courses, and many felt that it helped teach basic quadcopter control. It was also found that if the activity is repeated in later courses it should include some form of directed activity, instead of focusing on "free flight". Several students brought up another activity in the course, simulator flights, and these are briefly discussed. Future work should look to find if these activities have a positive effect on student flight performance.

Background

Unmanned Aerial Systems UAS programs offering Associate's or Bachelor's degrees are present at many universities and colleges across the United States. These programs vary in their requirements, and many mirror traditional professional pilot programs while others require manned flight certificates (Wentworth, 2017). The UAS program being discussed in this paper requires 120 credit hours, 24 of which are UAS specific, and 22 aviation courses related to UAS. The UAS program is the newest major at the school that includes professional pilot, aviation maintenance, and aviation management degree programs. There is a heavy lab focus in the UAS major with 16 of the 24 UAS specific credit hours are spent in various UAS labs. These labs teach UAS construction, manual flight, autonomous flight, data collection, data analysis, and problem solving with UAS. The learning activity investigated in this paper took place from the Fall 2017 to Fall 2018 semesters across two courses. For this paper these courses will be called UAS1 and UAS2. The course objectives for UAS1 is to introduce new major students to the UAS field and how UAS operate. This objective is met through construction of a Sig Kadet Mk II in small groups, individual micro quadcopter flights, and UAS flight simulators. The last two activities were added in the Fall 2017 semester because of the instructor's belief they would add learning value to the class, and as a way to mirror the professional pilot program present at the university.

The Blade Nano QX, will also be referred to as "Nano", was chosen for the micro-quadcopter flights due to their size, stability, and cost. These vehicles are 16.5grams in flight with a length and width of under 3 inches. The Blade Nano QX has an auto-stabilize feature, called "safe technology", that limits the vehicle pitch and bank angle and will auto level when pitch or bank control is neutralized (Flysaferc, 2013). This system is not the same type of auto-stabilize system that is present larger commercial vehicles, which will keep the current position of the vehicle without user input. The size of these vehicles makes them easy to fly indoors, and the auto-stabilization feature is intended to make the vehicles easy for new pilots to fly. A ready to fly package for the Nano can be found on Amazon which allowed a small fleet of 10. The vehicles and additional batteries are also able to be easily purchased.

UAS1 introduced the Nano flight activity, which required students to accumulate 4 hours of flight time with the vehicles throughout the semester. These flights could be accomplished during lab time or the instructor's office hours. The flights were performed without activity direction, i.e. the students were not told where to fly or what do while flying. For the flight area students determined the flight activities they would perform in a large open room. This activity was repeated in UAS1 Spring 2018 and UAS2. The same fleet of Nanos was used for all the classes mentioned in this paper with each vehicle receiving an average of 8.8 hours of flight time at the end of UAS1. Assuming all students correctly reported the hours they flew each vehicle would have begun UAS2 with 17.6 flight hours and ended with 26.4 flight hours. In addition to the flight hours each vehicle received they were also subjected to a variety of crashes as the students learned how to operate them effectively.

UAS2 added to the Nano flights with tasks such as; precision landing, orientation control, and precision flight. Precision landing involved landing pads that were slightly larger than the vehicle placed around the large lab table approximately 4 feet by 10 feet. The students would take-off from one landing pad and navigate to another approximately 4-5 feet away and land. Orientation control involved taking-off from a pad circling a pole and landing on the pad the flight started from. There were two tasks involved with flying around the pole; 1. Fly around the pole in any orientation without hitting the pole, 2. Fly around the pole keeping the nose of the vehicle facing towards the pole for the entire flight. Precision flight involved replicating a drone racing league (DRL) course. Students would fly around a predetermined course while being timed with the fastest time being considered the "winner". For all of these tasks students had as many tries as they wanted.

Analysis of Interview Transcripts

To evaluate student perceptions of micro UAS flights seven former students participated in 5-12 minute interviews focusing on their perceptions of the Blade Nano flights and the class. The participants had all taken UAS1 with four students having taken other courses in the program, in addition to the following course UAS2. The number of other UAS courses taken by each participant varies due to the freedom of course selection in the major, and class standing when UAS1 and UAS2 were taken. The first experience from the participants that became apparent is that the wide range of flight time concentrated at the extremes. The extremes varied from no prior flight experience to over 2,000 hours of flight time. These hours were self-reported and not logged, but the students who had a high amount of flight hours also said they had been flying for years. Due to the lack of flight logs it is impossible to know exactly how many hours they have, participants with a high hour time were assumed to be very experienced with UAS flight. One commonality between the inexperienced and experienced participants was DJI products. Participants who mentioned gaining flight experience outside of class, either before or after taking the course, stated that they gained their multirotor time from primarily DJI products. All participants mentioned the lack of fixed wing flight time in the class as a negative when asked about feedback regarding the Nano flights. This is mentioned as a drawback of the class with the classroom being too small for fixed wing flight to be performed adequately. Small fixed wings were attempted to be flown but lacked enough space to be operated.

When discussing the Blade Nano flights all participants stated that they had a positive view of the activity, but some negatives were also mentioned. The biggest negative mentioned was the flight characteristics of the vehicles. Flight conditions were broken down into three complaints; no auto-stabilize, worn condition, and battery life. The participants mentioned that the Nano's lack of auto-stabilize feature made the flights daunting to those who had no experience with quadcopter flight. As mentioned in the description of the Nanos, they do contain an auto-stabilize feature, but it is not as robust as what can be found on larger commercial platforms such as DJI. This is likely why the participants mentioned the lack of this feature. The lack of auto-stabilize was also stated as a positive by a participant stating that they felt it "reinforced manual flight control techniques" and another participant stated that the Nano performance gave confidence for later courses. The worn condition of the Nanos is brought up by participants who took UAS1. This used condition negatively affected vehicle performance and each vehicle performed differently due to the varying degrees of damages on the vehicles. One participant did mention that the vehicle discrepancies helped them to internalize problem diagnosis and solving. An example of this is on many occasions a propeller would not rotate on one of the vehicles. This could be due to a foreign object wrapped around a motor shaft or a dead motor. The corrective action for this problem can be removal of the propeller and removing the foreign object or completely exchange the motor. Multiple participants brought up the short battery life of the vehicles, while this is a limitation of the technology steps were taken in the course to address the issue. During the first few weeks of the initial offering of the course it became obvious that battery size would be a limitation of the vehicles and thirteen additional batteries were purchased doubling the amount of batteries available and leaving each quad with 3 batteries. While the short flight times were described as a negative of the vehicles, the short charge time of the batteries was noted as a positive. This short charge time and number of batteries available did not eliminate down time while all batteries were dead or charging but did reduce the frequency.

The Nano flights were repeated in UAS2 due to the assumed helpful nature and the fact that students enjoyed the activity. The feedback from the students that participated in UAS1 and

UAS2 provides interesting insight into developing Nano flights for multiple courses. The original idea was to repeat the activity in UAS2 so students could continue to familiarize themselves with multirotor flight characteristics at their own pace. The three participants who had taken both courses did not like this approach as the activity felt repetitive leading them to dislike it. This became apparent in the course quickly as the instructor began providing obstacle race courses for students to participate in. Three of the participants mention being engaged in these activities, and one mentions regretting not taking advantage of them in the class. Both courses were mentioned as being fun and engaging as well as helping to develop the students advance flying techniques.

One participant felt that the Nano flight activity provided a challenge as an experienced pilots however the experienced pilots were much "quicker to get the hang of it". This could suggest that even experienced pilots have an experience gap, possibly due to most of their flight experience being with a vehicle that has many pilot assist features. A participant also noted an increase in situational awareness while flying due to the number of vehicles that would be flying in the room simultaneously. This forced students who were flying to keep an eye out for other student vehicles and anticipate their movements in addition to controlling their vehicle. This was accomplished very well as the instructor only remembers a handful of mid-air collisions during the three times this activity was performed.

Discussion

This difference in experience makes engaging curriculum design for introductory courses difficult as some students are complete beginners and others are more experienced than the instructors. Even though the more advanced pilots initially struggled when they first flew the Nanos they gained control of the vehicle much more quickly than beginner pilots. One way to counter this would be to immediately introduce the more advanced techniques such as the landing pads or flying through and obstacle course, however if it is dominated by the advanced pilot or appeared to be made easy it can be very discouraging to beginner pilots. Through having multiple courses that focus on the various aspect of flying could be a way to encourage students to practice flying, such as having all three courses mentioned active at the same time and as students feel more confident allow them to attempt each course as they wish. By having no point values assigned to flying the courses allowing the students to not have to deal with the additional stresses of flying for a score.

One of the common complaints throughout the interviews were the lack of fixed wing experience. This was attempted to be mitigated through the fixed wing simulator flights, however this was mentioned, and the participants would have preferred actual flight experience in addition to the sims. This presents a challenge for the class as there are no classrooms that contain a large outdoor air field to fly on. This was attempted to be mitigated through having small indoor aircraft to fly, but it was soon found that the largest room available was still too small to provide adequate room for flying.

One item that was brought up during the interviews was the simulator flights that the students were required to do in addition to the Blade Nano flights. There were two set of challenges required for students to complete the fixed wing and the multirotor. These simulator flights were found to be beneficial to the participants although they were considered challenging. One of the positives that was mentioned about the simulator flights was that it provided the students with fixed wing experience, even if the only experience with fixed wing was on the simulator. Some of the drawbacks of the simulators involved lack of stations and difficulty of some of the challenges. In the down time of the Nanos charging students were able to attempt the simulator flights, however they felt that while the batteries were charging most would end up gathering around the computer and watching one person complete them due to their only being one seat available for the sims. With the challenges themselves students found that the fixed wing represented the most realistic course as it was replicated similar to the red bull air race. Students were required to fly through a set of pylons as quickly as possible for a score out of 100 points. For the quadcopter trials there were gates and landing pads that students needed to fly through or land on, this is where the dislike for the challenges came since to complete the challenges students had to fly their vehicles full speed into the ground or wall. This allowed completion of the challenge but enforced poor flying habits.

Conclusion

In conclusion this case study has found that the students who participated thought the micro-quadcopter flight activity to be useful and engaging. One student mentions that it helped build confidence for later courses, and multiple students stated that it helped teach basic control. It was also found that if this activity is to be integrated into multiple courses then it must be modified for each course. The students who had taken UAS2 mentioned that repeating the "free flight" activity was not engaging when done a second time, but the modifications made to the activity made it engaging. The interviews also suggest that the Nanos may not stand up to multiple semesters of novice UAS pilots, and may need to be replaced regularly due to about a third of the vehicles being used for spare parts. While not the focus of the study it would seem that the simulator activities had a large effect on student perceptions of the class. One participant was a very adamant supporter of the simulator activities, and after this interview questions about the simulators were asked in subsequent interviews.

Future Work

Future research should be done to determine the effect of these type of activities on student flight performance. This study suggests that students view the activity positively but it cannot determine the actual effect of the activity on their performance. The effect of the simulator activities should also be investigated. As UAS education becomes a regular part of higher education activities like this could provide a low cost introduction to flight, and help teach advanced flight skills. It should be determined early on how useful these methods are to that goal, and how they may be improved or if other methods should be used.

References

Flysaferc. (2013, May 28). *SAFE^{TM} Technology: A New Way to Reach for the Sky* [Video file]. Retrieved from https://www.youtube.com/watch?v=1JmWZVhnP_8

Wentworth, A. (2017). A comparative analysis of UAS crewmember collegiate curricula. *Journal of Unmanned Aerial Systems*, 3(1).

HAZARDOUS ATTITUDES IN US PART 121 AIRLINE ACCICDENTS

Bryan Nuñez, Carlos López, Jonathan Velazquez, and Oswart A. Mora
Inter American University of Puerto Rico
Bayamón, Puerto Rico
Kevin Román
Delta Air Lines
Atlanta, Georgia

The greater part of aviation accidents is often attributed to human error, with flight crew performance accounting for the majority of these mishaps. In 2016, the Federal Aviation Administration (FAA) published a rule to address pilot professionalism and to increase the likelihood that aviators adhere to standard procedures and prevent behavior that could lead to pilot errors in the airline domain. The FAA has identified 5 Hazardous Attitudes that afflict pilots: *macho, impulsivity, resignation, invulnerability, and anti-authority.* This study examined the FAA-defined Hazardous Attitudes and the regularity with which they occurred in the U.S. air carrier flight crew related accidents between 1991-2018. The top two Hazardous Attitudes were *anti-authority* and *invulnerability*, which were found in 92% and 68% of aviation accidents, respectively. The paper also explores the relationships among these Hazardous Attitudes and other contributing factors such as time of day, weather, flight conditions, and crew resource management, among others.

Literature Review

There are a multitude of factors that affect decision making and Crew Resource Management (CRM) within the cockpit. CRM is the proper use of all the available resources (human, hardware, and information) to conduct and complete a safe flight (FAA, 2004). Helmreich and Foushee found that the lack of CRM was responsible for more than 70% of the accidents during the period between 1959 and 1989 (1993). Furthermore, Wetmore and Lu studied general aviation (GA) accidents and found that Hazardous Attitudes have a great influence in the aeronautical decision making (ADM) process of pilots (Wetmore & Lu, 2005a, 2005b, 2006; Wetmore, Bos, & Lu, 2007).

An attitude is a predisposition to respond to an event in a given manner (FAA, 2009). Investigations have identified five Hazardous Attitudes, which interfere with the ability to make decisions and exercise authority properly (FAA, 2009). These Hazardous Attitudes are *macho*, *impulsivity*, *resignation*, *invulnerability*, and *anti-authority* (Table 1). Although they contribute to poor pilot judgment, they can be counteracted by saying the correct antidote (FAA, 2009).

Proposed Rulemaking (NPRM) by the Federal Aviation Administration (FAA) was published in 2016, to tipoff pilots to follow standards procedures and professionalism and to prevent behavior, which could lead to errors (81 FR 69908-Pilot Professional Development, 2016). Historically, most of the research on Hazardous Attitudes has focused on the GA population and the flight instruction environment (Hunter, Martinussen, Wiggins, and O'Hare

2011; Stewart, J. 2006, 2008; Wagener & Ison, 2014; Wetmore & Lu, 2005a, 2005b, 2006; Wetmore, Bos, & Lu, 2007). However, this study concentrated on the Hazardous Attitudes in the multi-crew environment and focuses on which ones were predominant in crew-related accidents. The specific research questions were:

1. Which Hazardous Attitudes are present, and with what regularity do these occur, in flight crew related accidents?
2. What relationships exist between the pilot Hazardous Attitudes and other contributing factors in these U.S. airline accidents?

Table 1.
Hazardous Attitudes Definitions and Antidotes as defined by FAA (2009, p. 2-5)

Hazardous Attitude	Definition	Antidote
Anti-Authority: (AA) "Don't tell me."	This attitude is found in people who do not like anyone telling them what to do. In a sense, they are saying, "No one can tell me what to do." They may be resentful of having someone tell them what to do, or may regard rules, regulations, and procedures as silly or unnecessary. However, it is always your prerogative to question authority if you feel it is in error.	"Follow the rules. They are usually right."
Impulsivity: (IM) "Do it quick."	This is the attitude of people who frequently feel the need to do something, anything, immediately. They do not stop to think about what they are about to do; they do not select the best alternative, and they do the first thing that comes to mind.	"No so fast. Think first."
Invulnerability: (IV) "It won't happen to me.	Many people feel that accidents happen to others, but never to them. They know accidents can happen, and they know that anyone can be affected. They never really feel or believe that they will be personally involved. Pilots who think this way are more likely to take chances and increase risk.	"It could happen to me."
Macho: (MA) "I can do it.	Pilots who are always trying to prove that they are better than anyone else are thinking, "I can do it – I'll show them." Pilots with this type of attitude will try to prove themselves by taking risks in order to impress others. While this pattern is thought to be a male characteristic, women are equally susceptible.	"Taking chances is foolish."
Resignation: (RE) "What's the use?"	Pilots who think, "What's the use?" do not see themselves as being able to make a great deal of difference in what happens to them. When things go well, the pilot is apt to think that it is good luck. When things go badly, the pilot may feel that someone is out to get me, or attribute it to bad luck. The pilot will leave the action to others, for better or worse. Sometimes, such pilots will even go along with unreasonable requests just to be a "nice guy."	"I'm not helpless. I can make a difference."

Methodology

The study used archival methods to explore the Hazardous Attitudes contributing to U.S. Part 121 flight crew accidents. The primary data source was the Embry-Riddle Aeronautical University (ERAU) National Transportation Safety Board (NTSB) database blue cover accidents reports. Research focused on NTSB Accident Reports periodical, which cataloged 37 accidents from 1991 to 2018. Excluded were all accidents with undetermined causes or those attributed to terrorism. All 37 accidents, that fit the above-mentioned characteristics, were analyzed by 5 subject matter experts (SMEs) so as to identify any Hazardous Attitudes and the contributing factors in the accidents.

The research team analyzed the accident reports to determine the presence of Hazardous Attitudes that may have been influential. The researchers also identified contributing situational factors, such as weather, CRM, airline management, flight rules, etc., that may have exacerbated the effect of the Hazardous Attitude. A priori codes were used; specifically, the FAA-defined Hazardous Attitudes. After the identification process was completed, the SMEs tried to find any connections between the Hazardous Attitude and the other contributing factors. All the relevant information from the accident reports was entered into NVivo (v. 12), a computer aided qualitative data analysis software. The use of such software allowed for a second stage of coding where themes began to emerge (e.g., additional contributing factors) in conjunction with the Hazardous Attitudes themselves.

Results

Descriptive Statistics

Table 2 and shows the regularity with which the Hazardous Attitudes were found in the analyzed accidents. The number represented under "Yes" means the number of accidents in which the Hazardous Attitude was found. Conversely, the number under "No" means the number of accidents in which the Hazardous Attitude was not found. The top two Hazardous Attitudes were *Anti-authority* and *Invulnerability*; these two were found in 34 and 25 accidents, respectively. In addition, each Hazardous Attitude was further analyzed in fatal accidents. These results are demonstrated in Table 3.

Table 2.
Frequency Count of Hazardous Attitudes in all Accidents

	Hazardous Attitudes Total Accidents				
	AA	IM	MA	IV	RE
Yes	34	15	9	25	12
No	3	22	28	12	25
Totals	37	37	37	37	37

Table 3.
Hazardous Attitudes Frequency Count in Fatal Accidents

	Hazardous Attitudes in Fatal Accidents				
	AA	IM	MA	IV	RE
Yes	13	3	3	10	4
No	1	11	11	4	10
Totals	14	14	14	14	14

Relational Analysis Results

Figure 3 represents a Cluster Analysis performed by NVivo. NVivo can attempt relational analyses through a dendogram such as this one. Cluster analyses are good visualization tools based on the frequency with which words or coding are shared in the text. This figure explores that relationship with word similarity. The dendogram indicates how sources of information have word similarities, which in turn could suggest relationships between two concepts. The proximity to, and color of codes within the diagram, suggests associations among the concepts.

Anti-authority and *Invulnerability* share the same color, and are near, to CRM issues. This result suggests that these Hazardous Attitudes are similar and are having an impact on the ability of crews to perform well together. Both Hazardous Attitudes (i.e., *Anti-authority* and *Invulnerability*) may lead to the bypassing of procedures and teamwork efforts, which ultimately affect CRM. No other relationships between factors and attitudes were established by the research team.

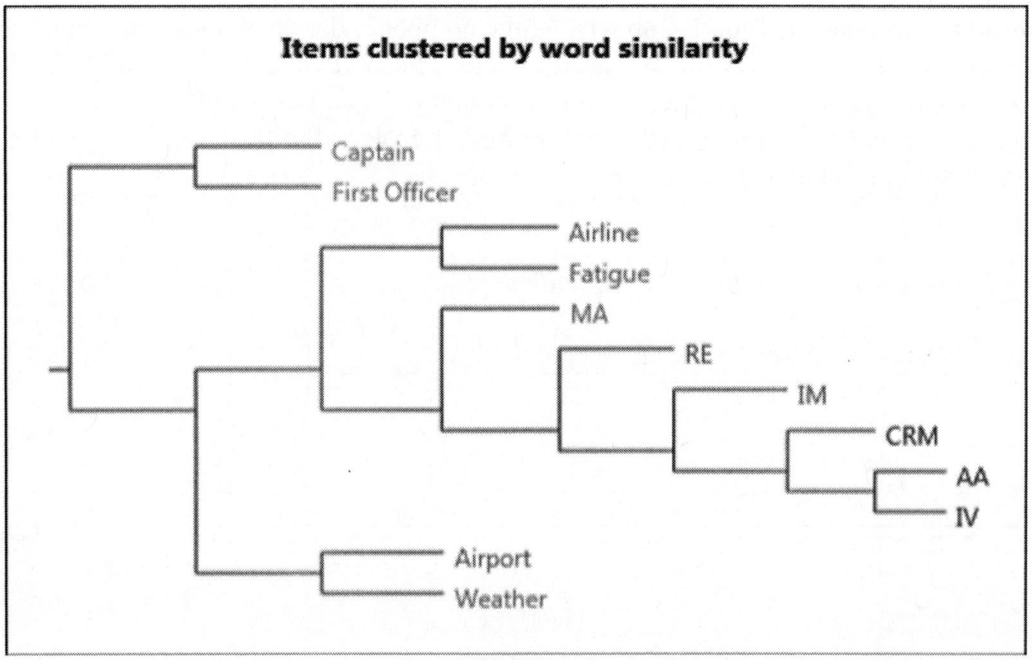

Figure 3. Cluster Analysis between CRM and the predominant Hazardous Attitudes.

Discussion, Conclusions, and Recommendations

All of the Hazardous Attitudes were found in the reports analyzed; however, the two more dominant ones were *Anti-authority* and *Invulnerability*. Anti-authority appeared in 92% of the accidents analyzed; meanwhile Invulnerability was found in 68% of the accidents. In addition, a relationship between these two Hazardous Attitudes and CRM was clearly established by NVivo. Thus, it is unsurprising that both Attitudes were the top two in total and fatal accidents.

As it was evidenced, the Hazardous Attitudes are affecting crew related operations and performance. The results of this study, where *Anti-authority* is the top Hazardous Attitude is aligned with Velazquez (2018) where he found *Neglect of Flight Planning, Preflight Inspections, and Checklists* to be the top Behavioral Trap. Also known as Operational Pitfalls, Behavioral Traps are accident-inducing attitudes that equally affect decision making (FAA, 2009). In both examples of negative behaviors, pilots bypass rules and procedures, fail to follow checklists, federal aviation regulations, and manufacturer recommended practices. Moreover, all of these studies justify the FAA's efforts to increase pilot professionalism in the U.S. Part 121 environment. Perhaps, it is time CRM training include a psychological element in which pilots identify and manage behavioral factors such as Hazardous Attitudes. This training can include scenarios and Hazardous Attitude modification techniques.

The FAA could implement more rigorously the NPRM (81 FR 69908-Pilot Professional Development, 2016). It is highly recommended that pilots be monitored, mentored, and well trained in CRM operations to avoid failures. After all, as Michael Huertas said, "We have some of the best pilots in the world and should take full advantage of our pilot's wealth of experience to raise professional standards and cockpit discipline" (FAA, 2016, para 2). Every time pilots enter an airline he or she should be briefed on how and why the majority of the accidents have occurred. In addition, besides observing the operations inside the cockpit, the pilot should be assigned to identify as many hazards as they can. This form of risk management should include behavioral hazards. Once the crew is on the ground, all pilots involved should be instructed on how to avoid them.

It will always be almost impossible to reduce human errors to 0% in an environment that relies heavily on humans for its operation. However, these studies help identify the shortcomings so that all involved can focus on improving safety and accident prevention.

References

Federal Aviation Administration. (2016). FAA to Boost Pilot Professional Development. Retrieved March 4, 2019 from: https://www.faa.gov/news/updates/?newsId=86647

Federal Aviation Administration. (2009). *Risk Management Handbook*. Washington, D.C.: Government Printing Office.

Federal Aviation Administration. (2004). Advisory Circular 120-51e: Crew Resource Management training. Washington, D.C.: Government Printing Office.

Helmreich, R. L. & Foushee, H. C. (1993). Why crew resource management? Empirical and theoretical bases of human factors training in aviation. In E. L. Weiner, B. G. Kanki, & R. L. Helmreich (Eds.), Cockpit resource management (pp. 3-45). San Diego, CA: Academic Press.

Hunter, D. R., Martinussen, M., Wiggins, M., & O'Hare, D. (2011). Situational and personal characteristics associated with adverse weather encounters by pilots. *Accident Analysis & Prevention, 43*(1), 176-186.

Stewart, J. E. (2006). *Locus of control, attribution theory, and the five deadly sins of aviation.* (Technical Report 1182). Fort Rucker, AL: U.S. Army Research Institute for the Behavioral and Social Sciences.

Stewart, J. E. (2008). Locus of control and self-attribution as mediators of hazardous attitudes among aviators: A review and suggested applications. *International Journal of Applied Aviation Studies, 8*, 263-279.

Velazquez, J. (2018). The Presence of Behavioral Traps in U.S. Airlines Accidents: A Qualitative Analysis. *Safety, 4,2.* https://doi.org/10.3390/safety4010002

Wagener, F., & Ison, D. C. (2014). Crew resource management application in commercial aviation. *Journal of Aviation Technology and Engineering, 3*(2), 2-13.

Wetmore, M., Bos, P., & Lu, C. (2007). Conflict management strategies in the flight training environment. *International Journal of Applied Aviation Studies, 7*, 256-271.

Wetmore, M. & Lu, C. (2006). The effects of hazardous attitudes on crew resource management skills. *International Journal of Applied Aviation Studies, 6*, 165-182.

Wetmore, M., & Lu, C. (2005b). *Reducing hazardous attitudes: The effects of pilot certification and flight experience.* White paper. Central Missouri State University.

Wetmore, M., & Lu, C. (2005a). The effects of pilot age on aeronautical decision making and crew resource management skills. White paper. Central Missouri State University.

PILOT WELLBEING & WORK RELATED STRESS (WRS)

Joan Cahill
Trinity College Dublin (TCD)
Dublin, Ireland
Paul Cullen
Trinity College Dublin (TCD)
Dublin, Ireland
Keith Gaynor
University College Dublin (UCD)
Dublin, Ireland

This paper presents the preliminary findings of an anonymous web-based survey addresing pilot work related stress (WRS) and wellbeing. The initial analysis indicates that pilots are under stress and experiencing wellbeing problems. Specific features of the job can result in wellbeing problems, spanning the three pillars of wellbeing. Critically, sources of WRS can increase a pilot's risk in terms of developing a mental health (MH) issue. Further, sources of WRS can impact on performance and safety. Considerable barriers still remain in relation to reporting MH issues at work. Coping mechanisms addressing sleep/fatigue, diet, exercise and communication/reporting, enable some pilots to thrive in an environment that has negative impacts for others. The vast majority of pilots indicated that issues pertaining to WRS and wellbeing are not being adequately managed in terms of airline safety management systems/processes. Potentially, airline interventions might focus on enhancing existing safety management system processes/technology to address risks associated with WRS and wellbeing, training pilots, and introducing new wellbeing briefing/reporting systems. Further, new digital tools might be advanced to support pilot self management of WRS/wellbeing and risk identification, both inside and outside work.

Work Related Stress (WRS) is defined as the response people may have when presented with work demands and pressures that are not matched to their knowledge and abilities, and which challenge their ability to cope (Leka, Griffiths & Cox, 2003). A high stress situation may not be detrimental to a person, once they have learned to cope with it in a healthy manner. As reported by Joseph (2016), stress coping is an important psychological construct which moderates/mediates the relationship between stressors and behavioural outcomes such as flying performance. Pilots experience many physiological, psychological and environmental stressors. Since the Germanwings 9525 accident (2015), the issue of pilot suicide and detecting/managing mental health issues amongst pilots has been gaining increased attention. Recent studies demonstrate that pilots are suffering with the same wellbeing issues as the general population (particularly those relating to mental health) and possibly to a greater extent (Pasha & Stokes, 2018; Wu et al, 2016). Overall these studies have attempted to measure the prevalence of wellbeing issues (and in particular, mental health issues), and to understand the factors that contribute to this. However, these studies fall short in terms of providing a rich picture of the lived experieince of pilots, and the complex relationship between individual wellbeing factors as conceptuallized in the biopsychoscial approach (Engel, 1977). In addition, there has been little emphasis on understanding (1) the relationship between WRS, pilot wellbeing and safety, (2) how pilots adapt to WRS and associated coping/self-management techniques, (3) the role of other stakeholders in relation to supporting pilots and managing this problem, and (4) potential solutions at different levels.

Prior exploratory interviews undertaken by the authors suggest that aspects of the job are impacting on pilot's physical, social, and emotional/psychological health (Cullen et al, 2017). Research indicates that aspects of the job present a potential threat to flight safety, given the ensuring impairments to task performance (Cahill, Cullen & Gaynor, 2018). In general, pilots try to normalize/adapt to the job and manage wellbeing issues. However, there is much variation in relation to coping ability. Overall, six impact scenarios were identified (Cahill et al, 2018). Of these, participants suggested that the primary focus of wellbeing interventions might be on the prevention of routine suffering, suffering which may degrade performance on the day, and suffering which ends in harm to the person. Following from the above research, this paper reports on the preliminary finding of the first wave of an anonymous web-based survey pertaining to pilot wellbeing. The survey and its analysis are both ongoing. Overall, the paper provides a preliminary descriptive analysis of the findings of the first wave of feedback (N=330, 67% completion rate). First, a brief background to this research is provided. The survey methodlogy is then reported. The high level results are then reported. These results are the discussed and some preliminary conclusions drawn.

Methodology

The objectives of the survey include: (1) to measure routine suffering amongst pilots, (2) to understand pilots experience of WRS/wellbeing issues, (3) to understand pilot attitudes to reporting wellbeing issues (including mental health), (4) to understand the relationship between work related stress, pilot wellbeing, pilot performance and safety, (5) to understand how pilots adapt to WRS and wellbeing issues, (6) to identify pilot coping/self-management techniques, and (7) to examine pilots perceptions regarding the role of their employers/airlines in terms of managing WRS/wellbeing issues. The is a cross-sectional descriptive study. An anonymous web-based questionnaire was developed which elicits feedback pertaining to the topics indicated above. The survey incorporates several standardised instruments to measure levels of common mental health issues. These are these Patient Health Questionnaire -9 (PHQ-9) (Kroenke, Spitzer & Williams, 2001), the Oldenburg Burnout (OLBI 8) (Demerouti, Bakker, Vardakou & Kantas, 2003), and the Oldenburg Burnout (Modified Instrument) (Demerouti, Veldhuis, Coombes & Hunter, 2018). Further, the survey design draws upon prior research undertaken by the authors pertaining to a biopsychosocial model of wellbeing, the factors that can positively and negatively influence a pilot's physical, mental and social health, and the ensuing impact on pilot performance and flight safety (Cahill et al, 2018, Cullen et al, 2017). Ethics approval was granted by the School of Psychology, Trinity College Dublin (TCD). The survey was completed by commercial pilots between 7th November 2018 and 28th February 2019. Using social media chanels, respondents were invited to participate in an anonymous online survey at a time of their choice. Advertising information informed participants that the survey elicits information of a sensitive nature and included a weblink to the survey. Prior to answering survey questions, respondents received background information about the study and completed the electronic consent. Following this, respondents completed questions for each of the nine sections. The survey concluded with a debriefing which included contacts information for relevant support groups. The survey was powered by the SurveyMonkey service and did not collect any identifying information about the person. Further, no internet protocol (IP) addresses were collected. It was assumed that each participant was a pilot and only completed one survey. Several questions in the survey required knowledge that would only be readily available to pilots. An active pilot (co-author in this study: PC) reviewed surveys for potential non-pilot participants. All surveys passed this screening. Descriptive statistics were used to describe the respondents and their responses on various survey items. We evaluated depressive symptoms via the Patient Health Questionnaire (PHQ-9) depression module. Tests for statistically significant group differences have not yet been undertaken.

Results

330 respondents participated in the survey, with 220 completing it fully (66.7% rate). 265 respondents completed the PHQ-9 (80.0%). Overall, the respondents can be described as male (84.5%), full time (91.8%), married (58.2%) and based in home country (80.3%). The repondents can be split into the following age brackets; <25 (4.2%), 25-35 (33.5%), 36-45 (27.8%), 46-55 (23.0%) and 56-65 (10.0%). Respondents had worked as a pilot for the following lengths of time; <2 years (8.5%), 2-5 years (12.6%), 6-10 years (17.1%), 11-15 years (15.7%), 16-20 years (14.7%), 21-25 years (7.2%), 26-30 years (12.0%) and >30 years (12.3%). 62% of respondents held the position of Captain. Over 3/4 (77.7%) of respondents rated their physical health as good/very good, while approximately 2/3 (67.7%) rated their mental health as good/very good. In general, the Pilots surveyed were a reasonably healthy population in terms of their health behaviours. The majority of participants reported obtaining between 7 and 8 hours sleep on non duty days (35.4% reported 8 hours of sleep, while 30.0% reported 7 hours). Respondents reported obtaining considerablly less sleep duirng duty periods (42.9% obtaining 6 hours, and 27.5% 7 hours). The vast majority exercise regularly (22.0% three times a week, 21.3% twice a week, and 16.8% once a week). Further, the majority reported eating a healthy diet (88.5%) while off duty, although a significant proportion (54.5%) reported that they ate an unhealthy diet while at work.

Just under half of the respondents (48.7%) reported that they had spoken to somebody about a MH issue they were experiencing or had experienced. 42.5% of respondents indicated that they have a close-friend pilot colleagues who has experienced MH issues. 12.8% of participants meet the threshold for Clinical Depression. 7.9%, had suicidal thoughts in the previous two weeks. However, although respondents reported experiencing wellbeing problems, the data suggests that Pilots are adapting and coping. Nearly half of respondents (48.1%) agreed to the statement 'Pilots are suffering, but they are also adapting and coping', while 8.7% strongly agreed.

45.6% strongly agreed that there are low levels of speaking out and/or reporting about mental health among Pilots, while 40.3% agreed. The vast majority of participants indicated that they would talk to a partner/spouse (79.5%) about a MH issue, closely followed by a friend (55.0%). Only 24.9% indicated that they would talk to a close friend colleague. 13.5% indicated that they would speak to a peer support group. A very small number (2.2%) indicated that they would speak with their line manager. Overall, participants indicated a considerable level of stigma in relation to reporting mental health issues at work. 78.0 % indicated they would not disclose a MH issue to their employer. 55.6% reported that if they were "unfit for flight" due to a mental health issue, they would provide a different reason. When asked about their reasons for this, the vast majority of respondents (68.6%) indicated 'fear of loss of license and loss of long-term earnings'. Other reasons included 'fear of stigmatisation by employer' (57.7%) and 'potential negative impact on career progression' (52.6%). On a more positive note, the vast majority agreed that they would look for help, if they had a MH issue (47.8% agreeing and 29.0% strongly agreeing). Further, 70.5 % strongly agreed with the statement 'Promoting mental health awareness (recognising problems in one's self or others) is important from a safety perspective', while 27.2% agreed.

Just over half of participants (51.0%) indicated that they find the job stressful 'now and again', while 23.5% indicated that the job is 'frequently stressful'. Pilots were asked to rate their ability to cope with WRS. The majority (69.6 %) agreed that they can tolerate the pressures of their work very well, while 13.8% strongly agreed. However, most participants (51.7%) agreed that 'they feel worn out and weary after work', while 22.9% strongly agreed. Respondents reported the top 3 most common sources of WRS as working irregular hours (70.2%), working ani-social hours (57.5), and the divergence of values between management and pilots (57.5%). Overall, the data indicates that sources of WRS have a negative impact on pilot wellbeing. Sleep difficulties (78.2%) were reported as the most common wellbeing issue that respondents either attributed to the job, or believed to be worsened by the job. This is followed closely by musculoskeletal symptoms (71.6%) and then digestive symptoms (53.8%). Other impacts include social isolation (42.2%), marital/family discord (36.9%), respiratory symptoms (32.9%) and psychological distress (31.1%). Although psychological distress was ranked the lowest in terms of wellbeing impact, the vast majority of respondents indicated that the environment in which Pilots work can contribute to the onset of, or worsen an existing a mental health issue (59.8% participants agreed, while 26.2% strongly agreed).

Data anlaysis suggests that sources of WRS impact on performance and flight safety. The vast majority of respondents (60.4%) agreed to the statement that 'certain sources of Work-Related Stress (WRS) have an impact on my performance', with 18.7% strongly agreeing. Further, 52.6% of respondents agreed to the statement 'Certain sources of WRS have an impact on my performance and by implication, have the potential to impact on flight safety', while 21.1% strongly agreed. Respondents were invited to identify specific performance impacts in relation to different sources of WRS. 82.4% of responedents reported 'working within the close confines of the cockpit' as the having the strongest impact, specifically, in relation to distraction and inability to focus on current task. Working irregular hours (73.6%) and working long duties (76.4%) were rated as having most impact on decision making. Over half of the respondents (52.4%) agreed to the statement that they are 'mostly coping well and that periodically, they may make a mistake but they will identify their own mis-take and correct their actions, thus ensuring that a safety event does not occur', with 7.8% strongly agreeing. Equally, the vast majority (56.7%) agreed to the statement 'if something were to give on the day, and I were to make a mistake, it is most likely that my fellow pilot would detect this and take a corrective action, thus ensuring that a safety event would not occur', with 12.6% strongly agreeing.

Pilots were asked to select from list of common methods of coping with (1) non WRS (stress outside work) and (2) WRS (stress inside work). 60.2% reported adopting coping strategies for non WRS, while 53.9% reported using coping strategies for WRS. In relation to coping strategies for non WRS, 30.8% reported using positive diet each day. Only 1.6% used relaxation devices/tools on a daily basis. On a several times per week basis, respondents reported using sleep and rest (54.6%), exercise (53.6%) positive diet (48.8%) and relaxation (13.0%). In relation to daily activities to manage WRS, the strongest focus appears to be on sleep and rest (28.0%), diet (27.6%) and exercise (14.0%). In terms of activities performed several times a week, respondents reported exercise (51.2%), positive diet (46.4%), sleep/rest (47.1%) . 21.9% respondents reported talking with colleagues while 17.6% reported talking with family and friends. The data analysis indicates that pilots do not use relaxation methods as frequently as other methods (3.1% every day, 11.8% several times a week and 8.1% once a week). In addition, it indicates that pilot use of professional supports is infrequent (2.0% several times a week, 0.7% once a week).

Overall, it seems that pilot engagement is quite low. Only 18.0 % agreed with the statement 'my employer and I share the same set of values', while 1.7% strongly agreed. 38.3 % of participants rated the level of engagement between themselves and their employer as very poor, while 39.6% rated it as poor. The majority of respondents indicated that ensuring and maintaining positive mental health for Pilots should be a key priority for all airlines (82.2% strongly agreed, while 16.9 agreed). However, it appears that this is not being taken seriously at an airline level. Only 10.2% of respondents agreed with the statement 'Ensuring and maintaining positive mental health for pilots is a key priority for my airline', while 7.6% strongly agreed. Most participants agreed that the process for supporting positive mental health and managing mental health problems in Pilots should be clearly defined at an airline level (62.5% strongly agreed while 34.8 % participants agreed). However, a very small number (8.5 %) agreed that this process is clearly defined at their airline, while 2.7% strongly agreed. Further, a small number of respondents (6.7%) agreed with the statement 'The Safety Management practices at my airline adequately address issues concerning the support & management of Pilot mental health & wellbeing', with 0.4% strongly agreeing.

Discussion & Conclusion

The wellbeing of pilots is being negatively affected by certain sources of WRS. Critically, wellbeing impacts span the three pillars of wellbeing, and are not limited to MH. Further, sources of WRS have implications from a human performance and flight safety perspective. In accordance with safety management system approaches, specific wellbeing issues and associated performance/safety risks need to be identified, measured and managed. Certain strategies enable some pilots to cope in a work environment that is detrimental for others. If these strategies can be better understood, lessons might be learned in terms of enabling pilots to increase their resilience to wellbeing challenges (including MH challenges). Also, these might be considered in relation to the design of solutions/interventions at different levels (for example, pilots, airlines and the regulator). Specifically, this research indicates that airlines are not adquately managing these issues. Overall, airline organizations might increase their support for preventative mental health treatment. Potentially, airline interventions might focus on enhancing existing safety management system processes/technology to address risks associated with WRS and wellbeing, training pilots (i.e. in relation to wellbeing awareness, coping strategies and self-assessment), and introducing new wellbeing briefing/reporting systems. In addition, future research might address the introduction of digital tools to support pilot management of specific sources of WRS both inside and outside work. The results of this study should be interpreted with potential limitations in mind. Next steps will involve detailed analysis of survey data. A further analysis is planned following a second wave of data collection (February to October 2019). Particpatory co-design activities will also be undertaken with different stakeholders to address wellbeing interventions at different levels.

Acknowledgements

The authors would like to thank pilots for their participation in this study. The views expressed in this study do not represent the views of the authors's employers.

References

Cahill, J., Cullen, P. & Gayor, K. (2018). Estimating the Impact of Work-Related Stress on Pilot Wellbeing and Flight Safety'. Paper presented at The 2nd International Symposium on Human Mental Workload: Models and Applications (H-WORKLOAD 2018), Amsterdam (September 20-21, 2018), Netherlands Aerospace Centre (NLR), The Netherlands.

Cullen, P., Cahill, J. & Gaynor, K. (2017). Pilot wellbeing. Paper presented at the Flight Safety Foundation 2017 – 70th International Air Safety Submit (IASS), Dublin, October 23-25, 2017.

Demerouti, E., Bakker, A. B., Vardakou, I., & Kantas, A. (2003). The convergent validity of two burnout instruments: A multitrait-multimethod analysis. European Journal of Psychological Assessment, 19(1), 12.

Demerouti,E., Veldhuis, W., Coombes, C. & Hunter, R. (2018): Burnout among Pilots: Psychosocial Factors related to Happiness and Score on Simulator Training, Ergonomics, DOI: 10.1080/00140139.2018.1464667.

Engel, G. (177). The need for a new medical model: a challenge for biomedical science. Science, 196, 126-9.

Kroenke, K., Spitzer, R. L., & Williams, J. B. W. (2001). The PHQ-9: Validity of a Brief Depression Severity Measure. Journal of General Internal Medicine, 16(9), 606–613.

Leka, S., Griffiths, A & Cox, T. (2003). Work Organisation & Stress: Systematic Problem Approaches for Employers, Managers and Trade Union Representatives. (Protecting Workers' Health Series: No. 3). Geneva, Switzerland: World Health Organization.

Joseph, C. (2016). Stress Coping Strategies in Indian Military Pilots: Preliminary Observations. International Journal of Aviation, Aeronauctics and Aerospace, Volume 3, Issue 4.

Pasha, T & Stokes (2018). Reflecting on the Germanwings Disaster: A Systematic Review of Depression and Suicide in Commercial Airline Pilots.

Wu AC, Donnelly-McLay D, Weisskopf MG, McNeely E, Betancourt TS, Allen JG. (2016). Airplane pilot mental health and suicidal thoughts: a cross-sectional descriptive study via anonymous web-based survey. Environ Health (2016) 15(1):1–12. doi: 10.1186/s12940-016-0200-6 9.

TRAFFIC FLOW MANAGEMENT FOR TRAJECTORY BASED OPERATIONS: SUPPORTING EFFECTIVE PREDEPARTURE REROUTES

Philip J. Smith, Mark Evans and Amy Spencer
The Ohio State University
Columbus OH

There are a number of tools and procedures identified as applicable to initial Trajectory Based Operations (iTBO). This includes Strategic Planning and Traffic Flow Management such as the use of Ground Delay Programs, Airspace Flow Programs and Collaborative Trajectory Operations Program initiatives (CTOPs). It also includes a focus on route management (including the Pre-Departure ReRoute (PDRR) and AirBorne ReRoute (ABRR) tools and airport surface management (as part of the Tower Flight Data Management program or TFDM). This paper focuses on enhancements to support the effective use of the Pre-Departure ReRoute tool (PDRR). These enhancements emphasize the importance of integration between the Traffic Flow Management System (TFMS) and TFDM.

PDRR provides access to displays which allow the traffic manager to view the flight lists (demand) associated with the different departure fixes for an airport. While this information on departure fix demand is very important, the decision about which flights to reroute and about which flights to reroute first requires real-time information about which flights have already pushed back and about the current lineup in the departure queue.

Multi-route Trajectory Options Sets (TOSs) provide a mechanism for the flight operator to communicate constraints on the reroutes that a particular flight is prepared to accept. By submitting a TOS for each flight, the flight operator is able to provide the FAA (through TFMS) with a prioritized set of alternative routes that a flight is prepared to accept. The dispatcher has reviewed this set prior to submission and the flight crew has a list on the flight release.

Thus, this TOS can provide the FAA traffic manager with information about what tactical reroutes (within 45 minutes of departure) a flight is prepared to accept without the need to further coordinate with dispatch or to return to the gate for additional fuel.

The availability of multi-route TOSs for FAA traffic managers offers the potential to reduce coordination and communication demands for the Air Traffic Control Tower (ATCT) controller, the flight crew, the dispatcher and the traffic manager. An additional benefit is that flights are less likely to have to return to their gates for additional fueling (or cancelation).

This paper describes an operational concept based on these considerations which has been recommended to the FAA by the Collaborative Decision Making Program Flow Evaluation Team (FET) and The Collaborative Decision Making Program Steering Group (CSG).

Operational Concept

This concept focuses on the integration of PDRR and TFDM (FAA, 2018) so that traffic managers can easily see the actual line-up of traffic on the airport surface in order to prioritize the order for making reroutes. It allow traffic managers to simply select a flight shown on the airport surface display in order to access information on the alternative routes in theTOS associated with that flight.

The scenario below assumes that multi-route TOSs are being submitted by the flight operators for pre-departure reroutes that can be made by a traffic manager at the Cleveland ARTCC (ZOB) using PDRR. (In the future, these reroutes might be made locally by the ATCT/TRACON at the affected airport Detroit (DTW) instead.

The goal of this scenario is to illustrate the functionality and procedures necessary to support effective use of PDRR to manage departures. The focus is an airport (DTW) using RNAV SIDS with departure fixes dedicated to that airport (see Figure 1).

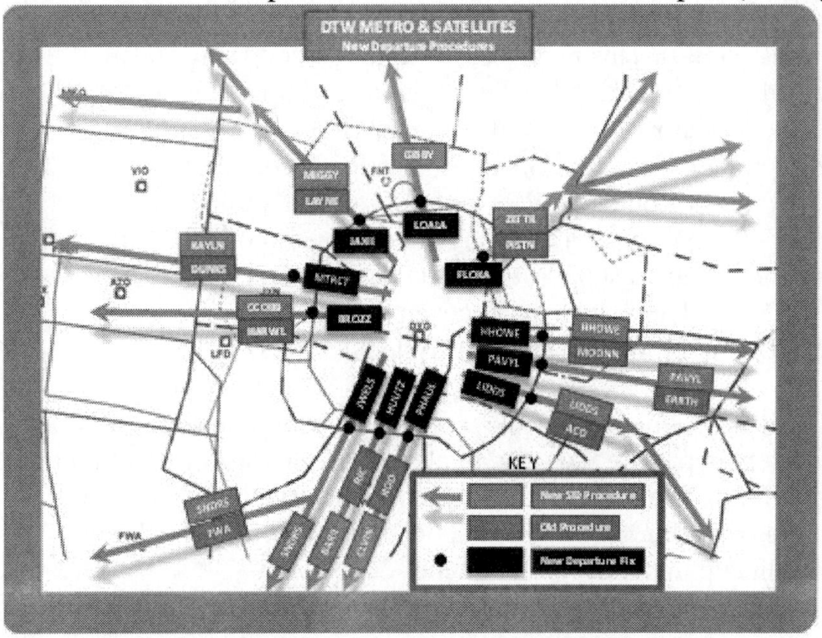

Figure 1. RNAV SIDS for departures out of DTW.

This scenario deals with a weather event in which a frontal system extends from just south of the Southgate departure fixes for DTW (see Figure 1) down into the Indianapolis ARTCC (ZID) airspace. This weather is moving from west to east and progressively impacts the Southgate departure fixes. Later it impacts departures to the southeast via LIDDS. During this period, storm cells to the south also are affecting ZID traffic, resulting in Miles-in-Trail (MIT) restrictions on the DTW departure fixes feeding ZID (JWELS, HUUTZ and PHAUL).

1800Z. Based on a forecast that this frontal system will impact Southgate departure fixes and the southern-most Eastgate departure fix sometime between 2000Z and 0030Z, ATCSCC sends out an FYI SWAP advisory for this time range indicating:

SOUTHGATE DEPARTURES OUT OF DTW (JWELS, HUUTZ and PHAUL) AND EASTGATE DEPARTURES VIA LIDDS AND CAN EXPECT CDRS/SWAP DUE TO WEATHER. USERS SHOULD FILE NORMAL ROUTES BUT CONSIDER SUBMITTING A TOS INCLUDING ALTERNATIVE EASTGATE, SOUTHGATE AND WESTGATE DEPARTURE FIXES (JWELS, HUUTZ, PHAUL, PAVYL, LIDDS and BROZZ) AND FUEL ACCORDINGLY IN ORDER TO EXPEDITE DEPARTURES.

1801Z. As recommended, many, but not all, flight operators begin to submit TOSs to the FAA for use in PDRR for flights included in this FYI advisory. These TOSs contain alternative routes using JWELS, HUUTZ, PHAUL, LIDDS, PAVYL and BROZZ as departure fixes. These flights are fueled for these alternatives that have been reviewed by the responsible dispatchers.

These alternative routes (and associated fuel requirements) are listed on the flight release for the pilots. The TOS that is submitted contains both the CDR (Coded Departure Route) eight letter code (if the alternative route is a CDR), as well as the full route string for that flight, and the prioritization for these routes is indicated by submitting the Relative Trajectory Cost (RTC) of each alternative route in the TOS for a given flight. For those flights for which no TOS has been submitted, the filed route is used for modelling purposes, but that route will not show up as a route in the TOS and will not appear in the list of TOS options in PDRR. In general, flight plans are submitted 60-120 minutes before the scheduled departure time of a flight.

Note also that, if a CTOP is in effect for a given flight, then the dispatcher is expected to file the route assigned by CTOP (and that route is used to assign ground delay if any). If the dispatcher determines that some other route needs to be filed, the flight is assigned a new delay based on the Flow Constrained Area (FCA), if any, that that route flies through. (Since this scenario assumes only PDRR is using the TOSs and that no CTOP initiative is in effect, the remaining discussion does not cover any additional considerations that could arise in the use of TOSs for CTOP alone or in conjunction with PDRR.)

2100Z. The frontal system moving west to east located just south of the DTW Southgate departure fixes is still west of the departure routes fed by those fixes (JWELS, HUUTZ and PHAUL). However, it is already impacting flights departing JWELS that are filed through SNDRS, resulting in a 15 MIT restriction for flights using SNDRS from 2100-2200Z.

2130Z. Cells associated with the front are beginning to significantly affect airspace in ZID making it necessary for flights to deviate from their routes. This results in a 10 MIT as one restriction initiated by ZOB for flights departing the Southgate departure fixes (JWELS, HUUTZ and PHAUL) from 2130-2230Z. This overrides the SNDRS restriction so it is canceled effective 2130. These fix restrictions are made available electronically to the flight operators in real time via some mechanism (This is a future enhancement that could be added to the National Operations Display or NOD.) PDRR is used by a traffic manager at ZOB to tactically reroute some flights within 45 minutes of departure that have been filed to depart via JWELS, HUUTZ or PHAUL so that they can depart using the a route in their TOS that includes the Eastgate departure fixes LIDDS and PAVYL and the Westgate departure fix BROZZ. To do this:

- The traffic manager looks at a surface management display to see how many flights are filed to depart using the restricted fixes (see Figure 2). This display indicates where these flights occur in the departure timeline and how many flights are included. It also indicates where they are on the surface (still at the gate, in the ramp area or in the active movement area). The traffic manager also looks at the display of departure fix loadings in the PDRR and looks at the timeline in order to help decide which departure fixes to consider for each reroute. The view of departures (organized by departure fix) also provides information on which of these flights have associated TOSs. Based on this awareness of activity on the airport surface and the upcoming demand for the departure fixes, the traffic manager decides which flights (if any) to reroute to LIDDS, PAVYL or BROZZ in order to ensure they are ready to depart when the aircraft reaches the runway threshold. Note that, without access to this surface information, the traffic manager would have had to talk with the Tower in order to determine which flights to reroute first.
- Based on the surface map indicating where flights filed to depart via the now restricted fixes (JWELS, HUUTZ and PHAUL) are located on the airport surface, the traffic manager proceeds to make reroutes as deemed necessary. (Note that just because a flight is filed to depart JWELS, HUUTZ or PHAUL does not necessarily mean it needs to be rerouted.)
- To reroute a flight with a TOS that is filed to depart JWELS, HUUTZ or PHAUL as indicated in PDRR, the traffic manager clicks on that flight on the surface map and opens a display in PDRR to look at its TOS (with the routes ordered in terms of their associated RTCs).

The traffic manager views the TOS route options for that flight that show both the CDR code (if applicable) and the full route string, and as appropriate selects the route using LIDDS, PAVYL or BROZZ as the departure fix. This is submitted to the En Route Automation Modernization system (ERAM) as a route amendment.

- PDRR also is used to reroute flights that do not have a multi-route TOS, but without the benefit of pre-coordination with the responsible dispatchers and flight crews.
- For many, but not all, flight operators, when this route amendment is submitted, the dispatcher receives an alert indicating that this flight will be cleared to depart on this new route. This alert indicates both the CDR code (if applicable) and the full route string. This alert also indicates whether the amended route was contained in the TOS and was therefore included on the release for that flight. In response to this alert, the dispatcher informs the flight crew that they should expect to be cleared by the ATCT controller for departure on this pre-coordinated route. To make it easy for the crew to identify which of the alternative routes on their flight release to expect, the dispatcher indicates the CDR code (if applicable) in this communication.
- The crew checks the fuel and weather conditions to make sure they can still accept this route. Assuming they can, they prepare for departure on this reroute.
- The ATCT controller receives a flight strip with the CDR code (if applicable) and full route string indicated.
- The ATCT controller clears the flight by voice to depart on the new route to depart via LIDDS, PAVYL or BROZZ (depending upon which flight amendment was made). At DTW, at present this likely will be done with a full route clearance. (At EWR, in contrast, 70-80% of the time this will be done using the CDR code for an abbreviated clearance. DTW could consider doing this in the future as well to increase efficiency and reduce workload.) Alternatively, Data Comm is used to electronically deliver the new route to the flight deck. In this case, the message includes both the CDR code (if applicable) and the full route string. The CDR code (if applicable) makes it easier for the pilots to quickly ensure that the cleared route is in fact on their flight release. (If not, they would need to clear it with dispatch.)
- The pilots accept the clearance and depart on schedule via LIDDS, PAVYL or BROZZ (depending upon which flight amendment was made).

This process eliminates the departure delay for those flights that have submitted TOSs that include routes using LIDDS, PAVYL and/or BROZZ. Some of the flights that do not have a multi-route TOS may have to pull over until they can coordinate with dispatch and ATC in order to find an acceptable reroute, or may even have to return to a gate to get additional fuel.

2200Z. The frontal system has moved east quickly enough to directly impact the Southgate departure fixes, resulting in significant airborne deviations for departing flights and responses from pilots prior to departure that they are unable to depart using these fixes due to the weather. Departures via those three fixes are therefore halted. These fix closures are communicated electronically to the flight operators in real time via the NOD or some future equivalent.

As with the previous step at 2130Z, the traffic manager identifies those flights filed to depart via the three Southgate departure fixes and when possible uses their TOSs as displayed in PDRR to make reroutes as appropriate.

Note however that, like today, even if a flight doesn't have a route option for a LIDDS, PAVYL and/or BROZZ departure in its TOS it could still be offered a reroute. However, without the pre-coordination provided by the TOS, there may be a much greater delay to accomplish the needed coordination involving the flight crew, dispatch and ATC. In some cases, the result for such flights is that they have to go back to the gate for refueling or cancellation.

2200-2300Z. As the weather progressively moves past the three Southgate departure fixes, flights are allowed to depart via each fix as it opens up.

2300Z. The restriction (closure) on the Southgate departure fixes is allowed to expire. Departures via LIDDS are now starting to deviate so a 20 MIT restriction is put into effect from 2300-0000Z. Similar to the management of reroutes at 2130Z, some of the flights filed to depart LIDDS are rerouted to depart via PHAUL using their TOSs in the RAD. This restriction is communicated electronically to the flight operators in real time via the NOD.

2330Z. Departures via LIDDS are halted due to significant airborne deviations for departing flights and responses from pilots prior to departure that they are unable to depart using these fixes due to the weather. Similar to the management of reroutes at 2130Z, some of the flights filed to depart LIDDS are rerouted to depart via PHAUL using their TOSs in PDRR when available. This restriction is communicated electronically to the flight operators in real time via the NOD.

0000Z. The front passes east beyond LIDDS and the restriction on LIDDS departures is allowed to expire.

Notional Display Illustrating the Integration of PDRR with Surface Information

To complement this scenario, below we provide a notional example illustrating how access to surface data could significantly improve performance (reducing communications and coordination between the ATCT and the Center traffic manager who is using PDRR, and streamlining the inputs required of the Center traffic manager using PDRR in order to make a route amendment). This display (shown in Figure 2) makes use of a previously developed prototype surface management simulation (Smith et al., 2012) to provide a concrete illustration using a hypothetical airport (KMJA) that includes associated departure fixes WICKR, WILEY and NOBLR.

The display below shows that two closed fixes to the west, WICKR and WILEY have been highlighted in the Selection Tool (left window). This results in the highlighting of those flights in the flight list (center window) filed to depart via one of the two closed fixes (WICKR and WILEY). Those flights also are highlighted on the surface map. In this display, the traffic manager has noted that the first flight in the departure queue that is filed to depart via one of the two closed fixes is DAL8889. The traffic manager has double clicked on that flight on the surface map and its data has appeared in a PDRR display. The traffic manager can now select the third option (a route using NOBLR, an open departure fix to the north) in order to select that TOS option as the reroute.

Conclusion

The scenario illustrates how, using an airport surface display that is linked to PDRR, a traffic manager can:

- Use a selection tool to highlight flights that were filed to depart using a departure fix that has been constrained by weather . They are highlighted in both the flight list and on the airport surface display (see Figure 2).

- View the impacted flights have been highlighted in the flight list and on the airport surface display.
- Use the visual display of the locations of these impacted flights on the airport surface in order to determine which are in the active movement area, which have pushed back and but are still in the ramp area, and which are still at their gates.
- Use this information to identify the aircraft that is closest to the departure runway.
- Clicking on that aircraft on the surface map (or in the flight list) to view the available pre-coordinated route options in the TOS in a PDRR display (see Figure 2).
- If judged appropriate, using PDRR to select one of these options as the reroute for that flight. (If a flight doesn't have a TOS, or if the route judged most effective by the traffic manager is not in the TOS, the traffic manager can still open up the PDRR display by clicking on the flight on the airport surface map and then making some other route amendment.)
- This process, supported by appropriate enhancements to PDRR and TFDM, offers an approach to significantly reduce the level of effort and coordination time involving the ATCT controller, Center or TRACON traffic manager, pilot and dispatcher when making a tactical pre-departure reroute. It further provides a mechanism to communicate and consider flight operator priorities and constraints when such a reroute is made.

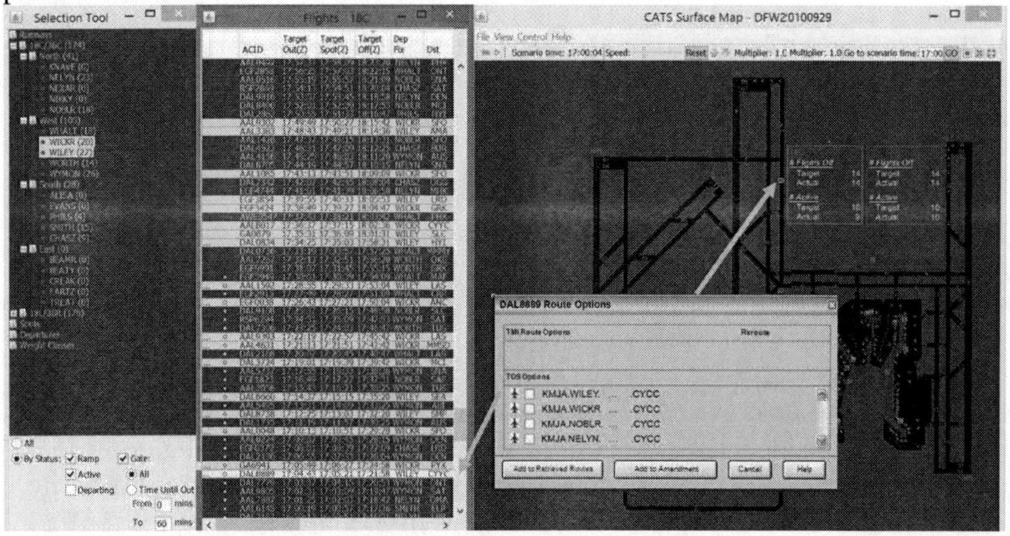

Figure 2. Notional display illustrating the linkage of PDRR with surface management displays.

Acknowledgements

This work was completed in collaboration with the Collaboration Decision Making Program Flow Evaluation Team.

References

FAA. (2018). *Tower Flight Data Management (TFDM).* Retrieved from https://www.faa.gov/air_traffic/technology/tfdm/.

Smith, P.J., Weaver, K., Fernandes, A., Durham, K., Evans, M., Spencer, A. and Johnson, D. (2012). Supporting distributed management of the airport surface. *Proceedings of the 2102 AIAA Digital Avionics Systems Conference*. Williamsburg VA: AIAA.

A FRAMEWORK FOR ASSESSING THE IMPACT OF PERFORMANCE-BASED NAVIGATION ON AIR TRAFFIC CONTROLLERS

R. Jordan Hinson, Michael W. Sawyer, Gary L. Serfoss
Fort Hill Group, LLC
Washington, DC

The air traffic control domain is undergoing significant modernization efforts through technology and procedure enhancements. Understanding the impact of these changes and ensuring these enhancements do not unduly increase controller cognitive workload is essential for success. This research provides a framework for assessing human performance impacts and cognitive workload associated with Performance-Based Navigation for use in an operational air traffic control environment. A panel of human factors and air traffic control subject matter experts assessed a broad set of measures of cognitive workload based on sensitivity, bandwidth, diagnosticity, selectivity, interference, controller acceptance, reliability, and implementation requirements. This resulted in a set of recommended operationally-viable measures of controller cognitive workload. Additionally, a series of potential human performance impacts associated with Performance-Based Navigation were identified. The benefits and limitations of each measure are summarized along with guidance for tailoring the recommended measures based on a research objectives and operational constraints.

The changes introduced by PBN procedures present a wide range of direct and indirect impacts to human performance for both air traffic controllers and flight crews. Achieving the potential benefits associated with PBN procedures requires that controllers and flight crews can effectively assign, execute, modify, and monitor the procedures. Considering the wide range of impacts to controller performance, of particular interest are those that may adversely impact a controller's cognitive workload. Cognitive workload represents just one of the elements of human performance that may be directly or indirectly impacted by PBN procedures. Excessive levels of cognitive workload have been shown to adversely impact human performance in air traffic control and many other similar domains. This paper presents one piece of a larger framework developed to equip the Federal Aviation Administration (FAA) to consistently assess and mitigate the effects of cognitive workload on controller performance. Measuring and managing controller cognitive workload may support the FAA in developing more effective PBN procedures, increasing PBN utilization rates, and ensuring that future technology and procedure changes reduce or do not unduly increase controller cognitive workload.

Methodology

A literature review was first conducted to identify a candidate set of cognitive workload measures for consideration. Each measure identified from the literature was categorized based on five potential measure types: Primary Task (Pri.), Secondary Task (Sec.), Physiological (Phy.), Subjective (Sub.), and Analytical (Ana.) ((Stanton, Salmon, & Rafferty, 2013) (Wilson &

Corlett, 2005). For each measure type, the related measures, source documents, and a brief measure summary were catalogued (Sawyer, Hinson, & Henderson, 2017). To identify which measures would be best given the defined project scope, researchers devised a system for assessing the measures to account for the following criteria developed from the literature: sensitivity (combined with bandwidth), diagnosticity, selectivity, interference, controller acceptance, reliability (combined with transferability), and implementation requirements (Wickens & Hollands, 1999) (Wierwille & Eggemeier, 1993). Scoring criteria definitions were also defined as presented below in Table 1. A workgroup consisting of air traffic control and human factors subject matter experts was then convened to review and rate each measure using the scoring criteria. A consensus approach was taken by the workgroup to assign a value of +1, 0, or -1 for each of the 7 scoring criteria.

Table 1.
ATC Measure Scoring Criteria and Definitions.

Criteria	**1**	**0**	**-1**
Sensitivity *(Sen)*	Measure distinguishes fairly rapid changing levels of cognitive workload, or task load without risk of the measures saturating.	Measure shows moderate variation in task / workload. Scale may become saturated but remains useful to a point.	Measure shows only sensitivity to extreme variations in workload. Measure reaches saturation quickly.
Diagnosticity *(Dia)*	Measure allows the cause of variation in cognitive workload to be identified, or indicates which cognitive resources are most affected.	Measure indicates minimal cause of variation in cognitive workload.	Measure does not indicate cause of variation in workload.
Selectivity *(Sel)*	Measure allows various confounding factors such as noise, physical workload, and emotional stress, to be distinguished from variations in cognitive workload.	Measure accounts for most causes of variation, but may not distinguish some confounding factors or noise.	Measure includes confounding effects which cannot be isolated.
Interference *(Int)*	Measure does not affect primary task performance.	Measure has minimal effect on primary task.	Measure significantly impacts primary task.
Controller Acceptance *(CA)*	Controllers likely accept measure.	Controller is neutral on measure.	Controllers likely reject use of measure.
Reliability *(Rel)*	Measure has documented research of use in ATC.	Measure has documented research with limited use in ATC.	Measure has very little development or validation.
Implementation Requirements *(Imp)*	Neither additional equipment nor specialized personnel are required. Training is minimal.	Minimal equipment or specialized personnel is required.	Significant equipment or specialized personnel are required.

Results

The 33 highest scoring, viable measures are provided in Table 2. Full details on measure identification, assessment, and prioritization are available by technical report (Sawyer et al., 2017).

Table 2.
ATC Measure Assessment Results.

Measure Name	Type	Sen	Dia	Sel	Int	Rel	Imp	CA	**Total**
NASA Task Load Index (TLX)	Sub.	0	1	1	1	1	1	1	**6**
Communications Data	Pri. Ana.	1	1	0	1	0	1	1	**5**
Coordination / Communication Rating	Sub.	1	0	0	1	0	1	1	**4**
Simplified Subjective Workload Assessment Technique (SWAT)	Sub.	0	1	-1	1	1	1	1	**4**
Trajectory-based complexity (TBX)	Ana.	1	0	0	1	0	1	1	**4**
ATC Tape Communication Analysis	Ana.	1	0	-1	1	0	1	1	**3**
Localized traffic density	Ana.	0	0	-1	1	1	1	1	**3**
Number of Handoffs	Ana.	0	0	-1	1	1	1	1	**3**
Simulator Test Score of Performance	Pri.	1	-1	-1	1	1	1	1	**3**
Subjective Workload Assessment Technique	Sub.	0	1	-1	1	1	1	0	**3**
ATC Complexity Measurement	Ana.	0	0	0	1	0	0	1	**2**
Bedford Scale	Sub.	0	-1	0	1	0	1	1	**2**
Checklist to Evaluate Airspace Complexity	Ana.	0	0	-1	1	0	1	1	**2**
Communication time, message length	Ana.	1	0	-1	1	-1	1	1	**2**
Communications Efficiency	Ana.	1	0	-1	1	-1	1	1	**2**
Handoff Acceptance Latency	Ana.	1	-1	-1	1	0	1	1	**2**
Hart & Hauser Rating Scale	Sub.	0	1	1	-1	-1	1	1	**2**
Mental Workload Index (MWLI)	Pri. Ana.	0	0	0	1	-1	1	1	**2**

Measure Name	Type	Sen	Dia	Sel	Int	Rel	Imp	CA	**Total**
Number of control actions	Ana.	0	-1	-1	1	1	1	1	**2**
Performance and Objective Workload Evaluation Research (POWER)	Ana.	1	-1	-1	1	1	0	1	**2**
Projective SWAT	Sub.	0	1	-1	1	-1	1	1	**2**
SME / Over-the-shoulder ratings	Sub.	0	-1	-1	1	1	1	1	**2**
Time required	Ana.	1	-1	0	1	0	0	1	**2**
Air Traffic Workload Input Technique (ATWIT)	Sub. Sec.	0	-1	-1	1	1	1	1	**2**
Behavioral Markers	Pri. Ana.	0	-1	0	0	0	1	1	**1**
Continuous Subjective Assessment of Workload (C-SAW)	Sub.	1	-1	-1	1	0	0	1	**1**
Number of aircraft under control per hour / traffic count	Ana.	0	-1	-1	1	0	1	1	**1**
Recall Ability	Sec.	1	-1	-1	-1	1	1	1	**1**
Respiration	Phy.	0	-1	0	1	0	0	1	**1**
Subjective Workload Dominance (SWORD) Technique	Sub.	0	-1	-1	1	0	1	1	**1**
Task Analysis Workload (TAWL)	Pri. Ana.	-1	1	0	1	-1	0	1	**1**
The Projective SWORD Technique (Pro-SWORD)	Sub.	0	-1	-1	1	0	1	1	**1**
Thermo-vascular activities	Phy.	1	-1	0	1	-1	0	1	**1**

PBN Human Performance Impacts

A list of identified human performance impacts related to cognitive workload impacted by PBN procedures were gathered from a review of operational safety reports from the Aviation Safety Reporting System (ASRS), research studies, industry guidance documents, and interviews with Human Factors and ATC subject matter experts. The resulting list of PBN human performance impacts are grouped into 11 categories listed below in Figure 1.

Aircraft Performance	Impacts caused by aircraft deviating from the expected flight path, altitude, and or speed.
ATC Automation	Impacts caused by ATC automation that supports controlling traffic and supporting tasks (map display, flight-plan processing, eligibility, etc.).
Acceptance	Impacts characterized by mistrust in a PBN procedure if it is perceived as less efficient, less safe, flawed, or otherwise inferior to previous or conventional routes.
Communications	Impacts characterized by the effect of PBN procedures on the coordination and communication among air traffic service users including Air Traffic Controllers, Flight Crews, Airport Operators, Traffic Management, etc.
Mixed Equipage	Impacts caused by aircraft using RNAV navigation in the same environment as aircraft using conventional navigational capabilities.
Nominal Operations	Impacts caused by the range of normal operating conditions that affect controller performance during day-to-day operations.
Design of Airspace Procedures	Impacts characterized by the design elements of PBN procedures (speed, course, altitudes, etc.) and interactions with other elements of the airspace (other routes, airspace boundaries, etc.)
Recovery	Impacts characterized by how PBN procedures affect a controller's response to an event that could lead to an adverse outcome.
Monitoring	Impact affecting how a controller monitors the airspace.
Training	Impacts relating to how training is conducted, including when it occurs, how often, what type, and its effectiveness.
Weather & Wake	Impacts to human performance caused by the presence and management of adverse weather conditions and the effects of wake turbulence.

Figure 1. PBN Human Performance Impact Categories.

Conclusion

The PBN cognitive workload assessment framework includes tools for assessing the impact of PBN procedures including recommended measures of cognitive workload. Further selection using the scoring matrix results of each measure resulted in the following recommended PBN Workload Measures for operational use: NASA Task Load Index (TLX), Trajectory Based Complexity Calculation (TBX), Communication Efficiency Rating, and Number of Handoffs. Alternative measures may be selected using the scoring matrix results to match specific research needs and constraints. Additionally, the framework recommends using interviews and impact surveys to assess potential human performance impacts associated with Performance-Based Navigation.

For more information on this framework and the tools supporting this research, see PBN Cognitive Workload Analysis Results Report (Hinson, Serfoss, & Sawyer, 2018b). For a detailed analysis of the science of Cognitive Workload, the pros and cons of the many various evaluation methods, and analysis and discussion of which cognitive workload tools seem most applicable and usable in the PBN air traffic controller environment, see PBN Cognitive Workload Analysis Plan (Sawyer et al., 2017). For a complete guide to the framework providing the appropriate tools and instructions for understanding, analyzing, and beginning to mitigate the impact of PBN procedures on controller performance including a full list of PBN Human Performance Impacts, see the report (Hinson, Serfoss, & Sawyer, 2018a).

Acknowledgements

Fort Hill Group would like to thank the FAA NextGen Human Factors Division (ANG-C1) for supporting and funding this research and Bill Kaliardos for providing technical guidance throughout the project. Additionally, we would like to thank the aviation safety subject matter experts who provided the valuable insight necessary to developing these results. The results presented herein represent the methodology and results of this research project and do not necessarily represent the view of the Federal Aviation Administration.

References

Hinson, R. J., Serfoss, G., Sawyer, M. W. (2018a). *Framework for Collecting and Utilizing PBN Human Performance Metrics.* Washington, DC: delivered to the Federal Aviation Administration with the intent of future publication.

Hinson, R. J., Serfoss, G., Sawyer, M. W. (2018b). *PBN Cognitive Workload Analysis Results Report.* Washington, DC: delivered to the Federal Aviation Administration with the intent of future publication.

Sawyer, M. W., Hinson, R. J., Henderson, A. S. (2017). *PBN Cognitive Workload Analysis Plan.* Washington, DC: Fort Hill Group, Delivered to the Federal Aviation Administration with the intent of future publication.

Stanton, N., Salmon, P. M., & Rafferty, L. A. (2013). Human factors methods: a practical guide for engineeringand design. *Ashgate Publishing, Ltd.*

Wickens, C. D., & Hollands, J. G. (1999). *Engineering psychology and human performance (3rd ed.).* New Jersey: Prentice Hall.

Wierwille, W. W., & Eggemeier, F. T. (1993). Recommendations for mental workload measurement in a test and evaluation environment. *Human Factors, 35(2)*, 263-281.

Wilson, J. R., & Corlett, N. (Eds.) (2005). *Evaluation of Human Work.* CRC Press.

ASSESSING HUMAN-SYSTEM RESILIENCE POTENTIAL THROUGHOUT THE DEVELOPMENT LIFECYCLE*

Amy L. Alexander
MIT Lincoln Laboratory (MIT LL)
Lexington, MA USA
Dan Herschler
Federal Aviation Administration (FAA)
Washington DC, USA

We worked with subject matter experts to create a human-system resilience checklist that can be utilized during Independent Operational Assessments (IOAs) of air traffic control systems as part of the system acquisition process. The checklist focuses on four key areas for evaluating human-system resilience characteristics: procedures, system use, workload, and training. A resilience scoring method indicates areas where a human-machine system under consideration does or does not have resilient characteristics. Overall resilience scores can be compared among design alternatives, or across different points in system development for a particular design. The ultimate intent is to provide guidance and metrics that will enable the FAA to address human-system resilience aspects in the implementation of NextGen capabilities in the National Airspace System (NAS). The goal of increased resilience is to reduce the risks and potential impacts of disruptive events, and to safeguard the efficiency, safety, and cost effectiveness of NextGen NAS operations.

The Federal Aviation Administration's NextGen program uses many complex systems and technologies to increase the efficiency, safety, and cost effectiveness of the National Airspace System. Although NextGen systems are designed to achieve defined system availability requirements, system degradation and failure are still a very real, if remote, possibility. Designing and assessing systems with resilience to failures in mind can reduce the risks or potential impacts of degradations. Looking to the literature, there are a variety of definitions of resilience (see Reason, 2000; Sheridan, 2008); however, a number of common characteristics emerge relating to anticipating adverse effects, withstanding unexpected conditions, maintaining control, sustaining operations, and recovering quickly when something goes wrong. Resilience is defined by the FAA as maintaining safety and a minimum level of service in reaction to system failures or degradations (FAA, 2016). The underlying goal is to prevent or mitigate impacts on air traffic operations.

Previous work (e.g., Hollnagel, Woods, & Leveson, 2006) has identified characteristics of resilient organizations and human-machine systems, and initial experimental methods for assessing resilience potential have been developed. However, these methods primarily apply to existing or well-prototyped systems. In an effort to assess the resilience potential of an

---

* DISTRIBUTION STATEMENT A. Approved for public release. Distribution is unlimited.
This material is based upon work supported by the Federal Aviation Administration under Air Force Contract No. FA8702-15-D-0001. Any opinions, findings, conclusions or recommendations expressed in this material are those of the author(s) and do not necessarily reflect the views of the Federal Aviation Administration.

operational capability earlier in the system development lifecycle, we worked with subject matter experts to create a human-system resilience checklist that can be utilized during Independent Operational Assessments (IOAs) of FAA air traffic control systems as part of the system acquisition process. The checklist focused on four key areas, identified through collaboration with subject matter experts in conjunction with review of the resilience literature, that should be considered when evaluating human-system resilience characteristics: procedures, system use, workload, and training. A resilience scoring method was developed to provide an indication of areas where a system under consideration does or does not have resilient characteristics. The overall resilience score can then be compared to design alternatives, or across different points in the system development lifecycle for that particular design and operational context. The checklist and scoring system has yet to be validated, but upcoming IOA testing is anticipated to provide insight and feedback about the utility of this approach for assessing human-system resilience.

Method

The first step in creating the human-system resilience checklist was to identify resilient characteristics of NextGen systems, including ways to build, enhance, and assess the resilience of complex human-machine systems. MIT LL conducted a literature review on characteristics of resilient systems, particularly focused on human-automation systems (Yenson et al., 2015). System reliability, system predictability, and operator engagement emerged as three key areas for examining resilience potential. The identified characteristics of resilient automation systems were then translated into a list of phrases (e.g., a resilient system is able to handle "unknown unknown" situations). These phrases formed the basis of a resilience job aid that was originally developed in reference to the safety risk management (SRM) process, without a specific target application or end user group. An excerpt from this job aid is presented in Figure 1. The job aid specifically pointed out questions to ask and actions to take, provided detailed explanations and rationales, references to SRM documentation, and included a basic scoring method for assessing resilience potential.

	Guidance	Phase/ Document	Reference	Scoring
	risk.			
8.	For this human-machine system, what are indicators of successful system performance? Of degraded system performance? Of a failed system? Can these indicators be specified by certain measurable qualities? Such qualities and indicators can be used in combinations as indicators of human-machine system resilience. Do the metrics appropriately measure system recovery and performance levels after adverse events?	PSP, IIA	4.2.2	□ 0: System resilience metrics not identified □ 1: System resilience metrics identified □ 2: System resilience metrics identified and considered appropriate
	Rationale: To determine if the system is resilient, practitioners should identify metrics that indicate the system's ability to adapt to, react to, and learn from abnormal conditions. These metrics can be found by selecting events or other indicators of degraded performance, such as a hazardous weather event, and monitoring a performance metric, such as throughput, before and after the event to determine how well performance returns after an event.			
9.	What are the minimum performance levels for the resilience metrics? (e.g. < 3 minutes to return to an acceptance rate of 25 aircraft per minute after an event)	PSP, IIA	4.2.2	□ 0: Baseline metric values not set □ 1: Baseline metric values set
	Rationale: To determine if the system is resilient, practitioners need to set a baseline for the nominal acceptable level of performance for each of the selected resilience metrics so that later performance data can be compared against the required performance.			

Figure 1. Original Resilience Job Aid Excerpt

Various discussions regarding resilience with the FAA led us to the Independent Safety Assessment Team (AJI-321) of the FAA Air Traffic Organization's (ATO) Safety and Technical Training office, which is responsible for conducting independent operational assessments (IOAs) of designated NextGen systems. IOAs verify new FAA systems or solutions are suitable, operationally effective, and safe prior to deployment in the NAS. Specifically:

- IOAs are independent from the FAA office responsible for deploying the new system/capability.
- IOAs are conducted at operational key sites during live NAS operations.
- IOAs are major structured assessments with the purpose of identifying safety hazards and operational concerns with new systems/capabilities.

AJI-321 agreed for IOA to be a focus area for our work, and we coordinated across seven working group meetings to review the original resilience job aid and customize it for use during IOAs. We determined that a more streamlined checklist would be most appropriate for the IOA context. Working group meetings then focused on carefully reviewing the overall checklist content, categorizing questions in a meaningful way, and revising the wording of the questions and their associated responses. Usability and usefulness of the checklist as well as a resilience scoring system were also discussed as our checklist development progressed.

Checklist

The final checklist contained questions broken down into four key categories for evaluating human-system resilience characteristics: procedures, system use, workload, and training. Example questions from each checklist section are presented in Figures 2-5. Questions were presented with up to four response options, each having a point value associated with it as well as a color-coded indicator of goodness (red: not indicative of a resilient system, yellow: resiliency needs improvement; green: indicative of a resilient system). The evaluator was instructed to select the most appropriate response for each question, and there were comment fields for any additional notes that would be helpful to capture.

4. Are detailed and appropriate procedures available for a wide range of situations, including:	See sub-question responses below:			
a. System usage under nominal conditions?	0: No	1: Yes, but most procedures need to be improved	2: Yes, but some procedures need to be improved	3: Yes
b. The most frequent and/or critical known off-nominal events?	0: No	1: Yes, but most procedures need to be improved	2: Yes, but some procedures need to be improved	3: Yes
c. Assessing system recovery and performance levels after adverse events?	0: No	1: Yes, but most procedures need to be improved	2: Yes, but some procedures need to be improved	3: Yes
d. Bringing the system down and back online for maintenance?	0: No	1: Yes, but most procedures need to be improved	2: Yes, but some procedures need to be improved	3: Yes
e. Certification of systems?	0: No	1: Yes, but most procedures need to be improved	2: Yes, but some procedures need to be improved	3: Yes
Comments:				

Figure 2. Example Procedures Checklist Questions

9. Does the system notify the controller if a degradation occurs?	0: No	1: For some critical functions	2: For most critical functions	3: For all functions
Comments:				
10. Are there design aspects within the system (e.g., alerts, warnings) that safeguard against controller errors and adverse conditions?	0: No	1: For some critical functions	2: For most critical functions	3: For all functions
Comments:				

Figure 3. Example System Use Checklist Questions

24. What types of tasks are performed by the controller under steady-state (i.e., nominal) conditions?	0: Tasks involve more passive monitoring than intended	0: Tasks involve more active engagement than intended	2: Tasks are the appropriate passive/active mix
Comments:			
25. Under steady-state conditions, does the system allow for an appropriate controller workload level?	0: Workload is too low – controller is disengaged	0: Workload is too high – controller is overloaded	2: Workload is appropriate
Comments:			

Figure 4. Example Workload Checklist Questions

33. Which of the following are provided as part of the human-machine system training protocol?	See sub-question responses below:			
a. Minimum training requirements?	0: Not addressed	1: Yes, but requirements need to be greatly improved	2: Yes, but requirements need to be somewhat improved	3: Yes, addressed adequately
b. Training on system vulnerabilities?	0: Not addressed	1: Yes, but training needs to be greatly improved	2: Yes, but training needs to be somewhat improved	3: Yes, addressed adequately
c. Operational aids (e.g., cheat sheet, help line) for less-experienced users?	0: Not addressed	1: Yes, but aids need to be greatly improved	2: Yes, but aids need to be somewhat improved	3: Yes, addressed adequately
d. Training sessions on contingency procedures?	0: Not addressed	1: Yes, but training needs to be greatly improved	2: Yes, but training needs to be somewhat improved	3: Yes, addressed adequately
e. Training sessions on novel events?	0: Not addressed	1: Yes, but training needs to be greatly improved	2: Yes, but training needs to be somewhat improved	3: Yes, addressed adequately
Comments:				

Figure 5. Example Training Checklist Questions

Checklist Scoring

A basic scoring system was developed to tally across responses and provide an ordinal resilence score for each of the four categories. An example resilience scorecard for the procedures category is presented in Figure 6. Total points possible are broken into three levels to provide a general assessment of low/moderate/high human-system resilience. Individual category scores can then be combined to provide an overall human-system resilience score, as shown in Figure 7.

Procedure Resilience Scorecard

Question	Response Score	Max Score Possible (Benchmark)
1.		3
2.		3
3a.		3
3b.		3
4a.		3
4b.		3
4c.		3
4d.		3
4e.		3
5.		1
6.		1
Total		29

0 – 9: Low Procedure Resilience
10 – 19: Moderate Procedure Resilience
20 – 29: High Procedure Resilience

Figure 6. Procedure Resilience Scorecard

Overall Resilience Scorecard

Category	Category Score	Max Score Possible (Benchmark)
Procedures		29
System Use		37
Workload		20
Training		47
Other (Q40)		1
Total		134

0 – 44: Low Resilience
45 – 89: Moderate Resilience
90 – 134: High Resilience

Figure 7. Overall Resilience Scorecard

This simple scoring system was developed so as not to imply any unwarranted precision in quantifying certain responses or categories over others. The notion here is that the checklist provides an indication of areas where a system under consideration does or does not have resilient characteristics, and a basis of comparison among design alternatives, or across different points in system development for a particular design, to determine if the design of a system is improving over time from a resilience perspective.

Conclusions

In an effort to assess the resilience potential of a system, we worked with subject matter experts to create a human-system resilience checklist that can be utilized during IOAs of air traffic control systems as part of the system acquisition process. The checklist and scoring method presented here have yet to be validated, but application of the revised checklist during upcoming IOA testing may provide initial validation and feedback about the utility of the checklist approach for assessing human-system resilience.

Acknowledgements

This material is based upon work supported by the Federal Aviation Administration under Air Force Contract No. FA8702-15-D-0001. Any opinions, findings, conclusions or recommendations expressed in this material are those of the authors and do not necessarily reflect the views of the United States Government or the Federal Aviation Administration.

References

Federal Aviation Administration (2016). *Performance Based Navigation NAS Navigation Strategy 2016.* US Department of Transportation.

Hollnagel, E., Woods, D. D., & Leveson, N. (2006). *Resilience Engineering: Concepts and Precepts.* Aldershot, UK: Ashgate, 2006.

Reason, J. (2000). Reducing human error through safety management practices. Presented at *The 14th Annual FAA/CAA/Transport Canada Human Factors in Aviation Maintenance Symposium.* Vancouver, Canada. Retrieved from: http://www.faa.gov/about/initiatives/maintenance_hf/library/documents/media/mx_faa_(formerly_hfskyway)/strategic_program_plan_(1998)/14th_symposium/reducing_human_error_through_safety_management_practices.pdf

Sheridan, T. (2008) Risk, human error, and system resilience: Fundamental ideas. *Human Factors, 50,* 418-426. doi: 10.1518/001872008X250773

Yenson, S. K., Phillips, S., Davis, A., & Won, J. (2015). Exploring human-system resilience in air traffic management technologies. Presented at the *2015 IEEE/AIAA 34th Digital Avionics Systems Conference*. DASC. doi: 10.1109/DASC.2015.7311403

THE NASA MATB-II PREDICTS PROSPECTIVE MEMORY PERFORMANCE DURING COMPLEX SIMULATED FLIGHT

Kathleen Van Benthem, Caitlin Shanahan, Chunyun Ma, Adam Fraser, Chris M. Herdman
Carleton University, Ottawa, Canada

Prospective memory is essential for flight, where failures can result in incorrect flight control settings, leading to loss of life and equipment. Furthermore, prospective memory is highly-sensitive to pilot age, cognition, and experience. This research reports on the relation of the NASA Multi-Attribute Test Battery-II (MATB-II) to prospective memory during simulated VFR flight (N=51). Prospective memory was indexed with specialized radio calls that were associated with non-focal visual cues. Linear regression models examined the relative association of MATB-II variables to prospective memory in low and high workloads. System monitoring, psychomotor tracking, and resource management, generally at higher difficulty levels, were the variables most predictive of prospective memory, r^2 =0.41. Pilot experience improved the model in the high-workload condition. Estimating risk for prospective memory failures via multitasking ability, with a focus on monitoring tasks, may inform cognitive assessment approaches to enhance aviation safety.

Prospective memory is a cognitive construct associated with remembering to perform critical tasks in the future and is a skill relevant to successful daily living, including aviation outcomes (Dismukes & Nowinski, 2007). A retrospective look at aviation incidents and accidents found that 74 of the 75 reports associated with memory errors were, in fact, prospective memory failures (Nowinski, Holbrook, & Dismukes, 2003). Failures of prospective memory in the cockpit, such as forgetting to adjust the flaps or to lower the landing gear, can have disastrous results and can occur with pilots at any level of expertise (Dismukes & Nowinski, 2007). Research has shown that prospective memory is sensitive to mental workload demands during flight and a pilot's ability to detect relevant memory cues in the environment (Van Benthem, Herdman, Tolton & LeFevre, 2015).

Despite its relevance to aviation safety, cognitive assessments designed for pilots have yet to explicitly measure prospective memory. CogScreen-AE (Kay, 1995) and CogState (CogState Ltd., Melbourne, Australia) are comprehensive neuropsychological tests with links to pilot performance, however neither test has demonstrated an association with the risk of prospective memory failures.

We selected the MATB-II as a possible predictor of prospective memory during complex simulated flight due to the similar demands for cue detection inherent in both the MATB-II and prospective memory. The NASA Multi-Attribute Test Battery-II (Santiago-Espada, Myer, Latorella, & Comstock, 2011) is a cognitive screening tool designed for aviators that incorporates planning, vigilance, and monitoring, with varying levels of multitasking requirements.

Method

Participants

Participants were licensed and medically certified aeroplane pilots (or those with permits) who had flown within the last 18 months (N=51). Pilots were recruited from local flying clubs, aviation interest groups, and flight schools via newsletters, posters, and social media. All research activities took place at a university flight simulation research laboratory and were part of a larger research agenda investigating the cognitive health screening and intervention for general aviation pilots.

Table 1. Description of Full Pilot Sample (N=51)

	Minimum	Maximum	Mean	Std. Deviation
Age	17	71	46.3	17.4
PilotLevel	1	6	4.1	1.3
Flighthours	2	12000	1311.0	2592.3
Recent Pilot-in-Command hours	0	582	49.1	96.3
YearsLicensed	1	70	14.4	14.5
Simulator Hours	0	1000	88.7	192.1

Procedure

The experiment took place over two sessions. At the first session, pilots were provided with a briefing of the flight tasks, which were based on the premise of a search and rescue familiarization flight. The flights were conducted in two segments, where pilots had access to an electronic navigational aid for one of the legs. Pilots were provided with a briefing regarding the flight simulator and completed three warm-up flights where they practised performing the required tasks, including the prospective memory task. After the warm-up session, pilots were given a second briefing on the flight plan, and material required to complete the flight and the tasks. Pilots wore a wristband, which collected biometric information such as heart rate, and a lightweight wireless 14-channel electroencephalography headset.

An ongoing peripheral detection task (PDT) was also undertaken by participants. The PDT was designed to measure mental workload. Pilots were provided with one break time to hydrate or rest briefly, if required. During this break pilots were queried regarding their situation awareness. The biometric, PDT, and situation awareness data were not analyzed for this report. The second session included completion of the three difficulty-levels of the MATB-II (counterbalanced presentation order) and a virtual reality flight simulation task using a custom flight control unit and graphics displayed using the Oculus Rift headset (Microsoft) (also part of the larger research agenda). The order of MATB-II versus flight simulation was alternated so that half the participants completed the virtual reality flight first.

Flight Simulation Apparatus

Pilots flew a converted Cessna 172 aircraft simulator. The cockpit displays flight information via a virtual '6-pack' (i.e., the primary flight controls) and was equipped with a yoke, throttle and flaps. The graphics were produced by Prepar3D (Lockheed Martin) "on the fly" and were modeled after Canadian aerodromes and their surrounding terrains. The flight graphics were displayed on a broad-angle display system consisting of eight theater-quality 1080p projectors and a 14-foot tall, 180-degree curved screen that provided 45 degrees of vertical field of view and 120 degrees of horizontal field of view. The time and the pilot's location, airspeed, heading, bank, pitch, and altitude were recorded at one hertz.

Measures

Prospective memory. The prospective memory task assessed participant's ability to perform radio calls in response to non-focal visual cues. Participants were instructed to alternate between two scripted calls each time they detected the appearance of the cue – a right-facing arrow presented on a screen mounted to the cockpit. This task required that participants remembered to check for the visual cue, and their previous radio call to form appropriate intentions. Prospective memory was measured as the ratio of the number of calls made over the time spent in each leg (leg 1=low workload and leg 2=high workload).

MATB-II multitasking measures. The Multi-Attribute Task Battery II (MATB-II) is a neuropsychological test designed to measure multitasking and mental load (Santiago-Espada et al., 2011). The MATB-II tasks were configured to provide three difficulty levels (low, medium, and high). As illustrated in Figure 1, the MATB-II subtests included system monitoring, psychomotor tracking, communication, and resource management. In the system monitoring module, participants were tasked with clicking on the green light when it went out, clicking on the red light when it appeared, and clicking on the scales when the central indicators deviated significantly from centre. For the tracking module, when manual mode was engaged, the participants used a joystick to keep the reticle inside the central square.

Participants were expected to monitor a task scheduling module to become aware of upcoming tracking tasks. During predefined periods, the communications module intermittently broadcasted messages. Each message was prefixed by a call sign. If the call sign matched the ownship call sign, then participants were to change the frequency of a specific radio. The resource management task required participants to route fuel into two main tanks (top), maintaining the fuel level at the target level indicated by the shaded blue region while circumventing blocked fuel-lines (in red).

For each level of difficulty, the outcome measures from the MATB-II included total occurrences where errors in the system monitoring displays were not detected and responded to (lapses in vigilance), root mean square tracking error, and for resource management, the average amount of time pilots did not maintain adequate fuel levels, and average units of fuel deficit. After the completion of the tasks at each difficulty level, the MATB-II presents a modified NASA Task Load Index (TLX) screen to measure subjective mental workload (Cao, Chintamani, Pandya, & Ellis, 2009). Communication scores were not used as most pilots scored 100% on this task.

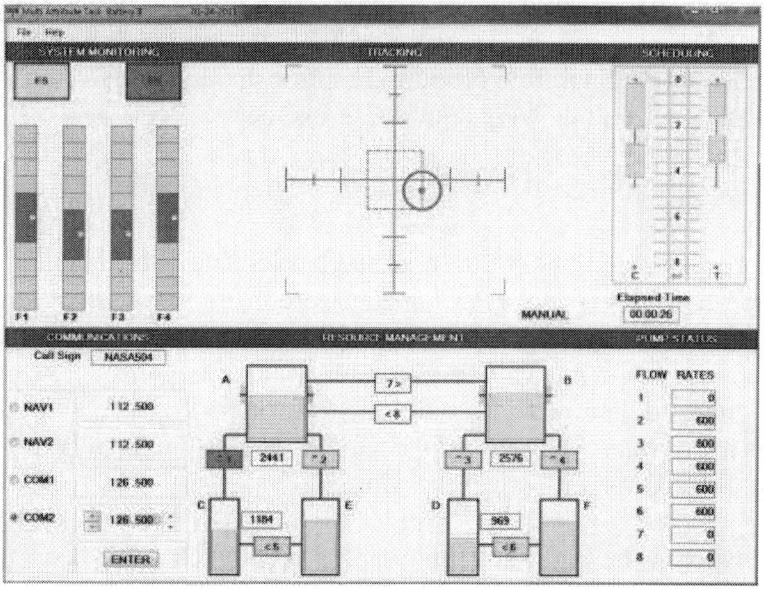

Figure 1. The MATB-II interface. Clockwise from the top-left: system monitoring (lights and scales), psychomotor tracking, scheduling (non-task), pump status (non-task), resource management, and communications.

Results

Pilot Attributes as Predictors of Prospective Memory and the MATB-II

As shown in Table 2, age was significantly associated with prospective memory in the low- but not high-workload condition. In contrast, pilot expertise positively correlated with prospective memory in the high- but not low-workload. The exception to the relationship of expertise and prospective memory was found for number of years licensed, where more years licensed was associated with lower scores. A post hoc analysis showed that the trend towards a deleterious effect of expertise on low workload prospective memory may be due to its conflation with age (e.g., age and years license were significantly correlated, $r = 0.634$, $p < .01$).

Table 2.
Pearson Correlations between Prospective Memory Scores and Pilot Attributes

PM	Age	Years Licensed	Hours Flown	Recent Hours	License Level
Low Workload	**-0.381***	-0.242	-0.231	-0.067	0.099
High Workload	-0.121	0.167	**0.282***	**0.329***	**0.321***

Note. * = $p < .05$ and ** = $p < .01$, two-tailed.

As shown in Table 3, older age was positively correlated with the error scores for all subtests (other than resource management time measure). Pilot level and hours flown did not correlate with any of the MATB-II subtests. However, recent pilot-in-command hours was negatively correlated with system monitoring errors in the medium- and high-difficulty levels.

Years licensed was positively correlated with errors in system monitoring, although this was most likely an artefact of the negative effect that age showed on performance.

Table 3.
Pearson Correlations between MATB-II Subtests, and Pilot Attributes and Prospective Memory

MATB-II Variable	Difficulty Level	Age	Years Licensed	Recent Hours	Low-Workload PM	High-Workload PM
System Monitoring Errors	Low	**.293****	0.165	-0.173	-.307*	-0.266
	Medium	**.247***	0.108	-.280*	-.346*	-.483**
	High	**.393****	0.252	-.255*	-.514**	-.401**
Tracking Deviation	Low	0.227	0.122	-0.201	-0.042	-.328*
	Medium	**.288***	0.157	-0.138	-0.143	-.331*
	High	**.274***	**0.335***	-0.152	-.379*	-0.161
Resource Management (Average Units Under)	Low	**.538****	0.217	-0.170	-.527**	-.314*
	Medium	**.478****	0.223	-0.171	-.463**	-.434**
	High	**.474****	0.189	-0.061	-.437**	-0.238

Note. PM = prospective memory. * = $p < .05$ and ** = $p < .01$, two-tailed.

Models of Prospective Memory Performance

We developed a hierarchical linear regression model for each level of workload. Each model consisted of five blocks (using the stepwise feature to account for the large number of potential predictors). The first block was age, then experience factors, then each of the three difficulty levels of tracking, system monitoring, and resource management. As shown in Table 4, the final model for low-workload prospective memory was comprised of the high-difficulty scores for tracking, system monitoring, and resource management, $F(4, 39)=7.53$, $p< .001$, $r^2 = 0.38$. The strongest effect was found with system monitoring.

Table 4.
Summary of Multiple Regression Analysis for Low-Workload Prospective Memory

MATB-II Subtest (difficulty level)	*B*	*SE(B)*	*β*	*t*	Sig. (*p*)
Tracking Deviations (3)	-0.030	0.015	-0.251	-1.971	0.056
System Monitoring Errors (3)	-0.191	0.074	-0.352	-2.573	0.014
Resource Management Deviation (3)	-0.002	0.001	-0.261	-1.941	0.059

The final model for prospective memory (high-workload) was comprised of similar-sized effects, ranging from $\beta= 0.279$ to -0.295, from the medium-difficulty scores for system monitoring and resource management and pilot license level, $F(3, 41) =7.54$, $p< .001$, $r^2 = 0.31$.

Discussion and Implications

In this work, we quantified the effects of pilot attributes and cognitive functions on prospective memory. Our results support the theory that prospective memory is strongly tied to executive functions, such as multitasking and cue detection (Dismukes & Nowinski, 2007; Kliegel, Martin, McDaniel, & Einstein, 2002; Van Benthem, Herdman, Tolton & LeFevre, 2015). We also found that pilot attributes, such as lower license level and older age, were associated with lower prospective memory and MATB-II subtests. In light of these findings, future tests designed to predict pilot prospective memory will benefit from a design that features executive cognitive functions and multitasking ability where monitoring the environment is integral to the task.

Acknowledgements

Research infrastructure was supported through the Canadian Foundation for Innovation and the Ontario Innovation Trust. We thank Anne Barr and Andrew Staples of the Carleton University, ACE Laboratory for their consultation and engineering expertise.

References

Cao, A., Chintamani, K. K., Pandya, A. K., & Ellis, R. D. (2009). NASA TLX: Software for assessing subjective mental workload. *Behavior Research Methods, 41* (1), 113–117. doi: 10.3758/BRM.41.1.113.

Dismukes, R., & Nowinski, J. (2007). Prospective memory, concurrent task management, and pilot error. In A. Kramer, D. Wiegmann, & A. Kirlik (Eds.) *Attention: From Theory to Practice*. (pp. 225-236). New York: Oxford.

Kay, G. (1995). *CogScreen Aeromedical Edition Professional Manual (Professional Manual).* Florida: Psychological Assessment Resources, Inc.

Kliegel, M., Martin, M., McDaniel, M. A., & Einstein, G. O. (2002). Complex prospective memory and executive control of working memory: A process model. *Psychologische Beitrage*, *44*(2), 303–318.

Nowinski, J. L., Holbrook, J. B., & Dismukes, R. K. (2003). Human memory and cockpit operations: An ASRS study. *In Proceedings of the 12th International Symposium on Aviation Psychology* (pp. 888-893). Dayton, OH: The Wright State University.

Santiago-Espada, Y., Myer, R. R., Latorella, K. A., & Comstock, J. R. (2011). *The Multi-Attribute Task Battery II (MATB-II) Software for Human Performance and Workload Research: A User's Guide*, 269. NASA/TM–2011-217164.

Van Benthem, K. D., Herdman, C. M., Tolton, R. G., & LeFevre, J.-A. (2015). Prospective memory failures in aviation: effects of cue salience, workload, and Individual differences. *Aerospace Medicine and Human Performance*, *86*(4), 366–373. doi.org/10.3357/AMHP.3428.2015

EXAMINING PILOT DECISION MAKING IN INFORMATION-RICH COCKPITS

Meredith Carroll
Florida Institute of Technology
Melbourne, Florida
Donna Forsyth Wilt
Florida Institute of Technology
Melbourne, Florida
Paige Sanchez
Florida Institute of Technology
Melbourne, Florida
Deborah Carstens
Florida Institute of Technology
Melbourne, Florida

Pilots are faced with making decisions based on a range of different information sources. One challenge pilots often face is the presentation of conflicting information between sources. This effort examined pilot decision making with conflicting information by conducting structured interviews with 13 pilots, including seven airline, three corporate, and three GA pilots. Pilots were asked questions regarding their experience with conflicting sources of weather, traffic, and navigation information on the flight deck or cockpit. Further, they were asked to describe how they responded to the information conflict, including which source they trusted, which source they ultimately acted on, and why they acted on that source. This paper describes the methods, results and implications for commercial and military aviation.

Pilots are faced with making decisions based on a range of different information sources. Whether commercial aviation, General Aviation (GA) or military pilots, increases in information sources on the the flight deck or cockpit have resulted in pilots having to determine which pieces of information are accurate and relevant, and integrate the information to create an accurate representation of the environment (Mosier & Fischer, 2010; Mosier, 2002). One challenge pilots often face is the presentation of conflicting information between sources. Conflicting information can significantly hinder decision making by reducing decision accuracy and decision confidence (Mosier et al, 2007; Chen and Li, 2015). Several questions remain regarding pilot decision processes during conflicting information events. Finding answers to these questions is imperative given the growing amount of redundant information available on the flight deck such as via both aircraft-installed avionics and Electronic Flight Bag (EFB) applications. The Federal Aviation Administration is particularly interested in this issue as they work to ensure that NextGen technologies and procedures being integrated on the flight deck are reliable and safe, and promote increased safety, capacity, and efficiency of the National Airspace System.

In order to better understand the decision-making processes that occur when pilots are faced with conflicting information, we conducted a series of studies including (a) a literature review, (b) pilot interviews, and (c) collection of a pilot questionnaire data. This paper describes the methods and results associated with the pilot interviews. Prior to the interviews, we performed a literature review to examine factors that influence how individuals make decisions when faced with conflicting and uncertain information. Results indicated that pilot decision making under these circumstances is influenced by a) system factors of reliability, transparency, and workload, b) individual factors of experience, system trust, and training, and c) task/environment factors of time pressure, risk, take action tendency, type of conflict, and task difficulty. After the literature review, we conducted pilot interviews to examine, among other things, information conflicts currently being experienced on the flight deck/cockpit and whether the factors identified in the literature review, are indeed the factors that influence aeronatical decision making with conflicting information. Based on results of the interview, a questionnaire was developed and administered to a large sample of pilots to obtain a more comprehensive view of the information conflicts being experienced by airline, corporate and GA pilots. The results of the questionnaire will be covered in a separate paper.

Methods

The research team conducted interviews with 13 pilots, including seven airline, three corporate, and three GA pilots. The goal of interviews was to obtain information from active pilots regarding what type of information conflicts are currently being experienced on the flight deck and what factors influence which source(s) of information pilots trust and ultimately act on. We utilized Florida Institute of Technology (FIT) College of Aeronautics (COA) alumni and faculty network to recruit airline, corporate and GA pilots via email and phone. We attempted to obtain participants who flew a range aircraft with various types of information sources on the flight deck or cockpit. We scheduled the pilots via email for two-hour interview blocks, and provided an electronic informed consent form and a short sample of questions to consider prior to the interview.

Interview Questions and Procedure

The research team prepared a standardized procedure and set of questions prior to the interviews, including a set of questions for airline and corporate pilots, and a slightly different set of questions for GA pilots. The questions targeted pilot demographics, experience with information conflicts, use of EFB applications, and experience with integrated displays on the flight deck. This paper focuses on the portion of the interview in which pilots described their experiences with information conflicts. The interview questions and associated procedures were submitted to, and approved by, FIT's Institutional Review Board (IRB).

Interviews lasted one to two-hour(s) and were conducted either via phone or in-person with each pilot. One or two researchers led the interview and another researcher acted as a scribe. Participants were initially asked to decribe the information sources on their current aircraft and we tailored questions during the interview based on the sources of information available to each pilot on their flight deck or cockpit. With permission from the interviewee, we recorded all interviews and transcribed the interviews for later analysis. The interviews were organized in a semi-structured format including a series of open-ended questions followed by prompts designed to elicit rich contextual responses, while keeping the interviews on track. The interviews commenced with a brief description of project background and ended with a request to follow up with additional questions in the future, and if the interviewee was interested, a commitment to share results of the research.

Data Analysis

First, the research team transcribed the interviews electronically using the Sonix online transcription services (https://sonix.ai/). Next, we reviewed interview transcripts against researcher notes for accuracy and used the transcripts to fill in any gaps in researcher notes. Then, we extracted question responses from the interview data and input responses into a spreadsheet where responses could be compared across participants. For each question, we analyzed participant responses to extract categorical themes. These themes were compared across different types of information conflicts (e.g., weather, traffic, navigation) for each type of question (e.g., why information was trusted/acted on) and converted into response categories. The research team then re-analyzed participant responses and classified each response within these categories. Finally, we summarized the interview results. Due to the small number of participants, only descriptive statistics were utilized in the analysis.

Results

Participant Demographics

Thirteen pilots were interviewed, including seven Part 121 pilots, three corporate pilots (based on operation of aircraft heavier than 6000 lbs. but not operating under Part 121), and three GA pilots (based on operation of aircraft lighter than 6000 lbs. and not operating under Part 121). Twelve of these thirteen pilots were male; one was female. Demographics are summarized in Table 1. On average, Part 121 and corporate pilots reported flying once a week to daily, while GA pilots reported flying between one to five times per month. Pilots also reported their highest pilot certificates held, which included: ten air transport pilot certificates, two commercial pilot certificates, and one private pilot certificate with an instrument rating. Pilots reported experience flying the following aircraft: Airbus (A320 and A321), Beechcraft Baron, Boeing (737-800/900, 747-400/800, 757-200, 767-300, and 787), Cessna (152 and 172), Challenger 650, Dassault Falcon (DA10 and DA20), Gulfstream G5, Lockheed Jetstar, Mooney, Piper (PA28-161, PA28-181, PA28-201, and PA44), and the Pilatus PC12-NG.

Table 1.

Demographic Data for Interview Participants

	Part 121	Corporate	GA	Overall
Number of Pilots	7	3	3	13
Gender (M/F)	6/1	3/0	3/0	12/1
Average Age, years	36 (SD = 9)	45 (SD = 7.4)	63 (SD = 14)	48 (SD = 3.4)
Average Total Flight Hours	7,800 (SD = 3,554.3)	11,059 (SD = 5,298.3)	6,417 (SD = 8,203.7)	8,233 (SD = 2,383.3)

Information Conflicts

Results indicated that pilots frequently experience conflicting information on the flight deck. When asked whether they had experienced an information conflict specifically associated with either weather, traffic or navigation information, all 13 pilots reported having experienced one type of conflict or another. In fact, pilots often gave answers such as "yes, all the time" or "yes, it is not uncommon".

Table 2 summarizes the number of Part 121, corporate and GA pilots who reported experiencing either a traffic-, navigation- or weather- information conflict on the flight deck. Weather and traffic were the most common types of information in which pilots experienced conflicting information. For weather, the onboard radar was the source most commonly found in conflict with another source such as Air Traffic Control (ATC), Next Generation Weather Radar (NEXRAD) or between the two onboard radars. Onboard radar was most commonly trusted over other sources. For traffic, the most common conflict was between Traffic Collision Avoidance System (TCAS) and ATC, with pilots equally trusting both sources. Navigation conflicts were less frequently reported in the interviews, but typically indicated that certified navigation-information sources in the panel (e.g., the Navigation Display (ND)) were trusted more than other uncertified navigation-information sources on their EFB or mobile devices, such as Jeppesen FliteDeck Pro or ForeFlight.

Table 2.
Number of Conflicts Reported by Part 121, Corporate, and GA Pilots During Pilot Interviews.

	Part 121	**Corporate**	**GA**	**Total**
Total Pilots	**7**	**3**	**3**	**13**
# Weather Conflicts	7	2	1	10
# Traffic Conflicts	6	2	2	10
# Navigation Conflicts	1	1	0	2
Total Conflicts	**14**	**5**	**3**	**22**

Factors Influencing Pilot Response to Information Conflicts

Further, results indicated that there were clear patterns regarding why pilots trusted information sources and ultimately acted on one source or another, and these align with findings from our previously conducted literature review. These patterns are consistent across the different types of conflicts, including conflicts related to weather, traffic and navigation information. The interview results indicated that pilots tended to trust, or distrust, an information source due to: 1) the recency of information on the source, 2) the source's reliability, 3) the pilot's knowledge of the source's strengths and weaknesses and when it is most trustworthy, 4) the source's accuracy, 5) the pilot's past experience with the source, and 6) the pilot's lack of knowledge about how the source's information is obtained. Table 3 summarizes the number of pilots who reported each reason as impacting their trust.

Table 3.
Reasons Pilots Trusted an Information Source and Number of Pilots that Reported Each.

Reasons Trusted	# Pilots
Recency of information on the source	16
Reliability of the source	13
Knowledge of the strengths and weaknesses of each source	11
Accuracy of the source	10
Better experience with this source in the past	5
Lack of knowledge about the sources	5

Note: Pilots typically reported more than one reason.

The interview results indicated that pilots ultimately acted on a source due to: 1) the source indicating a more hazardous situation, 2) their trust in the source, 3) being trained to use the source, 4) their knowledge that the source is certified, 5) the information being presented by the source requiring immediate action, 6) their experience with the source, and 7) being required to use the source. Table 4 summarizes the number of pilots who reported each reason as impacting their ultimate actions.

Table 4.
Reasons Pilots Acted on an Information Source and Number of Pilots that Reported Each.

Reasons Acted On	# Pilots
The source indicated more hazardous situation than the other source	16
I trusted the source the most	8
I am trained to use this source	4
I know that the source is certified	3
The information, as presented on the source, required immediate action	2
I have more experience with this source	1
I am required to use this source	1

Note: Pilots typically reported more than one reason.

Pilot Perception of Information Conflicts.

Results also indicated that pilots may not perceive the presence of conflicting information on the flight deck or cockpit as a problem. Each interview commenced by asking pilots if they had ever experienced conflicting information, in general, on the flight deck. When initially asked this general question, only nine of the 13 pilots (four Part 121 pilots, three corporate pilots and two GA pilots) reported having experienced a conflict. However, later in the interview when queried about specific conflicts such as conflicting weather, traffic or navigation information, all 13 pilots reported having experienced one type of conflict or another. It seems that although these conflicts occur somewhat frequently on the flight deck, pilots do not perceive them as a significant problem. Rather, pilots are accustomed to constantly evaluating and integrating information of varying levels of integrity, in order to hone in on ground truth and therefore perceive the conflict as a natural characteristic of the information-rich nature of the flight deck. Pilots appear to resolve these conflicts by collecting additional information from other sources, while considering the strengths and weaknesses of each source.

Conclusion

Results of this study indicate that pilots frequently experience conflicting information on the flight deck or cockpit and there are clear patterns as to which source a pilot will trust and ultimately act on. The patterns align with findings from our literature review and suggest that key factors that influence which source a pilot will trust and ultimately act on are influenced by a) system factors such as information recency, reliability and accuracy, b) individual factors such as system knowledge, experience and training, and c) environmental/task factors such as level of hazard.

Although pilots seem comfortable coping with information conflicts, these factors provide opportunities for better supporting pilots in making effective decisions. By optimizing system accuracy, recency and reliability, and by ensuring pilots have the training and experience necessary to understand the strengths and weaknesses of their systems, commercial, military and GA pilots can be better prepared to make decisions when faced with conflicting information.

References

Chen, K., & Li, Z. (2015). How does information congruence influence diagnosis performance?. *Ergonomics, 58*(6), 924-934.

Mosier, K. L. (2002). Automation and cognition: Maintaining coherence in the electronic cockpit. In *Advances in Human Performance and Cognitive Engineering Research* (pp. 93-121). Emerald Group Publishing Limited.

Mosier, K. L., & Fischer, U. M. (2010). Judgment and decision making by individuals and teams: issues, models, and applications. *Reviews of Human Factors and Ergonomics*, *6*(1), 198-256.

Mosier, K. L., Sethi, N., McCauley, S., Khoo, L., & Orasanu, J. M. (2007). What you don't know can hurt you: Factors impacting diagnosis in the automated cockpit. *Human Factors*, *49*(2), 300-310.

DEVELOPING A HIGH-FIDELITY SIMULATION LAB: CHALLENGES AND LESSONS LEARNED

Andrea M. Georgiou, Glenn E. Littlepage, Michael B. Hein, & Christopher R. Bearden
Middle Tennessee State University
Murfreesboro, TN

With a decade of experience, the Middle Tennessee State University (MTSU) NASA FOCUS (Flight Operations Center- Unified Simulation) lab is a vital part of the educational experience for senior aerospace students. The NASA FOCUS Lab is a high-fidelity simulation of a flight dispatch center in a collegiate setting. Students are trained in specific positions in the lab and must operate within a complex team environment to run the virtual airline within their shift. The purpose of the lab is to provide a learning platform for students to practice the requisite teamwork skills necessary to effectively work in airline operations. Designing effective simulation experiences and providing adequate performance feedback is complex. This paper discusses some of the challenges we encountered and lessons learned through a ten-year span of operation and refinement. By presenting this information, it may help future researchers in the design and development of high-fidelity simulation labs.

In 2008, Middle Tennessee State University (MTSU) developed the first high-fidelity simulation lab of a flight dispatch center in a collegiate setting. Originally funded by a NASA grant, the NASA Flight Operations Center- Unified Simulation (FOCUS) lab mirrors a Part 121 regional airline operations center. The simulation lab offers a unique, collaborative, and interdisciplinary opportunity for both faculty and students in the fields of aerospace and psychology. Additionally, and most importantly, the FOCUS lab provides MTSU undergraduate aerospace students the opportunity to participate in airline operation simulations in a nonconsequential environment. Simulation training has been identified as an effective approach to team training (Kozlowski & Ilgen, 2006; Salas, Cooke, & Gorman, 2010). The purpose of the lab is to provide a learning platform for students to practice the fundamental teamwork, communication, and coordination skills necessary to run an airline. A major component of the simulation is to integrate triggers during their shift that require them to effectively handle emergency and abnormal situations that have potential to disrupt operations.

The FOCUS lab consists of a simulated flight dispatch center of a regional airline and associated activities. The main lab room houses the following positions: flight operations coordinator, crew scheduling, weight and balance and fuel management, weather and forecasting, maintenance control, and passenger/cargo rescheduling. The dispatch center also interfaces with a flight crew in a CRJ-200 flight simulator, a pseudo pilot providing general direction for multiple flights, and a ramp tower controlling runway and gate activities. Participants are senior Aerospace majors in a capstone course from six difference aerospace concentrations. They are assigned to specific positions within the lab, receive training, and participate in up to three 2.5-hour simulation sessions. Since After Action Reviews (AAR's) are considered an effective component of team training, the teams participate in an after-action review (AAR) after each simulation that includes debriefing and discussing ways to improve their performance in their upcoming simulations (Tannenbaum & Cerasoli, 2013; Villado & Arthur, 2013).

From the initial onboard training to the hands-on training in the simulation lab, every training component was developed in-house and includes job-specific duties for roles within the lab. To encourage the consideration of downstream consequences prior to participation in the simulations, software training and exercises are provided to assist the students. In addition to learning a specific role within the lab, students learn the critical skills of communication and coordination in a learning environment that encourages making mistakes and learning from them prior to entering the industry. With over ten years of experience and expertise with the lab, there have been ample challenges and lessons learned along the way. Through weekly feedback and observations, the lab constantly evolves and changes to ensure student experiences are realistic and beneficial. By presenting a summary of the challenges and lessons learned with a decade of continuous improvement efforts, it is our hope that others can glean some beneficial information in the areas of simulation-based training. For additional information regarding the MTSU NASA FOCUS lab, refer to Littlepage, G. E., Hein, M. B., Moffett, R. G., Craig, P. A., & Georgiou, A. M. (2016).

Challenges Encountered

An initial challenge encountered during the genesis of the FOCUS lab involved assembling the expertise needed to design the lab. This required input from multiple aviation specializations including flight dispatch, maintenance, weather, professional pilot, as well as expertise in training, organizational behavior, and teams. Faculty members from the Aerospace and Psychology departments provided this knowledge and technological expertise was provided by consultants and a graduate student. The most effective method to gather everyone's input was in face to face meetings, rather than emails or phone conversations. It became apparent early in the process, that the researchers and staff members must work as an effective team in order to design an effective team simulation.

Funding was and is an ongoing issue. Initial equipment funding was obtained by two NASA grants and subsequent operational funding was cobbled from internal grants, departmental graduate student support, and from lab fees charged to students. Equipment maintenance is an issue as well, as there are often technological issues with the computers, LCD television screens, and communication headsets.

Faculty members provide the overall direction for both training and research and are involved in these activities and supervise other staff members. On the other hand, the actual operation of the lab and data management is conducted by graduate students, along with some assistance from undergraduates. Graduate students are funded through the two departments and undergraduates are paid as student workers. Due to the natural turnover amongst the student staff members with their eventual graduation, there is a challenge to ensure proper staff training for the incoming students selected to help run the simulation lab. A constant succession plan is imperative to ensure adequate transfer of knowledge and skills for the staff members.

Another challenge faced in the lab is the development and implementation of realistic simulations. This portion involved a great deal of planning and attention to detail. For example, flight routes and schedules needed to be developed, along with detailed information about each flight leg. This foundational information included the number of passengers per flight, pounds of cargo, and procedures for fuel calculation. Detailed weather information needed to be developed, along with procedures for determining alternate airports and additional fuel requirements. For crew scheduling, it was necessary to develop a list of crewmembers and their associated legal flight hour availability. An extensive listing of maintenance procedures for inoperative items prior to departure, in-flight mechanical issues, ferrying procedures, and more were included in the development of the maintenance control position. All these activities required substantial time commitments and ample revisions through trial and error.

A variety of training materials needed to be developed to include the following: onboarding to discuss the airline expectations, company policies, etc., orientation training, online positional training, hands-on positional training, and the development of job aids to assist during the simulations. Each training component was labor intensive and the difficult portion was to ensure we developed relevant training materials that aligned with the objectives of the lab.

As we encountered and tackled the various challenges in the development and operation of the MTSU NASA FOCUS lab, the simulations and training were meticulously refined over time. The collaborative efforts and dedication of the Aerospace and Psychology faculty and students resulted in simulations that were more involved, more realistic, and beneficial to student learning.

Lessons Learned

Student and Staff Training

We quickly learned that students needed more training than anticipated. Initially, we developed PowerPoint materials and conducted lecture-based orientation training to provide an overview of the lab and the duties of the various positions. Next, students were provided approximately 30 minutes of on-the-job training to learn their specific work duties in the lab. Over time, it was observed that more training would be needed to help them be successful during the simulations. The training was expanded to include an online overview module and position-specific online modules. Later, we required students to pass tests based on the online training prior to entering the lab for hands-on positional training. Job aids were developed to serve as training refreshers and include brief printed materials, online manuals (maintenance), and software to aid in calculations (fuel and weight calculations, duty-time calculations, etc.). The goal of the training is to ensure the students understand how to complete the technical aspects of their job and the duties and resources of other positions.

The lab is staffed with a team of faculty members, graduate, and undergraduate students. The consistency and realism of the simulations are greatly affected by how well trained the team of researchers and staff members are in the lab. New staff members are trained by the veteran members, along with annual frame of reference training to ensure that staff members make comparable ratings of team performance and fully understand the simulation objectives and procedures.

After-Action Reviews (AARs)

The simulations provide some natural performance feedback, but it was observed that students frequently were unaware of issues that were unresolved or handled improperly. For example, they might not be aware of an incorrect fuel load, passengers missing connecting flights due to delays, or a potential tarmac delay violation. As part of the educational component of the lab, it is imperative students receive feedback on their performance. This feedback allows them to learn from their errors, as well as learn from their successes. To close the loop from the simulation to learning, after-action reviews were implemented and it has been an extremely successful and vital part of the FOCUS lab. The after-action reviews (AARs) are facilitator-led discussions of positive and negative outcomes and specific team member actions that led to those outcomes. Participants are encouraged to discuss strategies that could be used to improve processes and performance. Because students are sometimes unprepared for the AARs, we require them to complete a form identifying effective and ineffective outcomes and the actions leading to each outcome. These are to be submitted within 48 hours and brought to the AAR for the round-table discussion.

Measures

A variety of measures are utilized from the simulation software, observer ratings, and instruments completed by students. As the simulation lab evolved, it became apparent that the measures obtained in the lab must also evolve. Careful attention to the measures is important to ensure the collection of valid and reliable data. Initially, the objective measures were primarily based on the team's on-time performance, such as delay loss and revenue. Eventually, the measures were supplemented with additional costs related to suboptimal performance (e.g. hotel vouchers) and penalties for safety and company violations (e.g. overweight flight, unsafe route). With the complexities present in the airline operation simulation, it became necessary to collect several types of data to measure their individual and team performance.

There are numerous participant-rated instruments administered after the completion of the simulations and there was concern for inattentive responses due to overload. Students consistently provided feedback stating they were tired of taking the surveys and felt it was excessive. This was addressed by spacing out the measures and administering them at various times such as onboarding, after positional training, and after each AAR. Additionally, certain measures were reduced in length and rotated between semesters.

Observer ratings tended to have relatively low inter-rater reliability and this seemed to reflect gaps in observer knowledge. Some observers served administrative roles (e.g. pseudo pilot and local maintenance facility) and were not in the main simulation room. For these staff members, they monitored the operations via electronic communication and would often be unaware of the verbal and nonverbal behaviors occurring during the simulations. The same was true for the staff members in the simulation lab in that they might be unaware of electronic communications. To remedy this issue, weekly staff meetings were held after the simulations to discuss the team's performance. In addition, we equipped observers with I-Pads and share an active document to enter immediate observations of the team performance as triggers are implemented throughout the simulation. Behaviorally anchored rating scales (BARS) were developed to assess team adaptation to non-routine events. These actions resulted in high levels of inter-rater reliability.

Maintaining a Professional Atmosphere

Efforts are made to ensure that students see the value in the lab and act in a professional manner. The professional benefits of the lab experience are highlighted and expectations communicated that the lab should be treated as a professional job. The necessity for regular attendance at the lab and AARs are emphasized and when a member of the team cannot attend the simulation, they are expected to give prior notice to the team leader (the flight operations coordinator). Cell phones and tablets are permitted for legitimate purposes in the lab, such as searching travel sites for flights that can be used to carry stranded passengers and cargo or researching contract maintenance services at off-site airports. When inappropriate use of these electronic devices is observed, corrective action is implemented by a staff member. Additionally, it is important the staff members are an example of professionalism and should not use their cell phones during down times or engage in non-task relevant conversations.

Balancing Standardization and Flexibility

The FOCUS lab has both training and research objectives and sometimes tradeoffs are needed. Training models suggest that simulation difficulty should increase as teams gain competence, but research models suggest consistency across simulations. This issue has not been fully resolved, but the movement has been for greater consistency. Initially, live weather was used which meant sometimes there were ample weather issues like icing and turbulence, and other simulations had calm winds and minimal weather effects on operations. To attempt standardization, previously recorded weather was captured and used in the simulations across the teams. Weather still varies across simulations, but the conditions were selected to reflect relatively comparable conditions.

A variety of non-routine events (triggers) are programmed into the simulations. The number of triggers increases with each simulation that a team faces, but there is some standardization with the creation of sets of non-routine events per simulation. Within each set, events were designed to represent similar levels of difficulty and similar actions to resolve the issue. For example, one team might encounter an unruly passenger, while another might be faced with a passenger with an in-flight medical issue. The use of sets of comparable triggers allows different teams to experience comparable events, but reduces the likelihood that communication across teams would make the triggers predictable. While the researchers plan the triggers in advance, some flexibility is needed about the specific flight that experiences the event. An event might have been planned to occur at a specific time on a specific flight, but if the team was slow in dispatching that flight, it becomes necessary to adapt the scenario.

Need to See the Big Picture

A major goal of the FOCUS lab is to enhance student understanding of the coordinated and nuanced nature of airline operations. This involves knowledge of how one's position relates to other positions within the team. The importance of situational awareness and how everyone works together is discussed in their training and reinforced in AARs; however, it is still difficult for students to see how their actions could affect other parts of the operation. Perhaps the most recurring problem is the failure to foresee downstream consequences. For example, if a plane must divert because of an in-flight emergency, teams almost always handle the diversion properly and generally get the affected passengers to their destination. However, they often fail to plan for another aircraft and crew to take over additional flight legs, along with other downstream issues. To correct this problem, we implemented a procedure borrowed from emergency medicine, known as the pause procedure. When an abnormal event occurs, the flight operations coordinator calls a "pause" and briefly explains the issue to the team. Next, one person is assigned to manage the problem and uses a checklist that evaluates potential implications and develops an appropriate action plan.

Need for Continuous Improvement

Throughout the evolutionary process of the development and implementation of the NASA FOCUS lab, there has been ample challenges faced and lessons learned for improvements. The initial semester involved only a round table description of the simulated airline and the various lab positions. In the following semesters, full simulations and AARs were conducted. Simulation triggers, training materials, and measures are continually developed and refined. While it does require continual funding and improvements, we have learned that dedication and passion for student learning are the two most important factors to keep the lab running.

References

Kozlowski, S. J., & Ilgen, D. R. (2006). Enhancing the effectiveness of work groups and teams. *Psychological Science In The Public Interest*, *7*, 77-124. doi:10.1111/j.1529-1006.2006.00030.x

Littlepage, G. E., Hein, M. B., Moffett, R. G., Craig, P. A., & Georgiou, A. M. (2016). Team training for dynamic cross-functional teams in aviation: Behavioral, cognitive, and performance outcomes. *Human Factors, 58*, 1275-1288. doi.org/10.1177/0018720816665200.

Salas, E., Cooke, N. J., & Gorman, J.C. (2010). The science of team performance: Progress and the need for more. *Human Factors, 52*, 344-346. doi:10.1177/0018720810374614

Tannenbaum, S.I. & Cerasoli, C.P. (2013). Do team and individual debriefs enhance performance? A meta-analysis. *Human Factors, 55,* 231-245. doi:10.1177/0018720812448394

Villado, A. J., & Arthur, W., Jr. (2013). The comparative effect of subjective and objective after-action reviews on team performance on a complex task. *Journal of Applied Psychology*, *98*, 514-528. doi:10.1037/a0031510.

DEVELOPMENT AND USE OF PERFORMANCE COMPOSITE SCORES IN DISPATCH TEAMS

Christopher Ryan Bearden
Michael B. Hein
Glenn E. Littlepage
Middle Tennessee State University
Murfreesboro, TN

Teams perform a variety of functions within organizations and should therefore be evaluated on multiple criteria. This paper argues for the use of a single value. We review the literature on team performance composites and briefly describe two approaches to developing evaluative performance composites in an academic setting by combining performance indicator data: A qualitative approach for performance feedback as well as an empirical approach for research purposes.

Over ten years ago, Mathieu, Maynard, Rapp, and Gilson (2008) suggested that because teams perform multiple functions, a best practice for evaluating teams is to include and combine multiple criteria dimensions to evaluate teams. Theirs' is not the first nor the only call to combine criteria to appear in the research literature (cf. Pritchard, 1990; Salas, Rosen, Held, & Weissmuller, 2009). An argument for a single index of performance can be made on the basis of parsimony. Additionally, a single value can be easily compared across teams, have motivational value, and convey performance data quickly to organizational stakeholders and management (Pritchard, 1990). When evaluating teams, the general recommendations in the literature are clear: criteria should be theoretically-based (Salas, Burke, Fowlkes, & Priest, 2003); criteria measurement should be designed keeping in mind the functions of the team (Mathieu et al., 2008), as well as the purpose and environment of the team (Kendall & Salas, 2004) and the desired outcomes (Rosen, Wildman, Salas, & Rayne, 2012); finally, differentiated criteria should be combined using a formal method (Mathieu et al., 2008).

Literature Review

The argument for a composite measure (i.e., a single criterion variable) begins with the idea that teams need to be evaluated on multiple criteria because they perform a variety of functions (Mathieu et al., 2008). Typically, these dimensions are examined one-by-one; however, it can be difficult to assimilate multiple pieces of information about a team's functioning (Pritchard, 1990). A single value quickly conveys a large amount of information to organizations, researchers, and the teams themselves. Additionally, it provides an evaluative advantage, demonstrating change efforts and allowing for easy between-team comparisons. A composite has motivational value to teams because it clearly demonstrates consequences of effort (e.g., performance increases or decreases; Pritchard, 1990). Differentiated criteria can always be examined for specific reasons (e.g., planning improvements). Table 1 contains example studies to illustrate the creation of various team performance composites.

Many different reasons exist for evaluating a team, such as research, training evaluations, team performance or process diagnostics, or determining team rewards. The reason for evaluation should drive the decisions for selecting the criteria and indicators of the criteria

Table 1.
A Sample of Example Studies Combining Disparate Evaluation Criteria.

Reference	Label	Indicator Dimensions	Theoretical Justification	Procedure
Hiller, Day, & Vance (2006)	Team Effectiveness	Planning and organizing (task) Problem solving (task) Support and consideration (relationship) Development and mentoring (relationship)	All four dimensions adapted from Managerial Practices Survey (Yukl & Lepsinger, 1990). For task versus relationship see Judge, Piccolo, and Ilies (2004)	1, "very ineffective" to 7, "very effective;" averaged; squared to reduce skew.
		Overall effectiveness	None	
Van Der Vegt & Bunderson (2005)	Team Performance	Efficiency Quality Overall achievement	Performance criteria (Ancona & Caldwell, 1992)	Item rated 1, "far below average" to 7, "far above average;" averaged
		Productivity Mission Fulfillment Overall team performance	None; suggested by company	
Mathieu, Gilson, & Ruddy (2006)	Quantitative Performance	Machine reliability Response time Parts expense	(Hyatt & Ruddy, 1997)	Converted to *z* scores; used as indicators for a latent variable
Komaki, Barwick, & Scott (1978)	Safety Performance	Department-specific safety item(s)	Analysis of archival incident reports	Behavior observation scale; score is the ratio of safe actions to total observed
Pearsall & Ellis (2006)	Team Performance	Offensive score Defensive score	(Ellis, Hollenbeck, Ilgen, Porter, West, & Moon, 2003)	Offensive and defensive scores were standardized and summed

86

(Salas, Reyes, & Woods, 2017). This ensures outcome-measurement congruence. That is to say, measures should capture what is needed to make the generalizations and draw the conclusions needed. Evaluators should not only rely on theory for criteria inclusion but for the justifying the composite itself as well. Choosing a theoretical model, such as Hackman's (1987) team effectiveness framework, will not only guide the measurement strategy but also provide conceptual clarity and lend credence to the approach (Salas et al., 2003). First, a systematic team task analysis should be conducted. Organizational leadership should be consulted for evaluation criteria during this process (Sundstrom, De Meuse, & Futrell, 1990). Consideration should also be given to the behavioral processes and performance criteria identified, defined, and organized through previously taxonomic efforts (e.g., Marks, Mathieu, & Zaccaro, 2001).

A criterion represents an objective or desired outcome, or product or service rendered. Each criteria included in the composite must have at least one indicator, and each indicator must be measurable/measured. When designing the measurement strategy, consideration should be given to the function, purpose, and environment of the team (Kendall & Salas, 2004). Pritchard (1990) makes several recommendations when selecting indicators: Indicators should meaningful to both the purpose of the evaluation and the intended audience; the long-term consequences of improving on the indicators should be considered; the indicators should be under the control of the team; and the indicator should not be contaminated by other units' performance. Additionally, indicators should not be selected if they do not vary between teams.

Indicators can be categorized as objective (e.g., points scored in a simulation game) or subjective (e.g., supervisor judgements). Meta-analytic findings have demonstrated the convergent validity of objective and subjective measures of performance (Bommer, Johnson, Rich, Podsakoff, & MacKenzie, 1995). Consideration should also be given to whether the indicators will be behavior-focused (esp. for training or feedback or rewards) *or* outcome-based. Kozlowski and Bell (2013) argue that team performance itself should be conceptualized as the action(s) the team takes as opposed to the outcomes, which is consistent with the distinction made by Beal, Cohen, Burke, and McLendon (2003) who argue that performance behaviors should be separate from performance outcomes. All decisions should be guided by the purpose of the evaluation. Finally, each indicator should represent the team as a whole not an individual.

Measures of objective outcomes have several advantages, such as possible automatization of data collection, and are also often the most intrinsically interesting to stakeholders. Teams, however, may not be able to control certain outcomes to the same extent they can control their own intrateam processes and behaviors. Subjective indicators are more widely used, in part, because data collection methods are typically easier to design and access. However, subjective measures have their own problems. For one, it can take numerous evaluators to effectively observe a team's performance. Ratings provided by evaluators can also be biased. If subjective indicators are used, raters should receive training and only assess four to five indictors of performance (Smith-Jentsch, Baker, Salas, & Cannon-Bowers, 2001). Ways to avoid this limitation include having raters assess only those indicators with which they are most familiar and increasing the rater pool to include self-report, peers, experts, and supervisors. This could lead to other issues, such as the inability to determine needed interrater agreement on indicators.

After data collection factor analytic methods (e.g., confirmatory factor analysis) can be used to establish construct reliability and provide evidence for construct validity. Often, highly correlated indicators ($r > .70$) are simply averaged. The problem with averaging indicators is that it assumes linearity. In other words, gains in the raw score at any level on any indicator contribute equally to the overall performance. Further, it assumes that deficiencies in one area of performance can be compensated for in another – which is not always the case in applied settings. Statistical methods (e.g., principal components analysis) can be used to inform how or whether to combine data on multiple indicators after the data are collected. Before being combined, indicators can and should be weighted based on their relative value. Weights can be determined through judgement or statistical methods. While this helps, weighting indicators does not solve the problem. Two examples are hereafter provided which do solve this problem.

Two Example Approaches for Developing a Team Performance Composite

Six objective indicators were identified for our team-training simulation of a regional flight dispatch center: number of flights dispatched, number of airline policy violations, total delay time, number of passengers missing connections, pounds of undelivered cargo, and number of airplanes with a tarmac delay fine. Two approaches were undertaken to combine data on the indicators. The first utilizes recommendations from the Productivity Measurement and Enhancement System (ProMES; Pritchard, 1990) to create a composite that provides students with actionable feedback and an ability to set goals. The second utilizes principal components analysis to maximize team differences on the indicators. Both approaches address the non-linear relationships among the indicators and with the criterion.

The ProMES recommends establishing three values for each indicator: a maximum value on the indicator, a lowest possible value, and an expected value on the indicator. See Table 2 for an example using airline policy violations. A raw score of zero represents the best performance. The purpose of this first composite is performance feedback for teams early in training.

Table 2.
Example using Archival Team Performance Data to Determine Indicator Values

		Indicator Value Label												
		Max.	Good		Expected			Poor				Min.		
Raw Score		0	1	2	3	4	5	6	7	8	9	10	11
Teams	*N*	5	9	6	6	3	4	3	1	0	1	3	1
	%	12	21	14	14	7	10	7	2	0	2	7	2

Therefore, using historical data, similar raw scores were grouped in such a manner that *most* teams would perform well or at least as expected (i.e., a negative skew). Some room was left for improvement. Similar values can be grouped considering the reliability of the measurement instrument or can be established using judgement. The points earned for *good* performance versus *expected* versus poor (etc.) can be established using a variety of ways. Here, maximum (Max.) performance is awarded an "A" or a 4.0 out 4.0 points, which uses a frame of reference with which undergraduate aerospace students are familiar. After new values are assigned to the raw scores, the indicator itself (i.e., the 4.0) is weighted relative to its contribution to overall effectiveness. Since policy violations are related to airline safety, and safety is our virtual airlines' number one priority, this indicator is weighted as the most important. This process is repeated for each indicator.

The second approach uses non-linear principal components analysis (PCA; see Linting, Meulman, Groenen, & van der Koojj, 2007). Non-linear PCA is suitable for all measurement levels, so indicators could be ordinal, nominal, numeric, or any combination thereof. Non-linear PCA reproduces more variance than traditional PCA – even in the unlikely cases where the relationships are linear. The second composite is to be used as a criterion for research purposes, therefore maximizing the differences between teams on the indicators is useful. The first step is to determine the overall number of components. Typically, this is one, but more may be needed. Second, the indicators are rescaled to account for non-linearity (this can be accomplished using the PRINQUAL function in SAS or SPSS's optimal scaling function). Third, a PCA is conducted on the rescaled indicator variables. The component score(s) produced maximizes the differences between teams while accounting for the non-linear relationships. Limitations of this approach include the need for large amounts of historical data, increased complexity of interpretations, and automation requires sophisticated information technology skills complex.

Conclusion

Several clear recommendations should aid in the design and interpretation of performance composites. Teams should be evaluated on multiple dimensions, which cover their functions and purpose (i.e., content validity; Mathieu et al., 2008); the specific criteria selected should fit within a theoretical framework (i.e., construct validity; Salas et al., 2003); the criteria must match outcomes (i.e., criterion relevance and criterion validity; Sundstrom et al., 1990). Each criterion must have at least one measurable, controllable, and uncontaminated indicator. When measuring behaviors and processes, these can be split into task and relationship (Judge, Piccolo, & Ilies, 2004), which could add more conceptual clarity and aid in interpretation. Indicators should be carefully combined using a formally articulated method. Methods of combining indicator data should account for the (potential) non-linear relationships among indictors and between the indicators and the evaluative criterion.

References

Beal, D. J., Cohen, R. R., Burke, M. J., & McLendon, C. L. (2003). Cohesion and performance in groups: A meta-analytic clarification of construct relations. *Journal of Applied Psychology*, *88*(6), 989. DOI: 10.1037/0021-9010.88.6.989

Bommer, W. H., Johnson, J. L., Rich, G. A., Podsakoff, P. M., & MacKenzie, S. B. (1995). On the interchangeability of objective and subjective measures of employee performance: A meta-analysis. *Personnel psychology*, *48*(3), 587-605. DOI: 10.1111/j.1744-6570.1995.tb01772.x

Hackman, J. R. (1987). *The design of work teams*. In J. Lorsch (Ed.), Handbook of organizational behavior (pp. 315–342). New York, NY, USA: Prentice Hall.

Judge, T. A., Piccolo, R. F., & Ilies, R. (2004). The forgotten ones? The validity of consideration and initiating structure in leadership research. *Journal of Applied Psychology*, *89*(1), 36. DOI: 10.1037/0021-9010.89.1.36

Kendall, D. L., & Salas, E. (2004). Measuring team performance: Review of current methods and consideration of future needs. In J. W. Ness, V. Tepe, & D. R. Ritzer (Eds.) *The science and simulation of human performance, Vol. 5* (pp. 307-326). Emerald Group Publishing Limited.

Kozlowski, S. W. J., & Bell, B. S. (2013). *Work groups and teams in organizations: Review update* [Electronic version]. Retrieved from Cornell University, School of Industrial and Labor Relations site: http://digitalcommons.ilr.cornell.edu/articles/927

Linting, M., Meulman, J. J., Groenen, P. J., & van der Koojj, A. J. (2007). Nonlinear principal components analysis: introduction and application. *Psychological Methods*, *12*(3), 336. DOI: 10.1037/1082-989X.12.3.336

Marks, M. A., Mathieu, J. E., & Zaccaro, S. J. (2001). A temporally based framework and taxonomy of team processes. *Academy of Management Review*, *26*(3), 356-376.

Mathieu, J., Maynard, M. T., Rapp, T., & Gilson, L. (2008). Team effectiveness 1997-2007: A review of recent advancements and a glimpse into the future. *Journal of Management*, *34*(3), 410-476. DOI: 10.1177/0149206308316061

Pritchard, R. D. (1990). *Measuring and improving organizational productivity: A practical guide*. Westport, CT, USA: Greenwood Publishing Group.

Rosen, M. A., Wildman, J. L., Salas, E., & Rayne, S. (2012). Measuring team dynamics in the wild. In A. Hollingshead & M. S. Poole (Eds.), *Research methods for studying groups: A guide to approaches, tools, and technologies* (pp. 386–417). New York, NY, USA: Taylor & Francis.

Salas, E., Burke, C. S., Fowlkes, J. E., & Priest, H. A. (2004). On measuring teamwork skills. In J. C. Thomas (Ed.), *Comprehensive handbook of psychological assessment, Vol. 4. Industrial and organizational assessment* (pp. 427-442). Hoboken, NJ, US: John Wiley & Sons Inc.

Salas, E., Reyes, D. L., & Woods, A. L. (2017). The Assessment of Team Performance: Observations and Needs. In *Innovative Assessment of Collaboration* (pp. 21-36). Springer, Cham.

Salas, E., Rosen, M. A., Held, J. D., & Weissmuller, J. J. (2009). Performance measurement in simulation-based training: A review and best practices. *Simulation & Gaming*, *40*(3), 328-376. DOI: 10.1177/1046878108326734

Smith-Jentsch, K. A., Baker, D. P., Salas, E., & Cannon-Bowers, J. A. (2001). Uncovering differences in team competency requirements: The case of air traffic control teams. In E. Salas, C. A. Bowers, & E. Edens (Eds.), *Improving teamwork in organizations: Applications of resource management training* (pp. 31-54). Mahwah, NJ, US: Lawrence Erlbaum Associates Publishers.

Sundstrom, E., De Meuse, K. P., & Futrell, D. (1990). Work teams: Applications and effectiveness. *American Psychologist*, *45*(2), 120. DOI: 10.1037/0003-066X.45.2.120

A CONTINUATION OF RESEARCH: STUDENT DECISION MAKING UNDER STRESS IN A FLIGHT CONTROL CENTER SIMULATION

Jessi Pope, Michael Hein, Meredith A. Russell, & Chelsea Burkholder
Middle Tennessee State University
Murfreesboro, Tennessee

Students enrolled in a capstone aerospace class participated in this study. The class involves the completion of simulations in a high-fidelity replication of an airline flight operations center called the FOCUS ((Flight Operations Center-Unified Simulation) lab. This lab also functions as a research center exploring individual and team-related attitudes, activities, and experiences. The current study builds upon previous research that suggested participation in the simulation lab resulted in improved self-efficacy towards making decisions under stress (DMUS). Additionally, data suggests that before the simulations, students' perceived fear of making the wrong decision (PFI) correlated with their perceived ability to make decisions under stress (Pope, 2018). Results of the current study showed that after completion of a full simulation in the lab, students reported a non-significant decrease in their personal fear of invalidity and a significant increase in their perceived ability to make decisions under stress.

Safety is an ever-present concern within the aviation industry. As some safety factors stem from the decision-making abilities of airline employees, this study seeks to further understand how stress can impact decision-making. The current study builds upon previous research that suggests students experienced increased confidence in their decision-making ability under stress as a result of participating in a high-fidelity simulation of a flight operations center (Pope, 2018). The relationships between the ability to make decisions under stress, perceived fear of invalidity, and stress are further investigated.

Stress has been defined in many differing ways: a stimulus, a reaction, or a hypothetical state (Sarason, 1984). However, researchers do agree on the effects of stress on an individual's cognitive states – leading to anxiety, poor decision-making strategies, cognitive interference, and decreased performance (Sarason, 1984; Johnston, Driskell, & Salas, 1997). Antecedents of stress include operating under time constraints, ambiguity (of problems encountered, environment, & goals), and high-risk situations (Cannon-Bowers & Salas, 1998**).** Within the context of this study, the primary stressors for participants include the pressure to dispatch flights on time, adhere to FAA regulations, balance productivity and efficiency, and solve unexpected problems in a safe and efficient manner.

Stress has been shown to influence an individual's ability to make decisions (Payne, Bettman & Johnson, 1988). Johnston & colleagues (1997) suggested that stress can increase the likelihood of maladaptive solutions in decision-making, such as employing the use of potentially inaccurate heuristics. Research suggests that people are sensitive to environmental changes (Payne, et al., 1988). This means that when a factor – such a stress – changes, an individual's adaptive strategies likewise change (Payne, et al., 1988). For example, if an individual's time constraints decrease or constraints increase, they may process information beginning with the most pertinent details and weigh this information more heavily to decrease decision time. These stressful situations can tempt the decision maker to oversimplify the situation they're assessing and they may fail to fully consider all contributing factors (Levi & Tetlock, 1980). Both maladaptive decision-making strategies and oversimplification can be detrimental to the outcome of the decision (Keinan, 1987). The present study builds on the work of Pope (2018), who found no relationship between stress and perceived ability to make decisions under stress. It is thought, however, that these results were due to research design constraints.

Personal fear of invalidity may be understood as a heightened "concern for making a mistake in the face of making a decision" (Pope, 2018; Thompson, Naccarato, Parker, & Moskowitz, 2001). Individuals high in personal fear of invalidity may take more time to ruminate before making a decision, which may be detrimental to performance related to tasks that require quick decision-making (Thompson, et al., 2001). Previous research suggests that after simulation-based training during which students solve real-life operations center problems fosters perceptions of an increased ability to make decisions that are high-risk and high-quality under stress (Pope, 2018). The same study also found support for a negative relationship between personal fear of invalidity and perceptions of decision-making ability under stress.

Methods

Participants

The participants of this study are 83 undergraduate senior aerospace students enrolled in a capstone aerospace class. This class involves completion of simulations within a high-fidelity simulation of a flight operations center and receive experience working together to run a simulated airline. Students are placed into one of nine different positions within the lab. The lab serves a secondary function as a research center exploring individual and team-related attitudes, activities, and experiences. After each simulation, the students attend an After-Action Review (AAR) where a facilitator and scribe discuss the student team's performance data with the students and facilitate discussion centered around behaviors to improve performance in the next simulation. Levels of participation fluctuated throughout the semester due to the attendance of students and general participation attrition. Previously collected data from 39 students who completed the lab in Fall 2018 was also used within the study.

Measures and Data Collection Sequence

Data was obtained with student consent through self-report surveys conducted confidentially online via *Qualtrics*. Personal fear of invalidity was measured using the 14-item Personal Fear of Invalidity Scale (PFI) by Thompson and colleagues (2001). Decision-making under stress was measured by 14 selected items from the Leadership Behavior Description Questionnaire (LBDQ; Stogdill, 1963; Brace, 2011). Stress was measured using a four-item scale developed by researchers in the capstone lab designed specifically to capture perceived stress (Pope, 2018).

DMUS, PFI, and stress measures were collected during the first half of the Spring 2019 semester. Archival stress data from Fall 2018 was also used in data analyses. These archival measures were collected after the team's AAR a week after the completion of their simulation. DMUS was measured three times over the semester: (1) after students were trained in their individual positions within the lab and participated in a mock simulation (training day); (2) after their first AAR (scheduled a week after the first simulation); (3) after they had completed the lab portion of their capstone course. PFI was collected on training day and after students completed the lab portion of their capstone course. Stress measures were distributed immediately after the completion of both simulations during Spring of 2019.

Results and Discussion

A Welch independent-samples *t*-test ($\alpha = .05$) indicated the average reported stress levels did not differ between Fall 2018 during Sim 1 (M=2.64, SD=0.76, n=39) or Sim 2 (M= 3.00, SD = 0.91, n = 39) compared to Spring 2019 during Sim 1 (M = 2.62, SD = 0.96, n = 31), $t(56.2)$= 1.37, p = .925, d = 0.02 or Sim 2 (M = 2.72, SD = 0.66, n = 31), $t(67.5)$=2.27, p =.140, d = 0.35. The test compared the stress levels of participants from Fall 2018 to stress levels of Spring 2019. The survey was administered a week after each simulation during Fall 2018, but the researchers administered the stress survey right after the end of each simulation during Spring 2019. Administration time was altered to more accurately capture real-time stress levels, as the stress reported by students a week after the simulation did not seem to reflect the level of stress students anecdotally reported experiencing during simulations. The lack of a significant difference in stress between the two conditions may be explained by the immediate relief experienced during survey administration when a stressful simulation is called to an end.

A Pearson's correlation indicated there was a significant negative association between DMUS (pre) and PFI (pre), r (56) = -.471, p = <.001. In other words, when perceived fear of invalidity increases, decision-making under stress decreases. Consistent with the literature (Pope, 2018; Brace, 2011), the students reported that when they begin the capstone course afraid of producing the wrong decisions, their perceived ability to make decisions under stress suffers. See Table 1.

Table 1.

Personal Fear of Invalidity and Decision Making Under Stress Before Simulations

		PFI (Pre-Simulation)	DMUS (Pre-Simulation)
PFI (Pre-Simulation)	Pearson Correlation	1	-.471**
	Sig. (2-tailed)		.000
	N	59	58
DMUS (Pre-Simulation)	Pearson Correlation	-.471**	1
	Sig. (2-tailed)	.000	
	N	58	58

Note. ** Correlation is significant at the 0.01 level (2-tailed).

A Pearson's correlation indicated there was a significant negative association between DMUS (post) and PFI (post), r (22) = -.547, p = .006. In other words, when perceived fear of invalidity increases, decision making under stress decreases. Researchers suspect this may be due to students' increased self-awareness of decision-making abilities after experiencing a full simulation and AAR. See Table 2.

Table 2.
Decision Making Under Stress and Personal Fear of Invalidity Post Flight Simulation

		DMUS (Post-Simulation)	PFI (Post-Simulation)
DMUS (Post-Simulation)	Pearson Correlation	1	-.547**
	Sig. (2-tailed)		0.006
	N	24	24
PFI (Post-Simulation)	Pearson Correlation	-.547**	1
	Sig. (2-tailed)	0.006	
	N	24	24

Note. ** Correlation is significant at the 0.01 level (2-tailed).

A paired samples t-test (α = .05) indicated PFI did not differ for measures taken before simulations (pre) (M=3.56, SD = .82, n =19) and measures taken after simulations (post) (M = 3.46, SD = .68, n = 19), t(18)=.780, p =.446, d = 0.13. However, the means between the pre and post conditions decreased, suggesting that as students gained experience in the simulations, they experienced less fear about potentially wrong solutions due their experiences in the lab. See Table 3.

A paired samples t-test (α = .05) indicated that DMUS scores differed for measures taken before the first simulation (M = 3.44, SD = .50, n = 19) and measures taken after the first simulation (M =3.70 , SD = .48, n = 19), t(18)= -4.54, p = <.001, d = -0.52. Additionally, the average reported DMUS after the simulations was higher than the average reported before simulations, indicating an increase in perceived decision-making ability under stress. The data shows that students feel they have learned from their experiences in the first simulation and AAR, helping them feel more confident that they can make better decisions under stress going forward. See Table 3.

Table 3.
Paired Samples Test

	M	SD	SEM	t	df	p (2-tailed)
Pair 1	0.1004	0.5611	0.1287	0.780	18	.446
Pair 2	-0.2545	0.2445	0.0560	-4.537	18	.000

Note. Pair 1: PFI (Pre-Simulation) - PFI (Post-Simulation)
Pair 2: DMUS (Pre-Simulation) - DMUS (Sim 1)

Conclusion

Results indicated that after experience with simulation-based training, students reported a decrease in their personal fear of invalidity and an increase in their perceived ability to make decisions under stress. Consistent with the literature, results also revealed a correlation between decision-making under stress and personal fear of invalidity – further supporting a connection between the two constructs (Pope, 2018). This suggests that individuals with a high fear of making incorrect decisions feel that they are less able to make good decisions during stressful situations. This finding is instrumental to the aviation industry as it suggests that by allowing individuals to experience job-relevant problem-solving opportunities while under stress – such as those offered by high-fidelity simulations – their decision-making self-efficacy can improve and personal fear levels can be lowered. As previous literature suggests, this may decrease rumination, maladaptive decision-making processes, and oversimplification – thus contributing to safer conditions in air flight operations (Pope, 2018; Johnston, et al., 1997; Thompson, et al., 2001).

Acknowledgements

We would like to thank Tyler Bos, Master's Candidate, for his advice in quantitative data analysis. This paper is based on the Master's Thesis of Jessi Pope, M.A. (2018).

References

Brace, B. R. (2011). Fear of failure and the effective leader. (Unpublished master's thesis). Middle Tennessee State University, Murfreesboro, TN.

Cannon-Bowers, J. A., & Salas, E. E. (1998). *Making decisions under stress: Implications for individual and team training*. Washington, DC: American Psychological Association.

Keinan, G. (1987). Decision making under stress: Scanning of alternatives under controllable and uncontrollable threats. *Journal of Personality and Social Psychology, 52*(3), 639-644.

Levi, A., & Tetlock, P. E. (1980). A cognitive analysis of Japan's 1941 decision for war. *Journal of Conflict Resolution, 24*(2), 195-211.

Payne, J. W., Bettman, J. R., & Johnson, E. J. (1988). *Adaptive strategy selection in decision making.* Duke University, Durham, NC. Office of Naval Research.

Pope, J (2018). "Examining change in self-assessments of decision making under stress in a flight operations center simulator." Unpublished master's thesis, Middle Tennessee State University, Murfreesboro, Tennessee.

Sarason, I. G. (1984). Stress, anxiety, and cognitive interference: Reactions to tests. *Journal of personality and social psychology, 46*(4), 929.

Stogdill, R. M. (1963). *Manual for the Leader Behavior Description Questionnaire-Form XII.* Columbus, Ohio: Ohio State University.

Johnston, J. H., Driskell, J. E., & Salas, E. (1997). Vigilant and hypervigilant decision making. *Journal of Applied Psychology, 82*(4), 614-622.

Thompson, M. M., Naccarato, M. E., Parker, K. C., & Moskowitz, G. B. (2001). The personal need for structure and personal fear of invalidity measures: Historical perspectives, current applications, and future directions. *Cognitive social psychology: The Princeton symposium on the legacy and future of social cognition* (pp. 19-39).

TOWARDS A META-MODEL TO SPECIFY AND DESIGN HUMAN-AGENT TEAMS

Michael F. Schneider and Michael E. Miller
Air Force Institute of Technology
Dayton, OH, USA
John M. McGuirl
Integrity Applications Incorporated
Dayton, OH, USA

This paper describes work towards developing a meta-model useful in the design and specification of Human-Agent Teams. The meta-model adapts components from the cognitive systems, human factors, software and systems engineering literature to form a model and language which can be applied early in the system design process. The resulting model provides a description of desired system behavior. More importantly, the model produces artifacts useful in deriving requirements for both the human and the artificial agents, as well as for the software/hardware human interface. Insight is also provided for manpower, training, and personnel requirements; as well as, requirements for agent sensing, processing, and actuating. This method has been developed to support student projects in a graduate human-agent teaming course at the Air Force Institute of Technology and has been useful in describing systems employing both embodied and disaggregated agents.

With the advent of agent-based software and multi-agent systems, agent-based modeling languages have been developed within the software development community to specify the intended behavior during development (Deloach & García-Ojeda, 2010). Generally these languages focus on the design and interaction of software agents and consider humans as entities external to the system. However, the terminology and the concepts applied within this domain mimic the terminology that is applied to describe the structure of human-human teams. The intent of the current paper is to examine the expansion of agent-based modeling languages to the description, analysis, and communication of systems which include human-agent teams to aid design and the communication of design to development teams.

Our interest is to explore agents as they apply to automation. Automation is defined as the process of substituting an activity originally performed by a human with an activity performed by a man-made artifact or system (Parsons, 1985). In the systems of interest, we are particularly interested in adaptive automation performed by agents. Expanding on existing definitions of agents (Weiss, 2013), we define an agent as a persistent entity that can: 1) perceive the environment to obtain state information, 2) apply this information to engage in non-deterministic reasoning relative to a set of goals, and 3) apply this reasoning to drive actions in the environment. Therefore, in human-agent teaming we consider humans and artificial agents who collaborate to fulfill a common set of goals. Our approach recognizes that automation may only be possible under certain circumstances, such that successful performance across a broad range of environmental conditions requires collaboration between humans and artificial agents.

In the current context, agents may be either a single physical entity (i.e., a robot) or disembodied and distributed where sensors or actuators are located remotely from the reasoning engine. These distributed components logically perform as an agent. An example of such a logical agent might be a tsunami warning system with distributed sensors to continually gather geological activity and changes in water level, reason about this information, and issue a warning to a human teammate or a broader population. For clarity, we will refer to agents as either human or artificial when it is important to distinguish human from man-made agents.

A meta-model to describe agent-based software systems was proposed by DeLoach and colleagues to describe the Organizational Multi-Agent Software Engineering (O-MaSE) modelling framework (Deloach & García-Ojeda, 2010) . In this model, an organization of software agents is comprised of one or more agents, with each artificial agent playing a role which is designed to achieve a goal. Agents within this model interact with external actors, for example humans. In this model the artificial agents' goals are not shared with the human. Instead, the artificial agents work alone to achieve goals and the humans are exterior to the agent model. While this arrangement may facilitate the design of a software system, is limiting when attempting to design human-agent interaction.

While human and artificial agent goals are not shared in most agent-based software engineering models, it is not common to all. Sterling and Tavateer propose that humans and artificial agents should have common, shared goals within an interaction model (Sterling & Taveter, 2009). In the model they propose, goals within an environment can be decomposed into a hierarchy of goals, similar to the hierarchy of goals used to analyze human activity. These shared goals can provide a starting point from which to design the interaction between humans and artificial agents. Such a design model should permit the designer to consider the activities, personnel selection criteria, and training of the human; the required capabilities of the artificial agents; and requirements for the user interface between the humans, agents, and any machine that they collaborate to control. Thus, we seek to define and explore an agent model that is more suited to the description and analysis of human-agent teams.

Overview of Proposed Meta-Model

The proposed meta model is designed to be integrated with Digital Engineering (*Department of Defense Digital Engineering Strategy*, 2018) techniques to develop multidiciplinary design models, while directly supporting cognitive analysis methods. The result of applying this meta model is a hollistic view of the system design focused on the cognitive elements that produce system performance.

Interdependence as a guiding principle of agent and team design

It is well understood that environmental impacts affect the ability of any system or human to perform any task. For example, environmental effects can deprive human or artificial agents from energy sources, degrade information sources, overwhelm computational capabilities, or change the interaction requirements. Teams are then structured to promote backup behaviors to improve system resilience by exploiting interdependence between agents. Therefore, as we explore meta-models for team design, it is critical that these meta-models include methods to understand and to design in interdendence (Johnson et al., 2014).

Proposed Meta-Model

The proposed structural meta-model is shown in Figure 1. This figure has been arranged so that items on the left side of the figure are heavily influenced by the user of the system prior to the introduction of automation and items on the right side are heavily influenced by the the designer during the design process. Items towards the center require both knowledge from the user and the designer. Elements are joined by associated relationships, indicated by arrows with descriptions; composition, indicated by filled diamonds, genearlizations, indicated by hollow arrows, and aggregation, indicated by hollow diamonds. It is important that the human in Figure 1 represents the system user(s) after the introduction of automation.

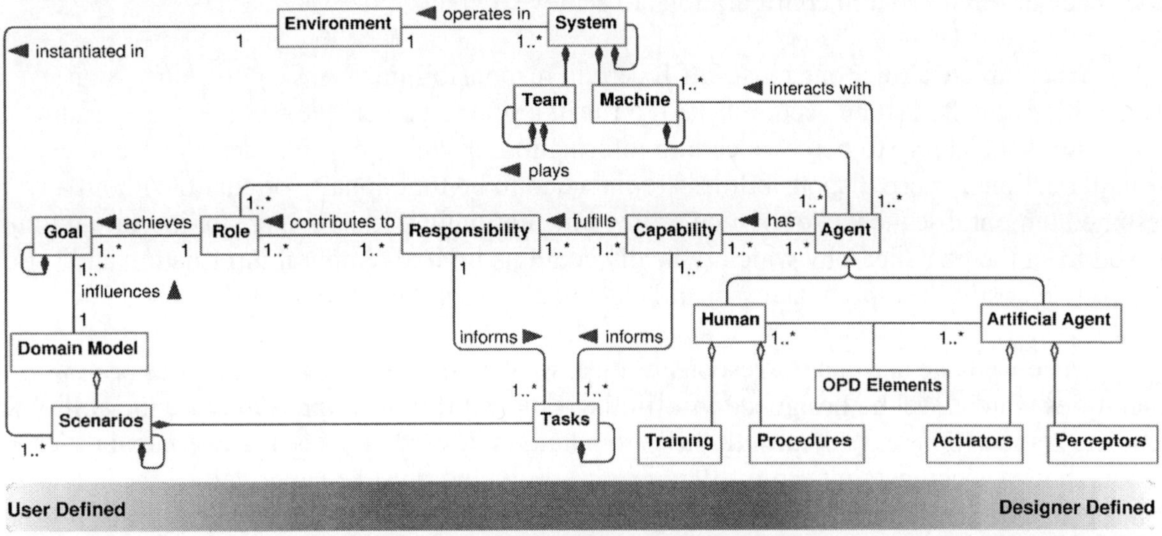

Figure 1. Proposed Meta-model of Human-Artificial Agent Team design in SysML notation

As shown, a *System* may be comprised of other systems, machines, and teams. Systems are operated in an *Environment* comprised of physical (e.g., weather), social (e.g., external collaborators), and informational (e.g., available knowledge) elements. We use the term *Machine* to denote the noncognative, supporting elements of a system (e.g., an airframe). In complex systems it is useful to loosly couple the hardware and supporting software from the agents. Any *Team* may be comprised of other teams and agents (e.g., air crew). An *Agent* is an abstract element which may be specified as either a human (e.g., pilot) or an artificial agent (e.g., autopilot). These agents interact with the one or more machines in the system in which the team is situated. These agents possess capabilities that can fulfill system functions. Of course the difficulty in system design is to understand the capabilities that are required and the interaction of team members to enable the most appropriate capabilities to be applied at the correct time. Understanding and defining these capabilities is central to the design of successful systems.

In our meta-model, understanding the necessary capabilities begins with defining goals. *Goals* are desirable states that must be achieved by the human-agent team (e.g., arrive on schedule)(Deloach & García-Ojeda, 2010). In the goal model, high level goals are decomposed into lower and lower level goals to provide a goal hierarchy. Importantly goals describe *what* needs to be achieved, but do not attempt to describe *how* these goals are achieved. This distinction is critical as high level organizational goals are unlikely to be changed significantly by technology. However, automation often changes how a goal is achieved or who accomplishes the steps to achieve the goals (Endsley & Jones, 2012).

Within this model, agents play *Roles*, which are defined as a distinct set of responsibilities necessary to fulfill one or more goals (e.g., flyer). Roles provide a mechanism to compartmentalize the responsibilities to reduce human training and the communication necessary between team members. These roles are designed to partition the high level activities necessary to fulfill the goals within the goal hierarchy for the human-agent team. At least one agent must play each role, however, multiple agents can contribute to responsibilities to fulfill a role. By abstractly mapping agents to goals, through roles, we avoid definitive allocations and provide the trade space of possible team configurations to achieve a goal.

To complete a role, one must fulfill certain responsibilities. A *Responsibility* is an abstract item which must be accomplished to fulfill a role (e.g., achieve and maintain heading). This requires the ability to perceive certain information, make appropriate decisions, and take particular actions. These responsibilities are intended to be functionally oriented, without any presumed temporal sequence, or particular implementation in mind. Responsibilities are initially derived from the goal hiearchy which typically contains limited temporal information and should be design agnostic. Temporal assessment is handled by scenarios and tasks discussed later.

Once we have defined the responsibilities, we can identify agent capabilities or new capabilities which must be designed to fulfill the responsibilities. *Capabilities* are the ability to complete some action (e.g., control throttles to maintain desired airspeed). These capabilities describe how a specific agent can fulfill a responsibility and must be supported by the agents to permit goal completion. Responsibilities and capabilities describe the possible methods for the team to achieve a goal, but during execution they are tied together by tasks. *Tasks* are temporal sequences of actions that, when executed, fulfill a responsibility. They trace a specific path through the trade space of possible actions that agents can take to accomplish a goal.

The *Domain Model* is a model that captures the policy, resource, value, and technique considerations that influence how the human agent team consider their operations (e.g., air track routes). The domain model influences goal definition and holds a set of scenarios that demonstrate and justify the elements of the domain model. *Scenarios* are specific examples of execution of temporally-arranged tasks by the team to accomplish some set of goals. Synonymous with use cases or user stories, they describe an instantiated form of the environment and are composed of individual tasks. Not all possible scenarios can be modeled, but common, critical, and special interest scenarios provide an oppportunity to identify capability or responsibility gaps when their tasks lack a relationship to the properties of the team.

Analysis Approach

While it is important to understand the capabilities of each of the agents, it is also important to understand that each agent will have certain constraints which limit the conditions under which these capabilities can be applied. Another key insight is that the agents' capabilities are context sensitive. That is, they are shaped, in part, by the environment and thus subject to change as the various constraints (both internal or external) impact the system. It is this context sensitivity that drives the need for an what Johnson termed an "interdependence analysis" as part of the Coactive Design process (Johnson et al., 2014). This process highlights the tasks and constraints that will impact a agent's capabilities and the potential need for multi-agent support or retasking. This analysis also supports the design process by identifying requirements to address observability, predictability, and directability (OPD) issues between agents. Johnson has applied interdependence analysis primarily to relatively simple robotic interactions usually between individual humans and agents rather than the design of multi-agent collections in complex systems as intended in our method. Therefore, the incorporation of this process into our model requires modification. In our model, this analysis is facilitated through scenarios. Scenarios also provide a method to assess new or novel situations and determine how the team could react based on the capabilities and responsibilities required by the tasks. Finding gaps and exploring potential automation surprises contributes to refining the training and procedures the human agents use to establish their capabilities. Complimentarily, the artificial agent's requirements for perceptors and actuators are refined during this same analysis.

Example Application

We applied this meta model to a hypothetical airborne recconaissance mission in a GPS denied/degraded environment. The system, environment, team, and machine elements were modeled using Systems Modeling Language (SysML) external and internal system context diagrams. A goal directed task analysis was applied to develope the goal hierarchy for the geo-locating targets portion of a mission. Six roles were developed by functionally grouping common goals. A total of 23 responsibilities were assigned to these roles with five contributing to multiple roles. These overlaping responsibilities related to commonality in the situation awareness necessary for tightly coupled roles (e.g., a flyer and a navigator). As stated earlier, this portion of the analysis indicates "what" the team needs to be able to accomplish.

In definining the team, a total of five agents, a human pilot, a human sensor operator and three artificial agents were identified. To fulfill the responsibilities, 27 functional capabilities were identified and associated with the responsibilities and agents. Figure 3 depicts the entire definition of the flyer role. Independently, narrative scenarios were developed to test the model. The tasks and domain model were derived from the scenarios with the tasks and capabilities forming the basis of an extended form of interdependency analysis. The analysis and modeling identified missing interfaces between an artificial agent and a human. We also identified sub capabilities that should have been explicitly modeled to provide a clearer understanding of the requirements. These are in addition to the OPD, training, and procedure requirements derived from interdependence analysis, sample in Figure 2.

Mission Phase	Role	Responsibility	Task	Required Capabilities	Primary Agent	Available Capabilities Primary Agent	Secondary Agent	Available Capabilities Secondary Agent	Observability, Predictability, Directability Requirements
Search AOI	Flyer	Fly aircraft	Fly the plane along a search pattern to cover the AOI	Control Aircraft Follow Route Understand Airspace Understand Current	AP	Control Aircraft Follow Route Understand Airspace Understand Current Situation Understand Terrain	PIC	Control Aircraft Follow Route Understand Airspace Understand Current Situation Understand Terrain Understand Threats	The PIC should have a method of placing constraints on the autopilot's execution such that it does not conflict with threats and or weather. There should be a method for the PIC and the AP to

Figure 2. Example of Extended Interdependence Analysis

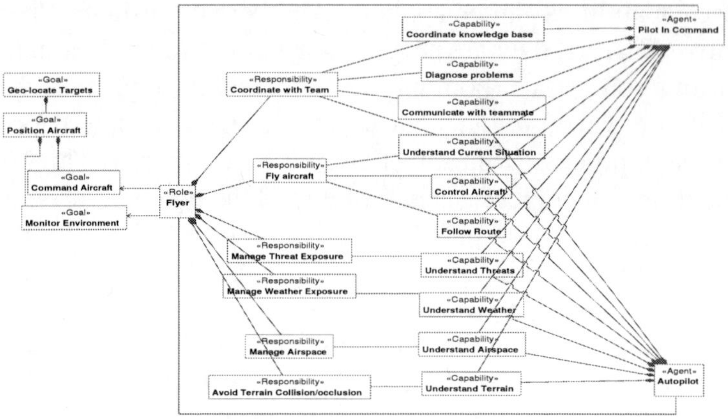

Figure 3. Flyer Role Definition Example

Conclusion

A meta-model for modeling human-machine development was discussed and illustrated through a short example application. It is believed that this approach provides a method for documenting a design for a human-machine team and provides a method for implementing interdependence analysis to the design of complex systems.

Disclaimer and Acknowledgement

The views expressed in this article are those of the authors and do not necessarily reflect the official policy or position of the Department of the Air Force, Department of Defense, nor the U.S. Government. The authors gratefully acknowledge the financial support of the Air Force Office of Scientific Research, Computational Cognition and Machine Intelligence Program.

References

Deloach, S. A. (2004). The MaSE Methodology. In F. Bergenti, M. P. Gleizes, & F. Zambonelli (Eds.), *Methodologies and Software Engineering for Agent Systems* (pp. 107–125). Springer.

Deloach, S. A., & García-Ojeda, J. C. (2010). O-MaSE: a customisable approach to designing and building complex, adaptive multi-agent systems. *Int. J. Agent-Oriented Software Engineering*, *4*(3), 244–280.

Department of Defense Digital Engineering Strategy. (2018). Washington, D.C. Retrieved from www.acq.osd.mil/se

Endsley, M. R., & Jones, D. G. (2012). *Designing for Situation Awareness: An Approach to User-Centered Design*. Boca Raton, FL: CRC Press, Inc.

Johnson, M., Bradshaw, J. M., Feltovich, P. J., Jonker, C. M., Van Riemsdijk, M. B., & Sierhuis, M. (2014). Coactive Design: Designing Support for Interdependence in Joint Activity. *Journal of Human-Robot Interaction*, *3*(1), 43. https://doi.org/10.5898/JHRI.3.1.Johnson

Parsons, M. H. (1985). Automation and the individual: Comprehensive and comparative views. *Human Factors*, *27*(1), 99–112. https://doi.org/10.1177/001872088502700109

Sterling, L. S., & Taveter, K. (2009). *The Art of Agent-Oriented Modeling*. Boston, MA: MIT Press.

SIMULATING HUMAN-ROBOT TEAMWORK DYNAMICS FOR EVALUATION OF WORK STRATEGIES IN HUMAN-ROBOT TEAMS

Martijn IJtsma, Sean Ye, Karen M. Feigh
Georgia Institute of Technology
Atlanta, Georgia

Amy Pritchett
Pennsylvania State University
State College, Pennsylvania

To foster resilience in teams operating in complex work domains, design should allow for a range of work strategies as appropriate to context. This paper describes how computational simulation and network visualization of a team's work can identify feasible work strategies and assess their appropriateness for different contexts. Network visualizations can identify constraints and dependencies that drive the feasible set of work strategies. After preliminary network analysis, these dependencies and inter-dependencies can be simulated in detail to better understand their impact. To illustrate, we describe a case study that explores two different work strategies that can each address the dependencies in a human-robot (rover) team in a manned space exploration mission.

Studies of expert workers in complex work domains teach us that much of a system's resilience orginates from its workers' ability to adapt their work strategies, both to manage performance and workload levels (Woods and Hollnagel, 2006). To support such adaptation in design of teams, designers can build in flexibility to allow team members to "finish the design" (Vicente, 1999). Many current design methods for teams, however, inherently prescribe normative work strategies through implicit assumptions about how the work ought to be done. Additionally, existing attempts to create a more formative approach to the design of the team (e.g. Ashoori and Burns, 2013) have applied static work models that cannot account for the evolving dynamics of the work itself, and the coordination and synchronization it requires within a team. This paper introduces computational work analysis as a means to creating designs that can support multiple work strategies.

Background

A team is composed of agents, which can be human or technological agents (i.e. robots or other forms of automated systems). Conceptual design of teams specifies team composition, work allocation, and mechanisms for coordinating activities (IJtsma, Ma, Feigh, and Pritchett, 2019). Cognitive Work Analysis (CWA) (Vicente, 1999) is a design framework that formalized the idea of designing for expert workers to adapt through the support for different strategies. Contrary to normative design frameworks, which often prescribe an "optimal" work strategy that in practice limits adaptation, CWA lays out "formative" methods that help designers in supporting experts workers. Formative analysis of teams can provide insight in the constraints and dependencies that need to be managed by a team, regardless of the context.

Earlier work has applied CWA methods to study work in teams (Ashoori and Burns, 2013; Miller, McGuire, and Feigh, 2017). However, the formative methods proposed in CWA have been based on qualitative methods that, relative to the needs of designers not versed in CWA, can be rather vague and do not model the temporal dynamics of the team's joint work. The methods additionally are manual and therefore labour and time intensive (Bodin and Krupenia, 2016). While our earlier work demonstrated computational work models to evaluate patterns of work (IJtsma et al., 2019), a remaining challenge is identifying and supporting-in-design multiple different feasible work strategies to support team adaptation.

Identifying Work Strategies Through Network Visualization

This paper proposes an analysis of work through graph network visualization, aimed to inform designers by identifying constraints and dependencies that define a set of feasible work strategies. The analysis examines a model of the work to be conducted in the team to identify the constraints and dependencies in the work that drive which work strategies are feasible. Such identification of feasible work strategies can then be used as input to a computational model of work, to further examine the temporal components of these dependencies.

First, a team's work within their given work domain is described at various levels of abstraction. At the highest level, the team's work is described in terms of one or a number of goals, further elaborated as values and priorities one level down in the abstraction hierarchy. These values and priorities can be further abstracted into work functions, i.e. the actions that can be performed by the team members. The lowest level describes the tangible aspects of the work environment, i.e. physical and information resources.

Each level in such an abstraction hierarchy provides a complete description of the team's work domain; multiple heterarchic linkages can then be identified between the levels. To identify dependencies within the team's work arising from the work environment, three types of linkages need to be identified between actions and resources: (1) Actions can require as input specific information resources, formalized as *get* relationships; (2) actions can serve to change the environment, described as changing aspects of the environment state through *set* relationships where actions manipulate pieces of information as output; (3) actions can require physical resources (such as tools), described through *use* relationships.

These three types of linkages identified in the work model then constrain the paths that can be taken: *get* and *set* relationships imply that certain actions are sequential, in which one action's output is input to another; *use* relationships imply that two actions that are both linked to the same physical resource cannot be executed together. These dependencies can help with methods of identifying feasible strategies, similar to classic methods in which strategies can be shown on the abstraction hierarchy as feasible paths through the work domain (Vicente, 1999). With these specific formalizations, however, a more systematic approach can be undertaken using two types of network visualization: one that contains the information resources and one that contains the physical resources.

When a computational form of these networks is available, the set of feasible action sequences can be formally identified. For example, one can identify all actions that need to be executed to reach a final state, or alternatively, from an initial state, all possible next actions can be identified by forward propagating through the network. Moreover, network theory has several constructs that can be used to analyze and characterize the work model. For instance, the connectivity of the networks provides insight into how constrained the work is.

The way these constraints are coordinated results in a pattern of actions that we refer to as the *work dynamics* (IJtsma et al., 2019). Thus, once the network has been analyzed and feasible action sequences have been identified, the work model can be extended into a computational form that can be used to simulate the work dynamics. Simulation can provide more detailed insight into the temporal aspects of each work strategy, where the interplay of the various constraints and dependencies might result in emergent behavior in the team.

Case Study: Manned Space Exploration with a Rover

Our earlier research has focused on simulation of work dynamics in human-robot teams for space operations (IJtsma et al., 2019). The case study presented here will further demonstrate how network visualizations can provide formative insight to designers, through analyzing constraints in the work and identifying feasible work strategies. Here we model a simple team consisting of a rover and two astronaut drivers on the lunar surface. Following a brief literature survey on lunar EVA and Mars Rover operations (Hooey, Toy, Carvalho, Fong, and Gore, 2017; Miller et al., 2017), as well as several informal discussions with space robotics and operations researchers, a work model was created in the form of an abstraction hierarchy, shown in Table 1.

Table 1.
Abstraction hierarchy of the work in a rover/astronaut team.

Functional purpose	EVA objectives			
Abstract functions	Resource consumption	EVA priorities	Rover safety	
Generalized functions	Life support system monitoring	Local navigation	Translation, control, and orientation	Imagery
Physical functions	1. Check battery levels	4. Plan path for rover	7. Select next waypoint	9. Change camera angle
	2. Check temperature	5. Estimate size of object	8. Move rover	10. Capture imagery
	3. Assess location and attitude	6. Localize obstacles		
Physical resources	a. Batteries	b. Rover vehicle	c. Rocks/obstacles	d. Camera
Information resources	A-B. (Observed) battery levels	Q. Goal location	T. Terrain map	W. Camera angles
	C-D. (Observed) subsystem temp	R. Planned path	U. Rock locations	X. Imagery
	E-P. (Observed) rover states	S. Next waypoint	V. Rock size	Y. Goal reached

Linkages between elements in this model identify constraints on the work. Figure 1 shows these linkages in a graph network representation. The nodes are actions (squares) and information resources (circles). The edges represent *set* and *get* linkages. The directional graph clearly shows the centrality of the rover dynamics (`Move Rover`), as well as the path planning

(`Plan Rover Path`). Several feedback loops are apparent in this representation, including the feedback loop for selecting the next waypoint and moving the rover iteratively, as well as a longer-timescale loop for replanning the path on a larger scale. A designer can further refine this model of the work iteratively by identifying new information requirements for actions and thereby creating new edges.

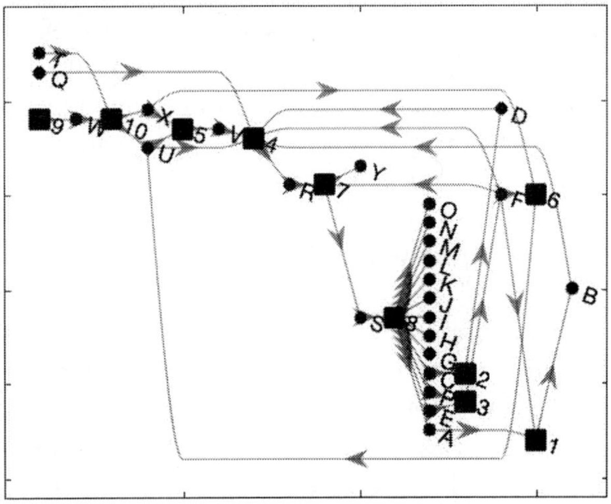

Figure 1. Network visualization of precedence relationships.

Once the network is fully flashed out, it can be used to identify feasible work strategies between two or more points. As an example, Figure 2 shows two feasible work strategies between setting the camera angle (`Change Camera Angle`) and moving the rover (`Move Rover`). In Strategy 1 the rover imagery is used straight away without formal analysis of obstacles, and is appropriate when high quality information on the "rock locations" is already available and does not require constant sampling. Strategy 2 steps through localization of obstacles and estimation of their size before doing path planning. This strategy would be appropriate when the quality of information for "rock locations" is low, and therefore the imagery should be frequently sampled and used to inform path planning. In many potential missions, the information quality can vary with location and terrain, and thus the team may need to adapt between these strategies.

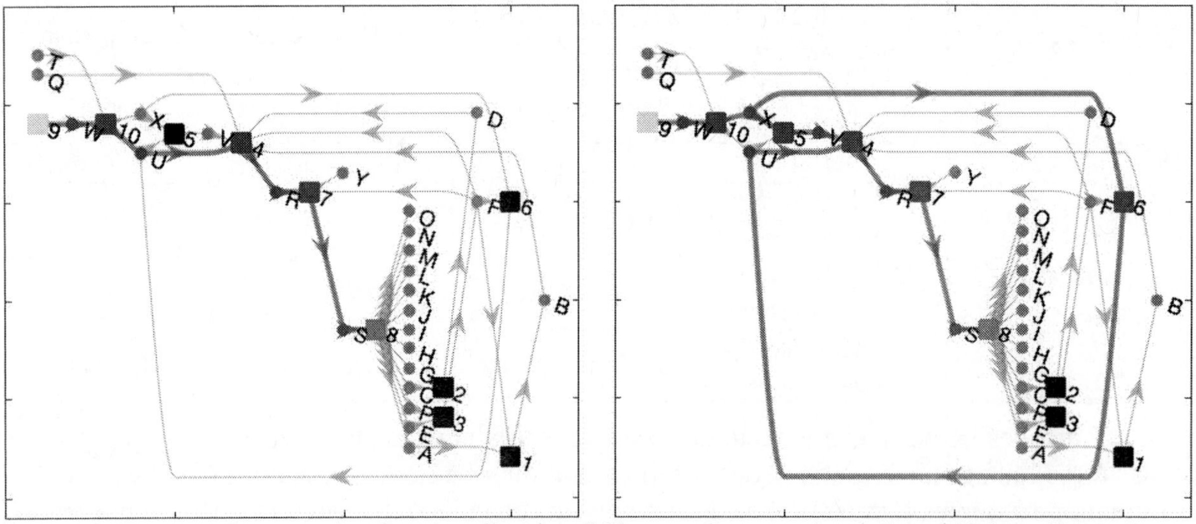

Figure 2. Two work strategies for coordinating "Change Camera Angle" and "Move Rover".

Once feasible action sequences are identified, more detailed evaluation can be performed through simulation of the work. A computational version of the work model was created to evaluate the timing of actions within each work strategy. From the graph networks it is clear that the rover dynamics play a significant role in the work dynamics. Thus, the computational model of the action `Move Rover` contains a dynamic model of the rover (updated at 2Hz), and models of how a human or automatic system would control speed and heading to track to a given waypoint. A lunar terrain model identified whether any location was open or has obstacles that the rover needed to maneuver around.. The `Localize Obstacles` action checks the surrounding area and updates the known obstacles in the environment. The action `Plan Rover Path` produces a path from the rover's current location to desired location. The `Select Next Waypoint` action provides the rover with the next desired waypoint to traverse to.

The two feasible work strategies in Figure 2 were simulated using this computational work model. In this case, all actions were performed by a single agent to illustrate the dynamics inherent to the work even before accounting for inter-agent coordination; the simulation can also examine a range of potential work allocations. For example, an astronaut might plan the paths based on an assumed database of terrain information and the rover then move through them autonomously (correspondingly closely with Strategy 1), or the human may actively steer the camera to confirm the feasibility of the planned path and interact more closely with the rover in path following (as may be necessary with Strategy 2).

Figure 3 shows the path that the rover traversed in each strategy. Clearly, having prior knowledge of the obstacles resulted in a more optimized path. With lower information quality in the right figure, the rover first attempts to drive straight to the goal position, only later finding out (through `Localize Obstacles` and `Estimate Size Object`) that this route is not a viable option, and then needing several iterations to find a path around obstacles.

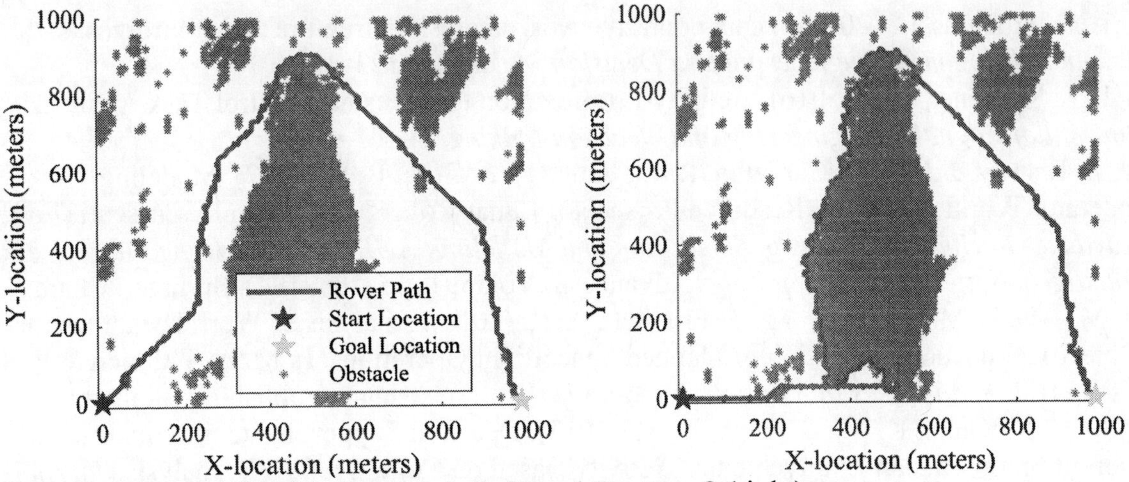

Figure 3. Rover path with Strategy 1 (left) and Strategy 2 (right).

Conclusions

This paper introduced a formative and computational work modeling approach that can support designers of teams for complex work domains by identifying and designing for multiple work strategies. Many further iterations are possible with this method. The network analysis and simulation described in this paper comprised a first step in examining primarily the task work that the team must collectively perform. Once the specific strategies (or attributes of a wide range of feasible strategies) are thus identified within the taskwork, subsequent design decisions can then examine the team composition, and the allocation of work within the team, by which this taskwork will be conducted.

Further, the method of coordination will impact the feasibility of different work strategies and, if they remain feasible, their timing and performance. In human-human coordination within teams, the humans are typically reasonably flexible and adaptive in their modes of interaction. On the other hand, in the case of human-robot or human-automation interaction, the machine is comparatively inflexible in the modes of interaction, requiring predictable patterns in the human's activity for commanding, controlling, monitoring and/or confirming the machine. These dynamics can be included in the network analysis as additional teamwork actions that require particular sequences of actions (e.g. the robot cannot perform an action until its human supervisor is free to command and monitor it – and the human may not be able to perform her/his own physical tasks in parallel). Further, these teamwork actions add a further temporal component to the collective dynamics of the team's work, which computational simulation can further predict.

Acknowledgements

This work is sponsored by the NASA Human Research Program with Jessica Marquez serving as Technical Monitor under grant number NNJ15ZSA001N. The authors would also like to thank the other WMC developers.

References

Ashoori, M., & Burns, C. (2013). Team cognitive work analysis: Structure and control tasks. *Journal of Cognitive Engineering and Decision Making*, *7*, 123–140.

Bodin, I., & Krupenia, S. S. (2016). Activity Prioritization to Focus the Control Task Analysis. *Journal of Cognitive Engineering and Decision Making*, *10*, 91–104.

Hooey, B. L., Toy, J. J. N. T., Carvalho, R. E., Fong, T., & Gore, B. F. (2017). Modeling Operator Workload for the Resource Prospector Lunar Rover Mission. In D. Cassenti (Ed.), *Advances in Human Factors in Simulation and Modeling. AHFE 2017. Advances in Intelligent Systems and Computing* (Advances i, Vol. 591, pp. 183–194). Springer, Cham.

IJtsma, M., Ma, L. M., Feigh, K. M., & Pritchett, A. R. (2019). Analysis of Work Dynamics for Objective Function Allocation in Manned Spaceflight Operations. In M. A. Vidulich & P. S. Tsang (Eds.), *Advances in Aviation Psychology* (Vol. 3). Ashgate Publishing Limited.

Miller, M. J., McGuire, K. M., & Feigh, K. M. (2017). Decision Support System Requirements Definition for Human Extravehicular Activity Based on Cognitive Work Analysis. *Journal of Cognitive Engineering and Decision Making*, *11*, 136–165.

Vicente, K. (1999). Cognitive Work Analysis: Modeling Behavior-Shaping Constraints. In *Cognitive Work Analysis* (pp. 113–136).

Woods, D. D., & Hollnagel, E. (2006). *Joint Cognitive Systems: Patterns in Cognitive Systems Engineering* (1st ed.). Boca Raton, FL: CRC Press.

RESILIENCE AND SAFETY FOR IN-TIME MONITORING, PREDICTION, AND MITIGATION OF EMERGENT RISKS IN COMMERCIAL AVIATION

Jon Holbrook
NASA Langley Research Center
Hampton, VA

Lawrence J. Prinzel III
NASA Langley Research Center
Hampton, VA

Michael J. Stewart
San Jose State University
Moffett Field, CA

Brian E. Smith
NASA Ames Research Center
Moffett Field, CA

Bryan L. Matthews
Stinger Ghaffarian Technologies, Inc.
Moffett Field, CA

Safety in aviation has been historically defined in terms of the occurrence of accidents or recognized risks; that is, safety is typically defined in terms of things that go wrong. An alternative and complementary approach is to focus on what goes *right,* and identify how to make that happen again. Focusing on the rare cases of failures attributed to "human error" provides little information about why human performance almost always goes right. Similarly, focusing on the lack of safety provides limited information about how to improve safety. This work builds upon a growing literature on resilience engineering and new approaches to safety (Hollnagel, 2014; Hollnagel, Woods, & Leveson, 2006). Data were collected from commercial airline pilots and air traffic controllers that illustrate the prevalence and value of resilient behaviors observed as routine in everyday operations. Results of data analyses as well as approaches to identify novel methods for data collection on resilient behavior for use in development of in-time safety monitoring, prediction, and mitigation technologies are described.

Every day in aviation, pilots, air traffic controllers, and other front-line personnel perform countless correct judgments and actions in a variety of operational environments. These judgments and actions are often the difference between an accident and a non-event. Ironically, data on these positive behaviors are rarely collected or analyzed. Data-driven decisions about safety management and design of safety-critical systems are limited by the available data, which influence how decision makers characterize problems and identify solutions. In the commercial aviation domain, data are systematically collected and analyzed on the failures and errors that result in infrequent incidents and accidents, but in the absence of data on behaviors that result in routine successful outcomes, safety management and system design decisions are based on a small sample of non-representative safety data.

NASA has proposed development of in-time safety monitoring, vulnerability prediction, and incident mitigation technologies for civil aviation (NASA, 2017). Ironically, a critical barrier to measuring safety threats and the impact of mitigation strategies in ultra-safe systems like commercial aviation is the lack of opportunities for measurement of beneficial events and conditions. Although it is common practice to relate safety to how many accidents or fatalities occur for a given number of flights, very safe systems have very few accidents. Therefore, accident data cannot be readily used to validate safety improvements for at least two reasons. First, the time necessary to observe the effect of a given safety intervention within accident statistics becomes excessively long, with estimates up to 6 years for a system with a fatal accident rate per operation of 10^{-7} (Amalberti, 2001). Second, attributing improvement to a specific intervention becomes intractable due to the many thousands of changes that a complex sociotechnical system would experience over that same time period (Nisula, 2018).

Historically, safety has been consistently defined in terms of the occurrence of accidents or recognized risks (i.e., in terms of things that go wrong). These adverse outcomes are explained by identifying their causes, and safety is assumed to be restored by eliminating or mitigating these causes. An alternative to this approach is to focus on what goes right and identify how to replicate that process. Focusing on the rare cases of failures attributed to "human error" provides little information about why human performance routinely prevents adverse events. Hollnagel (2014) has proposed that things go right because people continuously adjust their work to match operating conditions. These adjustments become increasingly important as systems continue to grow in complexity. Thus, the definition of safety should reflect not only "avoiding things that go wrong," but "ensuring that things go right." The basis for safety management requires developing an understanding of everyday activities. However, because few mechanisms to monitor everyday work exist in the aviation domain, there are limited opportunities to learn how designs function in real operational conditions.

This concept of safety thinking and safety management is reflected in the emerging field of resilience engineering. According to Hollnagel (2016), a system is *resilient* if it can sustain required operations under expected and unexpected conditions by adjusting its functioning prior to, during, or following changes, disturbances, and opportunities. To explore "positive" behaviors that contribute to resilient performance in commercial aviation, a range of existing sources of data about pilot and air traffic control (ATC) tower controller performance were examined, including subjective interviews with domain experts and objective aircraft flight data records. These data were used to identify strategies that support resilient performance and methods for exploring and refining those strategies in system-generated data.

Analysis of Operator-Generated Data

Pilot and ATC tower controllers were interviewed to elicit specific examples of resilient performance in routine operational situations. This approach focused on identifying behaviors and strategies based on the specific lived experience of the participants in an attempt to focus as closely as possible on work-as-done rather than work-as-imagined or work-as-designed.

Method

Participants. Twenty-one airline pilots and 12 air traffic controllers were recruited to participate. All pilot participants were employed by a major airline operating under Federal Aviation Regulations part 121 or its foreign equivalent. All controller participants were highly experienced, with an average of 33 years on the job. Interviews were conducted under approval from NASA's Institutional Review Board.

Procedure. Participants were interviewed individually, using a semi-structured protocol designed to elicit specific instances of unplanned or unexpected events experienced during routine operations, as well as their goals, motivations, pressures, and knowledge at the time of the described actions. Each interview lasted approximately 45 minutes.

Controller participants also completed a written questionnaire, in which they estimated the frequency of behaviors associated with resilient performance. In addition, controller participants took part in focus group discussions after completing their interview and questionnaire. Because pilots were interviewed during time-limited breaks between flights, they were not asked to complete the questionnaire nor participate in focus group discussions.

Results

The events and behaviors that participants described in the interviews were used to extrapolate strategies for resilient performance organized around four capabilities of resilient systems: anticipating, monitoring, responding, and learning (Hollnagel 2011). These strategies are shown in Table 1.

Table 1.
Identified Resilient Performance Strategies Employed in Routine Aviation Contexts.

Capability	Strategy
Anticipate	Anticipate procedure limits
	Anticipate knowledge gaps
	Anticipate resource gaps
	Prepare alternate plan and identify conditions for triggering
Monitor	Monitor environment for cues that signal a change from normal operations
	Monitor environment for cues that signal need to adjust/deviate from current plan
	Monitor own internal state
Respond	Adjust current plan to accommodate others
	Adjust or deviate from current plan based on risk assessment
	Negotiate adjustment or deviation from current plan
	Defer adjusting or deviating from plan to collect more information
	Manage available resources
	Recruit additional resources
	Manage priorities
Learn	Leverage experience and learning to modify or deviate from plan
	Understand formal expectations
	Facilitate others' learning

In responses to the administered questionnaire, all controller participants indicated that they exhibited resilient performance on the job as air traffic controllers, with 83% (N = 10) estimating that this occurs "at least once per session," where a "session" refers to each one of the multiple times that a controller works at their position during an 8-hour daily work shift.

Results showed that 75% of controller participants (N = 9) stated that they make traffic management decisions not explicitly specified within policies or procedures (e.g., FAA Order JO 7110.65, facilities standard operating procedures, letters of agreement) "at least once per week" with 58% (N = 7) estimating the occurrence to be "at least once daily." When asked, "How many of these decision would you categorize as 'resilient' decisions?", 75% estimated "more than 50%" (N = 9), and 58% indicated "more than 90%" (N = 7).

In focus group discussions with the controller participants, all stated that they had filed incident reports through one or more safety reporting systems. However, none of the participants stated that their narrative descriptions focused on detailing positive behaviors that demonstrate resilient performance. Participants noted "it was their job" to adapt to routine disturbances, and showing resilient behavior was "what they get paid to do." Participants believed that, in the current cultural environment, controllers might be reluctant to file positive incidents except in the case of extraordinary performance. Another barrier to positive event reporting is that most reporting systems are structured to capture negative events (i.e., when things go wrong). Participants suggested providing guided assistance for furnishing narrative details to ensure that filed reports focused on desired aspects or features of resilient performance.

Exploring Identified Resilient Strategies in System-Generated Data

Although operator-based data (e.g., structured interviews and self-reports) can provide rich data with regard to intentions, goals, pressures, or operator state, recollection-based approaches are subject to reconstructive attributes of human memory (Schacter, 1989). Examination of system-based objective data can substantiate subjective accounts and provide quantifiable details about events that are difficult or impossible to obtain from subjective data alone.

Based on the strategies and behaviors identified through operator interviews (see Table 1), the authors considered how these strategies might show up in aircraft flight data. For example, operators "anticipating resource gaps" might manifest in objective aircraft flight data as the pilot taking action to preempt an anticipated adverse state (i.e., a state indicating that one or more resources had reached their functional boundaries). These preemptive actions were identified using a machine-learning algorithm called deep temporal multiple instance learning (Janakiraman, 2018). This algorithm was designed to detect "precursor" states, ahead of a predefined known adverse event, that have a high probability of predicting that adverse event.

This method was demonstrated using Flight Operations Quality Assurance (FOQA) data. Commercial airlines with FOQA programs use data from flight data recorders to monitor daily operations. The adverse event used in this example was a high-speed exceedance at 1000 feet. A sample of 500 adverse event flights and 500 non-event flights were analyzed. Adverse event flights were analyzed to characterize those events based on 60 recorded variables, and non-event flights were then examined for high precursor probabilities.

An example of a flight that exhibited high precursor probability followed by the lowering of that probability is shown in Figure 1. The x-axis shows distance in nautical miles (NM) from the point at which the aircraft reaches 1000 ft. altitude. The solid blue line is the time series trace for the selected parameters that describe the precursor. The black dotted lines indicate the 10th-90th percentiles of the non-event data for each parameter for 0.25 NM binned distances to the event. Plot 4 shows the computed precursor score that the algorithm provided for each sample of the time series. Samples for which the precursor score was greater than 0.5 are marked with red dots in Plots 1-3 and are considered high-probability precursors of a high-speed exceedance at the end of the time series. The shaded green region in the precursor score plot represents the event of interest, in which a degraded state was identified and potential for a preemptive action was indicated.

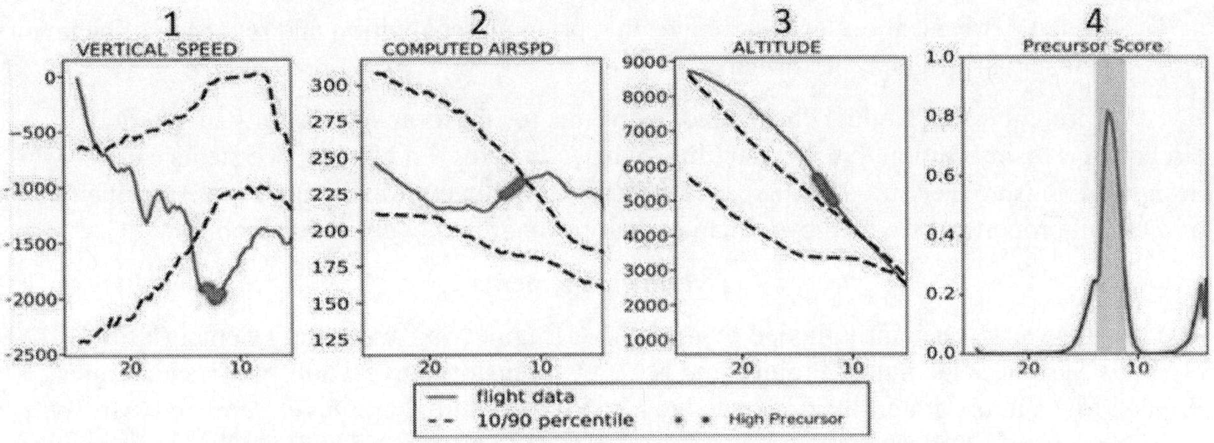

Figure 1. Time series plots for vertical speed, altitude, computed airspeed, and precursor score are depicted for a flight in which a preemptive action (i.e., slowing descent rate) was taken to avoid a high-speed exceedance at 1000 ft.

In Figure 1, the descent rate, inferred from vertical speed, was significantly faster than the normal distribution at that point in the flight (Plot 1). Simultaneously, the airspeed was trending upward toward the upper bound of the nominal distribution (Plot 2). At this point, the pilot slowed the aircraft's descent rate and the airspeed began to hold steady. Although the airspeed remained outside the normal distribution, the transfer of the aircraft's energy from potential (i.e., altitude) to kinetic (i.e., airspeed) reduced the probability of a high-speed exceedance adverse event. When aircraft energy is converted from altitude to speed, more tools available to the pilot to reduce kinetic energy, for instance through use of speed brakes, deploying flaps, etc.

FOQA data can provide many quantitative details about operator and vehicle performance, but cannot provide information about the knowledge state, motivation, or broader context for the event. Why was the pilot flying the arrival at a higher than normal airspeed? What contextual cues triggered the pilot to take action? If there were multiple appropriate actions that could have been taken, why did the pilot select that specific action? The answers to these questions could be obtained through observer- and operator-based data to supplement system-based data and provide a more complete understanding of work-as-done.

Discussion

This study highlighted the value and feasibility of learning from what goes right in addition to what goes wrong. To move forward in this area of research, the authors propose the following recommendations for the aviation safety community:

- Redefine safety in terms of the presence of desired behaviors and the absence of undesired behaviors.
- Leverage existing data to identify strategies and behaviors that support resilient performance.
- Develop tools to capture new operator-, observer-, and system-generated data on strategies and behaviors that support resilient performance.
- Develop a system-level framework for integrating insights from various data types to facilitate understanding of work-as-done.

- Develop organization-level strategies that promote recognition and reporting of behaviors that support resilient performance.

Through understanding "how" and "why" people perform successfully in a variety of circumstances, in addition to understanding "what," "where," and "when," systems can be designed to ensure the ultra-safe airspace system is not unintentionally made less safe due to loss of resilient properties provided by human operators yet are not well-understood.

Acknowledgements

This work was jointly funded by the NASA Engineering and Safety Center (NESC), NASA's System-Wide Safety Project, and NASA's Transformative Tools and Technologies Project. The authors gratefully acknowledge Mr. Viraj Adduru, Mr. Oliver Ammann, Mr. Ilya Avhrekh, and Ms. Colleen Cardoza, Mr. Gary Lohr, and Dr. Cynthia Null for their support in conducting and formulating this study. Additional detail is reported in NESC Technical Assessment Report, NESC RP-18-01304, which can be obtained by contacting the first author.

References

Amalberti, R. 2001. The paradoxes of almost totally safe transportation systems. *Safety Science, 37,* 109-126.

Hollnagel, E. (2011). "RAG – The resilience analysis grid." In: E. Hollnagel, J. Pariès, D.D. Woods and J. Wreathall (Eds). *Resilience Engineering in Practice. A Guidebook.* Farnham, UK: Ashgate.

Hollnagel, E. (2014). Safety-I and Safety-II: The Past and Future of Safety Management. Farnham, UK: Ashgate.

Hollnagel, E. (2016). Resilience Engineering. Retrieved from erikhollnagel.com/ideas/resilience-engineering.html.

Hollnagel, E., Woods, D. D. & Leveson, N. G. (2006). Resilience engineering: Concepts and precepts. Aldershot, UK: Ashgate.

Janakiraman. V.M. 2018. "Explaining Aviation Safety Incidents Using Deep Temporal Multiple Instance Learning". In *Proceedings of the 24th ACM SIGKDD International Conference on Knowledge Discovery & Data Mining*. ACM, New York, 406-415. doi: https://doi.org/10.1145/3219819.3219871.

Lay, E., Branlat, M., & Woods, Z. (2015). A Practitioner's Experience Operationalizing Resilience Engineering. In C. Nemeth & I. Herrera (Eds.). Special issue on Resilience Engineering. *Reliability Engineering and System Safety*. 141. doi: http://dx.doi.org/ 10.1016/ j.ress.2015.04.006.

NASA. (2017). NASA Aeronautics Strategic Implementation Plan. Retrieved from https://www.nasa.gov/aeroresearch/strategy.

Nisula, J. (2018). A risk management framework for a complex adaptive transport system. Doctoral thesis: Universite de Toulouse.

Schacter D.L. (1989). Memory. In Foundations of Cognitive Science, ed. MI Posner, pp. 683–725. Cambridge, MA: MIT Press.

PSYCHOLOGISTS ROLE IN BRAZILIAN AVIATION - PRESENT AND PERSPECTIVES

CABRAL, Lisia Maria Espinola da Silva Pacheco
RIBEIRO, Selma Leal de Oliveira
BARRETO, Marcia Regina Molinari
FAJER, Marcia
PEREIRA, Maria da Conceição Correia
SANTOS, Cesar da Silva
Brazilian Aviation Psychology Association - ABRAPAV
Rio de Janeiro, RJ, Brazil

This article describes a Group Discussion ocurred on the I National Congress of the Brazilian Aviation Psychology Association (ABRAPAV), in 2016. Among 158 participants on the event, 146 took part of this Group Discussion: 75 psychologists; 6 Psychology students; 25 other aviation profissionals; 40 professionals with unidentified formation. They chose one of the following subgroups to discuss about Aviation Psychology activities, facilities, difficulties and suggestions: Regular Aviation; Non-regular/General Aviation; Military Aviation; Regulator Authority/Aeronautical Industry; Clinics/Hospitals; Airclubs/ Aviation Schools/ Universities/Training Centers; Air Navigation/Airports. After the discussion, each subgroup representative presented the results of the main activities, facilities, difficulties and suggestions, respectively, as examples: Aeronautical Accidents Prevention; Managers Recognition, Support and Confidence; Reactive Organizational Cultures, Changes Resistence and Inflexible Manager; Professional Specialization and Specific Standard for Aviation Psychology. This enabled ABRAPAV to map relevant demands in this area and plan strategies for psychologists to minimize constraints and support improvements in their organizations.

Pilots selection, training and researches in aviation environment began in World Wars I/II, when Aviation Psychology started to have a great development with aviation technologies advances (KOONCE, 1984). Since then, several initiatives have emerged all over the world, aiming at strengthening the role of Aviation Psychology, such as: in 1956, the European Association for Aviation Psychology (EAAP, 2019); in 1981, the Australian Aviation Psychology Association (AAVPA, 2019); in 2011, the Journal of Aviation Psychology and Applied Human Factors (HOGREFE, 2019); since 1981, the International Symposium on Aviation Psychology (ISAP) and, in 1991, the International Journal of Aviation Psychology (ISAP, 2019); in 1998, the Spanish Aviation Psychology Association (AEPA, 2019); and in 2013, the Brazilian Aviation Psychology Association (ABRAPAV, 2019), which the Board members are the authors of the present article.

Brazilian Aviation Psychologists Innitiatives and ABRAPAV

In Brazil, the Aeronautics Ministry was founded in 1941 and assumed selection, training and research activities in aviation, but others were absorbed by the Selection and Orientation Service (SESO), formed in 1967: work analysis, psychological assistence, in 1970, when renamed as Nucleus of the Selection and Orientation Institute (NUISO); performance evaluation, organizational diagnosis, aeronautical accidents investigation, in 1980. NUISO was also renamed: in 1981, as Selection and Orientation Institute (ISO); and in 1988, as Institute of Psychology of Aeronautics (IPA) (COELHO et al, 2007).

In 1971, the Aeronautical Accidents Investigation and Prevention Center (CENIPA) was created to coordinate the Aeronautical Accidents Investigation and Prevention System (SIPAER). In 1986, IPA and CENIPA developed the Aeronautical Accidents Prevention Course/Human Factors (CPAA/FH) for civilian and military psychologists of aeronautical institutions to certify them as Certified Element/ Human Factors in Psychology (EC/FHP) for the aeronautical accidents prevention and investigation activity, and compose Aeronautical Accidents Investigation Comissions (BRASIL, 2017). Actually, besides this course, IPA developed the Extension Course on Aviation Psychology (CPAV) for EC/FHP (COELHO et al, 2007). In 2005, the Naval Aviation Center of Instruction and Training (CIAAN), subordinated to the Navy Command, developed the Special Course on Aviation Psychology for Official (C-Esp-PAVO), for military EC/FHP (BRASIL, 2007).

After ICAO regulation on Safety Management Manual (SMM), each member-state had to implement a State Safety Programme (SSP), and each aviation provider had to implement a Safety Management System (SMS) (ICAO, 2013). That's when increased the number of aviation psychologists certified as EC-FHP, because of both opportunity and necessity created for them to develop the aeronautical accidents prevention and investigation activity, in addition to selection, training and others.

Besides, some significant events and publications about Aviation Psychology happened, as well as several meetings, journeys, courses, promoted by different entities in Brazil. Two books were also edited, as follows: "Psychology Flights in Brazil: Studies and Practices in Aviation"; and "Scientific Articles Collection" (RIBEIRO, 2009).

In Brazil, the Federal Council of Psychology (CFP) does not recognize Aviation Psychology as an official specialization, so ABRAPAV main goals are to promote conditions for: this professional specialization recognition; researches and Brazilian aviation psychologists database creation; knowledges/experiences exchange by periodic activities; interdisciplinarity among aviation psychologists; study groups formation in this area; and Brazilian Aviation Psychology history preservation (ABRAPAV, 2019).

I National Congress of ABRAPAV of 2016 and its Group Discussion

Group Discussion Purpose and Methodology

The I National Congress of ABRAPAV, in 2016, at SP, promoted a Group Discussion (ABRAPAV, 2019), which main purpose was to raise issues about Brazilian aviation psychologists activities, facilities, dificulties and suggestions, in order to identify main demands in this area and project necessary actions. The questions of the form given to participants and subgroups for discussion and answering were: "Are you psychologist?"; "What are the main activities you perform at work?"; "Which are the main facilities to perform these activities?";

“Which are the main difficulties to perform them?”; “Do you have any suggestion to minimize or solve the difficulties cited?”.

The Group Discussion methodology was divided in five phases: Registration; Orientation; Execution; Presentation; Conclusion. On the Registration Phase, participants chose one of the subgroups: Regular Aviation; Non-regular/General Aviation; Military Aviation; Regulator Authority/Aeronautical Industry; Clinics/Hospitals; Airclubs/Aviation Schools/ Universities/Training Centers; Aerospace/Airports. The Orientation Phase consisted of instructions by the global coordinator to all participants in an auditory. The Execution Phase consisted of explanations by the specific coordinator of each subgroup to participants in different rooms: the subgroup will ellect a representative to present the discussion results; each participant will receive a form with questions to answer; each subgroup will discuss the individual answers and make a presentation with main results; each subgroup representative will present the answers in the first auditory, based on the discussion results. The Presentation Phase consisted of all discussion results presentations by each subgroup representative. The Conclusion Phase consisted of main points of all results comments by the global coordinator in the same auditory.

Group Discussion Results

The results were divided in: quantitative, based on the number of the global and the specific participation; and qualitative, based on the answers to the questions of the form distributed to each subgroup and each participant. Although the quantitative and qualitative results were classified in high, intermediate and low, this article will limit to comment on the high results, considered as main, to the detriment of the intermediate and low ones.

Quantitative results. Table 1 shows the quantitative results of the Group Discussion.

Table 1.
Global and Specific Participation per Subgroup.

Subgroups	**Global Participation**		**Specific Participation**		
	Signatures	**Forms answered**	**Psychologists**	**Students**	**Other professionals**
RegularAviation	30 (20,55%)	28 (26,42%)	18 (24,00%)	1 (16,67%)	9 (36,00%)
Non-regular/General Aviation	17 (11,64%)	15 (14,15%)	8 (10,67%)	3 (50,00%)	4 (16,00%)
Military Aviation	22 (15,07%)	19 (17,92%)	17 (22,67%)	0	2 (8,00%)
Regulator Authority/ Aeronautical Industry	4 (2,74%)	5 (4,72%)	1 (1,33%)	0	4 (16,00%)
Clínics/Hospitals	34 (23,29)	12 (11,32%)	11 (14,67%)	0	1 (4,00%)
Airclubs/Aviation Schools/Universities/ Training Centers	24 (16,44%)	17 (16,04%)	10 (13,33%)	2 (33,33%)	5 (20,00%)
Air Navigation/Airports	15 (10,27%)	10 (9,43%)	10 (13,33%)	0	0
Total	146	106	75	6	25
Percent (%)	100	72,6 (of 146)	70,75 (of 106)	5,6 (of 106)	23,58 (of 106)

Table 1 shows the quantitative results related to the global and specific participation of each subgroup, which, among 146 participants who signed the frequency list (100%), 106 returned the form fulfilled (72,6%): 75 (70,75%) psychologists; six (5,66%) students; 25 (23,58%) other professionals. Fourty participants with unidentified formations will not be cited. Besides, there was a high participation of subgroups: 34 (23,29%) by Clinics/Hospitals; 30 (20,55%) by Regular Aviation; 24 (16,44%) by Airclubs/Aviation Schools/Universities/Training Centers. Considering the total of 75 psychologists, there was also a high participation of: 18 (24,00%) in Regular Aviation; and 17 (22,67%) in Military Aviation.

Qualitative results. The qualitative results refer to the responses of Aviation Psychology activities, facilities, difficulties and suggestions by the subgroups, which represents a valious portrait of Brazilian Aviation Psychology practices to guide ABRAPAV Board Members in future actions. First, we will comment on Aviation Psychology activities with more than 50% of answers covered by subgroups.

Activities. Taking, as a reference, the total of seven subgroups, six subgroups (85,71%) indicated as main activities: Psychological Evaluation & Aeronautical Accidents Prevention & Training, Teaching etc., not cited only by Regulator Authority/Aeronautical Industry; and Work Health/Safety Programmes, not referred only by Air Navigation/Airports. Five subgroups (71,42%) cited Coach for Abilities Development as the main activity, only not mentioned by Non-regular/General Aviation and Regulatory Agency/Aeronautical Industry. Other main activities were considered by four groups (57,14%). All activities were cited by Regular Aviation and Military Aviation. Regulator Authority/Aeronautical Industry cited Work Health/Prevention Programmes as the main activity.

Facilities. Five subgroups - Regular Aviation, Non-regular/General Aviation, Military Aviation, Airclubs/Aviation Schools/Universities/Training Centers and Air Navigation/Airports (71,42%) - indicated Managers Recognition, Support and Confidence as the main atribute to facilitate psychologists performance. Three subgroups (42,85%) indicated other facilities as important to aid psycologists activities development in aeronautical context: Regular Aviation, Non-regular/General Aviation and Regulator Authority/Aeronautical Industry cited HF Multidisciplinary/Interdisciplinary Interface as a main facility. Military Aviation, Airclubs/Aviation Schools/Universities/Training Centers and Air Navigation/Airports indicated Continuous Knowlegde, Experience and Learning as a main facility.

Difficulties. Four subgroups (57,14%) indicated the following main difficulties in the psychologists performance at aeronautical environments: Regular Aviation, Military Aviation, Airclubs/Aviation Schools/Universities/Training Centers and Air Navigation/Airports cited Reactive Organizational Cultures, Changes Resistence and Inflexible Managers & Lack of Personell and Material Investiments on Prevention as main difficulties; Regular Aviation, Regulator Authority/Aeronautical Industry, Clinics/Hospitals, Air Navigation/Airports referred to Standards Deficiency as a main difficulty; Regular Aviation, Military Aviation, Clinics/Hospitals and Airclubs/Aviation Schools/Universities/Training Centers indicated Lack of Kowledge and Doubts about Psychologists Role & Lack of Specific Specialization and Professional Up-grade & Few Theorectic References, Research and Data-base in Aviation

Psychology as main difficulties; Regular Aviation, Military Aviation, Regulator Authority/Aeronautical Industry and Clinics/Hospitals indicated Lack of Instruments and Minimum Standardized Scores for Psychological Evaluation as a main difficulty.

Suggestions. Six subgroups - Regular Aviation, Non-regular/General Aviation, Military Aviation, Clinics/Hospitals, Airclubs/Aviation Schools/Universities/Training Centers and Air Navigation/Airports (71,42%) - cited Professional Specialization and Specific Standard for Aviation Psychology as a main suggestion to minimize or solve difficulties in psychologists activities. Four subgroups (57,14%) indicated other main suggestions: Regular Aviation, Clinics/Hospitals, Airclubs/Aviation Schools/Universities/Training Centers and Air Navigation/Airports indicated Qualified Aviation Psychologists Employment in the Organizational Staffs as a main suggestion; Military Aviation, Regulator Authority/Aeronautical Industry, Clinics/Hospitals and Air Navigation/Airports indicated ABRAPAV Exchange with Aeronautical Institutions for Psychological Evaluation Standards Definition & Improvements and Formation/Post-graduation Courses in Aviation as main suggestions; Military Aviation, Clinics/ Hospitals, Airclubs/Aviation Schools/Universities/Training Centers and Air Navigation/Airports indicated ABRAPAV Partnerships for Trainings/Improvement Courses for Psychologists and other Aviation Professionals as a main suggestion; Non-regular/General Aviation, Military Aviation, Regulator Authority/Aeronautical Industry and Airclubs/Aviation Schools/Universities/Training Centers indicated ABRAPAV Proximity to Academics Activities/HF Research Investiments/Aviation Psychology Dissemination as a main suggestion.

Conclusion

The Group Discussion of the I National Congress of ABRAPAV, in 2016, had the participation of different segments, as: students; psychologists; and other aviation professionals. The results enabled to map relevant Brazilian Aviation Psychology demands and plan strategies to assist aviation psychologists, for better performance and improvements projections, indicating that the main purpose of the Group Discussion was acbieved. Besides, all activities, facilities, difficulties and suggestions were considered for the adequate compreehension of this area.

Acknowledgements

ABRAPAV thanks to: all psychologists, who helped to found our Association; every participant of the I National Congress of ABRAPAV, who, in 2016, contributed to the Group Discussion results, realized on the referred event; all sponsers, who made possible and easier the event to occur; the speakers and presenters, who shared their knowledge and experiences on this occasion; the associate members, who trusted on ABRAPAV to increase the development of Aviation Psychology in our country; ISAP, which has approved this article; and you, who is interested on reading it.

References

ASOCIACIÓN ESPAÑOLA DE PSICOLOGÍA DE LA AVIACIÓN (AEPA). (2019). Website. Retrieved Feb, 18, 2019, from <www.aepa-spain.com/>.

ASSOCIAÇÃO BRASILEIRA DE PSICOLOGIA DA AVIAÇÃO (ABRAPAV). (2019). Website. Retrieved Feb, 25, 2019, from <www.abrapav.com.br>.

AUSTRALIAN AVIATION PSYCHOLOGY ASSOCIATION (AAvPA). (2019). Website. Retrieved Feb, 18, 2019, from <www.aavpa.org/home.htm>.

_______. CENTRO DE INVESTIGAÇÃO E PREVENÇÃO DE ACIDENTES AERONÁUTICOS (CENIPA). (2017). Training and Capacitation of the Human Resources of the Aeronautical Accident Investigation and Prevention System. [Formação e Capacitação dos Recursos Humanos do Sistema de Investigação e Prevenção de Acidentes Aeronáuticos (SIPAER)]. (NSCA) 3-10. Brasília, DF.: Comando da Aeronáutica (COMAER).

_______. MARINHA DO BRASIL. Comando da Força Aeronaval. (2007). Performance of aviation psychologist. [Atuação dos psicólogos de aviação]. NORMAERNAV 05-01. 10 Jul, 2007.

COELHO, E. C. et al. Psychology contribuitions to flight safety [Contribuições da Psicologia à segurança de voo]. In: BORGES, J. P. et al. (Orgs.). Collection of scientific articles [Coletânea de artigos científicos]. (2007) Rio de Janeiro: Instituto de Psicologia da Aeronáutica (IPA)/Sumauma Ed. e Gráfica.

EUROPEAN ASSOCIATION FOR AVIATION PSYCHOLOGY (EAAP). (2019). Website. Retrieved Feb, 18, 2019, from <https://www.eaap.net>.

HOGREFE. (2019). Website. Retrieved Feb, 20,, 2019, from <https://us.hogrefe.com/products/journals/apahf>.

INTERNATIONAL CIVIL AVIATION ORGANIZATIONAL (ICAO). (2013). Safety management manual (SMM). Doc 9859/AN 474: Montreal, Canada. Retrieved Feb, 23, 2019, from <http://www.icao.int/ safety/SafetyManagement/Documents/Doc.9859.3rd%20Edition.alltext.en.pdf>.

INTERNATIONAL SYMPOSIUM ON AVIATION PSYCHOLOGY (ISAP), (2019). Website. Retrieved Feb 18, 2019, from <http://aviation-psychology.org/>.

KOONCE, J. M. (1984). A brief history of Aviation Psychology. Human Factors, 26 (5), 499-508.

RIBEIRO. Selma Leal de Oliveira. (2009). Psychology in the aviation context: a brief retrospective. [Psicologia no contexto da aviação - breve retrospectiva]. In: Revista Conexão SIPAER, v. 1, n. 1, nov. 2009: Brasília, DF.: CENIPA. Retrieved Feb 23 2019, from <http:// conexaosipaer. cenipa.gov.br/index.php/sipaer/article/view/7/12>.

COMMUNICATING DATA-DRIVEN RISK INFORMATION TO PILOTS

Nicoletta Fala
Purdue University
West Lafayette, IN

Karen Marais
Purdue University
West Lafayette, IN

General Aviation safety is a pressing concern. In this research, we consider the factor that appears most often in accidents: the pilot. Newly-licensed pilots can fly without their instructor, potentially as the only or most experienced pilot in the aircraft. Commercial debrief products use technology in the flight deck to collect data and provide post-flight visualizations for performance reviews, but do not discuss flight safety. To manage risk, though, pilots need to perceive the risk associated with a situation before deciding whether they are willing to accept it. Safety-driven post-flight feedback may help address performance. However, it is not clear whether and how the way we present feedback affects how pilots perceive risk, or what the best way is. We designed and disseminated a survey to evaluate the communication factors that affect pilots' risk perception. In this paper, we evaluate whether different representation methods affect relative risk perception among pilots.

General Aviation (GA) consists of all civilian aircraft operations other than commercial air transport operations and it covers a range of activities, both commercial (business aviation) and non-commercial (recreational and flight training operations). In 2017, GA in the United States was responsible for 990 non-commercial fixed-wing accidents (AOPA, 2018a). Non-commercial GA, in particular, has been contributing disproportionately to the aviation accident rate, with an accident rate of 5.57 accidents per 100,000 flight hours, significantly higher than the rate of 2.33 accidents per 100,000 flight hours in commercial fixed-wing GA. Similarly, 20.3% of non-commercial fixed-wing accidents are fatal; almost double the 10.4% of commercial fixed-wing operations. Most GA accidents (~74%) are attributed to pilot-related causes—they occur because of the pilot's action or inaction.

Continuing to provide pilots with feedback on their flying even after they finish their training and are no longer flying with an instructor (and potentially flying as the sole pilot or the most experienced pilot in the aircraft) can improve GA safety. Rantz et al. (2009) evaluated how feedback and praise can be used to increase the extent to which pilots use checklists accurately, with some participants in their study showing abrupt improvements in performance after intervention. Commercial products that leverage the addition of technology in the cockpits of small aircraft to collect flight data and present pilots with a visualization of their flights, like CloudAhoy and CirrusReports, are becoming more prevalent. However, these products do not discuss risk or flight safety. O'Hare's Aeronautical Risk Judgment Questionnaire (ARJQ) suggests that pilots display low levels of risk and hazard awareness, and an optimistic self-appraisal of their abilities (O'Hare, 1990). If pilots do not identify risk that can be mitigated in their flying, we cannot expect that they will improve. To manage risk, pilots need to perceive the risk associated with a situation or hazard and decide whether they are willing to accept that level of risk in each situation (Hunter, 2002). Safety-driven post-flight feedback may help facilitate risk management in subsequent flights, by alerting pilots to potentially hazardous situations. However, no one has considered what the best way is to present risk-related feedback in the pilot population, and we do not know whether presentation format affects how they perceive risk.

Using flight data to proactively improve GA safety requires that we are able to (1) identify behaviors that may put the safe outcome of a flight at risk, (2) detect those behaviors in the available flight data, and (3) inform the pilot in a way that helps them improve in their future flights. We use a state-based representation of historical aviation accidents to define a list of undesirable events or behaviors that we need to communicate to the pilots, in the form of states and triggers. Each flight consists of states, which can be nominal or hazardous, and trigger events (Rao, 2016). A state is a period of time during which the system, consisting of the aircraft and the pilot, exhibits a particular behavior, and a trigger is an event that causes the system to transition between two states. We use flight data to retrospectively detect these states and triggers, upon completion of the flight, by mapping parameters or combinations of parameters that can be tracked in the flight data to the hazardous states and triggers defined. We then present any detected hazardous states to pilots in the form of post-flight debrief feedback, with the goal of using the information to improve safety on subsequent flights. To evaluate the effectiveness of feedback in different representation formats, we used an anonymous web-based survey where a sample of pilots self-debriefed flights with safety information presented in different ways, and assessed the risk of the flight in each case. We also asked the pilots how likely they are to make changes to their flying as a result of the information they reviewed, to evaluate feedback effectiveness in terms of motivation to change unsafe behaviors. We demonstrated this approach using the hazardous states that are specific to the takeoff phase of flight.

Cognitive Biases in Risk Perception Among Pilots

We hypothesize that pilots will perceive the risk of their flight depending on how information is presented to them. While research in the medical, education, and economics fields have established guidelines that designers can use when communicating risk to the general population, the pilot sub-population is understudied, so we do not know which cognitive biases affect their understanding. We consider three factors that may impact their risk perception: framing language, representation method, and parameter type. Framing language corresponds to whether we discuss risk in terms of safety-centric language or risk-centric language. For example, we might refer to a safe flight as being either 'very safe' or 'not risky'. Representation methods refer to how we present data: graphically or numerically/textually. For example, we can communicate how much runway distance was remaining at takeoff numerically (2,500 ft), or graphically on the airport diagram. Lastly, parameter type refers to how we frame the same metric. For example, there are two ways to measure deviation from the runway centerline: the distance between the aircraft's longitudinal axis and the runway centerline, or the distance from the aircraft's longitudinal axis to the edge of the runway. The former represents how close the pilot was to the ideal condition (with the aircraft's longitudinal axis aligned with the centerline); the latter measures how close the pilot was to being involved in an incident (runway excursion or collission with an object). While both versions represent the same thing mathematically, parameter type can affect how pilots perceive their own flight performance in terms of risk.

To investigate how these three factors impact safety-driven feedback effectiveness, we created a 2^3 full-factorial design experiment. Table 1 shows the eight resulting treatment combinations. Using de-identified flight data from a Garmin G1000 display and the CloudAhoy flight visualization software, we designed interactive prototype debrief screens that include safety-driven feedback. Figure 1 shows how we altered the commercial screens to add risk information in two popup windows; (a, top right) provides the pilot with a list of behaviors that could potentially appear in a flight, and (b, center) displays parameters that characterize the selected behavior. The survey is available at www.nicolettafala.com/survey and an example of an interactive prototype is available at www.nicolettafala.com/debriefexample. We repeated this process for three different flights, and survey respondents could respond to as many of the three flights as they wished.

Table 1.
The 2^3 full-factorial design evaluates main and interaction effects among the three factors.

Treatment Group	Framing Language	Representation Method	Parameter Type	Responses [Flight A]
1	safety-centric	graphical	performance	35
2	risk-centric	graphical	performance	31
3	safety-centric	numerical	performance	44
4	risk-centric	numerical	performance	33
5	safety-centric	graphical	safety	23
6	risk-centric	graphical	safety	34
7	safety-centric	numerical	safety	33
8	risk-centric	numerical	safety	35

To evaluate feedback effectiveness, we asked respondents six questions: (Q1) Given the information presented to you, how safe would you say this takeoff was? [5-point Likert scale] (Q2) In this takeoff, which of the following would concern you, if any? [Centerline deviation, Rotation airspeed, Engine RPM, Takeoff distance, Wind] (Q3) Optional Comments. (Q4) What changes (up to 5) do you think you could make to an upcoming flight after the information presented here, if any? [Freeform text, up to 5 changes] (Q5) How likely are you to make each of these changes to an upcoming flight? [5-point Likert scale for each change] (Q6) How important do you think each of these changes is to improving safety on takeoff? [5-point Likert scale for each change] To evaluate the impact of a risk-centric framing language, we reworded these questions, replacing safe with risky (How risky would you say this takeoff was).

We measure feedback effectiveness in two ways: (1) did the pilot understand how safe or unsafe their flight was based on their responses to Q1 and Q2, and (2) how motivated are they to change their behaviors to mitigate the risk in their flying activities based on responses to Q4, Q5, and Q6? Q1 captures how different treatment groups introduce cognitive biases in pilots and is subjective—we expect to see differences in the distributions of responses among the different treatment combinations. Q2 can identify whether pilots are perceiving risk in the correct categories more objectively; that is, if we deem a flight to be unsafe due to high crosswinds, did the pilots identify high crosswinds as an issue?

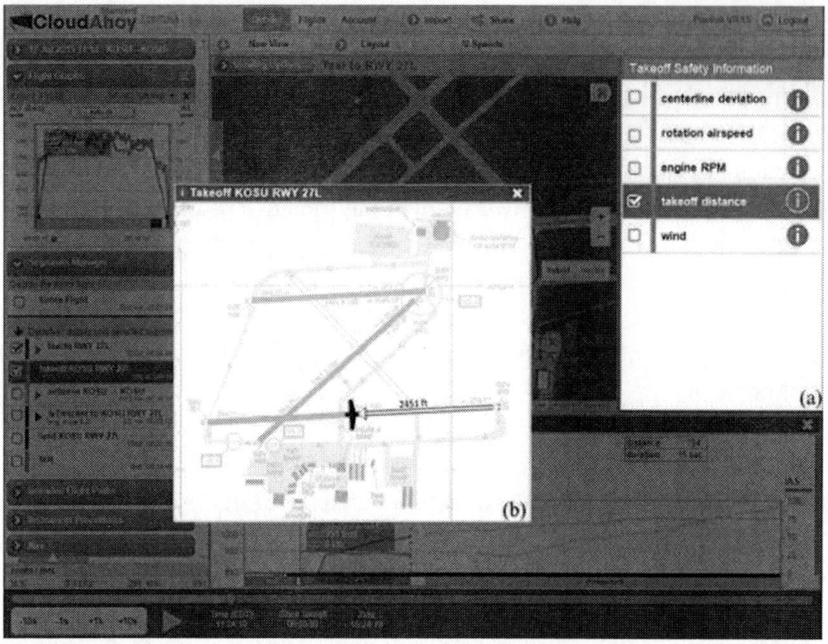

Figure 1. We supplemented the visualization of the flight data from CloudAhoy with information on the safety of the flight on five parameters.

Survey Results and Analysis

We used aviation mailing lists and groups to recruit participants and encouraged snowball sampling to maximize our responses. Approximately 70% of our respondents provided us with demographic information. Our sample consisted of 71% male and 26% female pilots, 76% of whom have completed at least a 4-year degree. Private pilots and commercial pilots made up the majority of the sample, at 49% and 30% respectively, with 58% of the pilots being instrument-rated. Most respondents (64%) fly primarily aircraft with steam gauges, fly at least weekly (59%) and have never used commercial debrief or flight visualization products like CloudAhoy (88%). We ended up with 268 responses for the first flight scenario, 195 for the second flight, and 189 for the third flight. Since the first flight in the survey got the maximum number of responses, we focus our initial data analysis on that flight before comparing results with the other two flights.

Figure 2 depicts the raw responses to Q1 that correspond to each group—the darker the marker, the higher the frequency of that particular response. The average response (marked in orange) in each treatment group tends to oscillate around the neutral response of 3 out of 5 on the risk Likert scale, but the spreads are also different, with the mean of the eighth group appearing lower than the means of the other treatment groups. We processed the data for the first flight to identify the response to Q1 and the treatment group it belongs to for each respondent and used ANOVA to test for the difference in treatment groups. The probability of the response means being equal for different treatment groups is < 2%. We also ran the Tukey procedure to identify which treatment groups had significantly different responses.

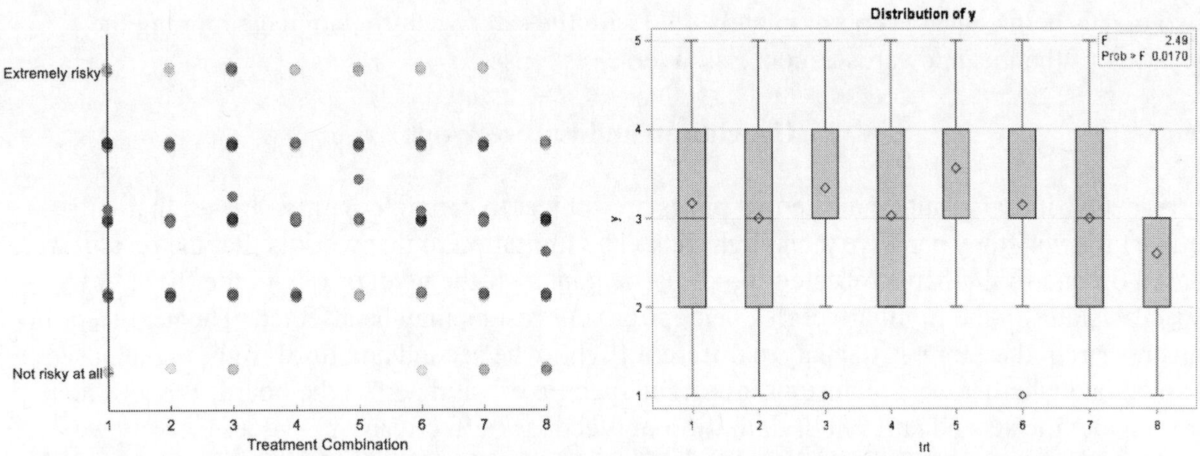

Figure 2. *The responses for the first flight show means and spreads that visually vary slightly among the eight treatment combinations. Treatment group 8 biased the respondents towards a less risk-averse response, and treatment groups 3 and 5 show a more risk-averse bias.*

Our ANOVA test indicates that there is a difference between the eight groups, with a p-value of 0.0170. Table 2 summarizes the ANOVA results. The box plot in Figure 2 suggests that the means for treatment groups 3 and 5 are higher than the mean for treatment group 8. Tukey's HSD test, used in conjunction with the ANOVA results, compared all possible pairs of means and identified which ones are significantly different from each other. The test identified that some treatment combinations—3 and 8 and 5 and 8—are different at the 0.05 significance level. Table 3 shows the difference between the means of these groups. Treatment combinations 3 and 5 have the safety-centric framing language in common, whereas treatment combination 8 is framed in terms of risk. This discrepancy suggests that asking a pilot how safe their flight was vs. how risky their flight was can potentially make the pilot more risk-averse.

Table 2.
We reject H_0 based on the ANOVA results. The means differ between the different groups.

Source	DF	Sum of Squares	Mean Square	F Value	Pr > F
Model	7	17.3332430	2.4761776	2.49	0.0170
Error	260	258.1443690	0.9928630		
Corrected Total	267	275.4776119			

Table 3.
Tukey's studentized range (HSD) test for the response variable at $\alpha = 0.05$. *The table only shows those treatment groups that are considered statistically different.*

Treatment Group Comparison	Difference Between Means	Simultaneous 95% Confidence Limits	
5 – 8	0.9652	0.1479	1.7825
3 – 8	0.7409	0.0513	1.4305

We ran a second ANOVA test to evaluate whether pilots responded differently to messages framed in a risk-centric language and safety-centric language. The probability of the two means being equal is approximately 1%, with the safety-centric language moving the location of the mean towards a more risk-averse response.

Conclusion and Future Work

Our initial results indicate that pilots are subject to certain cognitive biases that will impact the way they perceive their flight risk. The limitation to this work is that the results are based on one flight that is repeated for all participants—if the level of risk in the flight or the specific states in the flight affect risk perception, our results may be affected. The next step for this research, therefore, is to analyze different flights (the second and third flight in our survey) to evaluate whether the conclusions are flight-specific or valid across the board. We will then investigate the second part of our definition of "feedback effectiveness"—does the way we present safety information impact how motivated pilots are to change their behaviors? Our response variable metrics for motivation are (a) the number of changes suggested, (b) the average willingness to change among the suggested changes, and (c) the maximum willingness to change among the suggested changes.

Acknowledgments

We thank the Purdue Statistical Consulting Service for their help with the data analysis, and the people and organizations that were significantly helpful in our snowball sampling (the Ninety-Nines and the Cardinal Flyers in particular).

References

AOPA Air Safety Institute (2018a). GA Accident Scorecard. Retrieved from https://www.aopa.org/-/media/files/aopa/home/training-and-safety/nall-report/20162017accidentscorecard.pdf.

AOPA Air Safety Institute (2018b). 27th Joseph T. Nall Report—General Aviation Accidents in 2015. Retrieved from https://www.aopa.org/-/media/files/aopa/home/training-and-safety/nall-report/27thnallreport2018.pdf.

Hunter, D. R. (2002). Risk perception and risk tolerance in aircraft pilots. Retrieved from https://ntlrepository.blob.core.windows.net/lib/19000/19800/19856/PB2003100818.pdf.

O'Hare, D. (1990). Pilot's perception of risk and hazard in General Aviation. *Aviation Space and Environmental Medicine, 61(7),* 599-603.

Rantz, W. G., Dickinson, A. M., Sinclair, G. A., & Van Houten, R. (2009). The effect of feedback on the accuracy of checklist completion during instrument flight training. *Journal of Applied Behavior Analysis, 42,* 497-509.

Rao, A. H., & Marais, K. (2016). Comparing hazardous states and trigger events in fatal and non-fatal helicopter accidents. *16th AIAA Aviation Technology, Integration, and Operations Conference.* Washington, DC: AIAA AVIATION.

COLLEGIATE PROGRAM SAFETY CULTURE SURVEY – A COMPARISON OF DISCIPLINES

Wendy Beckman, Daniel Siao, Carlos Smith, Kevin Corns
Middle Tennessee State University
Murfreesboro, Tennessee

The use of safety culture surveys to determine constituent perception of an organization's efforts toward safety is widely accepted. However, the use of such surveys to compare disciplines within a collegiate aviation department has infrequently been employed. Students enrolled in flight, maintenance, and UAS programs participated in a safety culture survey, providing data on the safety culture within each program as well as the ability to compare programs. Subscales for safety values, safety fundamentals, and risk assessment were included. There were statistically significant differences in safety values between the maintenance group and both the flight and UAS groups, and between all three groups in safety fundamentals, but no statistically significant difference between programs in risk assessment.

In an effort to understand areas of strength and weakness, an inaugural safety culture survey was conducted of students in the flight, maintenance, and unmanned aircraft systems operations (UAS) programs of a collegiate aviation department. In addition to establishing a baseline perception of safety culture, differences between the responses of the students in the different programs were examined. While there have been previous studies comparing functional areas within companies, there has been little work examining differences between functional areas in an academic environment. Although all three programs are located within the same department, sub-culture development was felt to be likely, impacting student perceptions of safety culture.

Literature Review

Several studies (Clarke, 2006; McDonald, Corrigan, Daly, & Cromie, 2000; Varonen & Mattila, 2000) have confirmed the relationship between the employee perception of safety culture and the level of safety actually experienced within an organization. Thus, safety culture surveys serve as tools to direct efforts toward continuous improvement. There is now a significant body of research on safety culture in aviation; a 2011 study reviewed 23 safety culture surveys which had been utilized in military and commercial aviation organizations (O'Connor, O'Dea, Kennedy, & Buttrey, 2011). Although early studies tended to focus on a particular area of operations (i.e., flight, maintenance, cabin crew), several have concentrated on the variations found between operations within an organization. Patankar (2003) found significant differences between flight operations and maintenance personnel at a US company. Gao et al. (2015) found that while the overall safety climate across four groups (flight crew, cabin crew, mechanics, and ground operations staff) was positive, there were also significant differences between occupational groups.

As collegiate aviation programs embrace safety management system (SMS) concepts, they are looking to safety culture surveys to understand how well their efforts are being received.

In the largest study of safety culture at collegiate flight schools to date (Robertson, 2017), the Collegiate Aviation Program Safety Culture Survey (CAPSCUS) was completed by students and employees at 13 collegiate flight schools across the US. This study found a correlation between positive safety culture survey results and the degree of SMS implementation, level of safety promotion, and management commitment to safety at the collegiate level. Based on these findings, Robertson argued for the value of using safety culture surveys as both a baseline measure and as an ongoing indication of the effectiveness of a collegiate institution's SMS implementation efforts. There has been considerably less work done in areas outside flight training at the collegiate level. However, a 2016 study (Adjekum et al.) looked at aviation management, air traffic control, and UAS students' perception of safety culture. The study found that non-flight majors have "different areas of emphasis" (Adjekum, et al, 2016, p. 17) within the safety culture than do flight majors within the same department. It is this difference between programs that is of particular interest in this study.

Methodology

There have been numerous versions of safety culture surveys utilized over the past 15 years. These in large part have been developed from previously existing surveys, to meet the needs of the organization conducting the survey. This study is no exception to that precedent, with most survey questions based on the previously used CAPSCUS collegiate safety culture study (Adjekum, 2013; Robertson, 2017). The survey consisted of a series of Likert-type scale questions, divided into subscales of departmental safety values, safety fundamentals, and risk assessment of both participants themselves and their view of others. Note, three separate surveys were distributed, reflecting the appropriate program terminology (i.e., flight, maintenance, or UAS) although the questions displayed in Table 1 use generic phraseology. As a human subject research project, permission to conduct this survey was granted by the university's Institutional Review Board. In the spring 2018 semester, all students in the programs of interest received an e-mail from the department chair, requesting that they participate in the safety culture survey and providing a link to the electronic survey. Three days after the initial e-mail a follow up e-mail was sent, and four days later, the survey was closed. The survey was conducted anonymously, with no identifying information recorded.

Results

The 507 students enrolled in the flight, maintenance, or UAS operations program in the spring 2018 semester were the population for this study. The flight program had 370 students, maintenance had 81 students, and UAS had 56 students. A total of 188 responses were collected, representing 37% of the total population of students in the three programs of interest. There were 127 (67.5%) respondents enrolled in the flight program, 37 (19.7%) in the maintenance program, and 24 (12.8%) in the UAS program. The disparity in sample sizes is due to the imbalance in the number of students enrolled in each program. For instance, the flight program is the largest in the department, exceeding other concentrations by around 300 students. However, when the number of respondents in each program is compared to the population of each program, the response rates were 34% for flight students, 46% for maintenance students, and 43% for UAS students. Furthermore, the respondents comprised of 47 (25%) freshmen, 57 (30.3%) sophomores, 44

(23.4%) juniors, 39 (20.7%) seniors, and 1 (< 1%) graduate student. The means and standard deviations for each question, by program, can be seen in Table 1.

Table 1
Survey Questions and Descriptive Statistics

Scale: strongly disagree (1), disagree (2), neutral (3), agree (4) and strongly agree (5) **SAFETY VALUES**	**Flight** Mean (SD)	**Maint.** Mean (SD)	**UAS** Mean (SD)
1. Safety is a core value of the Aerospace Department.	4.57 (.89)	4.00 (.97)	4.71 (.69)
2. The Aerospace Department is more concerned about making money than being safe. (Reverse[1])	3.82 (.98)	4.06 (.86)	4.08 (1.0)
3. The Aerospace Department does not show much concern for safety until there is an accident or incident. (Reverse[1])	4.25 (.70)	3.97 (.88)	4.29 (.86)
4. The Aerospace Department does not cut corners where safety is concerned.	4.14 (1.0)	3.59 (.98)	4.25 (.99)
5. The Aerospace Department goes above and beyond regulatory minimums when it comes to issues of safety.	4.19 (.82)	3.54 (1.0)	4.52 (.73)
6. The Aerospace Department tries to get around safety requirements whenever the chance presents itself. (Reverse[1])	4.17 (.90)	4.08 (.76)	4.17 (.94)
7. Aerospace Department senior personnel view regulation violations very seriously, even when they do not result in any serious damage or injury	4.27 (.78)	3.68 (1.1)	4.33 (.96)
SAFETY FUNDAMENTALS			
8. Checklists and procedures are easy to understand.	4.36 (.71)	3.54 (1.2)	4.21 (.98)
9. The safety practices and procedures manual is carefully kept up to date.	4.21 (.66)	3.35 (1.3)	4.18 (.73)
10. The Aerospace Department is willing to invest money, resources, and effort to improve safety.	3.89 (.87)	3.75 (1.1)	4.42 (.83)
11. The Aerospace Department is committed to equipping aircraft with up-to-date technology.	3.91 (.94)	NA	NA
12. Instructors have a clear understanding of risks associated with specific operations.	4.29 (.70)	4.27 (.69)	4.87 (.34)
13. Safety is consistently emphasized during training.	4.41 (.65)	3.81 (.91)	4.78 (.42)
14. Instructors teach shortcuts and ways to get around safety requirements. (Reverse[1])	4.00 (.81)	3.97 (1.0)	4.45 (.91)
15. The Aerospace Department ensures that maintenance on aircraft is adequately performed.	4.00 (.88)	NA	NA
16. The Aerospace Department ensures that aircraft are safe to operate.	4.17 (.73)	NA	NA
17. The Aerospace Department safety reporting system is convenient and easy to use.	4.20 (.75)	3.97 (1.1)	4.26 (.92)
18. Students are actively involved in identifying and resolving safety concerns.	4.00 (.82)	3.38 (1.2)	4.43 (.95)
19. Safety is consistently emphasized in all stages of practical training.	4.41 (.65)	3.81 (.91)	4.78 (.42)
Risk Assessment of SELF (students asked to rate frequency they experienced the following):[2]			
Scale: never (1), once in the last six months (2), two to four times in the last six months (3), and five or more times in the last six months (4).			
20. Reported for a flight lesson when fatigued, ill, or under unusual stress because you felt you had no other choice?	1.48 (.82)	NA	NA
21. Were pressured to conduct a flight (or lab) in what you believed to be unsafe weather or environmental conditions?	1.17 (.42)	1.05 (.33)	1.04 (.21)
22. Were pressured to fly an aircraft you believed was in an unsafe mechanical condition (flight); unsafe equipment (Maint., UAS)?	1.24 (.63)	1.17 (.56)	1.00 (.00)
23. Failed to challenge more senior personnel (instructor or management personnel) on a safety issue for fear of being penalized in some manner?	1.07 (.26)	1.11 (.39)	1.04 (.21)
24. Made a hard landing that you did not report?	1.24 (.56)	NA	NA
25. Were aware of a safety issue but did not file a safety report?	1.26 (.48)	1.38 (.79)	1.09 (.29)
26. Were aware of another student acting in an unsafe manner but you did not file a safety report?	1.43 (.56)	1.28 (.83)	1.22 (.52)
27. Allowed an instructor or senior pilot's error to go unchallenged?	1.14 (.35)	NA	NA

Note: 1: "Reverse" indicates the question was worded in the negative to check and correct for student inattention to questions. Results indicated have been reversed to the positive so sub-scale statistics could be calculated. 2: The risk assessment questions were also asked as an assessment of "others" but are not repeated here due to space constraints.

Statistical Analysis for Research Question 1

Are students' perceptions of the safety value, safety fundamentals, and risk assessment scales consistent with their view of "safety is a core value"?

It was our objective to determine if the perception of safety was consistent across all scales. To accomplish this analysis, an ANOVA was conducted on the mean scores for each scale. The mean score was used because the number of items in each scale varied, negating the ability to use summated scores. The scores for "safety is a core value of the Aerospace Department" were not combined with any other scales and was analyzed independently. The data met the assumption of normality, but the assumption of homogeneity of variance was not met. The result of the Levene's test was $F(4) = 15.09, p < 0.05$. Welch's ANOVA was applied to the data to adjust for differences in group variances, as this test is not sensitive to unequal variances (Jan & Shieh, 2014). The result of Welch's ANOVA was statistically significant, Welch's $F(4, 441.91) = 41.59, p < 0.05$. A Tukey post-hoc test showed statistically significant differences ($p < .05$) among all scales when compared to the perception of safety as a core value ($M = 4.48$, $SD = .91$). The mean for the safety value scale was 3.71 ($SD = .48$), the mean for the safety fundamental scale was 3.93 ($SD = .51$), and the mean for the risk assessment scale was 4.06 ($SD = .26$).

Statistical Analysis for Research Question 2

Are there statistically significant differences in the perceptions of safety among students in different programs?

Our second objective was to determine any statistically significant differences in the perception of safety as a core value among the three programs selected for this study. The data met the assumption of normality and the assumption of equal variances. A one-way ANOVA was performed on the data, and the result was statistically significant: $F(2, 185) = 7.01, p < .05$. A post hoc analysis using the Tukey HSD test revealed that the mean for the flight group ($M = 4.57$, $SD = .89$) was statistically different than the mean for the maintenance group ($M = 4.0$, $SD = .97$). However, the difference between the flight and UAS groups was not statistically significant ($M = 4.71$, $SD = .69$). Additionally, there were statistically significant differences between the maintenance group and both the flight and UAS groups. Furthermore, the differences among the concentrations on the three other scales were tested. The result of the ANOVA for the safety value scale was statistically significant: $F(2, 184) = 5.49, p < .05$. The results of the post hoc analysis using the Tukey HSD test was similar to the previous result of safety as a core value, indicating that the mean of the maintenance group ($M = 3.82$, $SD = .67$) was statistically different than the means of both the flight group ($M = 4.13$, $SD = .54$) and the UAS group ($M = 4.28$, $SD = .72$). Similarly, there was no statistical difference between the UAS group and the flight group. For the safety fundamentals scale, the result of the ANOVA was also significant: Welch's $F(2, 49.6) = 11.61, p < .05$. However, the Tukey HSD post hoc test revealed statistically significant differences between all groups with the mean of the maintenance group as the lowest ($M = 3.76$, $SD = .70$), the UAS group with the highest ($M = 4.48$, $SD = .45$), with the flight group in the middle ($M = 4.18$, $SD = .48$). Finally, the risk assessment scale differed from the previous results in that no statistically significant differences were found among the three groups, $F(2, 182) = 1.97, p = .143$. The means for the groups are as follows: maintenance ($M = 4.77$, $SD = .40$), flight ($M = 4.80$, $SD = .29$), and UAS ($M = 4.92$, $SD = .13$).

Discussion

Overall, students in all programs of the department had a favorable perception of safety as a core value; however, their perceptions of aspects of safety across the other scales were lower. If safety is a core value of an organization, it should be reflected in all aspects of operation, with no statistical significance between safety as a core value and the other scales. The results also suggest that there is a disconnect between perceiving safety as a core value and understanding safety programs. Safety as a core value should be supported in practice. Two possible explanation for this disparity are posited. First, students may not have a fully developed understanding of SMS. They believe safety is a core value, but fail to make a logical connection to what they see in practice. Second, emphasis of safety through verbal and written communications to students may lead to higher perceptions of safety as a core value.

Additionally, among the three groups, maintenance had the least favorable perception of safety as a core value with statistically significant differences between maintenance and the other two groups. This is also true for the safety value and safety fundamentals scales. It is expected that students within the same department, under the same oversight, and operating with similar safety policies and requirements should have similar perceptions of safety, but this is not the case. Sub-cultures may explain the differences in perceptions of safety, as one of the factors of sub-culture development is geographical separation (Boisnier & Chatman, 2002). Maintenance, flight, and UAS programs all conduct a portion of their training on satellite campuses.

There is one scale in which there was no statistical difference among the three programs: risk assessment. Questions in the risk assessment scale are introspective—measuring one's own actions—while the other scales measure of the actions of the department. Taken together, this reveals that, irrespective of the program, students behave and act in similar ways regarding safety matters. However, the difference lies in how each program manages safety.

Conclusion

As discussed above, this study indicated that the safety culture in the department is perceived as positive by students in each program, but somewhat diverse results were obtained between the groups. Although all students in the department have 13 credit hours of aviation coursework in common, that core is so small relative to the total number of aviation credit hours required that different safety sub-cultures appear to develop within each concentration. The industry backgrounds of faculty in each program likely impact the development of these sub-cultures. Additionally, the time since instigation of SMS concepts within programs appears to have a strong impact on student perception of safety culture. The flight program was the first in the department to embrace SMS concepts, with that effort now over a decade old. The maintenance program SMS efforts began in earnest three years ago, and while progress is being made the perceptions of safety culture by students in this program currently lags slightly behind the flight and UAS program students. The UAS program is very different from the two other programs. As a young program which was developed after the tenets of SMS were widely understood in the industry, it was designed with these concepts built into every facet of operation. In summary, the longevity of efforts to implement SMS, as well as the impact of the faculty members within each program, appears to be greater than the influence of the department administration as a whole on student perception of safety culture.

Recommendations for Future Studies

Even though the response rate was relatively strong, it would be beneficial to repeat the study with stronger encouragement for students to participate. This study was a cross-sectional approach but in the future, it is anticipated that a slightly modified instrument will be utilized annually to provide a longitudinal approach. Finally, repetition of the survey by other collegiate aviation institutions is encouraged, as the further comparison of differences between programs within the same university may lead to an understanding of how to assist various programs in improving their safety culture.

References

Adjekum, D. K. (2013). *Safety culture: An assessment of a collegiate aviation program* (Master's thesis).

Adjekum, D. K., Keller, J., Walala, M., Young, J. P., Christensen, C., Demik, R. J., & Northam, G. J. (2015). Cross-sectional assessment of safety culture perceptions and safety behavior in collegiate aviation programs in the United States. *International Journal of Aviation, Aeronautics, and Aerospace, 2*(4), 1-34.

Boisnier, A., & Chatman, J. A. (2003). The role of subcultures in agile organizations. In R. Petersen, & E. Mannix (Eds.), *Leading and managing people in dynamic organizations*. Mahwah, NJ: Psychology Press.

Clarke, S. (2006). The relationship between safety climate and safety performance: a meta-analytic review. *Journal of Occupational Health Psychology, 11*(4), 315– 327.

Gao, Y., Bruce, P.J., & Rajendran, N. (2015). Safety climate of a commercial airline: A cross-sectional comparison of four occupational groups. *Journal of Air Transport Management, 47*, 162-171.

Jan, S., & Shieh, G. (2014). Sample size determinations for Welch's test in one-way heteroscedastic ANOVA. *British Journal of Mathematical and Statistical Psychology, 67*, 72-93.

McDonald, N., Corrigan, S., Daly, C., & Cromie, S. (2000). Safety management systems and safety culture in aircraft maintenance organization. *Safety Science, 34*, 151– 176.

O'Connor, P., O'Dea, A., Kennedy, Q., & Buttrey, S. E. (2011). Measuring safety climate in aviation: A review and recommendations for the future. *Safety Science, 49*, 128-138.

Patankar, M. S. (2003). A study of safety culture at an aviation organization. *International Journal of Applied Aviation Studies, 3*(2), 243–259.

Robertson, M. F. (2017). *Examining the relationship between safety management system implementation and safety culture in collegiate flight schools* (Doctoral dissertation).

Varonen, U., & Mattila, M. (2000). The safety climate and its relationship to safety practices, safety of the work environment and occupational accidents in eight wood-processing companies. *Accident Analysis Prevention, 32*(6), 761-769.

APPLIED USE OF SAFETY PERFORMANCE MONITORING IN GLOBAL AVIATION OPERATIONS

Timothy M. Wilky, Edward M. Austrian, R. Jordan Hinson, Michael W. Sawyer
Fort Hill Group, LLC
Washington D.C.
Mark Millam
Flight Safety Foundation
Alexandria, VA

For decades, aviation has been at the leading edge of safety and human factors data collection. These data have provided valuable insights into emerging trends and human-system performance needs. As industry continues to improve its data collection capabilities, stakeholders must develop a common understanding and use of safety performance monitoring (SPM) practices and terms governed by ICAO (ICAO Annex 19, ICAO Doc 9859). SPM is a critical component of Safety Management Systems and State Safety Programs. To understand industry's awareness and use of SPM in current operations, an SPM Survey was administered. Responses were received from 161 domain representatives in six ICAO global regions. Response data revealed the current state of industry SPM practices, SPM variability across domains and regions, and generalizable best practices. This paper will present top safety performance targets (SPTs), safety data analysis methods, and safety data sources utilized by respondents to track, analyze, and measure risk across five areas: Maintenance (n=120), Near Mid-Air Collision (n=95), Runway Safety (n=124), Loss of Control-Inflight (n=92), Controlled Flight into Terrain (n=109). Survey data revealed that the top SPTs set by respondents are: Unstable approaches (83.5%), Runway Excursions (70.1%). The top analysis methods used are: Causal Factor Analysis (68.1%), FDM/FOQA Software (59.7). The top data sources are: Voluntary Reports (93.8%), Mandatory Reports (86.2%). This paper will describe how SPM survey results may be used to develop a supplemental Safety Performance Monitoring Handbook in 2019, which will be intended to drive an industry-wide shift towards proactive and predictive safety risk management.

Introduction

A Safety Management System (SMS) is defined as a systematic approach to managing safety, including the necessary organizational structures, accountabilities, policies, and procedures. The framework of an SMS consists of four primary components: 1) Safety policy and objectives, 2) Safety risk management, 3) Safety assurance, and 4) Safety promotion. Safety performance monitoring (SPM) is defined as a fundamental element of the "Safety assurance" component. SPM involves selecting, defining, and monitoring multiple safety performance indicators (SPIs) with associated safety performance targets (SPTs) (ICAO, 2013b). These attributes are defined as follows (ICAO, 2013a):

Safety Performance Indicator: A data-based parameter used for monitoring and assessing safety performance.
Safety Performance Target: The planned or intended objective for safety performance indicator(s) over a given period.

In 2018, Flight Safety Foundation reported a need to provide the global aviation industry with safety performance monitoring guidance materials. This need was a response to identified differences and inconsistencies in global SPM terminology, standards, and practices (Flight Safety Foundation, 2018a). As a first step towards the development of data-driven guidance materials, Flight Safety Foundation initiated research to assess the current state of global aviation SPM. Flight Safety Foundation and Fort Hill Group executed the multi-year research as part of the Global Safety Information Project (GSIP). The United States Federal Aviation Administration (FAA) provided funding for GSIP through Cooperative Agreement 17-G-003.

Methodology

To develop a baseline understanding of global SPM practices, a panel of aviation, safety, and human factors subject matter experts developed an SPM survey. As part of the development process, panel members conducted literature reviews to inform the development of draft survey questions. Data from previous GSIP surveys and the results from 24 GSIP focus group sessions were also applied as part of the data-driven survey question development process. Panel members vetted each draft survey question during multiple panel review sessions and applied a consensus methodology to down-select a final set of 57 questions. The questions were then integrated into an online survey tool.

Respondents took the survey online using either a computer, tablet, or mobile device. Respondents first completed an initial set of survey questions to determine demographic criteria, such as organization type. Based on these initial responses, a tailored subset of questions was generated from the survey database. In this way, survey respondents were only asked questions germane to their demographic (e.g. air traffic controllers were not asked about aircraft maintenance operations).

Survey Outline

The SPM survey was organized into three distinct sections: Demographics, General Safety Performance Monitoring, and Risk Areas. Each section contained a set of questions in various forms, including multi-select, free response, and Likert scales.

Section 1, "Demographics," enabled the grouping of survey responses based on organization type, respondent role, and International Civil Aviation Organization (ICAO) region. This section contained 5 questions.

Section 2, "General Safety Performance Monitoring," included high-level questions about safety performance monitoring, such as familiarity with ICAO definitions/terminology and organizational practices. This section contained 14 questions.

Section 3, "Risk Areas," solicited information concerning the risk areas of Runway Safety, Controlled Flight into Terrain (CFIT), Loss of Control – Inflight (LOC-I), Near Mid Air

Collision (NMAC), and Maintenance. For each risk area, respondents were asked to identify the data sources used to gather information, to detail the tools and techniques applied to analyze the data, and to state the metrics used to monitor safety performance and set SPTs. This section contained up to 40 questions, depending upon whether the survey respondent indicated their organization tracks risk in each of the five risk areas.

Survey Results and Discussion

A total of 161 survey responses were received between March 26, 2018 and September 12, 2018. As shown in Figure 1, Airlines accounted for 41.9% of responses, followed by Other Aircraft Operators (30.4%). The "Other Domains" category included responses from consultants, auditors, and aircraft management companies.

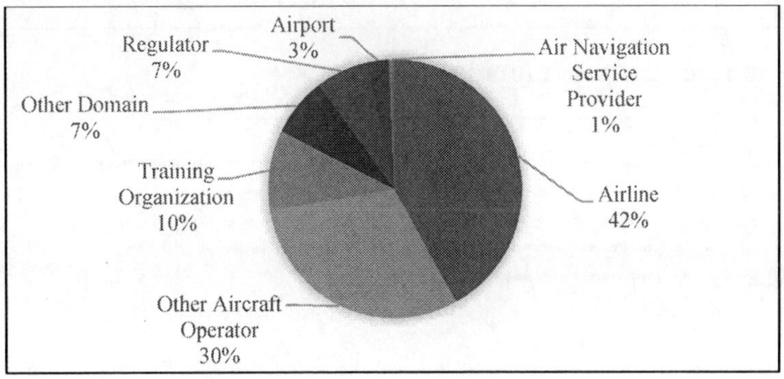

Figure 1. **Response breakdown by organization type**

Figure 2 displays the percentage of survey respondents (n=161) who reported that their organization measures risk in each risk area. Runway Safety was the most commonly tracked risk area, monitored by 77.0% of respondents. According to ICAO, the Runway Safety risk area accounted for the highest percentage of all accidents that occurred in 2017. Along with Runway Safety, ICAO identifies the Loss of Control – Inflight and Controlled Flight into Terrain risk areas as their top safety priorities in the 2017-2019 Global Aviation Safety Plan (ICAO, 2018).

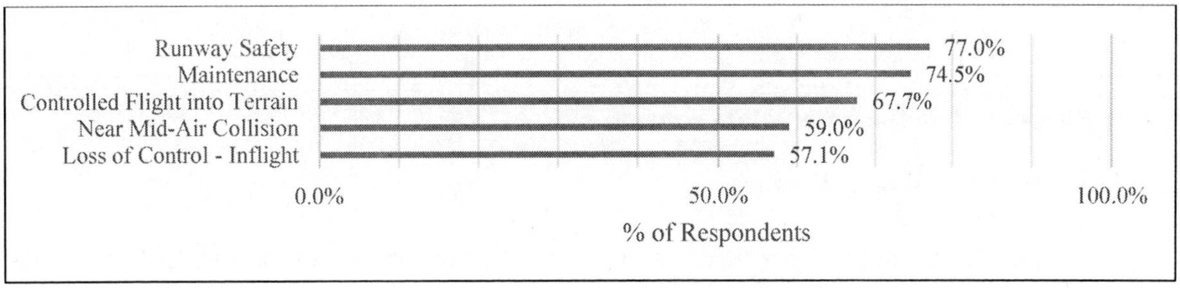

Figure 2. **Percentage of organizations measuring risk by area**

Figures 3, 4, and 5 respectively present the top five safety performance targets, safety data analysis methods, and safety data sources utilized by survey respondents across all risk areas (n=161). Safety performance targets for unstable approaches were set by 83.5% of respondent organizations. Unstable approaches are characterized by vertical, lateral, and/or speed deviations during approach and landing and can contribute to the risk of Loss of Control – Inflight and Runway Safety events (IATA, 2016). Causal Factor Analysis was the most commonly cited analysis method (68.1%). Flight Data Monitoring/Flight Operations Quality Assurance (FDM/FOQA) Software, Contributory Factor Analysis, and Safety Reporting Analysis Tools were each reported by slightly more than half of all respondents. Voluntary and mandatory reports were the top two most common data sources used to address risk, utilized by 93.8% and 86.2% of respondent organizations respectively.

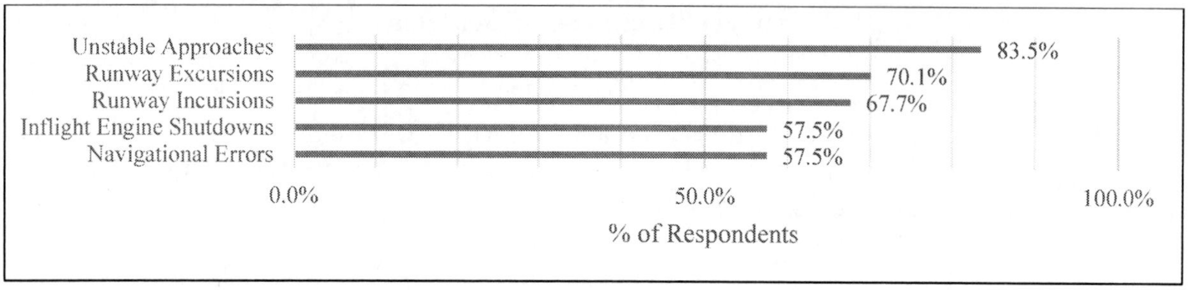

Figure 3. **Top five safety performance targets**

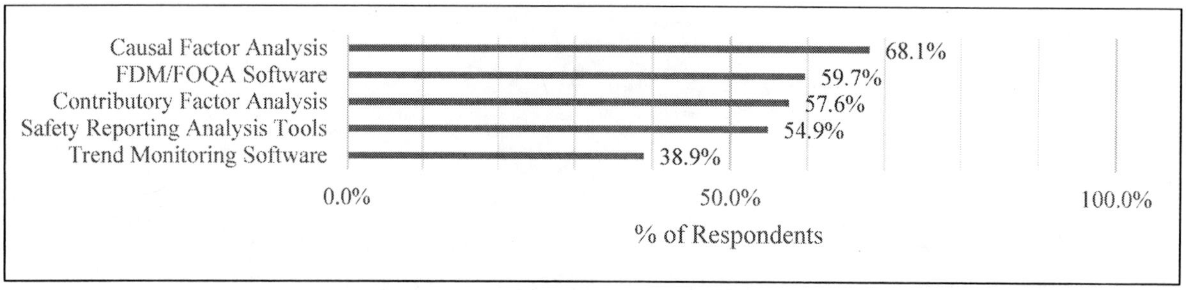

Figure 4. **Top five data analysis methods**

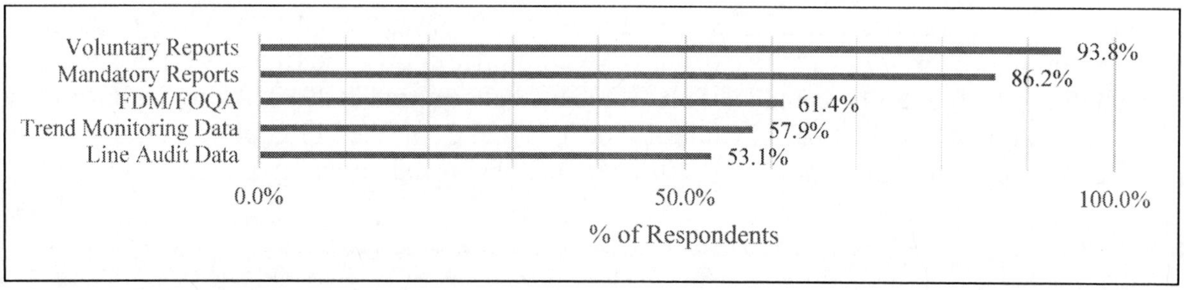

Figure 5. **Top five safety data sources**

Lagging versus Leading Indicators

A noteworthy finding from survey results was the prominence of lagging SPIs and lack of leading SPIs. SPIs are often classified as either leading or lagging. Lagging SPIs are reactive metrics that indicate historical performance and provide evidence of how effective an organization's interventions have been by observing whether the expected outcomes of changes are observable (Wreathall, April 2009). In contrast, leading indicators are "proactive, preventative, and predictive measures that monitor and provide current information about the effective performance, activities, and processes of a…system that drive the identification and elimination or control of risks" (National Safety Council, 2013).

The prevalence of lagging indicators is illustrated in Figure 3. The top five safety performance targets set by respondent's organizations are each reactive, lagging indicators of safety. Unstable approaches, runway excursions, runway incursions, inflight engine shutdowns, and navigational errors are all undesirable operational outcomes that have already occurred. Although some isolated examples of leading SPIs were found in the survey data, such as aircraft bank angle exceedances to indicate potential Loss of Control – Inflight risk, researchers identified an opportunity to encourage increased use of leading SPIs in future SPM guidance materials. A mix of both leading and lagging SPIs will equip aviation organizations with a broader range of metrics with which to measure risk in both current and future operations. "Safety can never be guaranteed by relying only on lagging indicators; rather it needs a continuous focus on lagging indicators of past deficiencies, leading indicators of current technical, organizational and human conditions and leading indicators of technical, organizational and human processes that drive safety forward" (Reiman & Pietikäinen, 2012).

Line Audit Data

Researchers identified opportunities to expand the use of line audit data in the safety risk management processes of respondent organizations. As shown in Figure 5, only 53.1% of respondents indicated their organization addresses risk using line audit data (n=161). Based on the Threat and Error Management (TEM) conceptual framework, programs such as Line Operations Safety Audits (LOSA) provide insights into the interactions that occur between humans, systems, and the operational environment during both nominal (routine) and off-nominal (non-routine) operations. In contrast to other data sources that describe adverse safety occurrences (e.g. mandatory reports), LOSA data describes the myriad operational factors present during routine flights (Klinect, 2005). As a result, line audit data often provides unique insights into the positive human-system resiliency factors that influence routine operational outcomes (Paulsgrove, 2018). Because this data describes the factors that can prevent undesirable outcomes, it can be key to establishing targeted leading SPIs.

Safety Performance Dashboard

To assist in the development of targeted SPM guidance materials, Fort Hill Group and Flight Safety Foundation developed a safety performance dashboard (Flight Safety Foundation, 2018b). The dashboard enables users to develop on-demand safety performance insights through the data analytics capabilities of Tableau Software. Users can create customized survey response data visualizations by applying sets of filters, including responses by region, risk type, domain, operational safety data sources, and others. This interactive filtering capability enables targeted deep dives into all meaningful aspects of the survey, such as understanding which risk areas are most important to different organization types in different ICAO regions.

Conclusion

To address industry's identified need for guidance, researchers identified an opportunity to apply the results of the SPM Survey to support the development of a Safety Performance Monitoring Handbook, containing example leading and lagging indicators appropriate for airlines, other aircraft operators, air navigation service providers, airports, and maintenance providers. By combining research findings with the results of a thorough cross-domain safety science literature review, a Safety Performance Monitoring Handbook would provide the aviation industry with actionable, data-driven information to enhance their safety performance monitoring capabilities and transition from reactive to proactive and predictive safety risk management.

Acknowledgments

Fort Hill Group and Flight Safety Foundation would like to thank the FAA for funding this research as part of Cooperative Agreement 17-G-003, as well as thank the subject matter experts who provided the valuable insight necessary to developing these results. The results presented herein represent the results of this research project and do not necessarily represent the view of the Federal Aviation Administration.

References

Flight Safety Foundation. (2018a). *Year Three Report of the Global Safety Information Project.* Alexandria: Flight Safety Foundation. Retrieved from https://flightsafety.org/gsip/gsip-reports/

Flight Safety Foundation. (2018b). *Safety Performance Monitoring*. Retrieved from https://flightsafety.org/gsip/safety-performance-monitoring

IATA. (2016). *Risk Mitigation Policies, Procedures and Best Practices 2nd Edition.* Montreal-Geneva: International Air Transport Association.

ICAO. (2013a). Annex 19 to the Convention on International Civil Aviation, First Edition. *Safety Management.*

ICAO. (2013b). Doc 9859 AN/474, Third Edition. *Safety Management Manual (SMM).*

ICAO. (2018). *Safety Report, 2018 Edition.* Montreal: ICAO.

Klinect, J. (2005). Line operations safety audit: a cockpit observation methodology for monitoring commercial airline safety performance (Doctoral dissertation). Retrieved January 9, 2019, from http://hdl.handle.net/2152/1967

National Safety Council. (2013). *Transforming EHS Performance Measurement Through Leading Indicators.* Retrieved from https://www.nsc.org/work-safety/tools-resources/campbell-institute/leading-indicators

Paulsgrove, D. (2018). A First for ACC: Line Operations Safety Audit. *Combat Edge Vol. 26 Issue 3*, p10, 6 p.

Reiman, T., & Pietikäinen, E. (2012). Leading indicators of system safety – Monitoring and driving the organizational safety potential. *Safety Science 50*, 1993-2000.

Wreathall, J. (April 2009). Leading? Lagging? Whatever! *Safety Science Vol. 47 Issue 4*, 493-494.

SAFETY ATTITUDE AND RISK PERCEPTION AMONG AIR PASSENGERS: A CROSS-REGIONAL STUDY

Abaid Ur Rehman[a]
[a]School of Psychology, Shaanxi Key Laboratory, of Behavior and Cognitive Neuroscience, Shaanxi Normal University, Xi'an, China.

You Xuqun[a]
[a]School of Psychology, Shaanxi Key Laboratory, of Behavior and Cognitive Neuroscience, Shaanxi Normal University, Xi'an, China.

The present study examined the safety attitude and risk perception among Air passengers at cross-regional levels. Moreover, the study also examined the differences in terms of safety briefing in the cabin. Although the Federal Aviation Administration and The international Air transport association has done much work on safety in cabin regarding air passengers, there is still the challenge to know to how to gain safety behavior of air passengers, particularly when they represent multicultural backgrounds. A sample of 700 air passengers with an average age of 26.5 was collected from three international airports in China. In this research, we used the two questionnaires (Safety attitude scale and Risk perception scale). The results of the study show significant differences at regional levels regarding safety attitudes and risk perception. The study provides a valuable discussion to move beyond the current research towards approaches that are more inclusive to the safety, behavioral context and provide new ways to develop safety strategies in the Aviation industry.

Keywords: Safety Attitude, Risk Perception, Air Passengers,

The current study aims to investigate passengers attitude and risk perception regarding cabin safety as past research concludes that an increase in passenger's safety knowledge will increase the chances of survival during an emergency, but during the past three decades, the issue has not gained due attention (Chang, Y-H., & Liao, M. Y., 2010). It is therefore important to know air passenger perceptions and safety attitude in flight. For this, we conducted an empirical in China to study this question. Research on safety behavior has recently gained attention after 9/11 to provide a safe environment for the air passengers, and cabin staff needs effective training strategies to cope with the uncertain situation during an emergency in the cabin.

Research on flight safety attitude on board in commercial airlines have not been well focused and nor air passengers been well instructed about aviation safety by the commercial airlines or the government institutes. Moreover, the literature is lacking data regarding risk perception and safety behavior of air passengers. Most of the passengers may believe that the commercial aviation incidence survivability rate are zero or low. Therefore, passengers pay less attention to what they should prepare for. Risk is part of life, its underestimation and overestimation can have unfortunate consequences (Burns and Slovic, 2012).

The attitudinal measures indicate that male passengers have greater likely to express undesirable cabin safety attitudes and findings provide very useful evidence for cabin safety-related issues (Parker, 2006). In addition, males express a higher level of confidence as compared to female (Roy Morgan Research, 2002). To define the differences, there must be a relationship between risk perception and demographic variables (Pidgeon, 1998). Individuals perceive as an active organizer in choosing what fear they have in order to bear in their life perspective. Therefore, such concepts relate to the present study as a participant's demographic, which may play a major role in defining the level and type of risk perception (Flynn, Slovic, & Mertz, 2006). When participating in travel activities uncertain consequences may exist as traveling includes moving abroad and interacting new environment (Yang, Khoo-Lattimore, & Arcadia, 2017a). Constraints for women travel were still present and compared to the European world, women are still facing restrictions in participating in leisure activities (Seow & Brown, 2018). Parker (2006) mentioned that attitudes and behaviors play an important role in shaping passengers perceptions towards safety briefing. Parker further states that if passengers express a more positive attitude towards safety on board may perceive safety information more useful in an emergency and if passengers have a less positive or negative attitude towards the safety on board will pay less attention to safety information.

In an airplane accident, the risk of being involved play a significant role in the public discourse in an air travel setting. However, safety risks have been mainly ignored by the air passengers in the selection of flight choice (Fleischer, A., Tchetchik, A., Toledo, T., 2015). Most studies have examined the impact of perceived risk on passengers online flights booking (Agag and El-Masry, 2016), However, there is little literature on how different aspects of safety attitudes which affect passengers' travel intentions. Although flight attendants are well trained, still they may not always be utilized to exercise their skills to respond to all passengers. In reality, air passengers may need to utilize their own abilities.

Aims of the study

Safety concerns have become the most important issue in the current travel decision process (Kozak, M., Crotts, J., & Law, R., 2007). One strategy for analyzing the passenger's behavior is to know the attitudinal and behavioral part of the passengers. This strategy is especially insightful for cabin staff when they interact with the background of multicultural passengers. To achieve these goals, we use a cross-sectional research design to collect the data from seven regions of the world as air passengers. Viewing the said literature review, we hypothesized that gender difference exists on safety attitudes and risk perception. Another hypothesis was that regional difference exists on both constructs. In the last, we also hypothesized that safety demonstration helps in shaping the passenger's attitude and risk perception.

Methods

Sample & Procedure

Accessible population in the study comprised of 700 air passengers who participated in the study. The data was collected at three international airports from China (Xian International Airport, Beijing International Airport, and Jinan International Airport). The survey intends at international travelers who were traveling from their country to another country. Most of the air passengers were students who were traveling to China for their study purposes. The average age of the participants was 26.5 years. Consent for the current study was taken from the respondents prior to handover the questionnaire. It was assured that the provided information would be used only in research purposes.

Instruments

Safety Attitude Scale. The safety attitude scale was developed by the researcher with the help of previously published literature and involving experts from China Eastern Airline and Pakistan International Airline. The safety attitude scale consisted of 30 items with five-point Likert-type scaled ranges from 1= strongly disagree to 5= strongly agree. Information was provided to respondents to choose the answer that best corresponded to their choice. The scale has been valid in Chinese and South Asian population. An alpha coefficient of the scale was .83. The scale development was part of the doctoral study.

Risk Perception Scale. The risk perception scale was also developed by the researcher to measure the level of perceived risk of air passengers. For the development of the scale research consulted with experts from China Eastern Airline and Pakistan International Airline. The scale consists of 13 sentences describing perception on five-point Likert type scaled ranges from 1= strongly disagree to 5= strongly agree. We briefed our respondents to select the best answer that corresponded with their level of agreement. The scale has been cross-validated in a sample of road travelers. Moreover, the scale was valid in a Chinese and South Asia population. An alpha coefficient of the scale was .93. This scale development process was part of a doctoral study.

Results

The analysis was conducted using a t-test, one-way ANOVA, and Post Hoc modules with the help of SPSS.

Table 1.

Mean, S.D, t-analysis of genders differences on the Post Traumatic Stress Disorder scale (N=100)

	Male (*n*=323)		Female (*n* = 318)			
Scales	*M*	*SD*	*M*	*SD*	*t*	*p*
Safety Attitude	161.84	18.04	159.04	17.85	1.97	.049
Risk Perception	37.72	13.48	37.92	14.13	.18	.857

Note. *df*= 639

Table 1 shows the mean differences between the male and female air passengers on safety attitude scale and risk perception scale. The mean scores of male and female air passengers show that there is a significant difference in safety attitude and nonsignificant differences in risk perception.

Table 2.

Mean, Standard Deviation and F-values for Video, Audio and Physical demonstration, Video & Audio and Physical & Audio on Study Variables (N = 641)

	Video (n = 90)		Audio (n = 34)		Physical Demo (n = 65)		Video & Audio (n=239)		Physical & Audio (n=213)				Post hoc
Variables	M	SD	M	SD	M	SD	M	SD	M	SD	F	η^2	
Safety Attitude	157.09	16.41	164.41	22.48	165.74	19.36	160.24	17.43	159.86	17.72	2.69**	.98	3 > 2 > 4>5>1
Risk Perception	37.17	12.58	44.35	16.07	41.26	13.14	37.09	13.67	36.82	13.95	3.46**	.97	2 > 3 > 5>4>1

Table 2 shows mean, standard deviation and *F* -values for Video, Audio and Physical demonstration, Video & Audio and Physical & Audio. Results indicate significant mean differences on safety attitude with {F (4,640) = 2.69**, $p < .01$}, and on risk perception with {F (4,640) = 3.45**, p < .01}.

Table 3.

Mean, Standard Deviation and F-values for Europe, East Asia, Africa, South Asia, USA, Gulf Countries, and Central Asia, on Study Variables (N = 641)

	Europe (*n* = 81)		East Asia (*n* = 114)		Africa (*n* = 110)		South Asia (n=150)		USA (n=61)		Gulf (n=64)		Central Asia (n=61)				Post hoc
Variables	*M*	*SD*	*M*	*SD*	*M*	*SD*	*M*	*SD*	*M*	*SD*	*M*	*SD*	*M*	*SD*	*F*	η^2	
Safety Attitude	158.07	15.78	164.34	17.82	160.02	19.31	*160.92*	*17.08*	158.44	14.85	159.14	21.74	159.36	18.86	1.37	.98	2 > 4 > 3>7>6>5>1
Risk Perception	34.47	15.41	41.20	13.67	37.88	14.85	36.39	12.57	37.98	13.11	39.67	14.03	37.07	12.04	2.46**	.97	2 > 6 > 5>3>7>4>1

Table 3 shows Mean, Standard Deviation and F-values for Europe, East Asia, Africa, South Asia, USA, Gulf Countries and Central Asia. Results indicate non-significant mean differences on safety attitude with {F (6,640) = 1.37 $p > .05$}, and significant differences on risk perception with {F (6,640) = 2.46**, p < .01}.

Discussion & Conclusion

It seems reasonable that safety attitude and risk perception influence passenger's behavior and their exposure to some future risks during their air travel. Passenger's overall helpfulness of safety attitude was to be significantly lower amongst female passengers when compared to that of male passengers (Table 1). Gender especially women seem to be influenced by the perception of risk during their travel. According to the results of the present study, a significant difference was seen for safety attitude as a male have a high attitude towards safety. Contrary to this, female's passengers, which, also shows that male, have more opportunities to interact with outside world and they have more information as compared to females about safety information as attitudinal measures indicate that male passengers have greater likely to express undesirable cabin safety attitudes and findings provide very useful evidence for cabin safety-related issues (Parker, 2006). In addition, males express a higher level of confidence as compared to female (Roy Morgan Research, 2002). Contrary to this, both found equal on risk perception variable. We have also analyzed safety finding regarding safety attitude and risk perception and it was observed that respondents who focus more on a physical demonstration during safety briefing have a high level of safety attitude and better perception of risk as compared to just video or audio demonstration. It can be observed that passengers were less interest in the video demonstration and have a low level of safety attitude as well as risk perception. As the past research shows that respondent who did not pay attention to think that paying attention to the safety briefing is a waste of time (Johnson, D.A., 1979). Air passenger who considers that the cabin staff is mainly responsible for the safety of passengers may pay less attention to safety briefings. Parker (2006) mentioned that attitudes and behaviors play an important role in shaping passengers perceptions towards safety briefing.

The study confirmed the third hypothesis which states that regional differences exist on safety attitude and risk perception constructs. The findings show that Asian as well as South Asian were higher on safety attitude construct as well as on risk perception as compared to Africans and European respondents. Our study appears to indicate higher feeling of safety attitude and risk perception be more careful in East Asian responded region and found high level of safety attitude but lower levels of risk perception in others. The study confirms that safety attitudes and risk perceptions that Asians have more careful about their safety and future travel risks and Europeans were less on the construct. It is known that an airplane cabin is like a sitting room where uncertain events may occur. Our study has several implications for aviation industry and research. although in all regions risk perception is perceived as a real risk, the level of safety appears to be rather low. Therefore, it can be suggested that passengers must be well informed and their Knowledge about safety-related issues should be broadened so that appropriate behaviors could be performed in an emergency. Furthermore, information programs must be launched at airport launches and waiting areas about safety in aviation. Key opportunities are identified to prove more effective was of safety in the cabin. It is also suggested that airliner should understand the importance of the relationship between passengers demographic characterizes, safety attitude and risks may be improved. More training at organizational and national levels can be launched to prove the knowledge of air passengers about safety issues and information can be improved about risk perception. Once it is recognized that the decisions of air passengers may be compromised with hazardous attitudes, airliners can improve more effective strategies to proved more safer journey

Acknowledgments

This research was supported by School of Psychology, Shaanxi Normal University, Xian, China. We are thankful to China Eastern Airline and Pakistan International Airline for their cooperation. We also thanks to our participants for their cooperation.

References

Agag, G., & El-Masry, A. (2016). Understanding the determinants of hotel booking intentions and moderating role of habit. *International Journal of Hospitality Management*, 54, 52-67, https://doi. 10.1016/j.ijhm.2016.01.007

Burns, W.J. and Slovic, P. (2012) Risk perception and behaviors: Anticipating and responding to crises. *Journal of Risk Analysis*, 32, 579-582. https://doi. 10.1111/j.1539-6924.2012.01791.x

Chang, Y-H., & Liao, M. Y. (2010). A comparison of cabin safety awareness among airline passengers in Taiwan and Mainland China. *Transportation Journal*, 49(1), 48-64. https://doi/10.2307/40904821

Fleischer, A., Tchetchik, A., Toledo, T., (2015). Does it pay to reveal safety information? The effect of safety information on flight choice. *Transportation Research*, Part C, 56, 210–220, https://doi.10.1016/j.trc.2015.03.039

Flynn, J., Slovic, P., & Mertz C. (2006). Gender, race, and perception of environmental health risks. *Risk Analysis an International Journal*, 14(6), 1101–1108. https://doi:10.1111/j.15396924.1994.tb00082.x

Johnson, D.A. (1979). An investigation of factors affecting aircraft passenger attention to safety information presentations. Report IRC-79-1, for the Federal Aviation Administration, Contract DOT-FA78WA-4095.

Kozak, M., Crotts, J., & Law, R., (2007). The impact of the perception of risk on international travellers. *International Journal of Tourism Research*, 9(4), 233-242. http://doi.10.1108/WHATT-07-2017-0032

Pidgeon, N. (1998). Risk assessment, risk values and the social science programme: Why we do need risk perception research. *Reliability Engineering & System Safety*, 59. https://doi 10.1016/S0951-8320(97)00114-2

Parker, A. (2006). Public attitudes, perceptions, and behaviours toward cabin safety communications. Australian Transport Safety Bureau Research and Analysis Report B2004/0238. Electronic version retrieved March 13, 2019 fromhttp://www.atsb.gov.au/publicationns/2006/pdf/B20040238.pdf

Roy Morgan Research, (2002). 'Survey of public attitudes to aviation safety 2002', Report prepared for System Safety and Human Performance Section, Aviation Safety Promotion Division, Civil Aviation Authority, Canberra.

Seow, D., & Brown, L. (2018). The solo female Asian tourist. *Current Issues in Tourism*, 21 (10), 1187-1206. http:// doi:10.1080/13683500.2017.1423283

Yang, E.C.L, Khoo-Lattimore, C., & Arcodia, C. (2017a). A systematic literature review of risk and gender research in tourism. Tourism Management, 58, 89-100.http:// doi: 10.1016/j.tourman.2016.10.011

DESIGNING MILITARY COCKPITS TO SUPPORT A BROAD RANGE OF PERSONNEL BODY SIZES

Summer Lindsey
Florida Insitute of Technology
Melbourne, FL
H. C. Neil Ganey
Northrop Grumman Corporation
Melbourne, FL
Meredith Carroll
Florida Institute of Technology

The Joint Primary Aircraft Training System (JPATS) body size cases provide aircraft designers access to representative Airmen anthropometric dimensions based on a database of air force personnel. To ensure new aircraft can support a broad range of pilot body sizes, designers can reference JPATS case numbers to assure adequate access to specific controls and clearances for ejection. JPATS cases 1 and 7 were added in response to the Air Force's goal to accommodate 95% of males and females. However, given that someone at the 90th percentile height may not have 90th percentile arm length, a far smaller percentage of accommodation than originally projected is actually achieved. This paper presents a discussion of the limitations of JPATS cases and past evaluation methods, and provides recommendations for methods to utilize during the design process to ensure next generation cockpits can accommodate a broader range of body sizes.

The Joint Primary Air Training System (JPATS) Program was developed to accommodate a large range of body types in military grade aircraft. This program sets requirements allowing a range of body sizes, from tall to short females, the ability to become military pilots. In 1996, United States Air Force (USAF) pilots were required to be within 64 to 77 inches tall – with a sitting height of between 34 to 40 inches. Due to these constraints, six percent of males, and over half of females were unable to become USAF pilots (Zehner, 1996). Anthropometric sizing is an issue for many military branches including USAF, the United States Navy (USN), and the United States Marine Corp (USMC).

The Cases

Zehner (1996) evaluated the range of anthropometric sizes of recent college graduates to derive a database of possible pilot body sizes. Based on his findings, JPATS cases 1 and 7 were built in response to the USAF goal to accommodate 95% of males and females. The addition of these cases expanded the range of acceptable height to between 58 and 77 inches tall, and sitting height to between 31 and 40 inches tall. The cases allow for an easy reference of body size. The USAF can reference these case numbers and require access to specific controls and specific head, shin and ejection clearances during Request for Proposal for new aircraft. The goal of accommodating 95% of the male and female population is good in theory; however, this creates a large amount of difficulties in the development phase for contractors and incoming extreme JPATS cases.

As the military includes more measurement requirements for the same individual e.g., thumb tip reach, sitting height and buttock-knee length, it limits the amount of the population that can actually fit into the percentages. Gordon, Corner and Brantley (1997) posit that if a requirement states that one individual must fit within the 5%-95% range for 5 different measurements, this may actually only accommodates 67% of the population. This issue arises due to the fact that an individual that falls within 90% height, may not necessarily fall within 90% thumb tip reach. Additionally, in a survey of USAF personnel, a pilot with 5th percentile height and weight is not actually a 5th percentile individual (Kennedy, 1986). Of this survey, only 1.3% of individuals were smaller than the 5th percentile in both dimensions. However, 9% of the population fell below 5% in either height or weight. These percentages can quickly become unrepresentative of pilots once percentiles are combined. As a result, a case 7 may be able to fly a particular aircraft; however, once an individual varies from these exact measurements in one or two categories, they are no longer accommodated. For instance, a case 7 may narrowly meet the requirements to reach critical safety controls, however an individual similar to a case 7 with longer legs would no longer be accomodated once the seat is adjusted back. Zenher (2002) discusses the concerns for larger pilots including overhead clearances, ejection envelopes, body interference with controls, body interference reaching for controls and shin clearance. On the other side of the spectrum, the largest concerns for smaller pilots include reaching emergency controls, external field of view, and full rudder and brake operating range. If a smaller case is in a harness locked position, it is important that all safety controls, primary flight controls and propulsion controls can be reached and that the operator has full operational range (MIL-STD 1333, 1987).

Gear being developed to support small female JPATS cases inside the cockpit also required modifications. For G-suits, women generally have larger hips and smaller waists then males. This would cause fit issues for incoming women, as many women must wear one size larger just to accommodate hip size. Modifications to three existing G-suit sizes are needed to support women pilots. These include reducing waist and calf circumferences, and shifting the abdominal bladder to the back instead of the front (Dooley, 1995). These reductions result in three new sizes which allow for 90% of the female population to be accommodated. However, this requires development of custom clothing in order to support new JPATS cases, and added cost. Further, considerations also must be given to these cases whenever additions are made in the cockpit. For instance, adding kneeboards for paper or an electronic flight bag (EFB) tablet has shown to have up to a 4.5% chance of a femur fracture for females in a T-38 during ejection compared to 2-3% for males (Perry, Burneka, & Stzelecki, 2015).

JPATS cases were developed over 30 years ago and their current applicability is in question. Over time and through new generations, average body sizes change. Research by Choi and Colleagues in 2011 found that while JPATS proportions still seem to fit the USAF population, individuals are becoming heavier overall. Tucker, Brattin, and Reason 2002 found that the reference databases for USN and USMC populations were non-representative of the actual population. It was found that the population consisted of mainly 4.7% female compared to the 40% female reference database. Furthermore, the population was shown to have some heavier individuals than the database stated. Twenty four percent of individuals were heavier than the highest weight in the database. Heavier individuals who are not considered during development can experience issues with safety gear, clothing, and even cockpit functionality such as center stick control (Tucker, Brattin, & Reason, 2002). The issues of body size variability, particularly when combining percentiles, as well as generational changes, makes it difficult for new aircraft designers. Further, significant body size differences were seen in Brazilian anthropometric data compared to U.S. data, as brazilian pilots were shown to have a shorter reach (Silva, Gordon, & Halpern, 2018). This is concerning as 28% of the USAF personnel are non-caucasian as of 2019, compared to 13% in 1996 when the JPATS cases were formed (Baseman, 1997; Air Force Personnel Center, 2018).

Legacy Aircraft

One of the largest concerns with allowing a larger percentile of males and females to become pilots, is the inability for legacy aircraft to support them. With respect to trainer aircraft, the T-38 is a common trainer aircraft for USAF pilots, however, only 47.1% of females will have a sufficient external field of view over the nose of the aircraft. The only way that many females can see over the nose is by adjusting their seat up all the way. However, this now prevents case 7 females from being able to reach the rudder pedals by multiple inches (Zehner, 2002). Small females are also unable to train in the T-38, due to the lack of reach to safety critical controls in a reel locked state and full rudder actuation. Considering all dimension categories together, only 27.2% of USAF female population and 86.9% of USAF male population can fly the T-38. However, the T-6 and T-37 allow a much higher percentage of the USAF population for safe flight. This includes full accomodation of female and male populations at the minimum size requirements or smaller. In addition, 86.3% of USAF females at the minimum size requirements or smaller are accomodated in the T-37 and 98.6% in the T-6 (Zehner, 2002). While these percentages support JPATS cases, they may not always be the appropriate training aircraft for the pilot's future career.

Zenher 2002 discusses the limitations with respect to fighter aircraft, the F-16 and F-15 still do not accommodate smaller females uch as case 7. This prevents case 7 females from pursuing a fighter pilot career path altogether. Such is also the case with respect to bomber and helicopter aircraft. It was also found that larger case pilots are also prevented from pursuing careers flying fighter aircraft or helicopters due to overhead clearances and/or leg clearances. Heavier pilots are able to fly bomber aircraft but are limited to training in the T-38. (Zehner, 2002).

Tucker, Brattin, and Reason (2002) discuss how USN and USMC legacy aircraft also encounter issues in an attempt to support 95% of the population. The T-44A, E-2C, and C-2A support less than 85% of the population. Additionally, the T-2C supports less than 54% of the population for the front cockpit. Each of these USN and USMC legacy aircraft exhibited issues supporting external FOV while also being able to reach controls in a locked hardness position. The T-2C also faced issues supporting heavier weighted individuals in order to safely eject. These legacy aircraft make it difficult for extreme JPATS case individuals to train, or safely operate in their particular aircraft. Due to the lack of accommodation of JPATS cases in existing aircraft, there is demand for new training and aircraft that can support extreme JPATS cases.

Moving Forward

Development for, and accommodation of, extreme JPATS cases cause an array of issues in existing and future aircraft. Numerous modifications have been made and current training aircraft are considered obsolete as they are unable to accomodate incoming JPATS cases. To truly allow 90% of the population to see and reach controls, have adequate external field of view, and have safe ejection clearances in future aircraft, there is a need to evaluate a large array of body sizes and variations of JPATS case measurements. Evaluation methods include measuring vision, clearances, reach tests and full range of motion for a large range of body sizes (Zehner, 2001). Ensuring these are researched appropriately and thoroughly during future aircraft test and evaluation can prevent the costs associated with issue correction during or after the production phase.

Manikins can be used in order to evaluate the impacts of ejection clearances as demonstrated in a study by Buhrman (1996). However, manikins that are representative of larger and smaller cases were shown to be less reliable for ejection tests. Particularly small JPATs manikins were shown to be less reliable for ejection impacts, showing different results under repeated tests. This can make it difficult to test the effectiveness of ejection seats using real-life ejection dummies that are representative of JPATS case 7. Furthermore, large case manikins were shown to exhibit larger seat forces than corresponding humans (Buhrman, 1996). The time to reach maximum velocity during ejection also varied from human data, which effects the calculations of injury possibilities. These variations indicate that manikin-tested ejection seats may be unsafe once used in a real-life situations.

Evaluation of the aircraft during development is vital to ensure accommodation of extreme JPATS cases. Crawford (2002) discusses how the USN utilizes 3D modeling in order to ensure that JPATS cases can be supported. A FaroArm is used to create a 3D rendering of the pilots performing certain tasks or reaching for particular controls. The FaroArm can be used in existing aircraft or used on a subject sitting in a seat similar to the one in the aircraft or a mockup of an aircraft in the development phase. The researchers can place the FaroArm rendering into CAD drawings in order to evaluate and predict possible issues during development. This method allows a controlled environment and does not require the actual aircraft in order to evaluate the reach and ejection envelopes which can allow more time and resources for evaluating a larger breadth of subjects (Crawford, 2000). However, the representativeness of digital human models requires repeated measurements, test, and evaluation. Whitestone, Hudson, & Rife (2018) evaluated the RAMSIS NextGen, a USAF tool that allows developers to test various cases, their reaches, positions and performance. The spinal data in the RAMSIS case sizes were not an accurate representation and failure to evaluate posture can lead to pain during prolonged flight. Through the use of 3D modeling and Luna fiber optic sensors, the researchers were able to correct the digital human models to a spinal position accuracy of 2mm. Robinette & Vietch (2016) stress the importance of still incoporating real human evaluations to evaluate not only accepted ranges and sizes, but to evaluate levels of pain, discomfort, pilot preferencess, and any potential improvements. The added benefit is that the anthropometric data from real human assessments can be added to virtual human databases to ensure the virtual anthropometric assessments are an accurate portrayal of the current military force.

Evaluating and discovering issues during the development phase can results in massive returns on investment. In past cases, these have ranged from -24% to 153% and even a 9,260% return on investment (Smith, 2015). However, when considering an aircraft that limits the population available to become pilots, it becomes more than just an investment decrement. This limits the pilots ability to pursue certain career tracks, and limits the manpower available to the military forces. The USN and USMC spend over one million dollars to train a jet aviator. Moreover, these costs can increase to three million if the aviator had to be reassigned, replaced, or was trained in a non-representative aircraft due to body size limitations (Tucker & Brattin, 2000). The JPATS program intends for a larger amount of the population to have the ability to become pilots. This comes with a large increase in test and evaluation resources and requires a much larger sample. For new aircraft in the developmental phase, evaluating an array of body sizes and types in large numbers is the most effective way to ensure 90% or more accommodation. However, pilots who approach the limits of JPATS extreme cases will always need to be fit-checked to see if they can be accommodated. While the possibility of all body type accommodation in aircraft is most likely improbable, the proper test and evaluation in early stages of development can help accommodate the more extreme percentiles.

References

Air Force Personnel Center. (2018). Military Demographics. Retreieved from https://www.afpc.af.mil/About/Air-Force-Demographics/

Baseman, L. (1997). Statistical digest. *United States Air Force.*

Buhrman, J. (1996). Impact testing of JPATS mankins. *Biodynamics and Biocommunications Division Wright-Patterson AFB.*

Choi, H., Coate, A., Selby, M., Hudson, J., Whitehead, C., Zehner, G., & Fleming, S. (2014). Aircrew sizing survey 2011. *Human System Integration Directorate, Wright-Patterson Air Force Base.*

Crawford, J. (2000). United states navy advanced crew station evaluation techniques. *Naval Air Warfare Center Aircraft Division Patuxent River MD.*

Dooley, J. (1995). Female acceleration tolerance enhancement. *Armstrong Laboratory.*

Gordon, C., Corner, B., & Brantley, D. (1997). Defining extreme sizes and shapes for body armor and load-bearing systems design: Multivariate analysis of U.S. Army torso dimensions. *U.S. Army Natick Research, Development and Engineering Center.*

Kennedy, K. (1986). A collation of United States Air Force anthropometry. *Harry G. Armstrong Aerospace Medical Research Laboratory, Wright-Patterson Air Force Base.*

N.a. (1987). MIL-STD 1333 Aircrew station geometry for military aircraft. *Department of Defense.*

Perry, C., Burnecka, C., & Strzelecki, J. (2015). Biodynamic assessment of pilot knee-board configuration during simulated T-38 catapult acceleration. *United States Air Force Research Laboratory.*

Robinette, K. M. & Veitch, D. (2016). Sustainable sizing. *Human Factors, 58*(5). 657-664.

Silva, G. V., Gordon, C. C., & Halpern, M. (2018). Comparison of anthropometry of brazilian and US military population for flight deck design. *International Journal of Industrial Ergonomics, 64.* 170-177.

Smith. N. (2015). Examining the return on investment of test and evaluation. *Air Force Institute of Technology.*

Tucker, H. & Brattin, L. (2000). Current USN/USMC aircraft anthropometric compatibility issues and the "street to fleet" proposal. *Naval Air Warfare Center Aircraft Division Patuxent River MD.*

Tucker, H., Brattin, L., & Reason, W. (2002). Revised anthropometric restrictions for U.S. Navy and Marine Corps trainer and fixed wing nonejection aircraft and U.S. Coast Guard HU-25. *Naval Air Warfare Center Aircraft Division Patuxent River MD.*

Whitestone, J. J., Hudson, J. A., & Rife, B. (2018). Determining spinal posture for encumbered airmen in crewstations using the luna positioning sensor. *United States Air Force Research Laboratory.*

Zehner, G., (1996). Cockpit anthropometric accommodation and the JPATS Program. *SAFE Journal, 26*(3).

Zehner, G. (2001). Prediction of anthropometric accommodation in aircraft cockpits. *Crew Systems Interface Division, Wright-Patterson Air Force Base.*

Zehner, G., (2002). Body size accommodation in USAF aircraft. *United States Air Force Research Laboratory.*

COLLABORATION INTERFACE SUPPORTING HUMAN-AUTONOMY TEAMING FOR UNMANNED VEHICLE MANAGEMENT

Elizabeth Frost[1], Gloria Calhoun[2], Heath Ruff[1], Jessica Bartik[2], Kyle Behymer[1], Sarah Spriggs[2], Adam Buchanan[1]
[1]Infoscitex, Dayton, OH, USA
[2]Air Force Research Laboratory, 711 HPW/RHCI, Dayton, OH USA

Advances in technology are leading to envisioned operational concepts that team a single operator with autonomy to manage multiple heterogeneous unmanned vehicles (UxVs). Several autonomy decision aids have been integrated into a prototype control station with innovative human-autonomy interfaces that allow multiple UxV management via high-level commands called "plays". Each play defines the actions of one or more UxVs, often in response to a mission event or task. This paper describes recent enhancements made to a Task Manager tool to better support operator-autonomy collaboration. After mission events are signaled in chat, corresponding tasks are communicated by an intelligent agent via pictorial icons designed to facilitate rapid retrieval of necessary actions. These icons also enable direct manipulation control functionality. The Task Manager supports shared awareness across human and agent team members by summarizing the relative priority, recency, and completion status of mission tasks.

Several autonomy advancements were integrated into a control station prototype referred to as "IMPACT" (Intelligent Multi-UxV Planner with Adaptive Collaborative/Control Technologies) to flexibly team a single human operator with autonomous decision aids performing a base defense mission (Draper, Calhoun, Spriggs, Evans, & Behymer, 2017; Draper, et al., 2018; Figure 1). To support human-autonomy teaming in IMPACT, a "play-calling" method is used that enables a single operator to develop and execute plans quickly for multiple heterogeneous unmanned vehicles (UxVs). This involved the design and implementation of a comprehensive suite of play-based interfaces to support calling plays, reviewing/revising the autonomy-generated play plan(s), and monitoring play execution. For example, when an IMPACT operator calls a play to achieve air surveillance on a building, an intelligent agent recommends a UxV (based on estimated time en route, fuel use, etc.), a cooperative control algorithm provides an optimal route to get to the building (taking into account no-fly zones, etc.), and an autonomics framework monitors the play's ongoing status (e.g., alerting if the UxV won't arrive on time).

Figure 1. IMPACT Control Station Prototype.

The operator is able to call plays through the selection of a corresponding play icon that represents both the vehicle type(s) that will be assigned and the high-level action of the vehicle(s) (i.e., play type; see Figure 2). IMPACT's play-calling interfaces also facilitate operator-autonomy communication on mission details to optimize play parameters (e.g., current visibility) as well as support operator/autonomy shared awareness (e.g., a display showing the tradeoffs associated with multiple agent-generated courses of actions). This adaptable extended play-calling approach is novel in its flexibility in providing fine-grained control whereby the operator can rapidly specify the level of automation along multiple dimensions, as well as seamlessly transition between control states. Additional details on the play-related interfaces are available (Calhoun, Ruff, Behymer, & Mersch, 2017; Calhoun, Ruff, Behymer, & Frost, 2018).

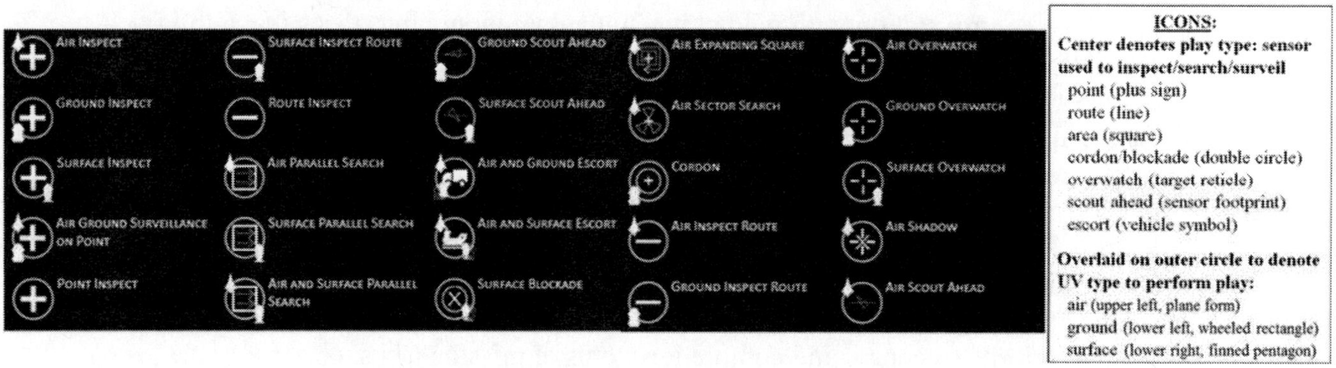

Figure 2. Icons that specify play and UxV type.

Play icons are accessible from several interfaces. A dedicated play-calling interface provides a categorized list of all 25 pre-defined base defense-mission related plays. Play icons can also be selected from a radial menu that is presented upon a right-mouse click of a target (i.e., location or other entity) or a UxV symbol on the map; the radial menu filters plays to present just those relevant to the selection. When utilizing these play-calling interfaces the operator has to specify, at a minimum, *what* type of play and *where* the play should be executed.

A third interface that includes selectable play icons is the Task Manager. This interface has been recently enhanced to facilitate operator-autonomy coordination. The Task Manager utilizes an agent that constantly monitors incoming communications (e.g., chat rooms) for mission events. For each identified mission event, the agent creates a corresponding task for the operator. Each task includes one or more subtasks that should be addressed in order to consider the higher-level task complete. The agent also suggests actions the operator could take to complete these subtasks, such as play calls that could be utilized (see Figure 3).

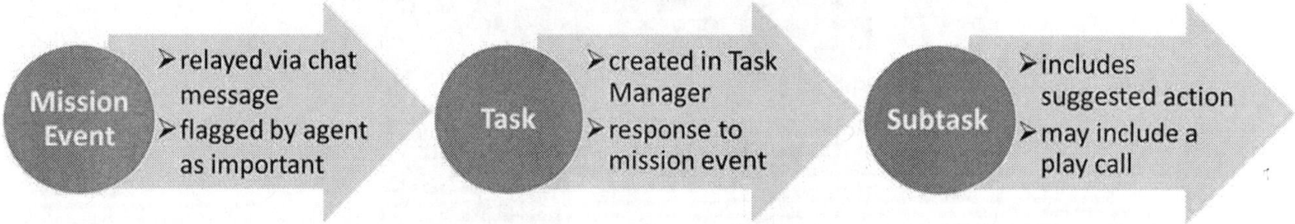

Figure 3. Representation of the relationship between mission events, tasks, subtasks, and plays.

Task Manager Interface

Left Pane of Task Manager

The left pane is persistent on the monitor and is where the high-level tasks are represented (see Figure 4). Each task has a corresponding mission-coded icon that has been determined in previous research to be intuitive and discriminable (Bartik, et al. 2017). The task icons utilized to date in support of IMPACT's base defense mission are illustrated in Figure 4.

The Task Manager uses rows (see Figure 4) to help organize multiple tasks that have varying priority since operational tasks will likely vary in priority. For IMPACT, each type of mission task has a pre-assigned priority in relation to the overall base defense mission, and this determines how the corresponding icons populate the four rows. The top row is for tasks that need immediate attention, specifically responding to events that indicate there is an active threat or attack on the base. The second row is for base defense activities that are performed as needed during normal operations (e.g., a specific threat has not been detected). Random Anti-Terror Measures (RAMs), or tasks that randomize base activities and help maintain base safety, are assigned to the third row. The question marks in the fourth row indicate that the agent detected queries in communications, each of which the operator can leverage the system in order to generate a reply. The magnifying glass icons are presented in experiments to task operators to provide mission information (as a measure of situation awareness). The number below each icon indicates how many of that task type have been identified (e.g., two queries in the fourth row). As tasks of the same priority are identified, they are added in respective rows from the right.

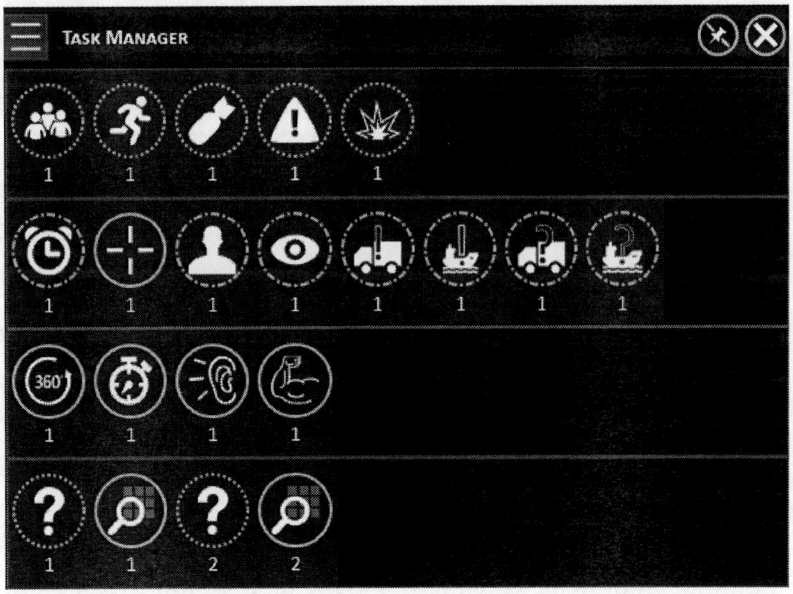

Figure 4. The left pane of Task Manager showing agent generated tasks in four rows, each row decreasing in priority. The following identifies the mission tasks, from left to right, for each row. Row 1: crowd forming, gate runner, mortar fire, perimeter breach, explosive device. Row 2: building/fence alarm, overwatch (provide air coverage), escort, eyes on (location), suspicious vehicle/watercraft, unidentified vehicle/watercraft. Row 3: 360 check, interval (temporal) check, listening post, and show of force. Row 4: queries and information retrievals (two each).

Each of the task icons is presented within a circle. The circle's line coding designates if that task needs to be completed by the human operator (solid line) versus the intelligent agent (dotted line). A dashed-dotted line for the circle indicates that the task (with multiple subtasks) requires action from both human and agent team members (e.g., the left most alarm icon in the second row of Figure 4).

The Task Manager's left pane in Figure 5 shows one of the task icons highlighted with a square outline and shade coding to indicate that the operator has selected the associated event, an activated alarm. (Thus these icons enable direct manipulation control functionality.) There are also two additional smaller icons to the right of the task icon by which the operator can delete the task (by selecting "X" the task is removed from the Task Manager and recorded in a separate log) or assign the entire task to the agent partner (by selecting the lightning bolt icon). The selection of a task icon in the left pane also brings up the right pane.

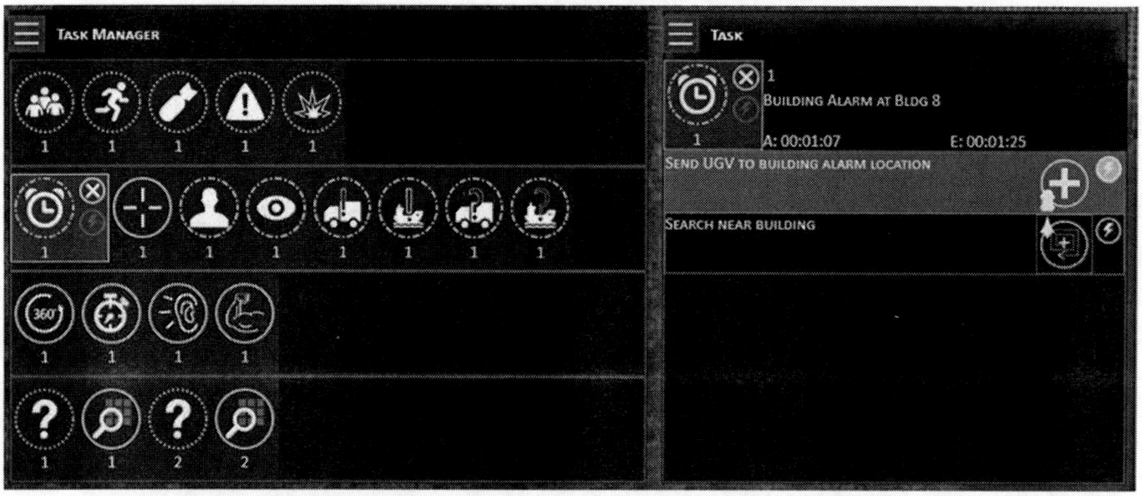

Figure 5. Illustration of Task Manager. Left pane shows a selected building alarm task. Right pane shows associated subtasks and suggested plays for the selected task.

Right Pane of Task Manager

The right pane displays additional symbology related to the selected mission task. At the top, the task icon and functions from the left pane are repeated along with the exact chat that triggered the creation of the task (e.g., "Building Alarm at Bldg 8") and two time fields (A: time when the task was added and E: time elapsed). Below this header, the agent's recommended subtasks are listed. The first subtask, as show in the right pane of Figure 5, is to send an unmanned ground vehicle (UGV) to the alarm location. A play that can be leveraged to complete this subtask, a ground point inspect, is represented on the right side of the row with a selectable play icon. The lightning bolt icon to its right allows the operator to assign that subtask to the agent. When selected, the lightning bolt will become highlighted and the row will be shaded. For example, the top row in Figure 5 indicates that the "Send UGV to building alarm location" subtask is being completed by the agent team member.

The second row in Figure 5 states that a search near the building with the activated alarm needs to be completed. In this example, the agent suggests an air expanding square search play as a means to complete this subtask. Also, the row is not shaded indicating it has not been completed by the operator. Once the operator completes an assigned subtask (i.e., calls a play in response), the row will become shaded (the lightning bolt will not be highlighted because it was completed by that operator and not the agent). When the task is completed via a play call, the initiated play and related task information is represented elsewhere (e.g., Active Play interface and map symbology). The right pane will remain open with all subtasks shaded until the operator closes the pane, deletes the task all together, or clicks on another task.

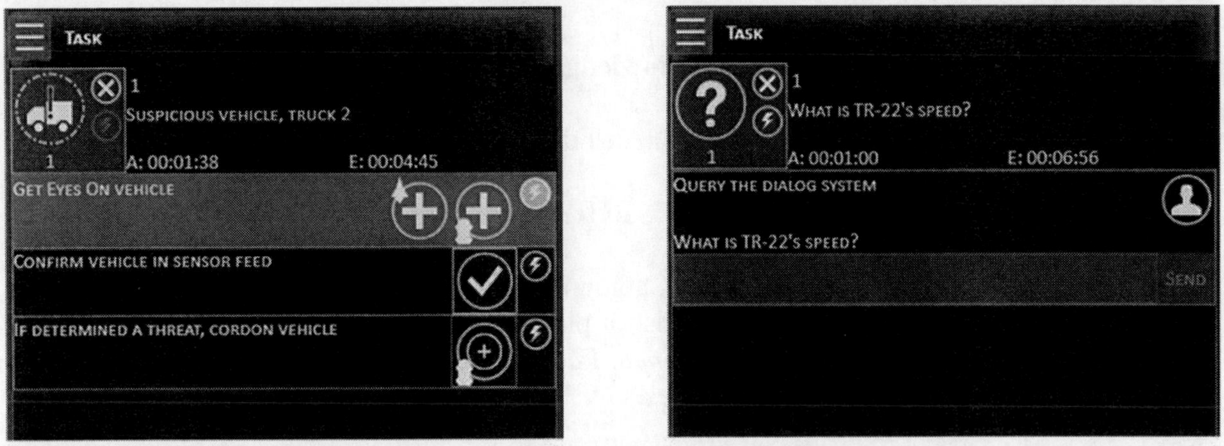

(a) (b)

Figure 6. (a) Right pane for a suspicious vehicle task. (b) Right pane for a query task.

Figure 6 illustrates the right panes of two additional tasks. The suspicious vehicle task has two subtasks that can be completed with play calls (the first and last subtasks; see Figure 6a). However, the second subtask requires communication with a sensor operator to confirm that the target is visible in the sensor feed. The operator can mark this subtask as complete by clicking the checkmark icon on the far right. Figure 6b shows the right pane for a query task. Instead of a list of subtasks, the query is shown along with a text box that can be used to send the response. By clicking the "person" icon, the operator can receive a response to the query from the agent.

Summary

In contrast to the other IMPACT play-calling interfaces in which the operator needs to specify *what* type of play and *where* the play should be executed, the intelligent agent, based on an ongoing analysis of mission events, proposes the *what* and *where* in the Task Manager and helps the operator team with autonomy in performing base defense. Moreover, it provides priority-based task organization and the control functionality by which the operator can assign tasks/subtasks to the agent for completion (Frost, Bartik, Calhoun, Spriggs, Ruff, & Behymer, 2018). With additional refinements, the interface could support coordinated distributed operations with other autonomy-aided human operators managing different UxV assets. The Task Manager is also a candidate interface to communicate and coordinate actions for envisioned autonomy advancements that result in the agent's ability to suggest pre-emptive tasks/plays to better posture the base for defense.

It should be noted, however, that the Task Manager does not function in isolation. The Task Manager's play recommendations are based on pre-established parameters for each type of task and these parameters can be refined (both prior and during play execution) to meet current mission considerations by employing other play-related interfaces. Also, upon play execution the Active Play interface is instrumental in maintaining a shared understanding between human and autonomy team members on the status of all ongoing tasks (see Calhoun, et al., 2018). Regardless, the Task Manager is paramount to play-based multi-UxV management because it facilitates rapid retrieval of necessary actions (and their relative priority), expedites execution of necessary mission-related actions, and provides a mechanism for sharing the workload between the human and autonomy team members.

Acknowledgement

This effort was supported by the U.S. Air Force Research Laboratory.

References

Bartik, J., Ruff, H., Behymer, K., Frost, E., Calhoun, G., Spriggs, S., & Hammack, T. (2017). Mission-coded map icon decision aids for play-based multi-unmanned vehicle autonomy delegation. In *Proceedings of the Human Factors and Ergonomics Society Annual Meeting, 61*(1), 237-241. Los Angeles, CA: SAGE Publications. 10.1177/1541931213601543

Calhoun, G., Ruff, H., Behymer, K. & Frost, E. (2018). Human-autonomy teaming interface design considerations for multi-unmanned vehicle control. *Theoretical Issues in Ergonomics Science, 19*, 321-352. 10.1080/1463922X.2017.1315751

Calhoun, G. L., Ruff, H. A., Behymer, K. J., & Mersch E. M. (2017). Operator-autonomy teaming interfaces to support multi-unmanned vehicle missions. In P. Savage-Knepshield & J. Chen (Eds.), *Advances in human factors in robots and unmanned systems: Advances in intelligent systems and computing* (Vol. 499, pp. 113-126). Switzerland: Springer International Publishing. 10.1007/978-3-319-41959-6_10

Draper, M., Calhoun, G., Spriggs, S., Evans, D., & Behymer, K. (2017). Intelligent multi-UxV planner with adaptive collaborative/control technologies (IMPACT). In *Proceedings of the International Symposium on Aviation Psychology.*

Draper, M., Rowe, A., Douglass, S., Calhoun, G., Spriggs, S., Kingston, D., ... & Reeder, J. (2018). *Realizing autonomy via intelligent hybrid control: Adaptable autonomy for achieving UxV RSTA team decision superiority (also known as Intelligent Multi-UxV Planner with Adaptive Collaborative/Control Technologies (IMPACT))* (AFRL-RH-WP-TR-2018-0005). Wright-Patterson Air Force Base U.S.: Air Force Research Laboratory.

Frost, E. M., Bartik, J., Calhoun, G., Spriggs, S., Ruff, H., & Behymer, K. (2018). Creating a well-situated human-autonomy team: The effects of team structure. In *Proceedings of the NATO Human Factors and Medicine Panel, Symposium on Human Autonomy Teaming*, STO-MP-HFM-300.

DEVELOPMENT OF AN INTEGRATED UNMANNED AERIAL SYSTEMS VALIDATION CENTER

Austin G. Decker
School of Aviation and Transportation Technology, Purdue University
West Lafayette, IN
John H. Mott
School of Aviation and Transportation Technology, Purdue University
West Lafayette, IN
Robert J. Connor
Lyles School of Civil Engineering, Purdue University
West Lafayette, IN
Ayman Habib
Lyles School of Civil Engineering, Purdue University
West Lafayette, IN
Darcy M. Bullock
Lyles School of Civil Engineering, Purdue University
West Lafayette, IN

Unmanned Aerial Systems (UAS) have the potential to drastically change how civil infrastructure is inspected, monitored, and managed. Deployment of UAS in areas such as bridge inspection and accident reconstruction will likely have far-reaching impacts and evolve over time, with new uses and users emerging as technology matures. However, with any new technology, limitations exist until new protocols are established, and industry must move forward with an appropriate level of caution. For example, statements regarding the ability of a UAS to replace a human bridge inspector are frequently observed in trade magazines, presentations, and in the literature, though no objective tests or standards exist in order to substantiate the claims. With no standard tests to verify such claims, agencies are left to rely upon vendors' promotional material when making decisions about UAS deployment. The Joint Transportation Research Program at Purdue University is working with the Indiana Department of Transportation and other state departments of transportation to develop an Integrated Unmanned Aerial Systems Validation Center that will create a structured validation process for civil UAS operators. This project will conduct a beta version of the validation center at Purdue University's Center for Aging Infrastructure (CAI) and the Steel Bridge Research, Inspection, Training, and Engineering Center (S-BRITE). Stakeholders including engineers, emergency response personnel, academics, and pilots will work together to determine the appropriate performance criteria needed to validate related civil UAS operations.

Literature Review

The Federal–Aid Highway Act of 1968 created the National Bridge Inspection Program to address growing concerns of aging infrastructure in the United States. Federal requirements mandated that states inspect public bridges that exceed 20 feet every 24 calendar months to collect data on the composition and condition of these structures (USDOT, 2007). A majority of these inspections are conducted by inspectors using non-destructive techniques to visually inspect the structures. To conduct these inspections, inspectors use Under Bridge Inspection Vehicles (UBIV) or commonly known as "snooper trucks" (MnDOT, 2018). UBIVs are expensive vehicles that can range in cost from $500,000-$1 million, and which require lane closures to operate. Closing lanes on roads and highways introduces a significant safety threat to inspectors and commuting traffic. While there are no statistics on the effect of UBIVs on traffic flow and accident rates, incidents and accidents from using these vehicles are not uncommon.

Unmanned Aerial Systems (UAS) have the potential to drastically change how civil infrastructure is inspected, monitored, and managed. In the context of this research project, a UAS is comprised of an Unmanned Aerial Vehicle (UAV), the imaging or scanning technology it carries, and the pilot or crew. Recent technological advancements in unmanned aviation and imagery capabilities have made these aircraft more useful than ever before. UAVs have the unique ability to fly into confined spaces that normally require bridge inspection vehicles and inspection crew. Implementing UAS for bridge inspections could reduce or remove the need for expensive UBIVs and the related operating costs.

Many state transportation departments are beginning to invest in research to determine the effectiveness of UAS for bridge inspection applications. The Minnesota Department of Transportation (MnDOT) has launched a multi-phase project to implement UAS for bridge inspections. MnDOT has used UAS to inspect 39 bridges throughout the state, with plans to expand UAS use. MnDOT has citied up to 40% in cost savings and an increase in inspection deliverables (Lovelace, 2018). Programs like these are committed to discovering the ability of UAS to replace a manual inspection of bridges and other civil structures.

Purpose of an Integrated UAS Validation Center

As UAS technology continues to grow, it is important to note that there are still limitations that exist until new protocols are established to ensure the industry moves forward with the appropriate level of caution. Many statements have been made in trade magazines, presentations, and literature that unmanned aircraft have the complete ability to replace humans in bridge inspection. Despite these claims, there are currently no objective tests or standards to substantiate them. Without any standards to verify these claims, agencies are left to rely on the promotional material when making decisions on UAS integration in bridge inspections.

Purdue University has created a Pooled Fund Study in collaboration with the Indiana Department of Transportation and six other state DOTs to develop an UAS validation center. Faculty and students from Purdue's School of Aviation and Transportation Technology and College of Engineering are creating the validation center. Each school is providing subject matter experts in aviation and civil engineering to develop performance criteria. The center is developing basic standards, protocols, and testing requirements applicable to all UAS utilized in the inspection of civil infrastructure. Currently, the use of UAS for civil engineering applications is completely unregulated. This project is in the process of creating validation criteria to set standards for UAV operators conducting bridge inspections. In order to identify performance criteria, Purdue has hosted stakeholder workshops, with participants consisting of engineers, owners, DOT representatives, and pilots. The ultimate goal of this validation center is to create an accredited course to properly train and certify UAS bridge inspectors.

Performance Criteria

As previously noted, Unmanned Aerial System includes the pilot, aircraft, and imaging capabilities. In order for a specific operator to be validated, that operator should demonstrate appropriate skills in all aspects of operating UAS for the particular bridge inspection application. Evaluation criteria for pilots include many of the same criteria for pilots of manned aircraft. Proper aeronautical knowledge of weather, airspace, air traffic control phraseology, aerodynamics, etc., should be expected of UAV operators. Additional technical subject areas such as crew resource management, aeronautical decision making, and flight planning will be applied to the UAS operation. Pilots will also be expected to demonstrate their ability to successfully conduct fundamental and task-specific flight maneuvers.

The aircraft itself must be able to handle the environmental factors related to operating in close proximity to structures (Figure 2). Electromagnetic interference from structures can disrupt GPS reception and signaling used by aircraft flight control systems. Aircraft that are heavily reliant on GPS for positioning information may be unacceptable for bridge inspection applications due to the frequent loss of GPS reception that can occur near large metal structures.

Safety must remain paramount for all UAS operations. Emergencies and abnormal flight situations will be introduced to operators to validate that appropriate contingencies have been considered and may be safely executed. In many urban environments, multipath interference can disrupt connections between aircraft controllers and the UAV. To demonstrate a UAS's ability to handle a loss of connection event, Faraday-cage devices are being developed that will simulate the effect such signal losses could have on a UAS.

Pilots and their aircraft must be able to demonstrate the ability of operating in adverse weather conditions such as high winds, extreme temperatures, and light-to-moderate precipitation. Wind moving over and under bridges creates turbulent airflow, and the ability to compensate for this air disruption must be compensated for by the UAS. All of these environmental factors are currently experienced by inspectors using traditional UBIV's; as such, UAS operators should be able to match their capabilities.

Throughout many inspection research projects, a common concern tends to be the variability of inspection results between inspectors. Previous research has highlighted this variability by comparing crack detection rates against various human factors for bridge inspectors (Campbell, 2019). The results of this research are shown in Figure 1. The goal of UAS for bridge inspections is to use sensors and algorithms to remove this variability. Operators will demonstrate their ability to use on-board sensing and imagery equipment to detect fatigue cracks, impact fractures, etc. When applicable, technology such as thermal imagery and LIDAR should be used in order properly detect and document structural defects beyond the visual spectrum. Data collection from inspections is crucial and the UAS should be able to properly identify and document structural deficiencies in the structure.

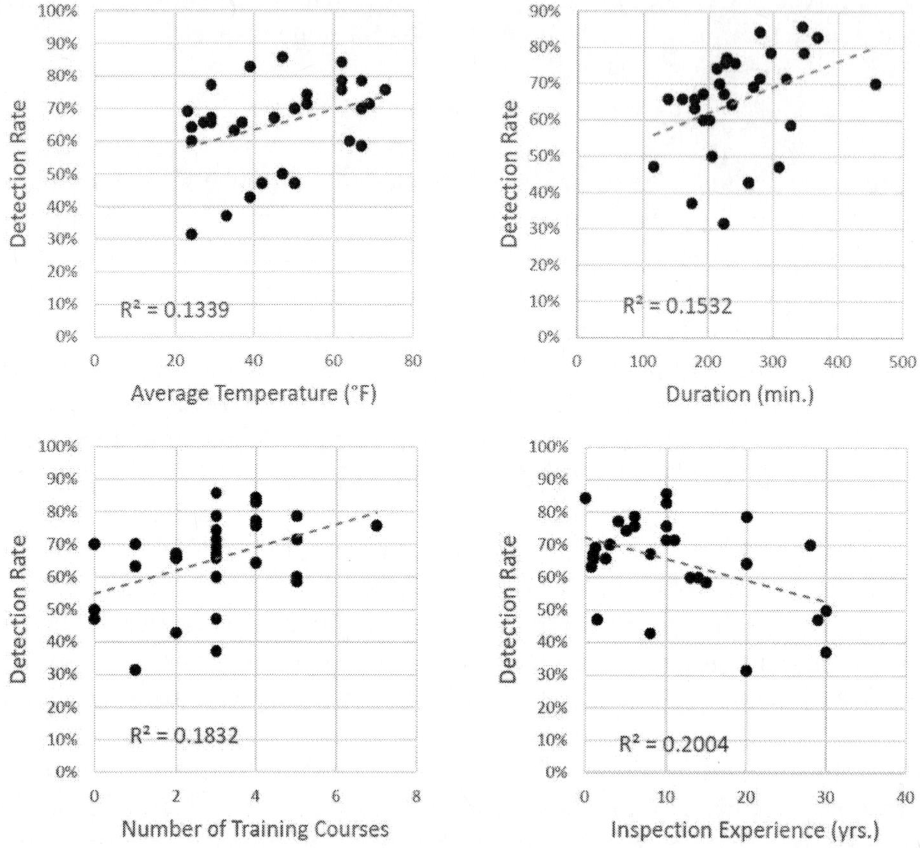

Figure 1. Human and environmental factors to detection rates (Campbell, 2019).

Figure 2. UAS bridge inspection training.

Facility for Validation Center

Purdue University's Center for Aging Infrastructure (CAI) and the Steel Bridge Research, Inspection, Training, and Engineering (S-BRITE) are being utilized in the creation of the UAS validation center. Purdue is uniquely equipped to perform the research described herein in part because of the existence of these facilities. S-BRITE is a multi-acre gallery of full-scale bridge structures, portions of complete structures, and individual components, with a host of common and uncommon details found among steel bridges (Figure 3). The S-BRITE Center provides the ability to inspect real world structures without complicating external factors such as costly traffic control requirements. Through CAI, engineers have documented all of the structural defects on each specimen. This information will be compared to the findings of UAS operators participating in the training and validation process to evaluate UAS performance.

S-BRITE is located 0.7 nautical miles from the Purdue University Airport. Purdue's airport is the second busiest airport in the State of Indiana and thus introduces many challenges to UAS operations at S-BRITE. Fortunately, Purdue has worked with the Federal Aviation Administration to obtain a Certificate of Authorization, giving the S-BRITE facility unique approval to operate unmanned aircraft for testing and validation purposes.

Figure 3. S-BRITE Facility.

Future Research

The utilization of Purdue University's facilities opens opportunities for future research in UAS applications. A major component of bridge inspections is the ability of the UAS to operate in urban environments near structures. Electromagnetic and multipath interference in these environments can restrict an unmanned aircraft's ability to operate. Further research into mitigation of multipath interference will be conducted through this center (Mott & Bullock, 2018). Turbulent airflow over bridges and other structures can produce many negative effects on aircraft stability; the extent of these effects will need to be researched, as well. To complete bridge inspections, aircraft design will have to be altered to accommodate capabilities such as inspecting directly underneath bridges and other structures. As the validation center continues to grow, so will Purdue University's commitment to further research UAS applications in civil infrastructure.

References

Gee, K. & Henderson, G. (October, 2007). Highway bridge inspections. *U.S. Department of Transportation*. Retrieved from https://www.transportation.gov/content/highway-bridge-inspections

Lovelace, B. (July, 2018). Improving the quality of bridge inspections using unmanned aircraft systems. *Collins Engineers, Inc.* Retrieved from http://www.dot.state.mn.us/research/reports/2018/201826.pdf

Bliss, L. (January, 2018). To care for aging bridges, Minnesota taps the power of drones. *Citylab*. Retrieved from https://www.citylab.com/transportation/2018/01/to-care-for-aging-bridges-minnesota-taps-the-power-of-drones/551339/

Campbell, L. (January, 2019). Inspection variability. TPF-5(387) Meeting. 9 January 2019. West Lafayette, IN.

Mott, J. H., & Bullock, D. M. (2018). "Estimation of aircraft operations at airports using Mode-C signal strength information," *IEEE Transactions on Intelligent Transportation Systems*, *19*(3), pp. 677-686. doi: 10.1109/TITS.2017.2700764

The Joint Tactical Air Controller: cognitive modeling and augmented reality HMD design.

Christopher D. Wickens, Colorado State University,
Gaia Dempsey, 7th Future, Los Angeles
Andrew Pringle, Trinity College Ireland,
Lucas Kazansky, Niantic Labs, San Francisco
Stefanie Hutka, Adobe Inc., San Francisco.

This paper describes the design and model based evaluation of DARSAD, an augmented reality head mounted display for the joint tactical air controller (JTAC), who manages and directs fire from air assets near the battlefield. Designs, based on 6 principles of attention, memory and information processing are produced for various phases of JTAC operations including target identification and airspace management. The different design candidates are evaluated and compared based on how they "scored" in adhering to model predictions, when those models were based on the above principles. Display designs, principles, models and the evaluation process are all described here.

The job of the joint tactical air controller (JTAC) near the battlefield is to integrate information about enemy attack units and nearby friendly forces and direct aircraft equipped with weapons to neutralize the enemy via close air support (CAS), while also safely coordinating and routing air traffic (USMC, 2014; Wickens et al., 2018). Thus, a substantial portion of the JTAC's job resembles that of the air traffic controller in a highly unstructured airspace. Because the JTAC must operate in a mobile environment and often on foot, in order to support such multi-tasking and information integration, we harnessed the technology and principles of head-mounted display design from aviation rotorcraft operations (Wickens Ververs & Fadden, 2004). Furthermore, becaused of the geospatial environment in which the JTAC operates, and the need to identify and locate objects within that 3D space, we have exploited augmented reality (AR), in order to provide pointers to, or attach labels to, entities within that environment. Our system is labeled DAQRI Augmented Reality Synthetic advanced display, or DARSAD HMD.

Design

The challenge of designing the HMD for the JTAC is that the typical mission must proceed through most or all of 12 different phases. While these are described in detail in USMC (2014, 2016) and in Wickens et al., 2018, in brief, a higher level description of multiple phases involves:

- A. Identifying targets and their locations on the ground
- B. Bringing in air support from more distant bases and "stacking" them at the ready for attack (many similarities with air traffic control)
- C. Developing and communicating a "game plan" of attack, to friendly ground forces and air assets.
- D. Coordinating the actual air attack, assuring that the pilot is focused on the correct target(s)
- E. Assessing the results of the attack
- F. Routing air assets back toward their base.

Each phase has somewhat different information needs, some focusing on the battlefield, some on air assets, some on digital data bases, and some on various aspects of all three of these. Thus it is not appropriate to design a single "one size fits all" display; but neither is it optimal to design 12 different formats, each optimized for the task performed during the phase. Such a design will lack overall consistency, and can also lead to possible disorientation, with discrete switches of displays from one phase to another. Instead, our approach was to try to find a compromise between these two extremes, and hence provide visual momentum (Woods, 1984) between succesively viewed formats.

Shown in Figure 1, is the Display for initial (A) target selection, illustrating the momentary field of view occupied by the display, in the center, and the information in augmented reality i.e., the AR 3D grid structure that can be viewed by rotating the head. The natural terrain is visible behind, so as to superimpose the 20 X 30 degree FOV

HMD over a distant part of the visual field. This will be described in more detail below. Shown in Figure 2 is a zoom-in view of the display, in this case, specifically designed for (B) airspace management. The features of the display variations shown in Figures 1 and 2 will be described in greater detail later. However one key feature, in both, and all other display variations is the central rectangle in the middle. This is an unobstructed "no clutter" region designed to contain no added symbology, that supports the most sensitive possible view of the far domain, in order to support far domain situation awareness and high acuity target search and confirmation. The SA support region will be viewed when the eye is scanning straight ahead. A rotation of the eyeball of 2 degrees to either side, from this straight ahead axis will be sufficient to gain HMD-displayed information.

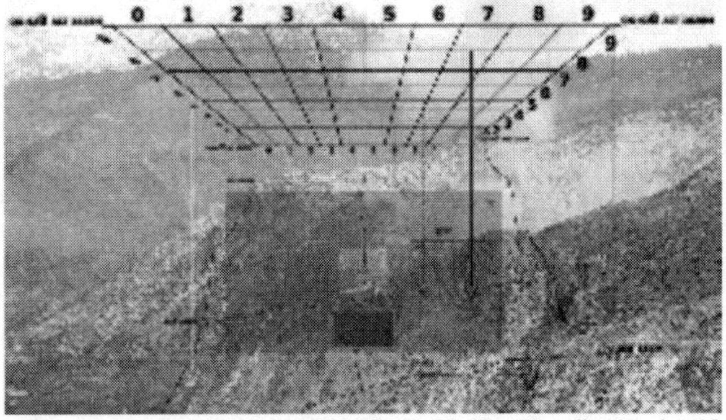

Figure 1. Example of the DARSAD HMD. The display itself is seen in the center rectangle, the grid imagery outside can be brought into view by head rotation. Vertical arrows point to ground location of targets (red arrows) or other entities. The AR gridlines in the actual display are rendered in low contrast white, to minimize clutter.

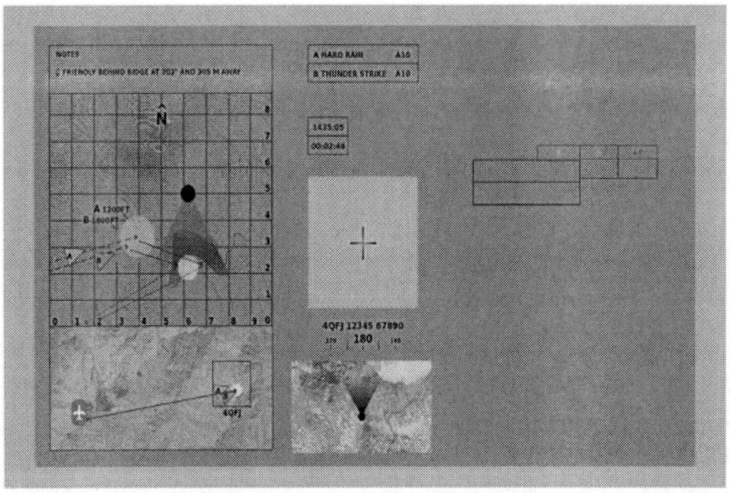

Figure 2. Airspace management design of the DARSAD. No far domain is visible. Note the no-clutter SA window in the middle. The three maps will be discussed below.

Our approach was twofold: (1) to design a single general format for a set of temporally adjacent phases; and to provide only minor modifications between phases within the set. For example item C in the bulleted list above, actually contains three of the 12 JTAC phases within it. (2) to endeavor to keep some general properties constant across all 12 phases (and hence all the different sets), in order to provide visual momentum. Thus it was necessary to design for the multiple tasks that must be performed by the JTAC, some concurrently (e.g., maintaining situation awareness of the battlefield, while communicating to air assets), and to consider the many different information processing demands of each task. Because of the multi-task nature of the DARSAD HMD, we chose to configure each display by adhering to multiple *principles* of display design, and then to evaluate a display's degree of adherence to the principle by a series of computational models, as described as follows.

1. *Situation Awareness Primacy*. As noted, a major rationale for using a transparent optical display in the first place is to keep the far domain in view, as this is the sole source for noticing dynamic changes in information on the battlefield. But our design expanded upon this feature to include the "protected zone" in the middle of the HMD (see square with crosshair in figure 2) that is never obstructed by display imagery, other than a center reticle that can be used as a component of digital target designation (center the reticle on a target and "click").

2. *Minimizing Scanning/Information Access Effort*. Scanning is effortful, and head movements are more so (Wickens, 2014). Hence our goal was to keep most information relatively accessible either on the display itself, or just outside its perimeter, accessible then by a short head rotation to look at a body-referenced location (e.g., as if mounted to a tablet attached to the shoulder). The visual angle distance of information sources from the center of the field of view (the reticle) was generally made proportional to its frequency of use and (Wickens, Vincow et al., 1997; Wickens, 2015).

3. *The Proximity Compatibility Principle*. We also endeavored to keep information sources that needed to be compared; such as a map and the forward view depicted in the map, or a commanded and actual aircraft altitude, as close together (proximate) as possible as dictated by the proximity compatibility principle (PCP: Wickens & Carswell, 1995; Wickens & McCarley, 2008). One direct derivative of the PCP is the very use of AR or conformal imagery, which creates the closest proximity possible between display information and its counterpart in the far domain (Wickens & Ververs, 2004). AR has the significant advantage (over pure spatial proximity) of creating maximum proximity while minimizing visual clutter (see below). Examples of AR to create proximity are seen by the virtual grid, and the target cue arrows, pointing to ground targets in Figure 1.

4. *Maximizing Legibility: The Tradeoffs*. One inevitable downside of superimposed imagery is the potential to mask information in the far domain by display symbology, and to mask information on the display by far domain scenes of high visual density or spatial frequency. We refer to both of these clutter sources as "overlay clutter" (Wickens, Hollands et al., 2013). The first type of clutter is mitigated by the SA protected zone, but is inevitable elsewhere across the display that may contain imagery. Furthermore, the close proximity of elements in a small space (i.e., small text designed to reduce overlay clutter) can create "density clutter" (Beck et al., 2010), by packing elements too closely together. Finally, both small display elements (i.e., reduced font and symbol size) and low intensity symbology (designed to reduce overlay clutter costs to the far domain), are the two elements most responsible for reduced legibility of critical display information (US DOD 1999). Collectively these inevitable costs of superimposition, must be balanced against the HMD benefits to situation awareness and reduced information access effort noted above, and we focus a great deal of design and test attention on the quantitative tradeoffs between them, exploiting the "sweet spot" in the tradeoff where possible.

5. *Frame-of-Reference Transformations (FORT)*. Much of the JTAC's operations require 3D spatial cognition: Where am I relative to my aircraft, and relative to the target? Where currently are and where will the aircraft be relative to the target, to terrain hazards, to ground hazards such as surface-to-air missiles and to each other? Each of these spatio-geographical elements may be represented in a different frame of reference, and the transformations between these can create high workload and error (Wickens, Vincow & Yeh, 2005; Wickens Thomas & Young, 2000); hence we seek ways to minimize these transformations, with the prototypical example being to superimpose the 3D grid directly onto the forward view, via AR imagery in Figure 1. The costs of such transformations in 3D displays can be quantified via a computational model (Wickens Keller & Small, 2011).

6. *Minimizing Working Memory Load*. In certain phases, much of the JTAC's tasks involve communications, often of somewhat arbitrary digits, codes or acronyms indicating geographic position, weapon selection, or codes representing target designation information related to, for instance, laser designators or IR sparkle. We endeavor to reduce the vulnerabilities of confusion and memory failures, by support from the visual display of information to be communicated and received. This is accomplished via voice-to-text automation that displays the codes spoken by the JTAC visually within the display interface.

Model based Display Evaluation.

Each of the principles above were leveraged in designing DARSAD phase displays (our efforts focused primarily on designs for A,B and D). However the principles were again leveraged to perform relatively low cost *evaluation* of the suitability of different designs for each phase; a great service performed by computational models (Byrne & Pew, 2005). To accomplish this for each principle, we either derviced, or used an existing computational model of how the level of performance would be driven by principle-relevant factors in the design. As four examples:

1. A model of working memory would predict that the quality of performance is directly proportional to the number of items to be retained (between encoding and response: communications or entry into a keyboard) times the length of required retention.
2. Some clutter models already exist (Beck et al., 2010), and the quality of search and reading performance can be inversely and linearly related to the computed abount of displayed clutter.
3. The proximity compatibility model penalizes a dispalty directly proportional to the number of items that need to be integrated or compared, times the average spatial separation between the compared items. Note that this separation is, by definition, = 0 for comparing a far domain item with its AR-displayed counterpart in DARSAD e.g., comparing a far domain target with a circular surrounding cue.
4. The SA support model penalizes displays proportional to the distance of displayed information from the central SA-window, weighted by the frequency of use of such items.

Greater details of the computational elements of all six models can be found in Wickens et al., 2018, or by contacting the first author.

Given the tasks that are performed in a given phase, each of the seven models can then provide a "score" for the displayXtask combination within that phase revealing the extent to which the display design serves the task in question (if more than one task is performed within the phase, the score can be averaged for the model). Thus a given model could generated either a penalty or its complement, a "figure of merit" of how poorly (or well) the display design serves the principle in question.

In principle then, an overall phase-related figure of merit for a design could be derived by summing (or averaging) the model scores across the six models. We refer to this as a "meta-model". One final step in this process is required however, because not all model principles apply equally to all phases. As an example, target identification (A above) places hevy demands on visual acuity through the SA window. So the SA model (Hooey et al., 2010) is quite relevant, but working memory considerably less so. In contrast, communications of an attack plan to ATC and to friendly ground troops places heavy demands on working memory (as do all communications; Morrow et al;, 2003), but far less on visual acuity and target search. Thus we add one more step of assigning an importance weighting (multiplier) of the degree of relevance of a model to a given phase. Here we chose three levels: 0. .5 and 1.0. From this we can create the figure of merit prediction of the weighted meta model.
With the above, we are equipped to derive a total figure of merit (FOM) to each designXphase combination.

However our interest was less in deriving an absolute FOM than it was in a *relative* FOM comparison between two display alternatives (e.g., AR versus non-AR presentation of geo-spatial information; or closely clustered versus more widely dispersed information). The reason that such comparisons are vital is because principles often trade-off against each other in design. For example a closely clustered display can impose penalties of clutter, but a widely dispersed one can impose penalties of information access and scanning. Computational models can inform the design/evaluation as to "who wins" a competition, for example between cluttered cluster display and the dispersed scan-intensive display, and by "how much".
We report below, the results of two such comparisons of alternative prototypes for a given display. First, for airspace management (see Figure 1b), we compared a version with a 2D plan view situation map (the large map on the left side of Figure 2) with a 3D perspective map. Figure 3 presents schematics of the perspective map designs. We also note the other two maps in the display of Figuren 2. At the bottom left is the small scale map of routing from the base to the battlefield, only displayed for phases A and F. At the bottom center is a rotating map, always depicting the momentary view forward at the top, with the cone subtending the momentary field of view of the HMD (Aretz, 1991; Olmos Wickens & Chudy, 2000).

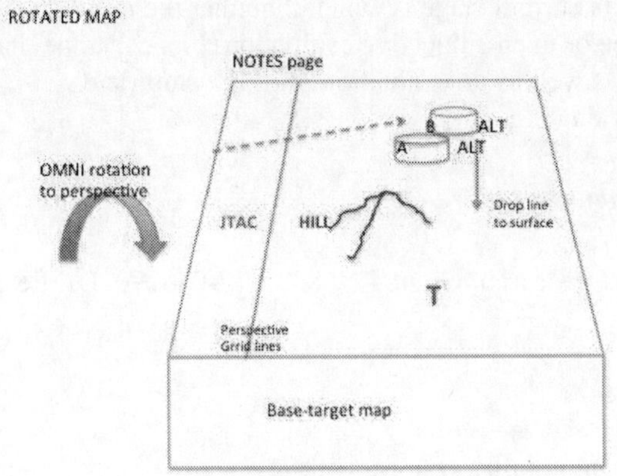

Figure 3. Schematic rendering of the 3D perspective version of the airspace management map Supporting phase B. The map can be rotated in depth, as shown by the blue arrow, to disambiguate distance and altitude of displayed aircraft, if required. The cylinders represent "stacks" where a given air asset may circle in a holding pattern.

Second, for target identification and acquisition, we compared the 3D AR grid rendering in Figure 1, with a second version in which there was no superimposed AR grid, where coordinates must be accessed from the screen perimeter. Figure 4 presents the two Tables in which the difference in six model scores, between the two designs are presented in the first column, the importance weights for each model are presented in the second, and the weighted model in the third. Shown in the bottom row is the meta-model score indicating the predicted percentage advantage of the candidate display (3D perspective airspace, gridline) over the less advanced display.

% advantage of 3D airspace map over 2D
For airspace management

Model	Importance weight of principle	% advantage
SEEV SA	.5	0
Proximity compatibility	.5	100
Legibility & contrast	.5	0
clutter	.5	0
Frame of reference	.25	12.5
Memory	.5	0
Meta-model		112.5%

% advantage of gridlines over no gridlines
for target ID phase

Model	Importance weight	% advantage
SEEV SA	.5	0
Prox compat	1	300
legibility	1	0
clutter	1	-50
Frame of reference	1	200
memory	.5	0
Meta model		450%

Figure 3: Relative percentage advantage of rotatable perspective airspace management display (left) and AR gridlines on target identification display (right).

The bottom line (literally and figuratively) of Figure 4 is that in both cases, the more innovative display is predicted to have performance advantages and particularly so in the case of the AR gridlines. It is noteworthy in the latter case that there does exist a penalty for clutter; but this is more than offset by the AR benefit to proximity compatibility, as described above.

Conclusions and Limitatations.

The value of computational modeling for evaluation of display prototypes in the complex JTAC support system is demonstrated here. Of course the limitation of this research, at its current stage is evident: neither the individual models nor the meta-model are validated, either via performance or even subjective evaluation. It is our hope that such evaluation can be accomplished for the DARSAD-HMD, as well as an evaluation, through empirical validation, of the overall meta-modeling approach, to system evaluation.

Acknowledgments

This research was supported by a grant from the Office of Naval Research, Grant # N68335-17-C-0290. Dr. Peter Squire was the scientific technical monitor.

References

Aretz, A. J. (1991). The design of electronic map displays. *Human Factors, 33,* 85–101.

Beck, M.R., Lohrenz, M.C., & Trafton, J.G. (2010). Measuring search efficiency in complex visual search t asks: global and local clutter. *Journal of Experimental Psychology: Applied, 16,* 238-250.

Byrne., M. & Pew, R. (2005) A history and Primer of Human Performance Modeling. In F. Durso (ed) *Reviews of Human Factors, Volume 5*. Santa Monica, Cal.: Human Factors Society.

Hooey, B. L., Gore, B. G et al (2010). Modeling Pilot Situation Awareness. In C. Cacciabue et al., (Ed.) *Human Modelling in Assisted Transportation*. Heidelberg: Springer.

Olmos, O., Wickens, C.D., & Chudy, A. (2000). Tactical displays for combat awareness: *The International Journal of Aviation Psychology, 10*(3), 247- 271.

Morrow, D. G., Ridolfo, H. E., Menard, W. E., Sanborn, A., Stine-Morrow, E. A. L., Magnor, C., ... Bryant, D. (2003). Environmental support promotes expertise-based mitigation of age differences in pilot communication tasks. *Psychology and Aging 18*, 268–284.

US Department of Defense. (1999). Department of defense design criteria standard: Human engineering. Retrieved f rom http://everyspec.com/

United States Marine Corps. (2014). TACP: TACSOP. Retrieved from http://www.mawts.usmc.smil.mil/

United States Marine Corps. (2016). JFIRE. Retrieved from http://www.alsa.mil/mttps/jfire/

Wickens, C.D. (2015). Noticing events in the visual workplace: The SEEV and NSEEV models. In R. Hoffman & R. Parasuraman (Eds). Handbook of Applied Perception. Cambridge, U.K.: Cambridge University Press.

Wickens, C.D. (2014), Effort in human performance and decision making. *Human Factors, 56(8),* 1329-1336.

Wickens, C.D., & Carswell, C.M. (1995). The proximity compatibility principle. *Human Factors, 37*(3), 473-494.

Wickens C.D., Dempsey, G., Pringle, A., Kazanski, L., & Hutka, S. (2018). Developing and evaluating an autmented reality interface to assist the joint tactical air controller by applying human performance models. Proceedings 2018 annual meeting of the Human Factors & Ergonomics Society.

Wickens, C.D., Hollands, J, Banbury, S. & Parasuraman, R.. (2013). *Engineering psychology and human performance* (4th ed.). Upper Saddle River, NJ:Pearson.

Wickens, C.D. Keller, J. & Small, R (2010) Development of the Frame of Reference Transformation Tool (FORT). *Proceedings of the 54th Annual meeting of Human Factors Society*. Santa Monica, CA: Human Factors.

Wickens, C.D. & McCarley, J.M. (2008). *Applied attention theory.* Boca-Raton, FL: CRC Press, Taylor & Francis.

Wickens, C.D., Thomas, L.C., & Young, R. (2000). Frames of reference for display of battlefield terrain and enemy information:. *Human Factors, 42*(4), 660-675

Wickens, C. D., Ververs, P., & Fadden, S. (2004). Head-up display design. In D. Harris (Ed.), *Human factors for civil flight deck design* (pp. 103-140). UK: Ashgate.

Wickens, C.D., Vincow, M.A., Schopper, A.W., & Lincoln, J.E. (1997). Computational models of human performance in the design and layout of controls and displays (*CSERIAC SOAR Report 97-22*). Wright Patterson Air Force Base, OH: Crew System Ergonomics Information Analysis Center.

Wickens, C. D., Vincow, M., & Yeh, M. (2005). Design applications of visual spatial thinking: The importance of frame of reference. In A. Miyaki & P. Shah (Eds.), *Handbook of visual spatial thinking*. Oxford University Press

Woods, D. (1984). Visual Momentum. *International Journal of Man-Machine Studies, 21,* 229–244.

HAPTIC SUPPORT FOR AVOIDING STATIC AND DYNAMIC OBSTACLES IN UAV TELE-OPERATION

Tomasz Piessens, M. M. (René) van Paassen and Max Mulder
Aerospace Engineering – Delft University of Technology
Delft, The Netherlands

Impoverished sensory input makes tele-operation of Unmanned Autonomous Vehicles a difficult task. Automation support can provide assistance to the operator, but may also produce automation surprise and a risk of loss of situation awareness, when the human operator fails to notice the actions of automation in high-workload situations. Previous work applied haptic feedback based on an artificial risk field to assist in avoidance of static obstacles with a small helicopter UAV. An off-line analysis of that solution shows that it would not be sufficiently effective for the avoidance of dynamic, moving obstacles. A new haptic assist algorithm, based on velocity obstacle theory was developed. This algorithm was first tested in off-line analyses, in scenarios with only static obstacles and in scenarios with a dynamic obstacle added. The off-line analysis showed the feasibility of the algorithm. The algorithm was then ported to a real-time environment and evaluated in pilot-in-the loop simulations, to verify the solutions and investigate the acceptability of the haptic feedback. Results indicate that the velocity obstacle approach works for both stationary and moving obstacles, in most, but not all scenarios.

Remotely operating of a vehicle in most cases is associated with limitations in sensory input (Custers, Oerlemans, & Vergouw, 2015). Operating beyond the line-of-sight operators often rely on camera images only, with limited resolution and field of view (FOV), and lack the natural multi-sensory environment as compared to when they would be on-board the vehicle themselves. To assist these operators in flying a UAV in a cluttered, obstacle-laden environment, a haptic interface for collision avoidance has been developed (Lam, Mulder, & van Paassen, 2007; Ho, Borst, van Paassen, & Mulder, 2018). The system uses a sensor to detect objects in the surrounding, and from this calculates an artificial force field to determine the appropriate haptic force feedback for the control stick (Lam, Boschloo, Mulder, & van Paassen, 2009). This proved helpful in obstacle avoidance, however, the method by which this support system is tuned makes it appropriate to avoidance of *static* obstacles only. Similar approaches, also for static obstacles, have been reported in literature (Murphy, 2012). As far as known, no support system has been reported that incorporates avoidance for *dynamic*, moving obstacles.

With trends in UAV operations it will be likely that multiple UAVs will be operating in close vicinity, making collision avoidance support with moving obstacles all the more relevant. Through an off-line analysis, it was determined that the avoidance logic as used in (Lam et al., 2007) does not properly support the avoidance of a dynamic obstacle. To support safe operation, the obstacle avoidance support should therefore be extended, which is the aim of the current project. To this end, elements from similar approaches to (visual) display design in air traffic control and airborne collision avoidance (Van Dam, Mulder, & van Paassen, 2008; Ellerbroek, Brantegem, van Paassen, & Mulder, 2013; Ellerbroek, Brantegem, van Paassen, de Gelder, & Mulder, 2013; Mercado-Velasco, Borst, Ellerbroek, van Paassen, & Mulder, 2015) were used.

The obstacle avoidance support in (Lam et al., 2007) was limited to the horizontal plane only, and was based on a range sensor that provided distance to obstacles around the UAV, with a resolution of 500 lines covering the full 360 degrees around the vehicle. Since it cannot be assumed that the position of all moving obstacles is known through other surveillance means, a similar (LiDAR) sensor input is also used here. Whereas previous work in airborne conflict resolution and air traffic control support could use surveillance data to determine the location of moving obstacles, and assume a circular Protection Zone (PZ) around each obstacle, here the location of surrounding traffic is to be inferred from LiDAR sensor-like input data.

The paper is structured as follows. First, a very brief explanation of relevant Velocity Obstacle (VO) theory is given, followed by an overview of our Collision Avoidance System (CAS) design. A pilot-in-the-loop experiment and its results are discussed, followed by final conclusions and recommendations.

System Architecture

Velocity Obstacles

The Velocity Obstacle method, here applied only in the horizontal XY plane, calculates permissible vehicle velocities that will avoid dynamic or static obstacles in the environment.

To this end, the *relative* velocities between a moving obstacle and the controlled UAV that would end up in a collision, respectively a path too close to the obstacle, are determined, and mapped to the UAV's absolute velocity space (Mercado-Velasco et al., 2015).

This is illustrated in Figure 1. Here, the vehicle that is under control, vehicle B (with velocity V_B), is crossing paths with another vehicle, A (with velocity V_A). Knowing the velocities V_A and V_B and the minimum distance that the two vehicles should remain separated – a circular disk referred to as the Protected Zone (PZ' in Figure 1) – allows the calculation of all velocities and headings of vehicle B (as it is the vehicle considered under control, but it would work similarly when vehicle A would be considered) that would end up with vehicle B entering the PZ of vehicle A (Van Dam et al., 2008).

What results is the dented gray circle drawn around vehicle B, which shows all possible *and safe* speeds and headings for this vehicle. Possible, as it shows the minimum and maximum speed range as a circle, and safe because *some combinations of heading and speed result in a conflict with vehicle A.*

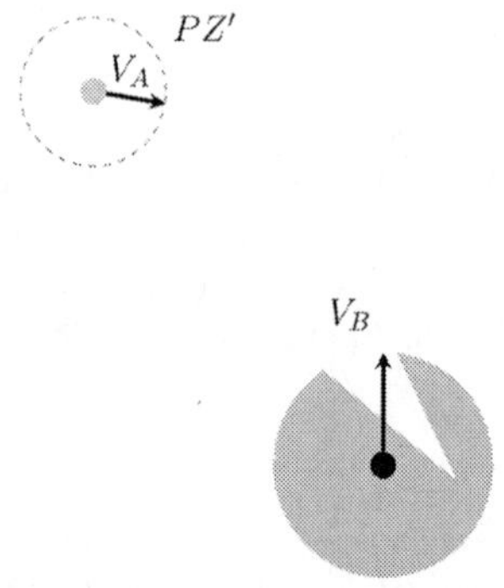

Figure 1: Bird's eye view of two moving vehicles in the horizontal XY plane.

Architecture

A scheme of the total system architecture for the Collision Avoidance System is given in Figure 2, it is composed of the following components:

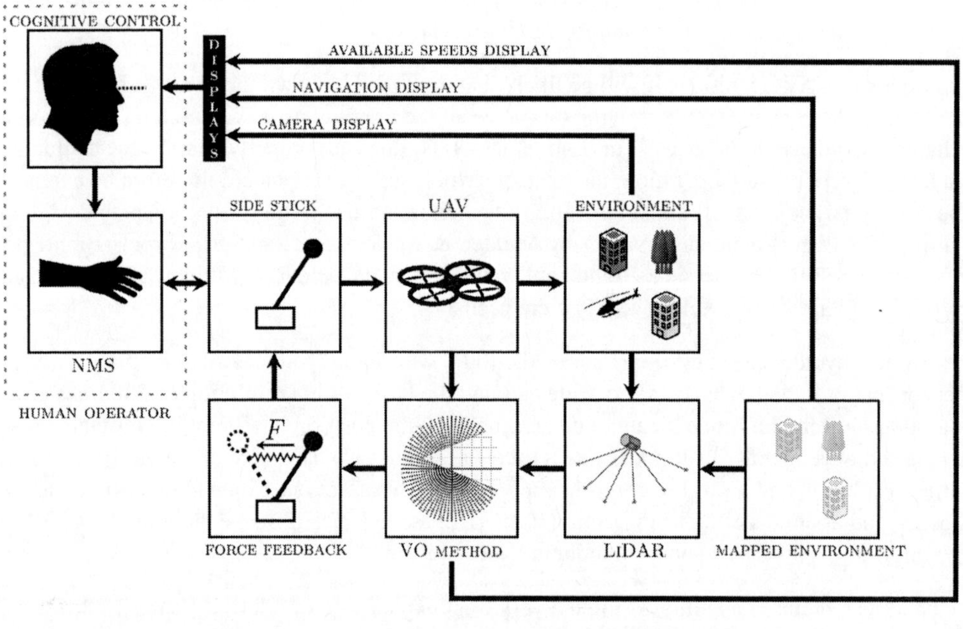

Figure 2: Schematic representation of the haptic Collision Avoidance System.

- The environment is sensed with a (simulated) LiDAR range finder, mapping dynamic and static objects onto a 2D plane, coded as range and bearing from the UAV. The environment is scanned with a resolution of 500 rays up to a distance of 100 $[m]$.
- Using a static map of the environment, measurements resulting from static obstacles are distinguished from the LiDAR range mapping. For static obstacles, the mapping as a velocity obstacle depends only on the vehicle's own velocity. A safety distance is added – as a circular buffer – onto the measurements. For this the super-positioning method for Protection Zones (PZ) from (Damas & Santos-Victor, 2009) is used.
- The static distances, with safety distance, are now translated into a maximum speed in the direction of the detected obstacle. A maximum acceleration/deceleration a_{max} of the controlled UAV in any given direction is assumed, then given a distance d_i to the obstacles in that direction, a modified form of the calculation given by (Damas & Santos-Victor, 2009) is used:

$$v_{max}(i) = \sqrt{f_s \cdot 2a_{max} \cdot d_i + (f_s \cdot a)^2_{max} \Delta T^2} - f_s \cdot a_{max} \cdot \Delta T \quad (1)$$

  Here ΔT is the discrete update time of the LiDAR system. The modification is the inclusion of a safety factor on the maximum possible acceleration, $f_s = 0.5$. This results in a permissible velocity map that accounts for static obstacles only.
- After the previous step, the measurements from the dynamics obstacles remain to be processed. It was originally attempted to determine the speed of dynamic obstacles from the measurements, as detailed for example in (MacLachlan, 2005), however, this proved to be impractical within the scope of the project. As a shortcut, the measurements were matched against a list of objects obtained from the simulation, producing the exact intruder velocity.
- The edges of the detected dynamic obstacles were determined, and using the detected distance, a cone of relative velocities that would bring the two vehicles too close together was determined. A radius of R_{total} is used there, composed of the radius of the own vehicle ($R_{UAV} = 1$ [m]) and a minimum safety distance $R_{pz} = 1.6$ [m], see Figure 3a.
- Using the respective intruders's velocities, the relative velocity cones are translated to the UAV's absolute velocity space, resulting in blocked zones in the permissible velocities, see Figure 3b.
- Given the current speed, the set of permissible speeds is searched for the closest permissible speed V_{safe}. The difference between the current speed and the closest safe speed is expressed as a speed magnitude change ΔV and a yaw angle change $\Delta\psi$, resulting in target normalized stick inputs for lateral and longitudinal control, respectively, x_{stick} and y_{stick}:

$$x_{stick} = \min\left(1, \max\left(\Delta\psi/0.32, -1\right)\right) \quad (2)$$

$$y_{stick} = \min\left(1, \max\left(\Delta V/4, -1\right)\right) \quad (3)$$

  This is rendered to a lateral and longitudinal haptic force with gains $K_{lat} = 2.5$ [Nm] and $K_{long} = 2.86$ [Nm] respectively.

Experimental Evaluation

A limited evaluation was performed with 5 participants, all right-handed, and none had previous experience in UAV tele-operation. The experiment was approved by the TU Delft Human Research Ethics Committee.

Apparatus and scenarios

The algorithm described above was programmed in the simulation used by (Lam, D'Amelio, Mulder, & van Paassen, 2006) in TU Delft's Human-Machine Interaction laboratory. A projected image was given of the front camera view of the controlled UAV, and head-down displays showed a map view based on the fixed obstacle database (Figure 4b) and a view of the available speeds calculation (Figure 4a). Figure 4c shows the set-up. The same vehicle simulation as used by (Lam et al., 2006) was used, and test tracks were generated from the same database.

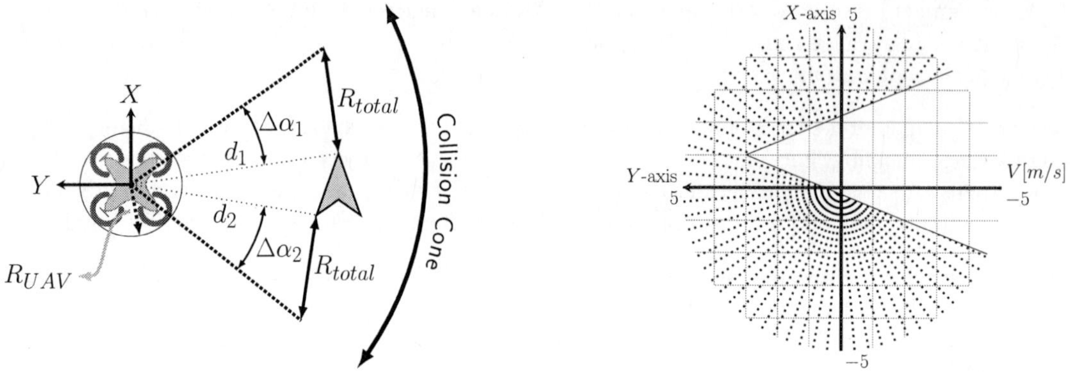

(a) The relative velocities blocked off by the intruder are constructed by adding a radius R_{total} to points detected at the edges of the obstacle.

(b) After using the intruder's speed to translate the blocked relative velocities to the UAV's velocity space, a set of permissible velocities remains.

Figure 3: Calculation of permissible velocities in the presence of a moving obstacle.

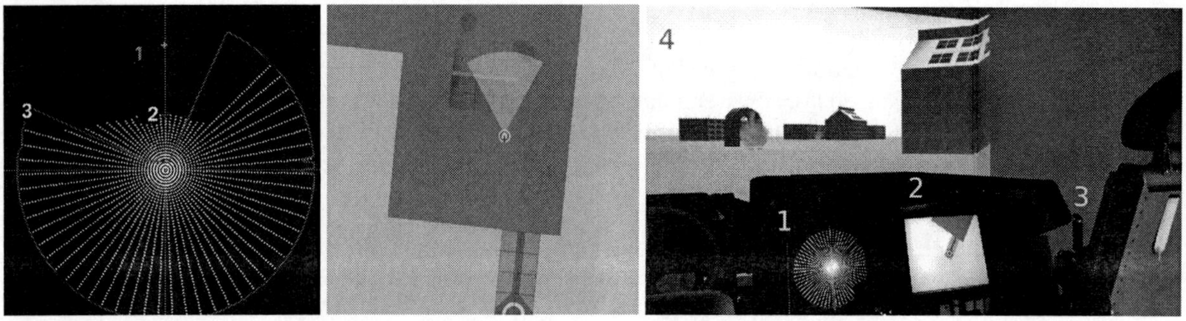

(a) Speeds display

(b) Map view

(c) Laboratory set-up

Figure 4: Experiment displays (left, middle) and laboratory set-up (right). In the right subfigure, '1' indicates the speeds display, '2' the map view, '3' the side stick, and '4' the outside view (simulated camera image).

Dynamic objects were added to the simulation and visualized as UAV's. These would appear and start moving when the operator controlled UAV would pass a trigger line in the database. Three simulated intruder traffic situation were used, illustrated in Figure 5:

Intruder 1 This obstacle appears around the corner when the operator is performing the maneuver of flying along a wall. The goal was to test the effectiveness of the CAS on obstacles that appear suddenly and become visible to the operator on the very last moment.

Intruder 2 This obstacle would cross paths with the UAV in the open field and would lead to a collision if the operator wouldn't adjust heading or speed. The obstacle was in the operators FOV the whole time.

Intruder 3 This obstacle would approach the UAV head on, while flying through a corridor. The operator would have ample time ($\pm$ 15 $[s]$) to get out of the corridor before the obstacle would actually cause a collision. This scenario was chosen to check the behavior of the CAS when a conflict is created in a confined space.

Experiment design

Participants were instructed to fly the given course and avoid collisions with buildings or moving obstacles. The course was marked with waypoints that were represented by smoke plumes through which the operator had to navigate the UAV (Ho et al., 2018). In addition, participants were asked to fly through the waypoints as close as

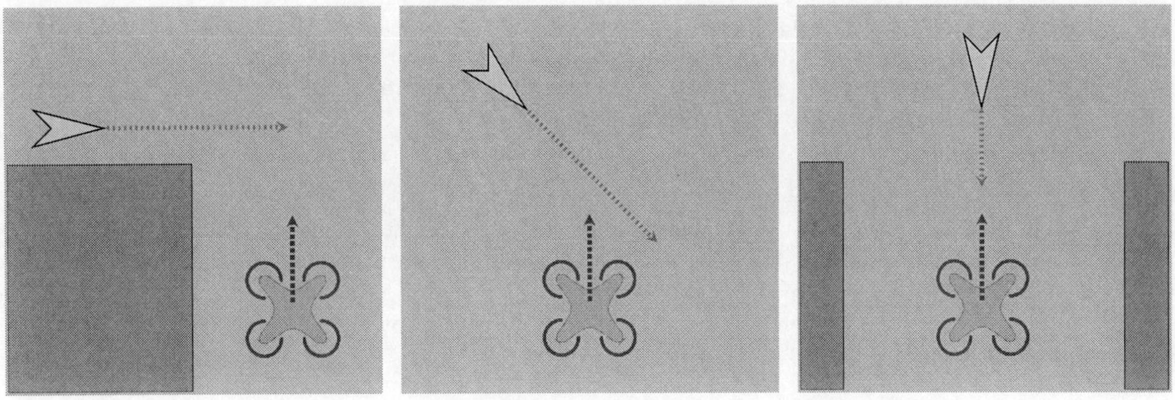

(a) Situation 1 *(b)* Situation 2 *(c)* Situation 3

Figure 5: The three flight situations with the intruder aircraft: (1) around the corner, (2) crossing in the open field, and (3) head-on in a corridor.

possible and to complete the course as fast as possible. If the UAV would have a collision, the screen would freeze for 5 seconds and the UAV would be placed back into the position just prior to the collision. Two training runs were performed prior to the start of the experiment. Each participant flew two measurement runs with each three sectors, with each sector having several static obstacles and one intruder aircraft. Runs with and without haptic feedback were balanced between subjects, the permissible speeds display was shown in all runs.

It was hypothesized that the haptic feedback would increase safety of the runs, to be verified from the number of collisions, that it would make the task easier, to be tested from a mental workload rating, but that the physical workload, to be determined from level of exerted force, might increase.

Results

Safety

Safety was measured by counting the number of collisions and measuring the minimum separation D_{min} between the UAV and the moving obstacles. Even with the small scale of the experiment, a nonparametric Friedman test indicated that the number of collisions differed significantly ($p < 0.05$). The collisions were further investigated, and it seems that all collisions due to the moving obstacle with the CAS active were with the third intruder scenario, in which the intruder was encountered in a small corridor, indicating that the CAS was not sufficiently supporting in that scenario.

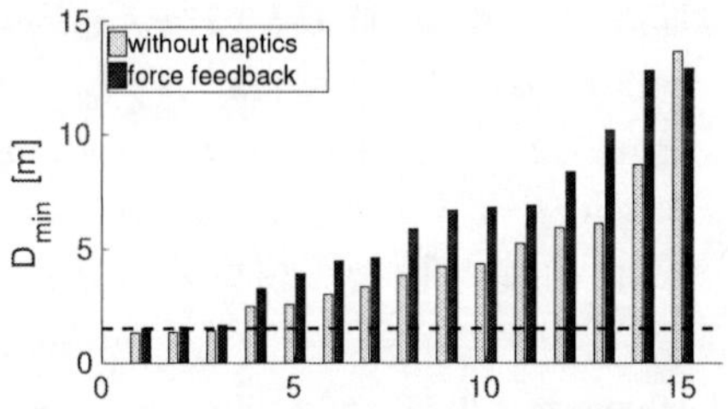

Figure 6: Minimum separation between UAV and intruder, sorted in size; not paired per intruder, for CAS active and inactive.

The minimum distance from intruder was investigated to determine the effect of the CAS on the flight, see Figure 6, which clearly shows that the CAS had a significant beneficial effect on avoidance of the intruder. Note that the figure shows the results of all fifteen runs (five subjects, three intruder scenarios).

Effort

The standard deviation of the exerted force in longitudinal direction and in lateral direction was determined and averaged for the runs with and without CAS. For the longitudinal input the stick force was increased by a considerable amount, 0.76 [Nm] versus 0.45 [Nm] without CAS. Workload was measured through the NASA TLX. No significant differences were found for the mental workload.

There was one "reverse parking" maneuver in each of the sectors of the scenario, and in the scenario with intruder #3 the UAV had to be reversed to avoid the intruder. These maneuvers in particular were associated with high stick forces, with the participants' input often conflicting with the input from the CAS, indicating that tuning and/or logic of the CAS were not appropriate for these situations.

Conclusions

The developed CAS supports operators in avoiding collisions with both static and dynamic obstacles, as apparent from the results of a limited evaluation. Operation in narrow spaces, particularly when travelling backwards, however, are not properly supported, and in these conditions the CAS tended to produce large and fluctuating lateral forces on the stick. No significant differences in workload were measured, however physical control forces and thus physical workload were higher.

Further optimization of the CAS, especially directed at maneuvers in narrow spaces, and investigation of scenarios where *both* UAV would be fitted with such a CAS, is warranted. Within the project there was no room to properly determine intruder velocity from the measured sensor values. Sources indicate that LiDAR sensors are able to perform this calculation, but these developments are not reported in open literature (MacLachlan, 2005).

References

Custers, B. H. M., Oerlemans, J. J., & Vergouw, S. J. (2015). *Het gebruik van drones*. Wetenschappelijk Onderzoek- en Documentatie Centrum (in Dutch).

Damas, B. & Santos-Victor, J. (2009). Avoiding moving obstacles: The forbidden velocity map. In *Proc. ieee/rsj int. conf. on intelligent robots and systems*.

Ellerbroek, J., Brantegem, K. C. R., van Paassen, M. M., de Gelder, N., & Mulder, M. (2013). Experimental evaluation of a coplanar airborne separation display. *IEEE Trans. on Human-Machine Systems*, *43*(3), 290–301.

Ellerbroek, J., Brantegem, K. C. R., van Paassen, M. M., & Mulder, M. (2013). Design of a coplanar airborne separation display. *IEEE Transactions on Human-Machine Systems*, *43*(3), 277–289.

Ho, V., Borst, C., van Paassen, M. M., & Mulder, M. (2018). Increasing acceptance of haptic feedback in UAV teleoperation by visualizing force fields. In *Proceedings of the IEEE international conference on systems, man, and cybernetics* (pp. 3023–3028). Myazaki (Japan).

Lam, T. M., Boschloo, H. W., Mulder, M., & van Paassen, M. M. (2009). Artificial force field for haptic feedback in UAV tele-operation. *IEEE Transactions of Systems, Man & Cybernetics, Part A*, *39*(6), 1316–1330.

Lam, T. M., D'Amelio, V., Mulder, M., & van Paassen, M. M. (2006). UAV tele-operation using haptics with a degraded visual interface. In *Proceedings of the IEEE conference on systems, man & cybernetics (IEEE-SMC)* (pp. 2440–2445). Taipei, Taiwan.

Lam, T. M., Mulder, M., & van Paassen, M. M. (2007). Collision avoidance in UAV tele-operation with time delay. In *IEEE international conference on systems, man and cybernetics* (pp. 997–1002). Montreal, Canada: IEEE.

MacLachlan, R. (2005). *Tracking moving objects from a moving vehicle using a laser scanner* (Technical Report No. CMU-RI-TR-05-07). Robotics Institute -Carnegie Mellon University. Pittsburg, PA.

Mercado-Velasco, G. A., Borst, C., Ellerbroek, J., van Paassen, M. M., & Mulder, M. (2015). The use of intent information in conflict detection and resolution models based on dynamic velocity obstacles. *IEEE Transactions on Intelligent Transportation Systems*, *16*(4), 2297–2302.

Murphy, R. R. (2012). A decade of rescue robots. *International Conference on Intelligent Robots and Systems,* 5448–5449.

Van Dam, S. B. J., Mulder, M., & van Paassen, M. M. (2008). Ecological interface design of a tactical airborne separation assistance tool. *IEEE Transactions on Systems, Man and Cybernetics, Part A*, *38*(6), 1221–1233.

MULTIPLE IDENTITY TRACKING AND MOTION EXTRAPOLATION

Ashley Buck and Esa Rantanen
Rochester Institute of Technology, Rochester, NY, USA

Abstract

Multiple Identity Tracking (MIT) is a research paradigm in which individuals track the location and individual identity information of several moving objects in the environment. The present study is an examination of how individuals are able to extrapolate the future movement of moving objects while they are masked. There has been conflicting research on the source of a decline in tracking ability; either the amount of time an object is occluded for, or the distance an object moved during an occlusion. Additionally, previous research has not included the use of a secondary visual search task in a mask. Our design was modeled after a task of a pilot, who has to divide his or her attention between flight information on a head-up display (HUD) and traffic information on a horizontal situation display (HSD), while maintaining good situation awareness on both sources of information.

The purpose of this study was to identify the determinants of performance in tracking multiple moving objects while maintaining their identity-location bindings in the visual short-term memory. This study expanded on past research by investigating the relationship between object displacement during masking of the objects and masking duration (simulating looking away from the HSD) on tracking performance. Isolating and identifying the aspects of tracking multiple objects that are most detrimental to rapid reacquisition of a given target object will help designers, engineers, and researchers identify solutions that would result in improved performance and a lighter cognitive load. No study to our knowledge has examined if poor performance is due to the displacement of an object or the duration of time that passes when a mask occludes the objects in a multiple identity tracking task.

This study also introduced the concept of task switching during an occlusion to an MIT scenario, making the experiment more realistic. Switching attention away from the MIT task may be brief but it requires processing additional information while maintaining the identity-location bindings in memory to quickly reacquire objects to be tracked when attention is again paid to them. Studying performance from this view point yielded results that are more aligned to what could be expected from people engaged in MIT tasks in realistic settings. Finally, this study is unique in that all objects in the task will be potential targets, adding another element of realism to the experiment.

We hypothesized that object displacement would result in poorer performance in identifying targets compared to the duration of the occlusion. In a realistic scenario, this would mean that if a pilot lost track of a moving object when they foveate to the HSD, it would likely be due to the objects moving far from the point they were last attended to, rather than the pilot spending more time on the HSD. Based on the outcomes of previous studies, we expected performance to be best in the lowest object speed, shortest occlusion task condition (smallest object displacement) and worst in the highest object speed, longest occlusion condition (largest object displacement). The results have important implications on estimating human performance in modern aircraft cockpits and for the design of both pilots' tasks and the displays helping them perform the tasks.

Method

Four combinations of object speeds and occlusion times were examined. These include a fast speed with a short occlusion (resulting in 2.54° visual angle (VA) displacement and 2 s occlusion), fast speed with long occlusion (5.08° VA displacement and 4 s occlusion), slow speed with short occlusion condition (5.08° VA displacement and 2 s occlusion), and slow speed with long occlusion (10.08° VA displacement and 2 s occlusion). The experimental conditions are presented in Table 1 below. We hypothesized that object displacement during the mask would drive performance in identifying targets rather than duration of the mask, based on the outcomes of previous studies (Keane & Pylyshyn, 2006).

Table 1
Object Displacement Calculations, Which Are Based on a 22.5-Inch PVD at a 500 mm Viewing Distance

Condition	Occl. Time	Obj. Spd (mm/s,	°VA/s)	Displ. (°VA)
Slow Speed, Short Occlusion	2 s	11.1	1.27	2.55
Slow Speed, Long Occlusion	4 s	11.1	1.27	5.08
Fast Speed, Short Occlusion	2 s	22.2	2.54	5.08
Fast Speed, Long Occlusion	4 s	22.2	2.52	10.08

Participants

Ten participants were recruited to pilot the study from Rochester Institute of Technology's undergraduate student population. All participants had normal or corrected-to-normal vision. All procedures were approved by Rochester Institute of Technology's Institutional Review Board.

Apparatus and Materials

Stimuli were shown in a program developed in Javascript and run in Java Runtime Environment. A Google satellite image (dark green forest) was set as the background for the HSD and the MIT task. The frame rate was set to 60 fps. Five small aircraft symbols were used as tracking stimuli. An alphanumerical call sign below each object, written in

12 point font, served as each object's identity. Each object moved in square paths at a predetermined speed of 11.1 mm/s or 22.2 mm/s. The objects moved for 7 s per trial before being masked and continued to move behind the mask.

An image of a head-up display (HUD) showing altitude and speed was used as the mask. The participants were tasked with determining if either the altitude or speed was safe (above 500 feet for altitude and below 1000 kts for velocity). The combination of altitudes and velocities was unique to each trial. Participants were prompted in the mask image to check either their altitude and velocity, and had 2 or 4 seconds to make a verbal "yes" (meaning safe) or "no" (meaning unsafe) response. Participants were given a reference sheet to use as needed and they practiced the associations as many times as they wanted to before performing the experimental trials.

Once the mask was removed, the objects froze in place, and the object identities were masked under black boxes. Participants were prompted to click on a specific object using a small pop-up window on the top left corner of the screen. Putting the cursor on an object revealed its identity. Once the participant clicked on the correct object, the objects resumed movement and the pop-up window disappeared.

Independent Variables

Object speed and mask duration were manipulated. The faster an object and the longer the mask duration, the larger the displacement during the mask. Object speeds were chosen to result in equal displacement in 2 or 4 seconds, to allow comparable conditions to determine if displacement from original position or occlusion time has a greater effect.

Dependent Variables

Response time, number object identity checks before clicking on the target object, and responses to the mask scenario were measured.

Design

A factorial 2 $\times$ 2 within-subjects design was used to compare 2 object speeds (11.1 mm/s and 22.2 mm/s) and 2 occlusion times (2 s and 4 s). A within-subjects design of this experiment accounted for individual differences in response time.

Procedure

All participants were seated 50 cm from a 22.5-in computer monitor. Participants were given a printed reference sheet and the researcher reviewed the instructions with the participants. The reference sheet included a simple representation of the mask scenario to help participants learn where to locate the altitude and speed on the HUD and what was considered "safe".

Participants completed 15 practice trials, or repeated the practice trials as many times as they wished until they felt comfortable with the task, followed by 100 experimental trials. The reference sheet remained in front of the participant during the experiment allowing participants to refresh their memory.

The experimental trials took approximately 20 minutes to complete. Five objects appeared on the screen, within a constraint of at least 1 degree away from the edge of the screen, and at least 1 degree apart. All objects on the screen moved at a consistent speed through the entire experiment, and all objects could potentially be targets (i.e., there were no distractors). The use of distractors in MIT tasks is not needed, because each object's identity is unique and distinct from all other objects (Oksama & Hyönä, 2008).

The participants tracked the moving objects for 7 s after which the mask screen occluded the entire tracking screen; the objects continued to move in the background. The mask appeared for either 2 or 4 seconds in a random order. Participants were asked to respond verbally if the altitude or speed of the object was "safe". Once the mask was removed, participants were presented with the tracking screen again, with the objects frozen in place and their identities masked. Participants were prompted to click on a target object as quickly as possible (Fig. 1). The software used to run the program logged responses and how many times the participant revealed an object identity by running their mouse over the label. Responses to the mask scenario were collected by the researcher by hand.

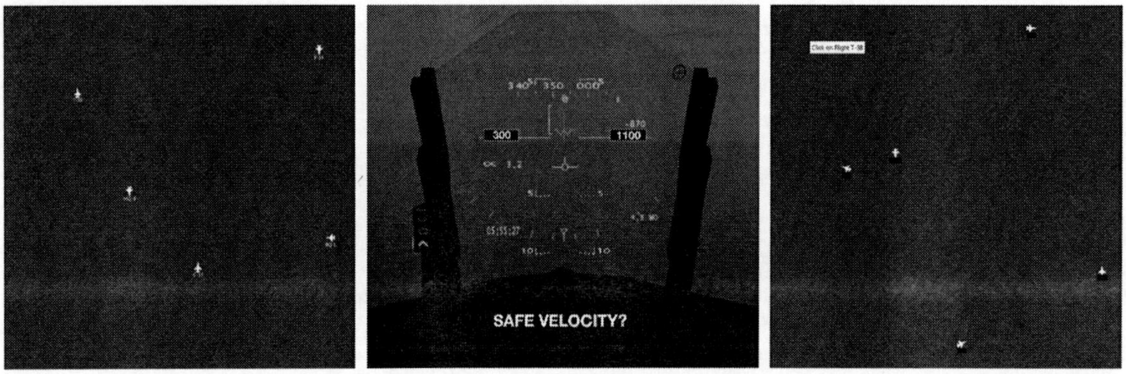

Figure 1. The experimental tasks and procedure. Participants tracked 5 moving objects with unique identities for 7 s (left) until the view of the object was blocked by a mask depicting a head-up display for 2 or 4 s, with the objects continuing to move on the background (center). Participants were required to see if the altitude or speed displayed was safe. After the mask was removed, the object reappeared frozen and with their identities masked. The participants were required to click on a target object queried in the pop-up box on the upper left corner (right)

Results

Data analysis was conducted using MS Excel and Minitab 18 software. The overall average response time to identify an object was 4.20 seconds (SD = 2.79). Table 2 shows these results by condition. A repeated measures ANOVA was performed; the differences in response times between conditions were not statistically significant, $F(3, 997) = 1.39$, $p = 0.246$, with $R^2 = -.42\%$.

Object label reveals were counted. If a participant only had one object reveal, this indicated that the participant knew the location of the object and this was considered to be perfect performance. Two or more means the participant revealed multiple object labels to be able to identify the correct object (Table 3).

Table 2
Average Response Times Per Condition, and by Overall Performance.

Condition	Average RT (s)	SD
Overall performance	4.20	2.79
Slow speed, short occlusion	4.44	2.94
Slow speed, long occlusion	4.16	2.46
Fast speed, short occlusion	3.94	2.29
Fast speed, long occlusion	4.19	3.18

Participants had perfect performance in about 50% of the trials per condition, and in 54% of the trials overall. Participants had two or less object reveals in 75% of the total trials, and participants revealed three or more objects in 25% of the trials.

Due to the nature of the pilot test, the performance on the task-switching scenario was monitored by two researchers and spot checked for performance. The researchers did not observe any errors in performance during the course of the study, indicating that participants were performing the task-switching task accurately.

Table 3
Percentage of Object Label Reveals Per Condition and Overall Performance.

Condition	1 rev.	2 revs.	3 revs.	4 revs.	5 revs.
Overall performance	54%	21%	12%	7%	5%
Slow speed, short duration	53%	24%	13%	5%	5%
Slow speed, long duration	55%	18%	12%	8%	6%
Fast speed, short duration	60%	19%	11%	5%	5%
Fast speed, long duration	49%	23%	11%	11%	5%

Discussion

The response times did not differ between the experimental conditions. There are several possible explanations for this finding. Participants who were unsure about the location of an object could have quickly uncovered the object labels nearby until the object was found (also meaning they had a quick visual search time). It is also possible that response time was not fully representative of performance in this task due to the high number of participants who were accurate on the first or second try (based on object label reveals). These findings suggest that participants were able to perform the task quite well without occlusion time or object displacement impacting performance.

Object label reveals indicated that the task was somewhat difficult for participants, but not impossible. Participants selected the correct item on the first try 54% of the time, and revealed two or less object labels 75% of the time. This suggests that participants generally could keep track of the objects. Revealing more than one object would indicate that the participant did not know where the object was located; choosing the correct object on the second try might suggest that participants had a general idea of where the target object was located, but confused objects that were in a close vicinity to one another.

Previous studies on the ability to extrapolate motion through occlusions have used shorter occlusion times that were less than half of the duration than the present study (Cohen, Pinto, Howe, & Horowitz, 2011; Fencsik, Klieger, & Horowitz, 2007; Franconeri, Pylyshyn, & Scholl, 2012; Keane & Pylyshyn, 2006; Zhong, Ma, Wilson, Liu, & Flombaum, 2014). The findings suggest that it is possible to track multiple moving objects and maintain identity-location bindings for longer periods than previously determined.

Conclusion

The present study was an examination of how object occlusions and object speeds impact tracking performance when the objects have unique identities. Additionally, the use of a task-switching scenario challenged participants to perform a second task during the occlusions. The results suggested that participants could perform a tracking task of this degree of difficulty, but participants had equal performances in each condition. The implication of these findings is that individuals have a limited ability to maintain the identity-location bindings of objects in their visual short term memory and switch to a brief alternative task without losing these bindings. The findings in the present study do not elucidate if tracking ability in a MIT task is impacted more by object occlusions and object speeds. This question should be further investigated, because knowing the limitations of performance in MIT tasks will aid in the design of appropriate systems in operational settings in the field of aviation.

References

Cohen, M. A., Pinto, Y., Howe, P. D. L., & Horowitz, T. S. (2011). The what-where trade-off in multiple-identity tracking. *Attention, Perception, and Psychophysics*, *73*, 1422-1434.

Fencsik, D. E., Klieger, S. B., & Horowitz, T. S. (2007). The role of location and motion information in the tracking and recovery of moving objects. *Perception Psychophysics*, *69*, 567–577.

Franconeri, S. L., Pylyshyn, Z. W., & Scholl, B. J. (2012). A simple proximity heuristic allows tracking of multiple objects through occlusion. *Attention, Perception, Psychophysics*, *74*, 691–702.

Keane, B. P., & Pylyshyn, Z. W. (2006). Is motion extrapolation employed in multiple object tracking? tracking as a low-level, non-predictive function. *Cognitive Psychology*, *52*, 346–368.

Oksama, L., & Hyönä, J. (2008). Dynamic binding of identity and location information: A serial model of multiple identity tracking. *Cognitive psychology*, *56*, 237–283.

Zhong, S.-h., Ma, Z., Wilson, C., Liu, Y., & Flombaum, J. I. (2014). Why do people appear not to extrapolate trajectories during multiple object tracking? a computational investigation. *Journal of Vision*, *14*, 1–30.

USING ENHANCED FLIGHT VISION SYSTEMS (EFVS) FOR LOW-VISIBILITY TAXI IN TRANSPORT CATEGORY AIRCRAFT

Dennis B. Beringer
FAA Civil Aerospace Medical Institute
Oklahoma City, OK

Andrea L. Sparko
U.S. DOT Volpe National
Transportation Systems Center
Cambridge, MA

Joseph M. Jaworski
Cherokee CRC, LLC
Oklahoma City, OK

Two studies (using Boeing 777 and 737 simulators) examined flight crews' use of an Enhanced Flight Vision System (EFVS) for low-visibility taxi operations. Twenty-five flight crews completed 21 short taxi scenarios under combinations of the following: Runway visual range (RVR: 300, 500, and 1000 ft); EFVS on head-up display (on/off); Airport infrastructure - 3 levels. The use of EFVS produced fewer route deviations, most at 300 feet RVR with edge lights and standard centerline or routes with LVO/SMGCS "enhancements" (without centerline lights). Larger turn angles and lower visibilities were associated with slower rates of travel. Crews detected the obstacle on the right-side most of the time and twice that of the left-side obstacle. Regardless of EFVS, crews had more route deviations on larger (>90 degrees) turns and right turns, possibly from loss of visual references in the turn. Recommendations are provided regarding benefits and limitations of EFVS for low-visibility taxi with suggestions for additional research.

The Federal Aviation Administration (FAA) Low-Visibility Operations/Surface Movement Guidance and Control System (LVO/SMGCS) voluntary program has supported safer taxi operations in low visibilities of less than 1200 feet runway visual range (RVR) since 1996. Approximately 70 U.S. airports have FAA-approved LVO/SMGCS plans, which comprise a combination of airport infrastructure and procedures as outlined in Advisory Circular (AC) 120-57A (FAA, 1996) and FAA Order 8000.94 (2012). The current LVO/SMGCS program has two levels: Level 1 is at visibilities from less than 1200 to 500 feet RVR and Level 2 is at visibilities from less than 500 to 300 feet RVR. A Level 3 (<300 feet RVR) is proposed once FAA/industry can jointly demonstrate that aircraft will operate safely with emerging technologies like an Enhanced Flight Vision System (EFVS), which displays a sensor image of the outside scene on a head-up display (HUD). EFVS technology combined with procedural mitigations could provide support for changing the existing low visibility taxi route program to include an EFVS low visibility taxi route at participating airports.

To gain a better understanding of how the proposed changes may be implemented, the FAA is interested in whether an EFVS can aid pilots in taxiing safely in low-visibility conditions when LVO/SMGCS infrastructure is reduced or not present. If so, it might increase access to airports that do not have an LVO/SMGCS plan because of current infrastructure costs. Although the FAA does not regulate taxi operations, the FAA is interested in understanding how to better support low visibility taxi operations without compromising safety with reduced airport infrastructure.

The focus of this examination was limited to a sensor-based display in light of (1) the fact that there have been numerous studies performed using database oriented (electronic map or synthetic vision) displays to facilitate low-visibility operations (maps: Lorenz & Biella, 2006; Battiste, Downs, & McCann, 1996; Yeh and Chandra, 2003, and perspective forward-looking displays: McCann, Andre, Begault, Foyle, & Wenzel, 1997; Beringer, Domino, and Kamienski, 2018), (2) a lesser number on use of EFVS (eg. Kramer, et al., 2013), and (3) some inherent limitations in displays generated from a database (accuracy of registration with the outside world, and obstacles or momentary obstructions unlikely to be in the database). The intent of this effort was to identify any potential safety decrements that might be encountered during the use of EFVS for taxiing in low-visibility conditions under likely airport infrastructure variations with less than that presently required for LVO/SMGCS. Data were collected for simulated wide-bodied aircraft (Boeing 777) and narrow-bodied aircraft (Boeing 737) to see if crews' anticipation of turns, being different in the two, might affect operations with the display.

Method

Participants

Twelve two-person B-777 flight crews from one carrier participated in Phase 1 and 13 two-person B-737 flight crews from various carriers participated in Phase 2. Both pilots in each crew were required to have at least 10 hours flight time within the past 30 days. The pilot flying was required to have at least 100 hours of head up display (HUD) experience. For the B-777 pilots, required HUD experience was as pilot-in-command in an aircraft equipped with an EFVS. At least one crewmember was required to be Category (CAT)-III ILS qualified for the previous five years. Each flight crew was comprised of pilots from the same company to minimize differences in standard operating procedures. On average, B-777 pilots had 17 years of CAT-III experience (SD = 10, Range = 0 to 35) and B-737 pilots had 12 years (SD = 9, Range = 0.5 to 30).

Simulation Environment

Phase 1 was conducted in a CAE B-777F level D full-flight simulator operated at the FedEx Flight Training Center in Memphis, TN, and Phase 2 was conducted in a CAE Boeing 737-800NG level D full-flight simulator operated by Flight Standards Flight Operations Simulation Branch at the Mike Monroney Aeronautical Center in Oklahoma City, OK. Both simulators used a version of the same Rockwell-Collins EP-8000 visual model for the EFVS image and airport simulation. The simulators were operated with the motion on to provide additional feedback (operational realism) to the pilots. The infrared (IR)-based EFVS was displayed on a Rockwell-Collins HUD in front of the left-seat pilot. The right-seat pilot did not have an EFVS. Pilots were able to control the pilot-adjustable settings (e.g., brightness) for the EFVS and HUD. All other EFVS settings were preset prior to the taxi scenarios. EFVS display features, characteristics, flight information, flight symbology, and sensor imagery were based on regulatory requirements (14 CFR §§ 91.176 and 25.773), minimum aviation system performance standards for EFVS (RTCA, 2011; FAA, 2016), guidance for EFVS operations (FAA, 2017), and/or as recommended by LVO/SMGCS subject matter experts (SMEs).

Design

Three independent variables were combined, as shown in Table 1, to form a 3x3x2 fully crossed within-subject factorial design: Runway Visual Range (RVR; 3 levels), Infrastructure (3 levels), and EFVS display (2 levels).

Table 1. *Experimental Conditions (taxiway edge lights were always present).*

RVR (ft)	Infrastructure	EFVS	
		On	Off
300	Standard centerline + edge lights Level 1 (L1)	300-L1-on	300-L1-off
	+centerline enhancement (L2)	300-L2-on	300-L2-off
	+ centerline lights (L3)	300-L3-on	300-L3-off
500	L1	500-L1-on	500-L1-off
	L2	500-L2-on	500-L2-off
	L3	500-L3-on	500-L3-off
1000	L1	1000-L1-on	1000-L1-off
	L2	1000-L2-on	1000-L2-off
	L3	1000-L3-on	1000-L3-off

Note. Heavily shaded cells with white text indicate the 18 cells of the 3x3x2 factorial design.

Task/Scenarios

Pilots performed taxi scenarios at a simulation of KSLC (Salt Lake) at night. Nighttime conditions were chosen based on SME input to represent the more commonly encountered difficult low-visibility condition, compared to worst-case dusk or dawn times. Because the study examined variable infrastructure, the KSLC simulator airport model was altered to remove LVO/SMGCS lights and markings along the taxi routes other than the specific LVO/SMGCS route used as a baseline reference. Twelve taxi scenarios were constructed such that (1) each contained at least one turn each of <90 degrees, 90 degrees, and >90 degrees (a sampling variable); (2) scenarios were balanced between left and right turns; (3) all began on a taxiway or runway. Some were repeated within an order, but those that were repeated were placed near the beginning and near the end of the counterbalanced orders. Three additional scenarios were included as supplemental conditions; two routes passed near a truck parked at the edge of the taxiway to examine crew obstruction detection/reactions, and a third used LVO/SMGCS centerline enhancements and designated route centerline lighting at 300 feet RVR, EFVS off, for a baseline.

Procedure

Upon arrival, crews completed the pre-experiment paperwork and briefing, including the Informed Consent Form, pilot experience questionnaire, and viewing of a PowerPoint briefing (Phase 2 only), along with a short briefing regarding the procedures to be followed. Pilots were told that they would be asked to traverse some non-standard routes (i.e. some that may be contrary to current SLC airport routings and requirements). These included some turns greater than 90 degrees. Additionally, pilots were also given paper route maps for each scenario that included printed copies of the ATC taxi instructions, EFVS setting (on or "hide"), and the aircraft's starting position on the airport surface.

Pilots next entered the simulator and completed a practice taxi scenario, to familiarize themselves with the simulator and EFVS setting, with the EFVS on in 500 feet RVR with LVO/SMGCS centerline "enhancements" (12-in wide with a 6-in black border), centerline lights, and edge lights. This was followed by the 21 experimental scenarios. Each scenario took approximately 5 to 10 minutes to complete. A 15 to 20-minute break was provided halfway through the scenarios. Two researchers sat in the simulator cab during the sessions, one acting as a simulated ATC controller, and the other as observer. When the scenarios were completed, the pilots returned to the briefing room where each completed a post-experiment questionnaire. After completing the questionnaire the test team solicited general comments or questions from the crews and presented an overview of the purpose of the study. The entire session required between 4 and 5 hours. As the results of both phases were similar, they will be presented together by type of performance measure.

Results

Performance Metrics

Centerline tracking. Although the means for centerline tracking were consistent and all averages were within 3.5 to 5 feet of the centerline, slightly more variability was evident when RVR was 300 than in the other visibilities, with 1000 RVR showing the narrowest variability. There was also slightly more variation with EFVS off than there was with it on, but the means were essentially the same. A similar pattern was observed for infrastructure but was somewhat anomalous with slightly less variation in conditions with the least infrastructure (Level 1).

Route Deviations. The majority of route deviations occurred at 300 feet RVR (Figure 1A). This was expected as a function of the reduced visibility. The maximum percentage of scenarios on which uncorrected deviations occurred was just over 15% for 777 at 300 RVR. About half as many deviations were detected soon enough to correct them. The 737 crews had roughly an equal proportion of uncorrected and corrected deviations in same visibility conditions (9% and 11.5% respectively). There were no uncorrected errors with either aircraft when EFVS was on at 300 RVR. This may also be related to the fact that crews taxied slightly slower in the lowest visibility than in the two higher visibilites. Overall, the percentage of deviations roughly linearly decreased as visibility increased.

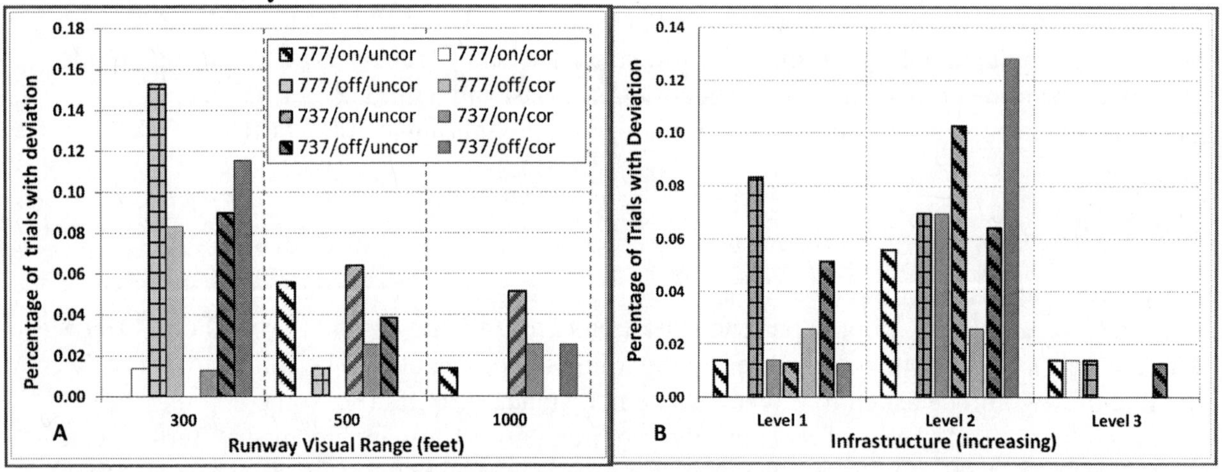

Figure 1. Percentage of route deviations by aircraft type, EFVS (on/off), and (A) RVR and (B) Infrastructure.

Interestingly, the pattern of deviations relative to increasing taxiway infrastructure was not entirely as anticipated. Level 3 did exhibit the lowest number of deviations (Figure 1B). The nonintuitive result was that Level 2 exhibited the most, with Level 1 having the middle frequency of deviations. One can also see across the two figures that the percentage of deviations with EFVS on (which averaged 2.2%) was smaller than when EFVS was off (4.3%).

Obstacle detection. Obstacle detection was defined as the pilot verbally indicating seeing the truck during the scenario. Flight crews in both studies detected the right-side truck the majority of the time, and about twice as often as they detected the left-side truck (Figure 2). In fact, the detection rate was approximately 90% for the 777 crews when the truck was on the right of the taxiway. If pilots did not verbally acknowledge seeing the truck, researchers asked the pilot about it during the post-test questionnaire and interview. However, for the purposes of analysis, only verbal indications of seeing the truck were included. The first officer frequently detected the obstacle, as the captain was often looking out the left-side window trying to keep the taxiway edge line in sight. The left-side truck was at a 90-degree turn to the left, and thus was not in the sensor field of view (30° x 15° in Study 1 and 32° x 15° in Study 2) during the turn, which possibly led to a lesser chance of being detected. This represented a situation where objects outside the sensor's field of view could potentially pose a hazard not immediately apparent in the EFVS. Descriptive statistics are presented here because the events were not independent due to the repeated-measures design.

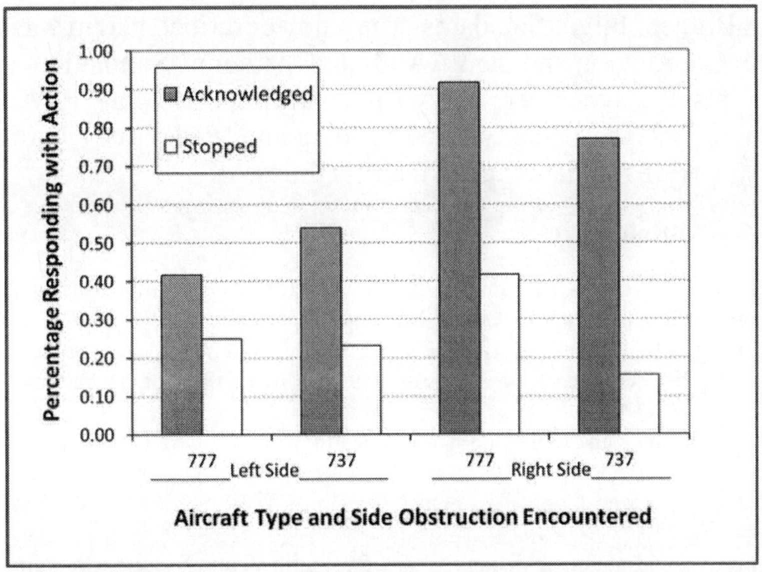

Figure 2. Pilot responses to an obstacle near the taxiway by aircraft type and object location.

Pilot Opinions

Boeing 737 pilots felt that reduced infrastructure contributed to increased workload. Moreover, Captains reported difficulty making right turns, particularly in the B-777, because they would lose visual reference to the centerline under the aircraft. Pilots did not feel that EFVS contributed to their position awareness above what their own direct observations provided. Although a moving map was not used in this study, pilots also felt that a moving map in addition to EFVS would provide improvements in position awareness. Some pilots had concerns about the use of EFVS in low-visibility operations, including the restricted EFVS FOV, limitations regarding EFVS visuals (e.g., parallax, blue lights showing up as green), and the limited effectiveness of EFVS under certain environmental conditions (e.g., precipitation or dense fog). Despite their concerns, both groups of pilots generally felt that an EFVS repeater should be made available to the First Officer.

Conclusions

EFVS provided a benefit to navigation performance at 300 feet RVR when there were no centerline lights. However, EFVS had no effect on navigation performance at 500 feet RVR and above. That is, with minimal taxiway infrastructure in visibilities of 500 feet RVR or greater, flight crews were generally able to navigate successfully with or without EFVS. Note that these results should not be taken to suggest that taxi operations are safe in these conditions without EFVS. Almost all of the wrong or missed turns observed in these studies were made on right turns. However, flight crews made very few wrong turns when centerline lights were available, suggesting that difficulties finding the centerline may be alleviated when the centerline is lit.

These studies also examined potential limitations on taxiing with reduced infrastructure. As mentioned previously, right turns were difficult and were observed to have more errors, notably without centerline lighting. It was also found that sharp turns greater than 90 degrees were associated with more route deviations and were described by pilots as being difficult, particularly without lights and markings. In wide-body aircraft, pilots may also find it difficult to oversteer on sharp turns. Additional research is needed to understand the impact of intersection complexity (using a more robust definition) during low-visibility taxiing. Taxi routes used in these studies were not designed to be complex. Although some complex intersections were noted, there were too few of them to make any conclusions about how effectively flight crews navigated them.

Acknowledgments

This research was completed with funding from the FAA NextGen Human Factors Division and the CAMI Aerospace Human Factors Division under the Flight Deck Program Directive in support of the FAA Flight Operations Branch, Flight Standards. The study described herein was conducted in coordination with FedEx Corporation and the FAA Flight Operations Simulation Branch, Flight Standards. Sponsor points of contact were Terry King and Bruce McGray, FAA Flight Standards. A special thank you to FedEx, especially Robert Riding and Mark Gouveia, for working with us to conduct the first half of this research using their facilities.

References

Battiste, V., Downs, M., and McCann, R. (1996). Advanced Taxi map display design for low-visibility operations. Proceedings of the Human Factors and Ergonomics Society 40th Annual Meeting, 997-1001.

Beringer, D.B., Domino, D.A. and Kamienski, J. (2018). Pilot performance using head-up and head-down synthetic-vision displays for SA CAT I approaches. In Proceedings of the 2018 Annual Meeting of the Human Factors & Ergonomics Society, 57-61.

FAA (1996). AC 120-57A, Surface Movement Guidance and Control System. Washington, DC: USDOT FAA.

FAA (2012). Order 8000.94, Procedures for Establishing Airport Low-Visibility Operations and Approval of Low-Visibility Operations/Surface Movement Guidance and Control System Operations. Washington, DC: USDOT FAA.

FAA (2016). AC 20-167A, Airworthiness Approval of Enhanced Vision System, Synthetic Vision System, Combined Vision System, and Enhanced Flight Vision System Equipment. Washington, DC: USDOT Federal Aviation Administration.

FAA (2017). AC 90-106A, Enhanced Flight Vision Systems. Washington, DC: USDOT Federal Aviation Administration.

Kramer, L.J., Bailey, R.E., Ellis, K.K.E., Williams, S. P., Arthur, Jarvis J. III, Prinzel, L. J. III, & Shelton, K. J. (2013). Enhanced flight vision systems and synthetic vision systems for NextGen approach and landing operations. NASA/TP-2013-218054.

Lorenz, B. & Biella, M. (2006). Evaluation of onboard taxi guidance support on pilot performance in airport surface navigation. Proceedings of the Human Factor and Ergonomics Society 50th Annual Meeting, 111-115.

McCann, R. S., Andre, A.D., Begault, D., Foyle, D.C., and Wenzel, E. (1997). Enhancing taxi performance under low visibility: Are moving maps enough? Proceedings of the Human Factors and Ergonomics Society 41st Annual Meeting, 37-41.

Pilot compartment view, 14 C.F.R. § 25.773 (2016)

RTCA (2011). DO-315B, Minimum Aviation System Performance Standards (MASPS) for Enhanced Vision Systems, Synthetic Vision Systems, Combined Vision Systems and Enhanced Flight Vision Systems. Washington, DC: RTCA, Inc.

Straight-in landing operations below DA/DH or MDA using an enhanced flight vision system (EFVS) under IFR, 14 C.F.R. § 91.176 (2017).

Yeh, M. and Chandra, D. (2003). Air transport pilots' information priorities for surface moving maps. Proceedings of the Human Factors and Ergonomics Society 47th annual meeting, 129-133.

WHAT IS SAFETY DATA?

Julia Pounds
Air Traffic Safety Oversight Service, Federal Aviation Administration
Washington, DC
Paul Krois
NextGen Human Factors Division, Federal Aviation Administration (Retired)
Washington, DC

This paper reports on our efforts to determine if the ubiquitous term *safety data* can be more specifically defined. That is, whether data can be categorized as safety data based on some unique characteristics such that other data not having these would be categorized as *not safety data*. FAA analysts rely on multiple sources of objective data for virtually all analyses supporting FAA's decision making. While profuse amounts of data are continuously collected twenty-four hours a day, only subsets are deemed useful for any particular purpose, such as assessing how well an organization conducts its safety or efficiency or security missions on a day-to-day or long-term basis. Therefore, safety data appears to be defined by whether it is used for safety activities, i.e., surveillance, compliance and verification. Conversely, data used exclusively for security and efficiency assessments could be defined as *not* safety data. We concluded that safety data are generally defined a posteriori by how the data are used rather than due to any intrinsic characteristics.

The Air Traffic Safety Oversight Service (AOV) monitors the Air Traffic Organization (ATO) using, in part, reports of ATO safety occurrences/data and, in turn, AOV shares safety data with ATO (FAA, 2006). The AOV order lists examples but does not specify what safety data are and are not.

Safety data has become such a ubiquitous term across safety studies that we questioned whether using this expression is essentially just a general convenience or whether it is actually being used as a specific construct in aviation as well as in other industries. Having well-defined data for analysis is the sine qua non - the gold standard - for safety professionals in the Federal Aviation Administration (FAA) and other industries around the world.

Researchers of all stripes learn early the need to define their work so that it is not open to nultiple interpretations; that is, to clearly define their constructs, methods, and data in terms of measurable and observable properties with operational definitions. Unambiguous and measurable data are essential for accurately identifying performance indicators, risk indicators, thresholds, tolerances, and so forth. Unambiguous results better support decision makers' interpretations. If the definition is determined by the industry using it, definitions of *safety data* would be expected to vary across different industries.

As practitioners striving to solve applied problems, we depend on clear and specific constructs and methods to reduce variability in our analyses. Otherwise, how would a researcher determine unambiguously what data were safety-related and useful for safety analyses versus "not safety data" that could be disregarded. Perhaps all data used for safety analyses are defined as "*safety data*." In that case, the term is merely a convenience.

According to the Miriam-Webster Dictionary online, *safety* is defined as *the condition of being safe from undergoing or causing hurt, injury, or loss,* which implies that it pertains to people, places, and things that can be hurt, injured, or lost. Perhaps data are labeled *safety data* if results of analyses are intended for protection of these entities. We collected some high-reliability industry examples to explore these hypotheses.

Nuclear Power Industry

The nuclear power industry uses safety data in its analyses and reports focusing on authorized activities and fundamental problems or hazards related to them, such as unintended conditions or events and radiological releases. The Nuclear Regulatory Commission (NRC) is responsible for regulation and its policies governing nuclear reactor safety relative to commercial use of nuclear materials. As an ultra-high reliability industry, safety data are developed by safety inspectors, reactor operators, equipment vendors, research laboratories, and other sources. Data and methods can include incident and investigation reports, probabilistic fast-time simulation models, equipment inspections, human-in-the-loop simulations, reactor operator training data, surveys, and evaluation checklists (NRC, 2018).

The NRC previously maintained a human factors information system reports database. The database provided annual summaries for each commercial reactor of human performance issues identified in Licensee Event Reports, inspection reports, and licensed operator examination reports. The information was a general overview of the types and approximate numbers of human performance issues documented in reports by either the NRC or licensees.

Other industry analyses focus on the industry's security and safeguards, which are linked but are also defined uniquely. Security analyses focus on intentional misuse to cause harm from external threats to materials or facilities. Safeguards analyses focus on unauthorized activities related to acquisition of materials and equipment or development of nuclear weapons.

The nuclear industry measures for safety, security, and power production tend to be independent of one another. Conceivably, undesirable occurrences in any of these categories might adversely influence persons, places, or things. If so, the idea of safety might be applicable to security and safeguards of people, places, or things. Testing this logic leads to the idea that safety data could pertain to all of these: authorized activities, materials/facilities, and unauthorized activities. However, the soundness of this idea is a bit strained and no instances of this industry using safety data for analyzing facility security, for example, was discovered during this activity.

Chemical Industry

Chemical safety involves a broad number of industries. These industries involve toxic chemical handling (e.g., chlorine and ammonia), hazards of combustible dust, reactive chemicals, oil and gas production, and hot work activities (e.g., welding). The Chemical Safety Board (CSB) has reponsibility to investigate significant chemical incidents and hazards and advocate for the implementation of recommendations to protect workers, the public, and the environment (CSB, 2017). The CSB mission has a cross-cutting relationship with federal and state regulations, industry standards, and other local policies and procedures. The CSB's focus is primarily on specific accidents and incidents especially those involving loss of life and injury.

As part of its work, the CSB may collect data to identify and analyze existing industry hazards and conduct broad safety studies of such hazards to examine commonalities among significant incidents and draw attention to key lessons learned. Their analysis can identify direct and proximate causes of equipment failures, underlying systemic and organizational causes such as inadequacies in corporate or facility-level safety management systems and organizational culture, and opportunities to improve operational practices, regulatory standards, and enforcement. The CSB issues safety recommendations to a variety of recipients, including Federal and state regulatory agencies, companies, industry and labor organizations, standard-setting bodies, and emergency responders.

An example of a CSB investigation was the West Fertilizer Company Fire and Explosion (CSB, 2013). The accident involved 15 fatalities and more than 260 individuals injured from the fire and detonation of fertilizer grade ammonium nitrate. The CSB report used data from multiple sources to assess contributing and causal factors. The report organized key safety findings as technical, regulatory, insurance, emergency response, emergency planning, and land use planning. Underlying many of these findings were the people involved in various roles, making decisions based on information available at that time, at varying points in time prior to the accident for which there could not have been any line of sight leading up to the accident such as a prescient warning.

Software Acquisition

Safety-critical software must meet both system specifications and safety-critical performance requirements. At the FAA, development of software is determined through policy in the Acquistion Management System (AMS) (FAA, 2019). The objectives of the AMS are to increase the quality, reduce the time, manage the risk, and minimize the cost of delivering safe and secure services to the aviation community and flying public. Data to support these objectives are collected across the acquisition life cycle, including an Operational Safety Assessment and Preliminary Hazard Analysis. Guidance for collecting human factors and human performance data during acquisition of safety-critical software is provided in the Safety Risk Management Guidance for System Acquisitions (SRMGSA) (FAA, 2018). Safety data are also obtained by operational testing of software prior to its implementation. Controllers and maintenance experts use representative operational scenarios to validate that the software is an effective and suitable design solution that meets operational needs (FAA, 2019).

Safety issues may also be identified after deployment in the field through post implementation reviews. These reviews provide an everyday operational perspective beyond what can be accomplished in a testing laboratory environment. Reviews can also draw on security, efficiency, environmental (e.g., noise abatement), and other important parameters and data sources as part of identifying and mitigating potential safety issues.

The FAA applies a safety risk management process to all acquisitions that impact the NAS. From development to operational deployment, data are important in determining whether an acquisition supports service that is at least as safe as what is currently being used. The addition of functionality through automation and procedures augments safety by providing more effective and efficient service. Data would be used to validate no decrease to safety while some of that same data could be used to validate efficiency gains.

Aviation

The International Civil Aviation Organization (ICAO) published the 2017-2019 Global Aviation Safety Program which defined safety data as:

> A defined set of facts or set of safety values collected from various aviation related sources, which is used to maintain or improve safety. Note. Such safety data is collected from proactive or reactive safety-related activities, including but not limited to: a) accident or incident investigations; b) safety reporting; c) continuing airworthiness reporting; d) operational performance monitoring; e) inspections, audits, surveys; or f) safety studies and reviews.

FAA

The continuing mission of the FAA is to provide the safest, most efficient aerospace system in the world. FAA established the national policy for safety management (FAA, 2016) which defines safety as the state in which the risk of harm to persons or property damage is acceptable.

Aviation has many stakeholder groups, the most important being the flying public. Each group monitors and improves safety levels in their areas of responsibility: commercial carrier companies, aircraft manufacturers, air traffic service delivery providers, and so forth. All depend on safety data collected by various means and from a variety of sources, pertaining to avoiding adverse outcomes.

Continuous improvements in aviation technologies, procedures, etc., have led to a level of safety such that accidents have become very rare events compared to traffic levels. To accomplish this, the FAA relies on multiple data sources to assess and monitor its programs, initiatives, plans and strategic goals.

The FAA collects and stores large amounts of data from real-time operations. However, FAA published system performance indicators that identified only three types as ATO safety metrics: number of runway incursions and surface incidents and en route losses of standard separation (FAA, 2018).

It's reasonable then to assume that the safety data label could apply to all types of aviation data, given that the goal is avoiding unsafe conditions and outcomes for its stakeholders. Here safety may be the primary use but with categories collected for other uses, similar to the nuclear industry, such as equipment outages, personnel training results, voluntary reporting systems, and oversight activities. Defining data primarily used for safety versus, for example, effeciency is helpful for several reasons, including, specificity of terms, support for efforts to standardize and harmonize data-based safety oversight tools and methods, and reducing conflicting interpretations by stakeholders.

AOV

The FAA Administrator established AOV in 2004 (FAA, 2006) making AOV responsible for independent oversight of the ATO. The order directs AOV to use safety data to fulfill its mission and responsibilities and gives examples of data types. The order also makes ATO responsible to collect, track, and analyze safety data, and to report safety data to AOV upon request. Examples identified include: ATC incident and accident rates, NAS equipment maintenance issues, flight inspection issues, results from safety risk assessments, and results from ATO internal oversight, evaluation, and quality assurance programs. Collected from different types of ATO operations, these data may contribute to the three FAA reported safety metrics.

In 2011 Press undertook to operationally define ATO safety data as it might pertain to AOV activities, e.g., surveillance, compliance and verification. In 2004 Press had advised that any definition should prevent multiple interpretations. For example, it must be representational, unique, and meaningful of the target. He identified several challenges and prescribed solutions based on the idea that safety data cannot be defined without also including its ulterior use.

ATO

The ATO performance indicators published by the FAA characterized air traffic operations (2018) and included three to indicate the safety of the NAS based on number of runway incursions and losses of airborne separation. The metrics for runway incursions were limited to the Core 30 Airports showing counts from Fiscal Year (FY) 2013 through 2017. Incursions were categorized by type of incident. The Loss of Standard Separation Count was shown for en route centers from FY 2013 through 2017.

Other categories of data could potentially intersect to influence safety and thus could be used in safety analyses and so also considered as safety data. For example, the report also includes numbers representing system efficiency, such as NAS delays, diversions, go-arounds, and cancellations, and numbers from traffic management initiatives, such as those used to manage traffic volume, excess demand and airport acceptance rates. As part of research on human performance in provision of air traffic services, Cardosi & Yost (2000) reported using safety data in a study of controller and pilot errors. They used incident and accident reports from ATO operations, the Aviation Safety Reporting System (ASRS), and the National Transportation Safety Board (NTSB).

In 2013 Kimble proposed to improve integration of ATO safety-related data. Examples listed included: ATC operational data from automated NAS systems; ATC personnel data, e.g., training, certifications, and proficiency; voluntary safety reporting programs; radar, voice, and facility communications data; weather; and facility logs (ATO, 2013).

Discussion

After considering how the term *safety data* was used in these high-reliability industries, it seems that the the common denominator is safe human performance but the industry's goals and respnsibilities seem to determine whether data related to safety is the primary goal, an equivalent goal or a secondary concern. For example, a business would use one set of data to track workplace safety and another set to evaluate product safety. The goal of analysis would determine which data are used as primary, equivalent, and secondary safety data rather than safety data vs. not safety data.

So, is *safety data* a ubiquitous term, an unnecessary term, or a useful term? Certainly, its use is pervasive. It's short, to the point, and satisfying in a sentence. Synonyms seem more clumsy, wordy or redundant, e.g., well-being, secure, safe-keeping. The answer to the question: "What is not safety data?" seems to be that it depends on the industry, the industry's production and outcome goals, and the industry's responsibilitues for avoiding harms to persons, places and things.

Disclaimer

The views expressed herein are those of the authors and do not reflect the views of the Federal Aviation Administration.

References

International Civil Aviation Organization. (2016). Global Aviation Safety Plan. Doc 1004 (pp. ix). Montréal, Quebec, Canada. Retrieved from https://www.icao.int/safety/Pages/GASP.aspx

Kimble, T. (2013). Operational Analysis and Reporting System (OARS) Program. Presented at Integrated Safety Risk Assessment Advisory Committee Meeting. Washington, DC.

Cardosi, K & Yost, A. (2000). Controller and pilot error in airport operations: A review of previous research and analysis of safety data. DOT-VNTSC-FAA-00-21 Office of Aviation Research, Washington, DC

Chemical Safety Board. (2017). Strategic Plan, 2017-2021. Washington, DC: U.S. Chemical Safety and Hazard Investigation Board. Retrieved from https://www.csb.gov/assets/1/6/csb_strategic_plan1.pdf

Chemical Safety Board. (2013). Chemical Safety Board Investigation Report, West Fertilizer Company Fire and Explosion, Final Report 2013-02-I-TX. Washington, DC: U.S. Chemical Safety and Hazard Investigation Board.

Federal Aviation Administration. (2019). Acquisition Management Policy. Washington, DC: Federal Aviation Administration. Retrieved from https://fast.faa.gov/docs/acquisitionManagementPolicy/AcquisitionManagementPolicy_4.pdf

Federal Aviation Administration. (2018). Safety Risk Management Guidance for System Acquisitions (SRMGSA). Washington, DC: Federal Aviation Administration.

Federal Aviation Administration. (2018). Air Traffic by the Numbers. Washington, DC: Federal Aviation Administration.

Federal Aviation Administration. (2016). Safety Management System (Order 8000.369). Washington, DC: Federal Aviation Administration.

Federal Aviation Administration. (2006). Air Traffic Safety Oversight Service (Order 1100.161). Washington, DC: Federal Aviation Administration.

Nuclear Regulatory Commission. (2018). Research Activities 2018-2020, NUREG-1925, Rev. 4. Washington, DC: Nuclear Regulatory Commission. Retrieved from https://www.nrc.gov/docs/ML1807/ML18071A139.pdf

Press, J. (2011). A data definiton framework for air traffic safety oversight. Washington, DC: Federal Aviation Administration.

Press, J. (2004). A measurement framework for air traffic safety metrics. DOT/FAA/CT-TN04/10. Washington, DC: Federal Aviation Administration.

A SYSTEMS-BASED MODEL AND PROCESSES FOR INTEGRATED SAFETY MANAGEMENT SYSTEMS (I-SMS)

Diogo Silva Castilho

Massachusetts Institute of Technology

Boston - MA

The study of the vulnerabilities of a system is often organized in a hazard analysis. Methods based on systems thinking are relevant tools to analyze the operation of modern products. The purpose of this research is to develop, implement, and validate a systems-based model for aviation Safety Management Systems (SMS) incorporating the treatment of collected data to foster the effectiveness of mitigating measures over time. The model uses data monitoring systems, management of change reports, flight inspections, voluntary reports, and other sources as input messages to an Active Hazard Analysis. The new requirements, constraints, and the preventing and mitigating measures are organized and delivered timely to the operators. The analysis on unstable approaches found contributions to documentation and procedures in practice. In accordance with SMS standards, the new framework provides organized safety information for management, fostering better planning on the use of workforce and resources.

Complex operations defy cognitive limitations. Safety-critical systems face accidents when these limits are unknown or ignored. A careful hazard analysis provides the knowledge that is necessary to reduce the risks. However, in dynamic systems, experience could generate negative learning, and even a thorough knowledge of the initial condition in which the system is delivered is not enough to guarantee safe operations.

The lifetime of an aircraft is expected to be long. A new airliner might endure more than four decades of operation. Throughout its lifetime, different generations of pilots, flight attendants, and mechanics will operate all the equipment developed for the system. This system comprised of hardware, software, and operators with different cultures will change over time because the environment and the mindset of operators will change. Technology will impact operations as upgrades of components, new functionalities, and different levels of automation are implemented. The challenge becomes to assure safety for operations when assumptions made at the beginning of the project are no longer valid.

The first efforts for hazard analysis should start during ConOps (Concept of Operations). In this phase, engineers need to make assumptions about how operators will interact with the product, and some of these assumptions will become obsolete. For example, the Boeing 777 entered into service in 1995. Back then, it would be impossible to imagine that airlines would be using electronic flight bags (EFB) or tablets[1]. It is easy to believe that this is a natural evolution

---

[1] EFBs and off-the-shelf tablets are accepted to be integrated to the dashboard to substitute all paper charts and manuals (FAA InFO, 2011)

in hindsight, but there was no smartphone when the aircraft was certified. The operational lifetime needs to be used to update the assumptions previously made, and consequently, the hazard analysis.

Safety Management Systems

The concept of an SMS (Safety Management System) was introduced in commercial aviation as a formal, top-down, organization-wide approach to manage safety risk and assure the effectiveness of safety risk controls. This perspective aims to make aviation even safer, but the processes within SMS leave room for improvement. It does consider software controlled systems and higher levels of automation, but it fails to proactively monitor the impacts of human factors and changes in the environment. It focuses on risk assessment (accident prediction) (FAA, 2016) rather than using a hazard analysis for accident prevention.

SMS is a new requirement for air operations, maintenance, and air traffic services. Annex 19, the document that formalized this initiative, is the first new ICAO (International Civil Aviation Organization) Annex to come out in over thirty years. All aviation organizations must show compliance with Annex 19 before November 2019, but there are a variety of ways to do it. The ICAO and the FAA offer manuals to guide Safety Risk Management (SRM) and Safety Assurance (SA), but there is no orientation on the use of hazard analysis at the organizational level.

There are tools based on Systems Theory that could be added to SMS. These techniques consider the operator's behavior to be the result of social, psychological and even environmental conditions. The mapping of actions applied to a controlled process and the analysis of the feedback that the operator is receiving provide a qualitative understanding of the real issues behind the unsafe behavior.

This research links systems engineering and management actions that are necessary to comply with SMS. In this context, we answered the following research question: How to apply systems-based concepts to collect aviation operational data and update a hazard analysis? The solution was the introduction of the Integrated Safety Management System (I-SMS) as a model to guide safety managers using concepts from Systems Engineering.

The purpose of this research was to develop and implement a systems-based model of safety management incorporating the treatment of collected data to foster the effectiveness of mitigating measures over time. This model has a general framework that the safety manager can adjust to each specific system.

This model work as a method to monitor safety in operations to maintain a higher level of safety by using an active hazard analysis. The foundation for the I-SMS is STAMP (Systems-Theoretic Accident Model and Processes), which is based on Systems Theory. STAMP is a modern model of causation that has proven to be successful in aviation. I–SMS is the model that improves the completeness on the application of STAMP techniques.

Systems-Theoretic Process Analysis (STPA) is the hazard analysis technique based on STAMP (Leveson, 2011). STPA covers not only the accidents caused by component failures but also the ones caused by a faulty interaction between components of a system that are each functioning properly, as a consequence of system design flaws. It recognizes safety and security as emergent properties of a complex system caused by the interaction of its components. The main characteristic of security is the malicious intentions behind control actions. However, safety is a more general term, and it is affected by both well-intended operators and the ones attacking the system. The STPA is complemented by organizing assumption-based leading indicators (Leveson, 2015) to register the reasoning behind performance indicators.

Integration of Hazard Analysis and SMS

Proactive management requires effective communication and monitoring activities. The proposed solution is the use of a structure that puts the hazard analysis at its core to feed the decision making of higher hierarchical levels with new indicators and their trends.

The hazard analysis performed during system development becomes the structure that will be in constant evolution as it is revisited during the whole lifetime of a system. The output of this active hazard analysis adapts the organization to a dynamic reality. The general framework of the I-SMS is presented in figure 1.

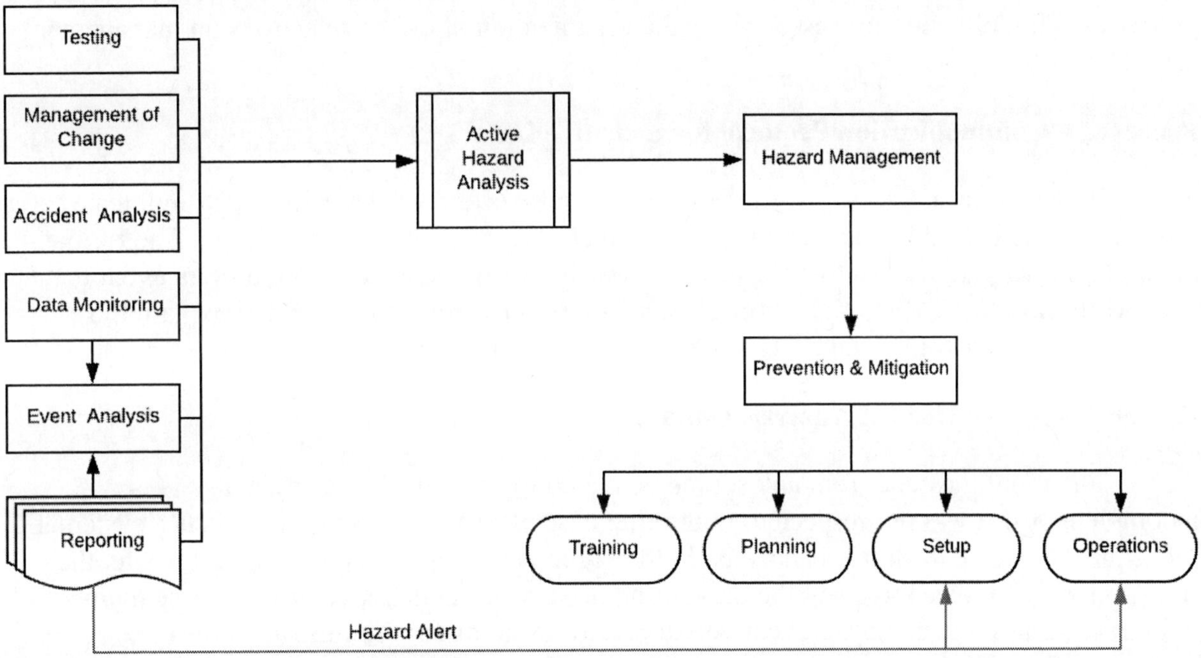

***Figure 1*. I-SMS general framework.**

On the left side, there are many different sources of input to the Active Hazard Analysis. The verification and validation tests become opportunities to add to the hazard analysis the details that developers did not consider when the product was just an abstract idea. When the system is already delivered and operating, changes initiate a process that requires revisiting the hazard analysis to avoid surprises. Incidents and accidents are seen as potential learning events. Also, the system must be open to voluntary contributions. The stakeholders' participation regarding hazardous conditions works both to enhance the system and to foster a safe attitude, keeping the organization awareness aligned with its culture.

On the right side, the output of the active hazard analysis feeds the hazard management and its preventive and mitigating actions that update the system's information flow, bringing it to a safer state. However, changing the documentation without understanding the operator's needs and difficulties is not effective management. It is essential to apply the mitigating measures considering how and when critical information is delivered to operators because it will be better assimilated if presented at a proper time.

A manager must guarantee that the information generated by the active hazard analysis will arrive at the desired destination, communicated to and understood by everyone who should have it, and applied to the system. Those tasks demand an observant manager to assure that all previous work is effective. Without monitoring the information flow, an accident could occur due to a causal factor that was identified and treated, but the measures to prevent it were not correctly followed over time.

The proper implementation of the model to a specific system requires tailoring the general framework. That means that each box of the general framework will need another particular label. Once this structure is drawn, the tasks are divided into the application of four processes. The following processes guide the organization of effective actions on management activities.

Process 1 - Communication Protocol for Sensitive Data

Proactive measures to prevent accidents require effective channels to communicate safety information. This goal is obtained only if all stakeholders use the same language. The format of the input message for the Active Hazard Analysis has a complete description of an event that starts with the context, lists all actions of each controller chronologically, and explain the reasoning for the decisions taken as reported by each controller.

Process 2 - Active Hazard Analysis Update

The input message from any source is treated to verify if the hazard analysis is incomplete or if it was not respected. In the first case, the safety manager conducts a systematic procedure to update the hazard analysis. In the second case, the analyst investigates why the rules were not followed to adapt the prevention or mitigating measures or to enforce them. In both cases, management acts preventively to avoid future losses. The list of actions and controllers from the previous process become a reference to relate the event with the correct part of the hazard analysis. The identified mental models are discussed to reason on assumptions previously made. The assumptions are updated, followed by the scenarios and measures derived from it.

Process 3 - Hazard Management

Currently, most aeronautical product development organizations use risk assessments to decide how to prioritize mitigating measures and to judge if it is worth taking action. This risk management is the evaluation of both on-going and new initiatives in a systematic attempt to address areas with the potential to pose a risk to safety during operations.

The concept of an "acceptable level of safety" is expressed by two measures required by ICAO: Safety Performance Indicators (SPI) and Safety Performance Targets (SPT). These solutions are in place because the company top management requires measurable safety targets that are acceptable to regulators and other stakeholders, and consistent with the SMS.

The problem is that, without a structured hazard analysis, the selection of SPIs and SPTs is subjective and relies only on the experience of a few managers. Systems Engineering provides qualitative ways to manage hazards and the analysis result on the elaboration of SPIs that will explain if the system is drifting to a more hazardous status. In other words, the new set of SPIs diagnose the safety culture of the organization and verify if communication channels are effective.

Process 4 - Prevention & Mitigation

The strategy for mitigation involves a range of possible actions including:

- Revision of the system design with changes to the functional control structure.
- Modification of operational procedures.
- Re-arrangements of staffing.
- Training of personnel to specific scenarios.
- Development of emergency and contingency plans.
- Ceasing operation.

All updates of the hazard analysis will result in changes in the company's documentation. Most safety critical organizations have standard procedures that are taught during training and enforced throughout the operation. They culturally become rules to avoid blame.

The desired safe behavior requires building mental models to facilitate proper actions when specific conditions are detected and recognized. It also requires responding signs of those conditions that alert about the proximity to hazard. That becomes necessary as humans are affected both by an excess of information, that causes high workload and stress, or lack of information, which leads to low situation awareness and distractions. Most systems have hazards related to both extremes, but prevention is possible if human factors are properly considered.

The solution is to organize the required safety knowledge into communication events that occur at different moments. Each communication opportunity has specific characteristics. The four categories and their vehicles presented in the general framework are a proposed reference that should be adjusted to the desired system:

- **Training:** The first opportunity to teach and to present limitations and rules has the benefit of a mind clearer of biases and preconceptions. The study of manuals must have a set of information regarding safety reasonably complete. That will be used to form the mental models and to serve as a consultation source during operation.
- **Planning:** The time dedicated to plan a set of actions (e.g., a mission in military activities) is opportune to communicate safety concerns to operators. The addition of safety information during the planning activity reduces surprises and the variability of improvisations.
- **Setup:** When the task is complex and requires a fast and accurate response, the operator prepares himself or herself recalling the mental models and remembering the responses for off-nominal situations. In many systems, checklists are tools that deal

with memory limitations. New technology solutions provide multiple ways to feed up-to-date information in dynamic systems.

- **Operation:** In dynamic phases of operations, there is no time to search for the manual or to read an order. The solution for the communication of safety information is the use of cues that can be aural or visual. They must be simple, recognizable and unequivocal.

Alerting System

In the I-SMS framework, the voluntary reports are classified into two different types: Hazard Analysis update and Condition Alert. In the hazard analysis update, the safety officer receives the description of the situation observed by the operator to perform the reasoning described in Process 2. In the second case, time-critical observations, such as a drone crossing the runway final approach, potentially dangerous environmental phenomena, or even criminal actions, require extra instant communication. Instant messages called Condition Alerts are transmitted using software solutions and connectivity to alert other operators.

Case Study and Conclusion

A complete STPA on unstable approaches was used as basis for the I-SMS. This project had the participation of major airlines in the USA, Brazil, Europe, and Asia. To avoid the correlation of companies with unsafe events, all data was condensed in one single database and analyzed altogether. The airlines provided flight monitoring data, pilot reports, observation flights, and investigation reports on unstable approaches for landing.

The outcomes of the project included more robust documentation for training, a better understanding of the vulnerabilities of airline operations, a more complete hazard analysis, a more explicit allocation of responsibilities, optimized enforcement mechanisms, and the observation of new trends that gives the feedback that is required for proper management.

References

FAA, FAA InFO, 2011. The Apple iPad and Other Suitable Tablet Computing Devices as Electronic Flight Bags (EFB).

FAA, 2015. Advisory Circular 120-92B (Safety management systems for aviation service providers) 1–4.

ICAO, 2013. Annex 19 to the Convention on International Civil Aviation, Annex.

Leveson, N., 2015. A systems approach to risk management through leading safety indicators. Reliab. Eng. Syst. Saf. 136, 17–34. doi:10.1016/j.ress.2014.10.008.

Leveson, N., 2011. Engineering a Safer World: Systems Thinking Applied to Safety, Vasa. doi:10.7551/mitpress/8179.001.0001.

Acknowledgments

This research was possible due to the scholarship provided by CNPq/Brazil.

DRONE ACCEPTANCE AND NOISE CONCERNS - SOME FINDINGS

Hinnerk Eißfeldt & Verena Vogelpohl
DLR German Aerospace Center
Institute of Aerospace Medicine
Department of Aviation and Space Psychology
Hamburg, Germany

Drones are becoming ever more present in public perception. Ranging from parcel delivery to wildlife protection, from precision farming to law enforcement, and from industrial inspection to digital fireworks, many applications are said to have market-changing potential. Against this background, nations and institutions around the world are trying to keep up with the dynamic development concerning rules and regulations. Since all of the parties involved anticipate a strong increase in both the number of drones and their range of uses, there is a rising interest in the acceptance of civil drones in the public. Widespread public acceptance can promote the dissemination of new technologies. Conversely, concerns among citizens about the use of drones in their daily environment could pose potential barriers to the further proliferation of civil drones, especially in urban areas. The psychoacoustic properties of drones have repeatedly been discussed as being one such limiting factor. This paper reports results of a representative national study on the social acceptance of civilian drones, taking a closer look at noise considerations. Therefore the results help improve understanding of the perception of civil unmanned aerial vehicles.

Drones – understood here as unmanned aerial vehicles (UAV) of a civilian nature – are becoming increasingly visible among the public. Applications range from parcel delivery to animal welfare, from the production of live images of major events to the fight against crime, and from the inspection of industrial facilities to the design of artificial fireworks. Almost monthly, the media reports on new uses for drones and patent applications. Thus drone technology is often regarded as having a disruptive quality in certain markets and industries. On a global level, the International Transport Forum of the OECD (ITF 2018) and the World Economic Forum (WEF 2019) have described opportunities and challenges for future drone usages in recent reports. National and international institutions are trying to establish rules and procedures to keep up with the dynamic development. With a continued strong increase in the use of drones expected by all of those who are involved, there is also an increasing interest in the public's perception of this new element. As airport planning has repeatedly shown, a lack of public acceptance can be a limiting factor for further growth in aviation (e.g. Suau-Sanchez, 2011). Similarly, certain concerns among the public regarding the use of drones could restrict their wider dissemination: "One potential outcome of scaled-up drone operations is an increase in urban noise volume exceedances above legal or desired limits" (ITF 2018, p.39).

Method

The study on drone acceptance was conceptualized at DLR German Aerospace Center and fielded by infas Institute for Applied Social Sciences as a Computer Assisted Telephone Interview (CATI). Using a dual frame technique with 70 % landline and 30 % mobile phones, a random digital dial design was used with the aim of reaching conclusive results representative for the German population.

The questions were asked in a standardized manner by specially trained employees in a telephone interview of approximately 20 minutes in length. After each call, the answers were entered into an online database using an appropriately designed template. For quality assurance, online supervision was performed by senior staff who occasionally listened in on the calls. The study fully adhered to the professional code of conduct for telephone interviews agreed upon in Germany (ADM 2016). 832 respondents took part in the study, which was conducted between March and May 2018, and answered all questions. Respondents were 51.8% male, 48.2% female; their age ranged from 14 to 94 years (mean 51.5, SD 18.2); the mean size of household was 2.5 (SD 1.3). Further information on the response rate and sampling procedures, as well as detailed results, can be found in Eißfeldt et al. (2018).

Results

The study was planned as a telephone survey to measure the public acceptance of civil drones in Germany. Only a few questions contained information about noise and will be referenced in the following in order to assess the effect of noise concerns on the acceptance of civil drones.

Associations with the term drone

After explaining the purpose of the study and gaining consent for participation, at the beginning of the interview, the respondents were asked whether they knew the term "drones" in aviation. All of the 97% participants who answered that question in the affirmative were subsequently asked in an open question to indicate what they associate with a drone. A total of 794 participants gave answers ranging from a single word to several complex sentences, all of which were protocolled onsite by the interviewer. Later these qualitative data were coded into 6 categories: espionage/surveillance/observation (32%), film/video/photography (27%), leisure/hobby (21%), parcel delivery/transport/air taxi (21%), danger/accident/threat (20%), and military/weapon (19%). About 18% were coded "other," indicating a wide range of associations not covered by these categories. Among the 715 different associations with the term, drone noise was among the least mentioned, only 6 times in total. In one of these cases, noise was explicitly considered unproblematic as drones would fly with electric engines making no sound.

Attitude towards civil drones in Germany

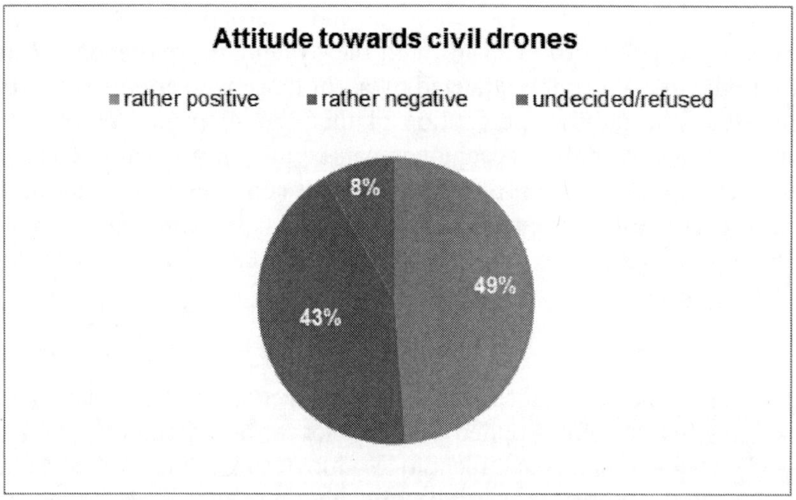

Figure 1. Attitude towards civil drones

After being asked for their associations with the term drone, study participants were informed that the drones referred to in the remainder of the interview were unmanned aircraft that look like small helicopters with several rotors, typically four or more, and that only civil

applications were relevant for this study. They were then asked how they would describe their general attitude towards civil drones, specifically, whether it was rather positive or rather negative. If they could not decide, the answer was coded as "undecided." Very few respondents refused to answer certain questions. For the sake of simplicity, those reactions were combined with "undecided" into one category, "undecided/refused."
Although there was a somewhat even distribution of negative and positive responses to civil drones, there was a slight advantage on the positive side (43% rather negative, 49% rather positive, and about 8% undecided, see *Figure*1). The results vary in accordance with with several sociodemographic factors such as gender, age, income, and place of residence. Male respondents have a more positive attitude toward civil drones compared to females. Younger study participants show higher acceptance than older participants.

Areas of concern with civil drones

Later during the telephone interview, 7 different areas of concern that had been identified from the literature were asked about in randomized order to avoid sequence effects. When asked to what extent they are concerned about aspects of civil drone usage, most of the respondents confirmed their concern about the possibility of misusing drones for criminal purposes (91%, see also Figure 2), followed by privacy concerns (86%). Concerns connected with mishaps all raised concerns in the range of 72% - 75% followed closely by concerns about animal welfare. Concerns about noise were confirmed less frequently (53%).

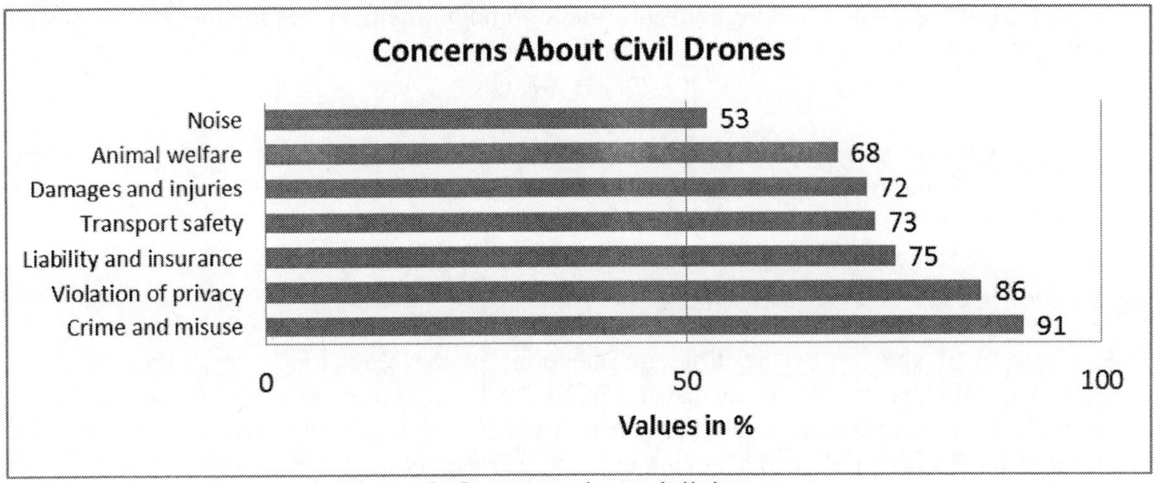

Figure 2. Concerns about civil drones

As a whole, a large majority of respondents named at least three or more subjects of concern regarding civil drone usage (91%). However, the number of aspects mentioned varied with respondent age and gender, with women and older respondents more concerned than younger or male respondents.

Experience and concerns. About half of the participants (47%) reported having experiences with drones in their personal lives (36.4%), on the job (4%), or in both contexts (6.1%). Looking into the concerns expressed by this group reveals that those who have some kind of experience with a drone have significantly less concern about potential accidents, animal welfare, or transportation risks than those who have no experience. Chi-square tests at the 10% level reveal significant values for concerns about damages and injuries χ^2 (1) = 3.09, p = .08, OR = .76, animal welfare χ^2 (1) = 4.29, p = .04, OR = .73, and transport safety χ^2 (1) = 3.39, p = .07, OR = .75. As shown in Figure 3 throughout all areas asked about the amount of concern is higher for participants reporting no experience with civil drones all areas of concern.

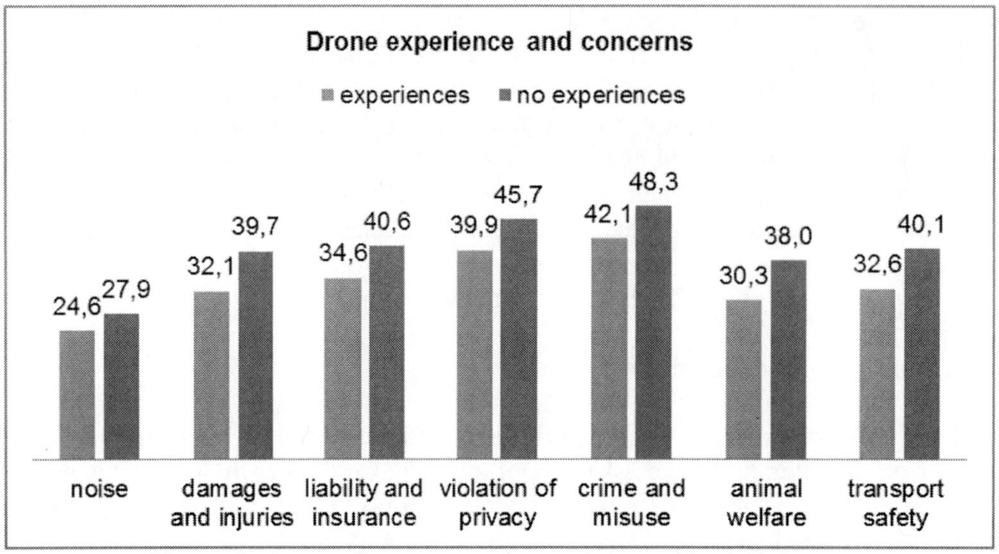

Figure 3. Areas of concern and experience with civil drones

Noise concerns and direct experience. Somewhat surprising was the rather low level of concern about drone noise (53%), as this had been discussed as being a potential barrier before. However, when looking into information about whether a respondent has or has not reported having heard a drone yet, for those having heard a drone, a higher percentage of noise concern was revealed: χ^2 (1) = 3.29, p = .07, OR = 1.45.

Concerns about civil drones and acceptance. The influence of the various concerns about civil drones on the public acceptance thereof was analysed using Chi-square Automatic Interaction Detection (CHAID). According to Perreault & Barksdale (1980), the CHAID method partitions a contingency table produced from cross-tabulation by using a semihierarchical, sequential procedure. One of its advantages is that it can be used with non-parametric survey data. In our case, the attitude towards civil drones was the parent group variable to be split up by the different categories of the predictors – the various areas of concerns. Of all areas of concerns listed in Figure 2, being or not being concerned about noise explained the attitude towards civil drones best χ^2 (2) = 38,6, p = .000, OR = .41. On the next level of the decision tree model, among those concerned about noise, concerns about transport safety explain the most variance, and among those not concerned about noise, their concerns about the violation of privacy are the major factor.

Knowledge about drones

Towards the end of the interview respondents have been asked to what extent they felt informed about drones in general. Answers were given on a 4-point-Likert-scale ranging from 1 = very well informed to 4 = not informed at all. 11.7% described themselves as "very well" informed, 40.6% were informed "a bit," 33.2% indicated being "only a little" informed, and 13.9% "not at all."

In a first step, the subjective level of information about drones was tested against the attitude towards civil drones. As can be seen in Figure 4, subjects who describe themselves as better informed about drones in general have a more positive attitude towards civil drones.

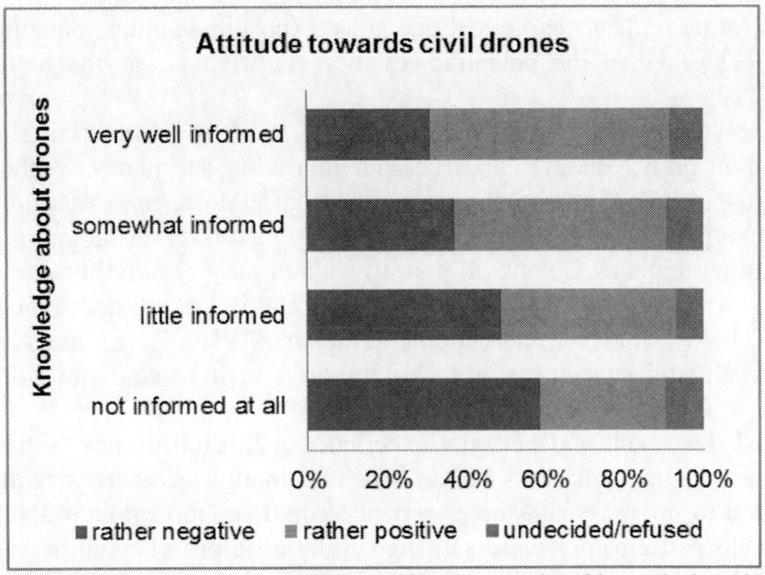

Figure 4. Attitude towards civil drones at different levels of knowledge about drones

Information about drones comes through various channels and could be biased; for example, information on noise levels could be exaggerated. Therefore in a second step, the subjective level of information about drones was tested against concerns about noise. As can be seen in Fig 5, subjects who describe themselves as being better informed about drones in general are less concerned about noise.

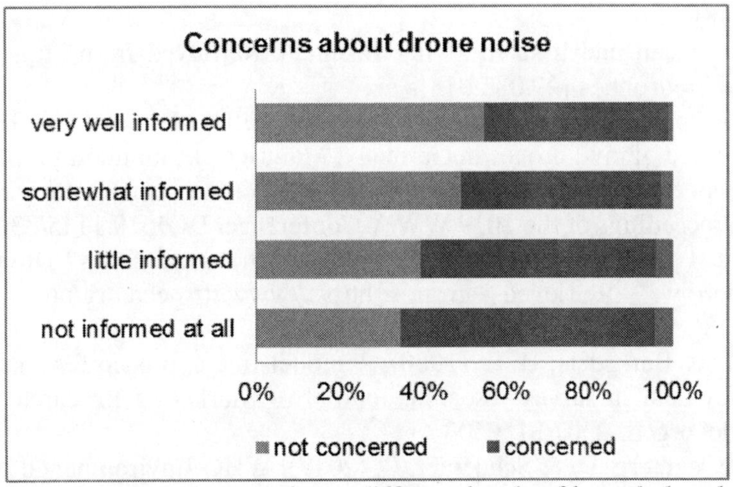

Figure 5. Concerns about drone noise at different levels of knowledge about drones

Discussion

Similar to comparable studies, a somewhat consolidated pattern of acceptance was found with slightly more than four out of ten respondents being rather negative about civil drones, about five out of ten indicating rather positive attitude towards drones, and the rest being undecided. A more detailed look revealed that the attitude towards drones in a civil context has a complex pattern of origins. Among other things, it depends on gender and age, but also on the individual level of information about civil drones. This is well in line with an online survey published by German Industries Aerospace Association (BDLI 2016), which showed acceptance concerning the civil usage of drones to be evenly split among participants, with 42% positive and negative each and about 15% stating they do not know. Also this study

found that 53% of participants expressed that noise exposure would be potential risk of drone usage, and also found that the potential violation of privacy was the highest concern of participants (84%).
The results presented here have shown that a good level of information about drones has positive effects on both reducing concerns and improving acceptance. Although not in the focus of the initial study and not prominent on first glance, noise concerns could be confirmed as being an important factor for the acceptance of civil drones. Although reported by only about half of all participants, among all concerns about usage of civil drones noise concerns have the strongest impact on acceptance. Environmental noise and annoyance is targeted by recent studies (Guski 2017) and international guidelines (WHO 2018). Stakeholders of drone usage thus are well advised to invest at maximum on reducing sound emissions to the lowest level possible.
Increased knowledge about and personal experience with civil drones both comes together with a decrease in noise concerns. To conduct information campaigns tailored to specific target groups and to provide hands-on experience could support drone usage in general. For metropolitan areas participatory noise sensing (Eißfeldt, in press) could be another approach supporting the development of urban air mobility. Further research should focus on such measures to further increase the public acceptance of civil drones and the successful development of the U-space and its applications.

References

ADM Arbeitskreis Deutscher Markt- und Sozialforschungsinstitute e.V. (2016). Richtlinie für telefonische Befragung. Retrieved from https://www.adm-ev.de/standards-richtlinien/ [29.08.2018].

BDLI (2016). Chancen und Risiken ziviler Drohnen. Retrieved from https: //www.bdl.aero /de/results/?q=drohnen [29.08.2018].

Eißfeldt, H., Vogelpohl, V., Stolz, M., Papenfuß, A., Biella, M., Belz, J. & Kügler, D. (2018) The acceptance of civil drones in Germany. Manuscript submitted for publication.

Eißfeldt (in press). Supporting urban air mobility with citizen participatory noise sensing: A concept. Proceedings of the 2019 WWW Conference. DOI: 10.1145/3308560.3317059.

ITF International Transport Forum (OECD) (2018). (Un)certain Skies? Drones in the World of Tomorrow. Retrieved from https://www.itf-oecd.org/uncertain-skies-drones [29.08.2018].

Perreault, W. D., & Barksdale, H. C. (1980). A model-free approach for analysis of complex contingency data in survey research. Journal of Marketing Research, 17(4), 503-515. http://dx.doi.org/10.2307/3150503

Guski, R., Schreckenberg, D. & Schuemer, R. (2017). WHO Environmental Noise Guidelines for the European Region: A Systematic Review on Environmental Noise and Annoyance. International Journal of Environmental Research and Public Health 14(12):1539. DOI: 10.3390/ijerph14121539.

Suau-Sanchez, P., Pallares-Barbera, M., & Paül, V. (2011). Incorporating annoyance in airport environmental policy: noise, societal response and community participation. Journal of Transport Geography, 19(2), 275-284. doi: http://dx.doi.org/10.1016/ j.jtrangeo.2010.02.005.

WEF World Economic Forum (2019). Advanced drone operations toolkit: Accelerating the drone revolution. Geneva, WEF.

WHO (2018) Environmental Noise Guidelines for the European Region. World Health Organisation Regional Office for Europe.

FEEDBACK ON THE USE OF MATB-II TASK FOR MODELING OF COGNITIVE CONTROL LEVELS THROUGH PSYCHO-PHYSIOLOGICAL BIOSIGNALS

Yannick Daviaux
AKIANI - Bordeaux, France
Christophe Bey
AKIANI - Bordeaux, France
Laurent Arsac
Univ. Bordeaux, CNRS IMS UMR 5218 – Bordeaux France
Olivier Morellec
Dassault Aviation – Paris, France
Sami Lini
AKIANI – Bordeaux, France

Modeling individuals' cognitive control levels in operational situations is a major challenge for safety in aeronautical industry. Standardized experimental tasks - as the Multi-Attribute Task Battery II (MATB-II) - are dedicated to such a challenge that can be faced using psycho-physiological biosignals. These biosignals are known to be sensitive to cognitive worload, performance, and expertise that are intricate features of MATB-II subtasks. Thus, it remained necessary to investigate whether these features could be set to ensure controlled experimental conditions. Two groups (15 experts in time-pressured decision making and 13 novices) completed 3 MATB-II sub-tasks (tracking, monitoring, and resource management tasks). Biosignals accounting for autonomic nervous system activity were measured continuously, as objective markers of cognition.
Confrontation between performance data and (objective and subjective) cognitive markers reported contrasting perspectives regarding the exploitation of MATB-II as a pertinent tool to insure controlled experimental conditions in the context of cognitive control characterization.

Designing adaptive human-machine interface is a major challenge in aeronautics, where the stakes relate to security. To this aim, we were looking to characterize experts' cognitive states in operational context, using psycho-physiological objectification tools. This paper will set out the details of the approach chosen to take up this challenge using the Multi-Attribute Task Battery (MATB-II, Santiago-Espada et al. 2011) computer-based task.

Theoretical framework of this study: Hollnagel's Extended Control Model
This study is based on Hollnagel's cognitive control theory (Hollnagel 1998) to address cognitive resource management. Although cognitive resources are usually seen as a form of "fuel" for cognitive processes – a fuel that could be assessed to determine the operators' margins and limits (Yerkes & Dodson 1908) – the specificity of this model is that it considers a principle of cognitive resource saving, which is more of a mean of optimization, than a simple cognitive resource consumption reduction process. Its main advantage is therefore to approach cognitive processes at an integrative level: there would be cognitive shortcuts to face familiar situations, and means to protect oneself against the unknown. Mental representation, abstraction ability, sufficiency principle and anticipation could be some of these means. The suggested ECOM model (Extended COntrol Model, Hollnagel 1998) mentions 4 identifiable levels of cognitive

control, from long term planning with the highest level of abstraction to short adaptative loops with short available time. It was thus required to propose a human-machine interaction environment in which these levels of cognitive control could be simulated, granting access to the operator's performance as well.

Experimental simulation: MATB-II

To this end, the Multi-Attribute Task Battery computer-based task developed by the NASA team (MATB-II in revised version, Santiago-Espada et al. 2011) was identified as a favorable simulation environment. Although MATB-II is originally a multi-task environment, it offers isolated subtasks that are resembled levels from the ECOM model. Among the 5 proposed tasks, 3 stood out: a "tracking" task (Track), which consists in holding a sight in the center of a target using a joystick (short adaptative loop with short available time) ; a "monitoring" task (Monit), which consists check for abnormal gauge oscillations on 4 gauges and correct it as quickly as possible, pressing the corresponding key on the keyboard (short-term planning); and a "resource management" task (Manag), which consists in maintaining the level of 2 tanks consuming resources, by activating/deactivating pumps that enable the transfer of resources from different tanks (highest planning level). To satisfy the stakes of this project, the cognitive states needed to be characterized during the realization of these tasks.

Objectification of the cognitive states in operational situation and interpretations

Heart rate variability (HRV), electrodermal activity (EDA), and pupillary dilatation are known for cognitive state objectification means in operational situations (e.g., Wilson 2002). As indirect markers of autonomic nervous system's activity, these physiological indicators are also considered representatives of workload, involvement in the task, emotional states, and waking indicators. Given their ubiquitous nature, we needed to ensure the nature only was a dependent variable (Track *vs.* Monit *vs.* Manag). Confounding variables which are workload, operator's involvement and emotional states, therefore needed to be controlled. Since confounding variables and the nature of the task are closely intertwined, the only way to guarantee control of the experimental conditions would be to check the confounding variables' stability between tasks. Since the MATB-II enables to 1) program difficulty levels, 2) measure the operators' operational performance, and 3) subjectively assess the workload, it was the perfect tool to obtain feedback on the participants' involvement through their performance in each task, and on the workload perceived, thanks to subjective assessment scales (NASA-TLX, Hart 2006).

Experimental conditions programming

The MATB-II computer-based task offers an environment allowing event programming for each task, independently from each other. To our knowledge, no gold standard already exist to ensure standardized difficulty. Pre-testing experimental conditions being a common approach in human sciences, pre-experimentation on 5 participants allowed to adjust the difficulty of the tasks, according to subjective feedback on the perceived difficulty.

For Track, level 2 on 3 (pre-programmed medium level) was chosen as default level. For Monit, adjustments during the pre-tests have led us to consider faulty gauge scheduled every 10 seconds, randomly made the difficulty similar to that of the Track task. For Manag, pre-tests led us to program one faulty pump event every 10 to 20 seconds for a duration of 15 seconds, to match Track's and Monit's perceived difficulty.

Hypothesis

Considering the pre-experimentation efforts, we expected no observation of inter-condition effects on the confounding variables, be it on the subjectively declared workload or the performance. Considering the ubiquitous nature of the psycho-physiological variables, we were expecting to observe significant correlations with the cognitive load and performance levels, indicating the necessity for data correction.

Method

Participants

A group of experts in high time-pressured context task management (high-level handball players, N = 15;16.6 ± 1.1 years old ; 8.2 ± 2.9 years of practice) was compared with novices (N = 13; age 20.6 ± 2.0; years of practice < 2 years), for a total of 28 participants. After they had been informed of the experimentation conditions, adult participants and legal representatives of underage participants signed a written consent to participate in accordance with the Helsinki Accords (General Assembly of the World Medical Association, 2014).

Experimental design

Experimental visit

Participants were checked for enough sleep the night before and no energy drinks in the last 6 hours. The experimental session was held in a temperate (20°C), constantly lit, soundproof room. Participants then trained for each condition for 1 minute as a habituation session. We made sure performance instructions were understood for each condition, if not, a second attempt was realized. Participants then performed each experimental condition randomly. Each condition lasted 5 minutes, with at least 3-minute rest between conditions.

Experimental conditions

As presented in the introduction, 3 out of the 5 MATB-II subtasks leaded to 3 distinct experimental conditions. These mono-task conditions create a human-machine interaction relevant with the ECOM model's definition of control levels.

Measurements

Physiological measurements

The participants' electrodermal activity (EDA) and cardiac activity (ECG) were measured continuously at 1000-Hz and amplified with a dedicated acquisition chain and an A/D 24-bit converter (MP150 and BioNomadix system, Biopac, California, USA). For EDA, 2 Ag/AgCl electrodes were placed respectively on the index's and the middle finger's first phalanx, on the non-dominant hand [1]. For the ECG, 3 Ag/AgCl electrodes were placed in conformity with the representation of Einthoven's triangle.Pupillary dilatation data were collected continuously in 60-Hz by a dedicated system (T60 XL Eyetracker, Tobii, Sweden), after individual calibration. All physiological data were recorded on a shared computer for synchronization.

Subjective measurements (cognitive load)

At the end of each experimental task, the level of cognitive load perceived by the subjects was assessed using NASA-TLX. Participants assign a score of 0 to 100 (one score every 5 points) to 6 sub-scales including mental demand, physical demand, time demand, global effort, frustration level, and estimated performance.

Performance measurements

For Track, the performance indicator chosen was Root-Mean-Square Deviation (sampled at 1-Hz). For Monit, reaction times sampled at 100-Hz (maximum allowed by MATB-II software) were collected as dependent variable. For Manag, the difference between both of the tanks' target filling and actual filling was collected at 0.1-Hz.

Data processing

Data were processed in Matlab programming environment (Matlab 2017a, The MathWorks, Natick, MA, USA).

Electrodermal activity

Time-frequency analysis for 0.08 to 0.24-Hz bands was applied to the EDA signal using complex demodulation, according with Posada-Quintero et al. (2016). Spectral power was averaged over time to provide an indication on the sympathetic autonomous activity (EDAsymp, no unit).

Heart rate variability

The ECG raw signal's was preproced according with Pan and Tompkins (1985, QRS complex detection) and Dos Santos et al. (2013, correction of abnormal values in R-R values tachogram). Spectral analysis was performed for each condition on the entire time window (5-min) for low frequencies (0.04 Hz < LF < 0.15 Hz) and high frequencies (0.15 < HF < 0.4-Hz)(Task force paper, 1996). Spectral powers LF_{pow} and HF_{pow} are expressed in s^2/Hz. The LF_{pow}/HF_{pow} ratio reports on the sympatho-vagal system (ratio, no unit).

Pupil dilation

Time series of left and right pupillary diameters were linearly interpolated, merged, then averaged over time and normalized by the time series' standard deviation to provide the EyeT index (no unit).

Cognitive load (subjective)

Scores from the 6 NASA-TLX subscales were averaged to obtain a global score out of 100 (Hart, 2006).

Performance measurements

For Track and Monit, the data have been averaged over time to provide performance indexes, respectively $Track_{perf}$ (in millimeters, mm) and $Monit_{perf}$ (in milliseconds, ms). For Manag, data from both managed tanks have been averaged between themselves, then averaged over time to provide the $Manag_{perf}$ index (no unit).
Subsequently, each data series has been transformed into z-scores to be compared and merged as needed.

Statistics

Statistic tests were carried out with XLSTAT (XLSTAT 2018.1, Addinsoft, France). A non-parametric variance analysis to compare 2 expertise modalities [Intra-subject comparison: Exp vs. Nov] by 3 experimental conditions [Inter-condition comparison: Track vs. Monit vs. Manag] was applied to compare the different variables.

Results

Each variable's averages and standard deviations are presented in table 1a.

Table 1 : Results of data processing.

Table 1a. Mean and standard deviation of scores and dependant variables for tracking condition (Track), Monitoring condition (Monit) and Management condition (Manag).

	Track		Monit		Manag		Effect
	Experts	Novices	Experts	Novices	Experts	Novices	
Performances' scores (mm / ms / no unit)	25,1 ± 7,3	21,9 ± 3	1279 ± 194	1166 ± 243	-745 ± 455	-444 ± 238	*Expertise*
Scores for Workload NASA-TLX (sur 100)	48,3 ± 11,9	51,8 ± 15	50,2 ± 10,5	59,9 ± 12,3	47,9 ± 12,5	47,3 ± 11,2	*Condition*
EDAsymp (index, no unit)	0,52 ± 0,05	0,54 ± 0,1	0,54 ± 0,08	0,61 ± 0,21	0,56 ± 0,07	0,56 ± 0,1	*Condition*
HRV LFpow (s2/Hz)	103,9 ± 26,9	92,8 ± 33,7	105,8 ± 22,5	101,1 ± 31,8	107,5 ± 36	95,3 ± 30,2	
HRV HFpow (s2/Hz)	13,1 ± 7,5	10,4 ± 4,3	12,9 ± 6,3	12,8 ± 6,4	13,9 ± 8,1	11,3 ± 4,7	
HRV LFpow/HFpow (ratio, no unit)	9,1 ± 2,6	9,3 ± 1,8	9,2 ± 2,3	8,8 ± 2,6	8,9 ± 2,5	8,9 ± 2,1	
Pupil dilation (index, no unit)	13 ± 4,9	12,7 ± 2,7	14,8 ± 4	13,9 ± 2,1	18,6 ± 3,3	17,4 ± 4,6	*Condition*

Discussion

This work relates to a project whose global goal is to model the cognitive states of operators in operational situation, using psycho-physiological tools. To this end, MATB-II was suggested as the appropriate experimental environment, because it offers a range of standardized experimental tasks addressing situations that mobilize distinct cognitive states as defined in Hollnagel's cognitive control levels theory. Psycho-physiological variables measured during these tasks being indirect witnesses of the autonomic nervous system's activity, interpretation of the experimental effects can therefore be tricky, considering their ubiquitous nature. This work's aim was thus to make sure that only the nature of the tasks differed between experimental conditions (i.e., that cognitive load and/or involvement in the task (via performance) should not be confounding factors). Precautions were taken in the form of pre-tests to adjust the experimental tasks' difficulty. Consequently, we made the hypothesis that the experimental condition would have no detectable effects neither on the subjectively declared workload, nor on the performance.

Interpretation of effects

Contrary to our first hypothesis, a major effect of experimental conditions was detected on subjectively scores of cognitive load. Although adjustments were carefully made prior to the experimentation – in accordance with common experimental approaches – it is however not possible to confirm that cognitive load has been standardized, and thus that only the nature of the task differed between experimental conditions. This is particularly problematic as far as psycho-physiological data interpretation is concerned, since we observe a similar effect. The difficulty of cognitive load standardization resides in the fact that, to our knowledge, there is no existing gold standard to normalize difficulty between cognitive tasks of different nature. The multiple setting parameters used to adjust the difficulty of a given task makes it particularly difficult to implement similar experimental conditions. As an example, the Track condition offered only 3 spatial difficulty levels (i.e. the more difficult the level, the wider the sight's random movements). Comparatively, the Manag condition offered dynamic (flow management), spatial (which pumps?) and time (when? for how long?) setting parameters. It was therefore difficult to imagine establishing configuration rules to come close to difficulty level standardization. Regarding performance measurements, variance analysis has allowed to detect an interaction effect between the expertise level and experimental conditions. Performance being one of the behavioral witnesses of the involvement in a task, it is once again tricky to confirm that participants' involvement is similar from one task to the next, and that only the nature of the task differed between experimental conditions. We should also note that task performance and cognitive load are closely linked. Poor performance can indeed reflect both an overload and an underload, according to Yerkes and Dodson's law (Yerkes and Dodson 1908). Difficulty level configuration therefore also plays a crucial role, to make sure that poor performances are not linked to a form of boredom in case of too simple a task, or to a total disengagement from a task which is too complex.

About the MATB-II task

Despite the difficulties mentioned in this paper, we should note that MATB-II was used in an unusual way during this study. This experimental environment has indeed originally been developed to simulate a given level of cognitive load in multi-task experimental conditions (e.g., Fairclough et Venables 2006). To our knowledge, no experimentation had tried comparing MATB-II experimental sub-tasks between themselves, attempting to control / normalize confounding factors such as cognitive load and involvement in the task via performance. Even though the solution put forward to address demands like this project's still seems relevant to us today, it is good practice to use this feedback to take precautions when treating and interpreting data, systematically measuring the levels of subjectively assessed cognitive load and the task performances. This would allow to adjust dependent variables (psycho-physiological) to model the operators' cognitive states as accurately as possible.

References

Boucsein, W., Fowles, D.C., Grimnes, S., Ben-Shakhar, G., Roth, W.T., Dawson, M.E., & Filion, D.L.; Society for Psychophysiological Research Ad Hoc Committee on Electrodermal Measures. Publication recommendations for electrodermal measurements. Psychophysiol (2012) 1017-1034.

Dos Santos, L., Barroso, J. J., Macau, E. E. N., & de Godoy, M. F. Application of an automatic adaptive filter for Heart Rate Variability analysis. Med Eng Physics (2013), 1778–1785.

Fairclough, S. H., & Venables, L. Prediction of subjective states from psychophysiology: a multivariate approach. Biol Psy (2006) 100–110.

Hart S. Nasa-Task Load Index (NASA-TLX); 20 Years Later. Proc Hum Factors Ergon Soc Annu Meet (2006) 904–908.

Hollnagel, E. Context, cognition and control. In Y. Waern (Ed.), Co-operative process management, cognition and information technology. In Collection, London: Taylor & Francis (1998) 27–52.

Posada-Quintero, H.F., Florian, J.P., Orjuela-Cañón, Á.D., Chon, K.H. Highly sensitive index of sympathetic activity based on time-frequency spectral analysis of electrodermal activity. Am J Physiol Regul Integr Comp Physiol (2016) 582-591.

Santiago-Espada, Y., Myer, R., Latorella, K., Comstock, J. The Multi-Attribute Task Battery II (MATB-II) software for human performance and workload research: a user's guide (2011).

Task Force of the European Society of Cardiology and the North American Society of Pacing and Electrophysiology. Heart rate variability. Standards of measurement, physiological interpretation, and clinical use". Eur Heart J (1996) 354–381.

Wilson, G. F. An analysis of mental workload in pilots during flight using multiple psychophysiological measures. Int J Aviation Psych (2002) 3–18.

Yerkes, R. M., & Dodson, J. D. The relation of strength of stimulus to rapidity of habit-formation. J Comp Neurol Psychol, (1908) 459–482.

EVALUATION OF UAS OPERATOR TRAINING DURING SEARCH AND SURVEILLANCE TASKS

Pratusha Reddy
[1] School of Biomedical Engineering, Science & Health Systems, Drexel University
Philadelphia, PA, 19104, USA
Dale Richards
[2] School of Mechanical, Aerospace & Automotive Engineering, Coventry University
Coventry, UK
Kurtulus Izzetoglu
[1] School of Biomedical Engineering, Science & Health Systems, Drexel University
Philadelphia, PA, 19104, USA

Unmanned aircraft system (UAS) sensor operators are typically required to execute search and surveillance tasks. Brain-in-the-loop measures during such tasks can help evaluate expertise development and cognitive capacities of the operator, which can be an important asset in designing adaptive and personalized training systems. Emergence of functional near infrared spectroscopy (fNIRS) has enabled monitoring of operators' prefrontal cortex (PFC) area, which is associated with higher level cognitive functioning such as decision-making, problem-solving, working memory and attention in everyday working environments. In a previous sensor operator training study, we investigated and reported preliminary evidence suggesting that fNIRS measures acquired from the left prefrontal cortex were associated with the development of scanning efficiency. Here we extend these findings by exploring skill acquisition in terms of changes in the functional brain activity correlated with the improvement in target search task. During each target search task participants were required to engage in route scanning, and target identification. Neurophysiological measures via fNIRS were found to be positively correlated with behavioral results suggesting that those who were actively engaged in finding targets, had significant changes in both left and right prefrontal cortex.

Unmanned aircraft systems (UAS) afford operational flexibility, including but not limited to effective and efficient surveillance tasks. Although the use of UASs presents a great opportunity, their associated mission success rate is greatly dependent upon human-system interaction and platform capability. In 2004, a comprehensive assessment of UAS accidents in different sectors of the US military, reported that both system and human factors contributed to the likelihood of an incident (Williams, 2004). While technological advances have been able to decrease the contribution of system-related failures, those attributed to human factors still remain at 70%. This issue is somewhat confounded when we consider wider human factors issues to do with the lack of perceptual cues available to the operator, the issues of lag, and the associated issues with operating within the wider national airspace system (NAS) with other air users.

The composition, designated roles and responsibilities of a UAS flight crew is somewhat dependent on the platform and nature of operations (Mccarley & Wickens, 2004). In some instances, several roles within the UAS crew may be shared across several crew members, and alternate between them, or be carried out by the same individual. Regardless of the crew composition, the operator in control of the asset is a focal point for ensuring not only the safe flying of the UAS, but also the operational effectiveness associated with the mission. Specifically, mission effectiveness is very much driven by the payload or sensor operator (SO). In the case of surveillance and search uses, the pilot (or a specific SO crew member) is expected to operate a number of sensors such as a camera during different phases of flight. To accomplish the manipulation of the sensor in relation to the surrounding mission constraints and specialist instructions, it is not surprising to assume that a SO must undergo an effective training program that allows the individual to develop appropriate cognitive skills that best suit the aforementioned SO tasks. The design of an appropriate training methodology must ensure appropriate and effective SO performance related to improved search behaviors, tracking, and classification accuracy to reduce incidents of false identification.

In order for the SO to reach a particular level of operational competence, it is expected that significant cognitive effort and activity will be elicited in specific regions of the brain, particularly the prefrontal cortex (PFC) (Izzetoglu et al., 2014; Menda et al., 2011). Cognitive effort assocaited with regions of the PFC produce a metabolic demand, which in turn causes an increase in blood flow to the specific regions of the human brain taxed by the assigned task and/or specfic target search mission. Recent advances in optical brain imaging techniques, in particular functional near infrared spectroscopy (fNIRS), have allowed portable application of monitoring brain activity of operators within their normal working environments. fNIRS exploits the optical properties of biological tissues and hemoglobin chromophores in assessing changes in brain activity. It does so by deploying wavelengths between 700 to 900nm, where the chromophores of oxygenated and de-oxygenated hemoglobin (HbO_2 and HbR, respectively) are found to be the main absorbers. The changes in HbO_2 and HbR are directly associated with changes in brain activity. Therefore, fNIRS can offer a direct method to assess a SOs' brain activity, via metabolism of oxygen, during task execution.

Methods

Participants

Eleven participants between the ages of 18 to 42 ($\bar{X}$= 22; SD=8) voluntarily consented to participate in the Institutional Review Board (IRB) approved study. Two participants were not included in the analysis due to incomplete sessions. All participants had no prior UAS piloting simulator experience, had normal or corrected to normal vision, were verified as right handed via the Edinburgh Handedness assessment.

Experimental Protocol

A Ground Station simulator, the Simlat's C-STAR (Simlat Inc., Miamisburg, Ohio), was used in this study as it offers a simulator training apparatus that implements SO's tasks and presents a realistic representation of their role (Reddy et al., 2018; Izzetoglu & Richards, 2019). The simulator allows for two trainees and one instructor to operate a generic tactical unmanned system (G-TAC UAS) simultaneously with designated roles. The participants took part in five sessions, of which only the first three training sessions are reported within the scope of this study. The route and scanning area of the map were identical for all the sessions, however location of the target per each sub-area was random. Additionally, to reduce task complexity, task variability, the piloting of the UAS was set to auto-pilot mode following a pre-determined route via pre-designated waypoints.

During each session, the screen display was orientated to show a map (left display) and a sensor payload screen (right display), as per Figure 1. The trainee was provided with real-time field of view (FOV) measures on the map screen in the form of differently sized polygons. On the other hand, the sensor screen displayed the simulated model of the landscape of Mallorca, Spain. It had a crosshair located in the center and zoom level gauge located to the left, which were utilized by the operator to complete their missions.

Figure 1. Trainee's Screen. Left side is showing the map screen with the route that the UAS will be following, while the right is showing the payload screen, with the zoom level indicator placed at the left edge of this screen.

The trainee was expected to move the position of the camera while varying the camera's FOV to search the sub-area (scan task). While an area was being scanned by the UAS camera, participants engaged in a target identification task where they screened the area for a pre-identified target (in this instance a singular *red civilian bus*) located in each sub area. To accomplish this, the participant was instructed to zoom in as close as possible when they located the target. The participant was then instructed to position the camera's crosshair onto the target and lock on. Throughout this task the participant was attached to a 16-Channel fNIRS device that acquired data from the left and right PFC.

Data Analysis

Behavioral data processing. The Performance Analysis & Evaluation module (PANEL) of the C-STAR system outputted sub-area start time, time elapsed since the sub-area started, zoom angle at which the scan was being conducted, scan polygon vertices, region of interest (ROI) polygon vertices, target location within the ROI, and lastly true or false indicating whether a target fell within the camera FOV or not. Target found or NOT found was assessed by extracting all the polygons that happened at a zoom angle less than 20 and had "true" tag in the target in FOV column, an example of this is show in Figure 2.

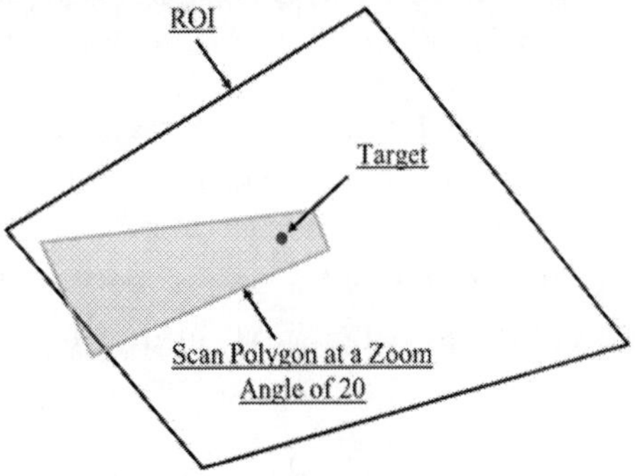

Figure 2. Target was classified as found if there existed a scan polygon (yellow) that happened below a zoom angle of 20 and the target was within the polygon.

fNIRS Signal Processing. fNIRS signal can be corrupted by instrument noise, physiological noise and motion artifacts (M. Izzetoglu et al., 2005). Therefore, to improve the sensitivity and spatial specificity of neuronal activity, a finite impulse response low pass filter, a linear detrending algorithm and a motion correction method (temporal derivative distribution repair) (Fishburn, Ludlum, Vaidya, & Medvedev, 2019) were applied. Then, modified Beer-Lambert Law was used to calculate the oxygenated (HbO_2) and deoxygenated-hemoglobin (HbR) changes at each channel (Villringer & Chance, 1997). Using HbR and HbO_2 measures, oxygenation ($Oxy = HbO_2 - HbR$) and total hemoglobin ($Hb_{total} = HbO_2 + HbR$) were derived. Lastly, samples that were three standard deviations above the expected values were classified as outliers and removed from further analysis.

Statistics. Due to small sample size, non-parametric Independent-Samples Mann-Whitney U Test was used to determine differences between groups.

Results

Each task per sub-area, trial and participant received a Found or Not Found label according to the criteria described previously. There were six targets to be found per trial, making a total of eighteen targets per participant, which sums up to 162 tasks. Investigating total number of targets found per participant, resulted in the identification of two groups. The first group, determined as a high-performance group, consisted of 6 participants who had medium to high counts of total finds, as shown in Figure 3a. Consequently, Figure 3b shows the low-performance group, which comprised of 3 participants who had minimal number of total target finds. On average, as shown in Figure 3c, the total number of finds in the high-performance group was significantly greater than that of the low performance group. Furthermore, significant differences were also observed when average number of finds per trial was assessed, as shown in Figure 3d.

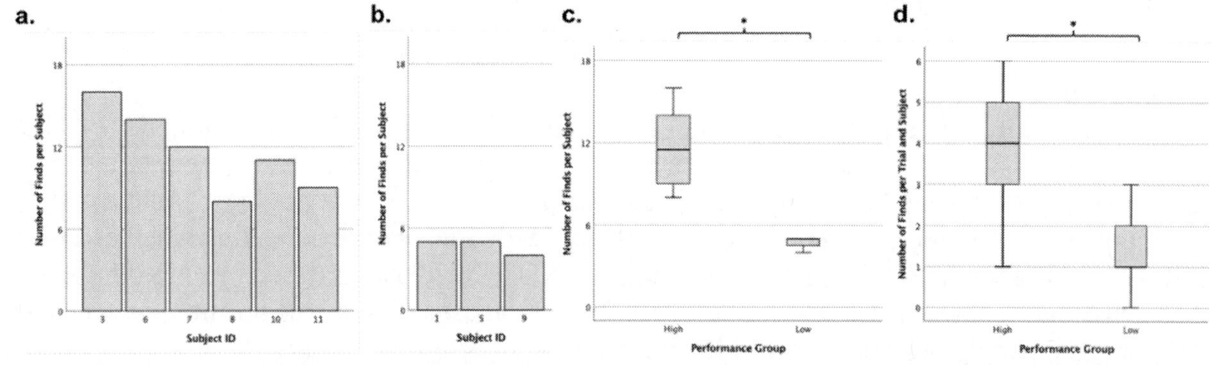

Figure 3. a. Total number of finds per subject in the high-performance group ***b.*** Total number of finds per subject in the low-performance group ***c.*** Average target finds per subject ($p = 0.024$) ***d.*** Average target finds per trial and subject ($p = 0.001$)

Our hypothesis for the target task was that oxygenation levels in the prefrontal cortex, especially in left PFC which is part of the working memory network, should be higher in engaged or high performer than that of a low performer (M. Izzetoglu et al., 2005). Comparison of HbO_2 measures from Optodes over middle frontal gyrus between performance groups support our hypothesis that high performers on average have higher activation during target finds than low performers, as indicated in Figure 4a. If a participant is engaged in finding a target, then he/she should also be actively scanning. Therefore, we also expected that the high performer group on average will have more activation in right PFC. fNIRS measures from the right PFC, have been known to be associated with sustained attention (M. Izzetoglu et al., 2005). Results from Figure

4b support this statement, where a significant difference can be seen between HbO_2 measures from high and low performers.

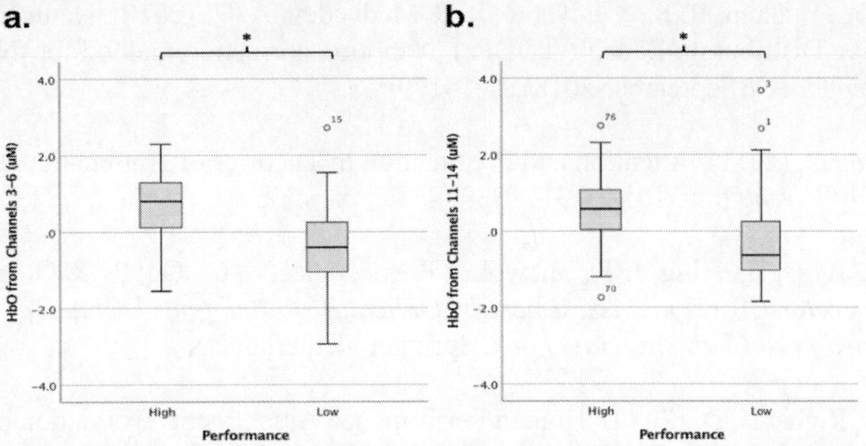

Figure 4. Significant differences were observed in HbO_2 measures between high and low performers in both ***a.*** left PFC ($p < 0.01$) and ***b.*** right PFC ($p < 0.01$) regions

Discussion

The cognitive demands placed on the SO play a crucial role within the context of UAS operation. Understanding the effect of these cognitive demands on the performance of a SO has implications for mission effectiveness, and the associated evaluation of training. Traditionally the assessment of an operator's ability would identify behavioral performance assessment parameters, such as task performance, task load assessment, etc. However, we suggest how direct neurophysiological measures through wearable sensors can provide complementary quantitative assessment methods.

In our previous preliminary studies, we reported separately that as a participant actively scanned or found targets, they had higher oxygenation in attention and working memory PFC regions (Izzetoglu & Richards, 2019; Reddy et al., 2018). In this current study, we sought to further explore the evolving relationship using target search tasks. Our behavioral results from target task demonstrated difference between high and low performers. The fNIRS results were positively correlated with behavioral results suggesting that those who were actively engaged in finding targets, had significant changes in both left and right PFC regions.

Although these behavioral and neuro-physiological measures are in line with previous studies, and provided supporting evidences to the previous findings, a comprehensive analysis on relationship between tasks and their influence on fNIRS measures acquired from all the PFC regions need to be assessed with large sample size. Specifically, we plan to study whether advances in cognitive engagement lead to faster response times in identifying a target and reduce incidents of false positive identification rate as shown in prior studies (Gordon D. Logan, 2011).

Acknowledgement

The authors express their gratitude to Shahar Kosti for his valuable review and support for scenario design, UAS simulator setup and behavioral data. The authors also would like to thank Simlat, Inc for providing access to their UAS training simulator.

References

Fishburn, F. A., Ludlum, R. S., Vaidya, C. J., & Medvedev, A. V. (2019). NeuroImage Temporal Derivative Distribution Repair (TDDR): A motion correction method for fNIRS. *NeuroImage*, *184*(September 2018), 171–179.

Gordon D. Logan. (2011). Attention and Preattention in Theories of Automaticity. *The American Journal of Psychology*, *105*(2), 317–339.

Izzetoglu, K., Ayaz, H., Hing, J. T., Shewokis, P. A., Bunce, S. C., Oh, P., & Onaral, B. (2014). *UAV Operators Workload Assessment by Optical Brain Imaging Technology (fNIR). Unmanned Aerial Vehicles Handbook.* Springer Netherlands.

Izzetoglu, K., Richards, D. (2019). Human Performance Assessment: Evaluation of Wearable Sensors for Monitoring Brain Activity. In M. Vidulich & P. Tsang (Eds.), *Improving Aviation Performance through Applying Engineering Psychology: Advances in Aviation Psychology* (1st ed., pp. 163–180). Boca Raton, FL: CRC Press.

Izzetoglu, K., Bunce, S., Onaral, B., Pourrezaei, K., & Chance, B. (2004). Functional Optical Brain Imaging Using Near-Infrared During Cognitive Tasks. *International Journal of Human–Computer Interaction*, *17*(2), 211–227.

Izzetoglu, M., Izzetoglu, K., Bunce, S., Ayaz, H., Devaraj, A., & Onaral, B. (2005). Functional Near-Infrared Neuroimaging. *IEEE Transactions on Neural Systems and Rehab. Engineering*, *13*(2), 153–159.

Mccarley, J. S. & Wickens, C. D. (2004). *Human Factors Concerns in UAV Flight.* Technical Report.

Mccarley, J. S., & Wickens, C. D. (2005). *Human Factors Implications of UAVs in the National Airspace*. Techical Report

Menda, J., Hing, J. T., Ayaz, H., Shewokis, P. A., Izzetoglu, K., Onaral, B., & Oh, P. (2011). Optical Brain Imaging to Enhance UAV Operator Training , Evaluation , and Interface Development. *Journal of Intelligent & Robotic Systems*, *61*, 423–443.

Reddy, P., Richards, D., & Izzetoglu, K. (2018). Cognitive Performance Assessment of UAS Sensor Operators via Neurophysiological Measures. In *2nd International Neuroergonomics Conference*.

Villringer, A., & Chance, B. (1997). Non invasive optical spectroscopy and imaging of human brain function. *Trends in Neurosciences*, *20*(10), 435–442.

Williams, K. W. (2004). *A Summary of Unmanned Aircraft Accident / Incident Data : Human Factors Implications*.

CONTROLLER-PILOT COMMUNICATION AS AN INDEX OF HUMAN PERFORMANCE IN THE NATIONAL AIRSPACE SYSTEM

Mustafa Demir, Sarah Ligda, Nancy Cooke, Megan Seeds, Mariah Harris, Mary Niemczyk
Arizona State University, Polytechnic School,
Mesa, Arizona, 85212, USA

New capabilities to modernize the U.S. National Airspace System (NAS) include support of real-time information streams derived from many data sources across the NAS. This provision allows for system risk prognostics originating from sets of diagnostic health information. The current exploratory paper presents how to model human performance with the larger purpose of developing NAS risk prognostics. We explore ways in which human performance relates to communication and coordination among controllers and pilots in the context of their objectives, technologies, and environment. A literature review shows communication is often associated with controller performance in both experimental simulations and safety reviews. We gathered controller and pilot verbal communication data from two incidents and one accident and examined them using a dynamical systems method—discrete recurrence quantification analysis—to visualize and identify stability and flexibility between controller and pilot during the failures. From our findings, we conclude that controller-pilot need effective and timely interaction in order to overcome fatal incidents.

The U.S. National Airspace System (NAS) is a vast and complex system, comprised of macro and micro level components, such as airports, control centers, airlines, aircrafts, pilots, and passengers, that are nested within one another (Laskey, Xu, & Chen, 2012). According to the Federal Aviation Administration (FAA) report in 2018, more than 26 million flights carrying nearly 972 million passengers were operated in the NAS in 2017 (Meilus, 2018). In order to meet future growth rates of about 2% per year, advanced technologies, services, and procedures are being developed and implemented in the NAS under the Next Generation Air Transportation System program (Joint Planning and Development Office, 2010). With these capabilities, new and existing sources of real-time data will be available and provide opportunities for system-wide diagnostic health information and prognostic risk assessment via data fusion.

As an emergent property, safety of the NAS arises from interactions between many elements and different levels, ranging from those attributable to humans, technology, and the environment. NAS selectively open systems, each component needs to interact with other components, exchange resources and information, and operate under broad regulations to achieve overall system objectives (Harris & Stanton, 2010). Sometimes incidents and accidents result from insufficient interaction (communication and coordination) between humans (e.g., pilot-controller). The 2012 review by Edwards and colleagues centered on nine human factors constructs and reported that the leading contributors to incidents were communication, teamwork, and attention-related measures. In research of controller-pilot verbal communication, content-based evaluations have shown two consistent themes: (1) controller transmissions that are lengthy and (2) those with more than one piece of information correlate with more frequent pilot readback errors (Morrow, Lee, & Rodvold, 1993; Cardosi, 1996; Prinzo, Hendrix, &

Hendix, 2009). Controller-pilot communication often corresponds to phase of flight activity. Cardosi (1996) analyzed 48 hours of communication from eight Terminal Radar Approach Control (TRACON) facilities, 24 from high controller workload and 24 from moderate workload. Despite typical increases in pilot workload during departure and approach phases of flight, the authors found less than 1% of messages resulted in communication errors. Moon, Yoo, & Choi (2011) suggest that the level of air traffic density impacts verbal errors by controllers when operating terminal airspace sectors that service large Korean airports. The authors documented elements in controller-pilot verbal transmissions that indicated difficulties in interactions, such as "wrong call sign used" and "non-standard phraseology". Results revealed controllers made 1.37 verbal errors when controlling 2-3 aircraft per 15 minutes, while 6.30 verbal errors were committed when controlling 30 aircraft per 15 minutes (Moon et al., 2011).

The content of communication will continue to provide value and support understanding with a multitude of team, individual, and data sets within air traffic research. In addition, another dimension to communication with a potentially rich source of understanding is everything but its explicit meaning. Cooke and Gorman (2009) describe methods of communication flow between teams (considered as a system) that have proven insightful. The first is a ratio of team members speech quantity, which can indicate the degree of influence one member has over others. Another is the communication required and passed score, or how much variation there is in actual team communication from expectations. Flow quantity represents how much speech each member of the team produces. In another study, Gorman, Amazeen, & Cooke (2010) underline the importance of coordination dynamics, and explain that "systems with different material substrates can exhibit the same dynamics", which is known as dynamical similitude, which can be used to "guide selection of appropriate dynamical systems methods for a system that has not been previously analyzed using a dynamical approach" (Gorman et al., 2010, p. 285). Gorman et al. (2012) study applied discrete Recurrence Quantification Analysis (RQA) method on team communication flow data as a measurement technique for coordination dynamics Unnamed Vehicle (UAV) teams, wherein mixed teams (i.e., team members changed) or intact teams (i.e., stayed the same over successive experimental sessions. Interestingly, mixed teams were better able to adjust to unexpected perturbations, and this ability was linked to team level coordination dynamics. That is, mixed teams adopted a globally stable pattern of communication while exhibiting strong temporal dependence (Gorman, Cooke, Amazeen, & Fouse, 2012). Demir, Cooke, & Amazeen (2018) found that metastable team coordination (not too stable nor too flexible) between team members is important to successfully overcome novel events (i.e., team situation awareness) in a dynamic task environment. In the current study, we investigate the potential of dynamical systems perspectives to capture the differential dynamics of three cases between controller-pilot communication flow during incidents and accidents.

Aviation Incidents and Accidents

Three cases of controller-pilot audio transmissions were obtained from "Cockpit Voice Recorder Transcripts," (2019), see Table 1, and analyzed via discrete RQA. The cases represent situations of particular interest, communication and coordination.

Table 1.
Flight Incidents and Accidents and Their Description

Flight	Description
Southwest Airlines Flight 1380 ("CVR Transcript Southwest 1380," 2018):	Boeing 737-700 enroute from New York–LaGuardia Airport to Dallas Love Field on April 17, 2018. Parts of the engine broke off and struck a window on the plane, causing rapid cabin depressurization and prompting the flight crew to conduct an emergency landing in Philadelphia International Airport. One passenger sitting adjacent to the failed window received fatal injuries and eight passengers received minor injuries. The aircraft sustained substantial damage.
US-Bangla Airlines Flight 211 ("US Bangla 211 CVR Transcript," 2018)	Bangladeshi airline (Bombardier 8-Q400) in Katmandu, Nepal in 2018. Airport tower cleared plane to land on runway 02, but confusing communication led the pilot to approach runway 20. Crashed due to runway confusion: 52 people died and 19 survived.
Aer Lingus Flight 12C ("CVR Transcript Aer Lingus 12C," 2018)	A flight from Dublin Airport to O'Hare International Airport in Chicago, Illinois. On takeoff, they reported a landing gear issue. They landed safely.

Discrete Recurrence Quantification Analysis Results

One of the approaches for investigating interaction patterns between the system components (in the controller-pilot case) and their change over time involves looking at communication flow using discrete Recurrence Plot (RP) and corresponding Recurrence Quantification Analysis (RQA) that quantifies how many recurrences with a given length are present by multidimensional space (phase space) trajectory in a dynamical system (Marwan, Carmen Romano, Thiel, & Kurths, 2007). The basis of discrete RQA is the RP (Eckmann, Kamphorst, & Ruelle, 1987) which is a visual tool for demonstrating a system's recurrent structure in the phase space when a system revisits specific states or sequences of states within a region of phase space over a period of time. In the case of two or more systems, discrete RP displays the times when two or more separate dynamical systems show a recurrence simultaneously (Marwan et al., 2007).

In this study, we used discrete RP to measure two or more behavioral dynamics of dyad communication flow encoded in discrete codes. The input to the discrete RP consisted of an ordered sequence of dichotomous codes: ''0" for ''ground/controller", and ''1" for ''flight", or if there was a third party involved "2" for "Rescue". Therefore, there is a series of discrete speaker states represented by a sequence of codes. Discrete RQA quantifies not only the effect of interventions (such as unexpected events) on instability, but also the dyad interaction processes and the dynamics that contribute to that process. The RQA was used to produce several measures, including: percent recurrence rate, percent determinism (DET), longest diagonal line, longest vertical line, entropy, and laminarity. Of these, the focal variable was determinism (DET; depicted in formula (1), Marwan et al., 2007), which indicates the amount of organization in the communication of a system. DET is derived from the recurrence plot by examining how the recurrent points are distributed: dyads with high determinism tend to repeat sequences of states many times—producing many diagonal lines (see Figure 1)—while controller-pilot with low determinism rarely repeat a sequence of states, producing few diagonal lines. The numerical value of DET comes from considering the upper triangle of points in the recurrence plot and then computing the proportion of points that form diagonal lines (see Figure 4) (Marwan et al., 2007).

$$DET = \frac{\sum_{l=l_{min}}^{N} lP(l)}{\sum_{l=1}^{N} lP(l)} \qquad (1)$$

where l is the diagonal line length considered when its value is $\geq l\text{min}$ and $P(l)$ is the probability distribution of line lengths. For instance, a 0% means that the time series never repeats; 100% means the time series repeats perfectly. DET essentially measures the degree to which two or more components (in this context ground and flight) are interacting in sync and how much they influence each other. This works by considering communication flow between controller and pilot, i.e., each dyad's communication for the duration of the task. For instance, in Figure 1a, we show discrete RPs between ground (i.e., controller) and flight (i.e., pilot) for SWA 1380 failure event. In the figure, the plot demonstrates a number of observations based on the dyads' (i.e. ground and flight) frequency of communication around a 17-minute events (79 communication sequences).

In Figure 1, we give three real case RPs based on ground and flight (and for Figure 1c also rescue) for two incidents and one accident (either two or three-code sequences) for (a) *Southwest Airlines Flight 1380 incident* (DET = 50%), (b) *US-Bangla Airlines Flight 211 accident* (DET = 64%), and (c) *Aer Lingus Flight 12C incident (*DET = 28%). These three examples of discrete recurrence plots demonstrate three different synergies among the ground and flight personnel during their own novel events. In Figure 1, the top of each RP depicts the communication flow between ground and flight (and the third party, if relevant; see Figure 1c: ground, Flight 12C, and rescue team). As a reminder, if ground sent a message, it was coded as 0, otherwise it was coded as 1 on the *y-axis* for the flight or it was coded as 2 for the rescue. The *x-axis* indicates the sequence of the communication flow (the number of communications sent).

According to Figure 1a, dyad communication shows metastable behavior (i.e. neither stable nor flexible: DET= 50%). In this specific case, air traffic control (ground) anticipated the issue in a timely manner and solved the issue regarding landing. The pilot and the controller were both aware of the situation. First, pilot pulled the nearest airport information from the controller, but quickly decided on Philadelphia. Then, the controller provided flight information to the Philadelphia airport in a timely manner. Even though loud sounds and other aircraft distractions caused initial communication issues between the controller and the pilot, as the aircraft stabilized, communications improved. In the middle of the event (Figure 1a), ground communicated and coordinated with the pilot (SWA 1380) and other aircrafts which were ready for landing. The controller's anticipation of the pilot's and airport's needs continued until the flight ended in a safe landing ("CVR Transcript Southwest 1380," 2018). This incident case shows the difference that timely awareness of the situation (aware of the technological failure by pilot and the controller), communication (timely anticipation between controller-pilot), and metastable coordination can make.

From the RPs, Ground-UBG 211 (see Figure 1b) had more synchrony than the other two cases. However, having more synchrony within ground and flight (i.e., controller-pilot) does not equate to successfully overcoming a novel situation; it can even create a novel situation based on the communication behavior. In this case, Ground-UBG 211 was one of the deadliest aviation disasters in aviation history and it was caused by confusion from conflicting communications between the controller and the pilot. In the beginning, the controller (i.e., ground) gave information to land on runway 02, but then the confusion about runway numbers start (between runways 02 and 20). Later on, the control tried to fix the confusion (see Figure 1b: between 30 to 40 black dots on the diagonal with repeated communication pattern). However, the confusion continued until the controllers gave a last try (long diagonal dot at the end of Figure 1b), and then the plane crashed ("US Bangla 211 CVR Transcript," 2018). The communication between controller and the pilot was full of confusion. Therefore, an argument can be made that the quality and effectiveness of the communication is more important than the quantity or frequency of communication. Most importantly, the controller and the pilot were not aware of the problem in a timely manner (i.e., lack of situation awareness).

Finally, in Figure 1c, three roles were considered to create the discrete RP and extract the DET measure. Overall, interactions in this incident were more flexible in comparison to other novel events, which may be partially explained by the mere presence of the third party (i.e., the third party increases the number of possible communication patterns: DET= 28%). When looking at the substance of the communication, the controller was coordinating and communicating with several flights and rescuers about the runways. In the beginning through the middle of the diagonal of the RP, the situation was routine in terms of landing and coordination was random across the three roles (*Ground, 12C, and Rescue*). Later, 12C noticed the landing gear issue and let the controller know in a timely manner. After that, both the controller and pilot anticipated each other's needs in order to land safely while rescue was preparing the runways ("CVR Transcript Aer Lingus 12C," n.d.). Overall, interaction dynamics and situation awareness indicate that effective interaction between the controller and pilot is crucial to effect situation awareness, successfully overcoming the failures.

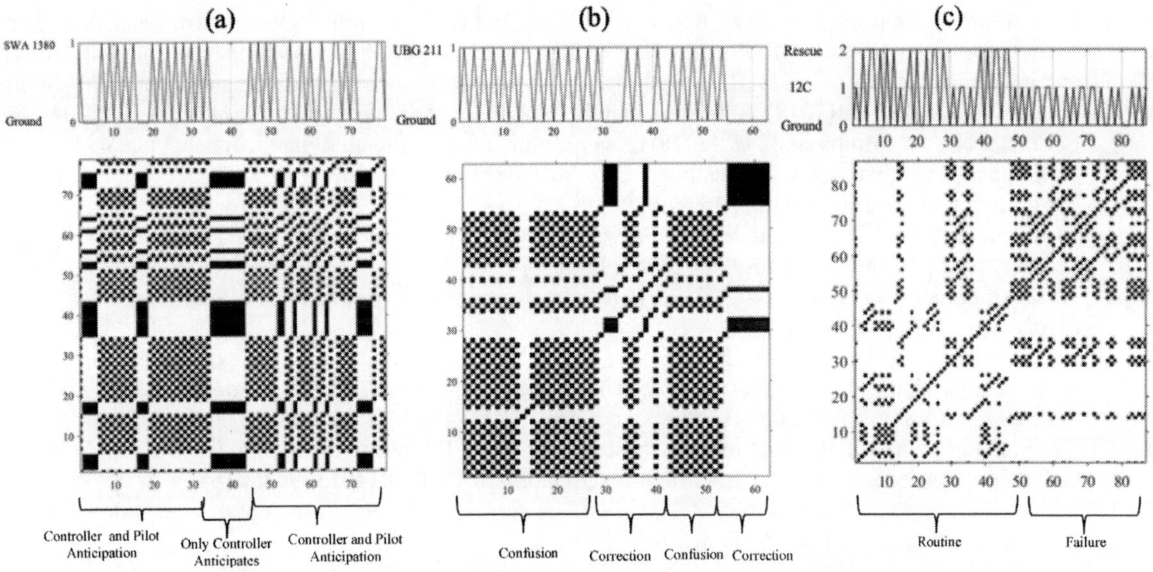

Figure 1. Discrete RPs based on number of communication events from: (a) Ground and SWA 1380 dyads (%DET= 50%), (b) Ground and UBG 211 (%DET= 64%), (c) Ground, Flight 12C, and Rescue (%DET= 28%).

Conclusion

We have presented three controller-pilot communication flows via discrete RP and RQA methods that differentiate three real cases based on discrete interaction sequences. The measures extracted from the RQA and visualizations of the interaction patterns show that effective communication and coordination is needed for effective situation awareness, i.e., overcoming the failures. Based on the previous studies (Demir et al., 2018), we expected the rigidity of the coordination dynamics between controller and pilot in the UBG 211 case would cause the fatal accident as well as lack of communication (confusion during the landing) and in turn lack of situation awareness. On the other hand, two other incidents demonstrated more flexible behavior across the roles (controller-pilot) to adapt to dynamic environment. In this case, the key lies in the dynamic transition between interaction and the environment. That is, controller-pilot are compelled to adjust their interaction patterns (flexibility) to adapt to changes in the environment and maintain a stable trajectory toward meeting their goals, such as safely landing. Thus, there are three crucial states for effective interaction in both temporal and spatial states "what needs to be communicated", "when it needs to be coordinated", and "how it needs to be communicated and coordinated".

Acknowledgment

This work is supported by a cooperative agreement award (NNX17AJ86A) from the National Aeronautics and Space Administration (NASA). The views expressed are those of the authors and do not reflect the official policy or position of the National Aeronautics and Space Administration, U.S. Government, or Arizona State University.

References

Cardosi, K. M. (1996). *An analysis of TRACON (Terminal Radar Approach Control) controller-pilot voice communications*. Available through the National Technical Information Service.

Cockpit Voice Recorder Transcripts. (2019). Retrieved March 11, 2019, from https://www.tailstrike.com/index.htm

CVR Transcript Aer Lingus 12C. (2018). Retrieved March 11, 2019, from https://www.tailstrike.com/180218.html

CVR Transcript Southwest 1380. (2018). Retrieved March 10, 2019, from https://www.tailstrike.com/170418.html

CVR Transcript US Bangla 211. (2018). Retrieved March 11, 2019, from https://www.tailstrike.com/120318.html

Demir, M., Cooke, N. J., & Amazeen, P. G. (2018). A conceptual model of team dynamical behaviors and performance in human-autonomy teaming. *Cognitive Systems Research*, *52*, 497–507. https://doi.org/10.1016/j.cogsys.2018.07.029

Eckmann, J.-P., Kamphorst, S. O., & Ruelle, D. (1987). Recurrence Plots of Dynamical Systems. *EPL (Europhysics Letters)*, *4*(9), 973. https://doi.org/10.1209/0295-5075/4/9/004

Gorman, J. C., Amazeen, P. G., & Cooke, N. J. (2010). Team coordination dynamics. *Nonlinear Dynamics, Psychology, and Life Sciences*, *14*(3), 265–289.

Gorman, J. C., Cooke, N. J., Amazeen, P. G., & Fouse, S. (2012). Measuring Patterns in Team Interaction Sequences Using a Discrete Recurrence Approach. *Human Factors: The Journal of the Human Factors and Ergonomics Society*, *54*(4), 503–517. https://doi.org/10.1177/0018720811426140

Harris, D., & Stanton, N. A. (2010). Aviation as a system of systems: Preface to the special issue of human factors in aviation. *Ergonomics*, *53*(2), 145–148. https://doi.org/10.1080/00140130903521587

Joint Planning and Development Office. Next Generation Transportation System: Concept of Operation v.3.0. Government Printing Office, Washington D.C. (2010)

Laskey, K. B., Xu, N., & Chen, C.-H. (2012). Propagation of Delays in the National Airspace System. *ArXiv:1206.6859 [Cs]*. Retrieved from http://arxiv.org/abs/1206.6859

Marwan, N., Carmen Romano, M., Thiel, M., & Kurths, J. (2007). Recurrence plots for the analysis of complex systems. *Physics Reports*, *438*(5–6), 237–329. https://doi.org/10.1016/j.physrep.2006.11.001

Meilus, A. (2018). *Air Traffic By The Numbers*. Federal Aviation Administration. Retrieved from https://www.faa.gov/air_traffic/by_the_numbers/media/Air_Traffic_by_the_Numbers_2018.pdf

Moon, W.-C., Yoo, K.-E., & Choi, Y.-C. (2011). Air Traffic Volume and Air Traffic Control Human Errors. *Journal of Transportation Technologies*, *01*, 47. https://doi.org/10.4236/jtts.2011.13007

Morrow, D., Lee, A., & Rodvold, M. (1993). Analysis of problems in routine controller-pilot communication. International Journal of Aviation Psychology, 3(4), 285-302.

Prinzo, O. V., Hendrix, A. M., & Hendrix, R. (2009). The outcome of ATC message length and complexity on en route pilot readback performance (NTIS No. ADA494551). Oklahoma City, OK: FAA Civil Aerospace Medical Institute.

Evaluating Team Dynamics for Collaborative Communication Alignment Tasks

Adam Jones, Nia S. Peters*
University of Cincinnati, Cincinnati, Ohio
711th HPW, Battlespace Acoustic Branch, Air Force Research Laboratory, Dayton, Ohio*

The primary focus of this work is in exploring human teaming dynamics within goal-oriented communication alignments tasks. A communication alignment task within the context of this work is one in which two teammates have the exact same target information, but from differing perspectives and must communicate in an effort to align their knowledge and agree on the target output. Such an interaction within aviation could occur between a pilot and air traffic controller or ground troop personnel and Unmanned Aerial Vehicle (UAV) controller. The goal is to compare the task performance and time of completion of a communication alignment task between various team compositions: novice-novice, novice-expert, and expert-expert teams. In this work a novice team member is defined as one who is new to the experimental subject panel and has no experience with the simulated communication alignment task at the onset of the data collection. An expert on the other hand has over 6 months of experience on the experimental subject panel and was trained on the task 9 months prior. It is hypothesized that there is a positive correlation between the number of experts within the team composition and the task performance and a negative correlation between the number of experts and the time of completion. The results indicate a decline in task performance and increase in task completion time with less experts on a team. This work aims to better inform the impact that teammate expertise could contribute to performance outcomes within goal-oriented collaborative knowledge alignment interactions.

Effective communication within a team setting is vital and serves as an essential backbone in tasks that require collaboration. In a world where communication is moving rapidly towards electronic mediums, it is important to understand the dynamics that produce the most effective method of communication to maximize the efficiency and performance on collaborative tasks. One such dynamic is that of expertise. This dynamic is important for communication in the domain of aviation, particularly for pilots and air traffic controllers where proper collaborative alignment of information over electronic mediums is imperative and room for error is low. Whereas previous research has focused on the effect of expertise and communication on the individual, this paper serves to better inform team composition in the future by analyzing the importance of expertise level in individuals of a team using accuracy and average time of completion of the task as markers of performance.

Background

In regards to the study of expertise, much of the current literature is focused on the performance on experts and novices outside of a team setting. However, in a recent study it was shown that novices experience collaborative inhibition when working collaboratively on a task, and experts experience collaborative inhibition when working on a simple task they have previously completed alone (Nokes-Malach, Meade, & Morrow, 2012). This implies that successful collaboration depends on both prior knowledge and experience. It has been shown that experts in a task outperform novices by both quantitative measure of skill and qualitative measure of communication method in normal task settings (Adelson, 1984). Communication methods used by experts have been shown to rely on abstract representations (Hinds, Patterson, Pfeffer, 2001) and greater precision in their descriptions (Solomon, 1990). It was also shown that experts better perform in collaborative tasks when receiving information from other experts as opposed to novices (Solomon, 1990). More recently, experts were shown to give more attention to relevant task information than novices (Sheridan & Reingold, 2014). With these trends in mind, it is the interest of this paper to analyze the effect of expertise level of a team composition on performance on collaborative communication alignment tasks.

Method

A communication alignment task is simulated via a teaming endeavor in which two teammates are tasked with identifying a target from two different perspectives. Within the context of this task, randomly selected team compositions are evaluated based on: 1) team performance, and 2) average task time of completion (TOC). This work explores the composition of expert and novice team compositions. An *expert teammate* is one who has over 6 months of experience on the designated subject panel and was trained on the task 9 months prior. A *novice teammate* is new to the designated subject panel and has no experience with the simulated communication alignment task at the onset of the data collection. The overall assumption is that teams with more experts will outperform teams with less expertise within a collaborative task not only because of task knowledge, but also potential rapport with other subjects that have done the same task. This is a loose assumption because task knowledge does not always guarantee useful teaming collaboration. Other factors such as personality, communication skills, individual performance on the task are all factors that can influence performance outcomes. In this work, the primary focus is on evaluating task performance as a function of teaming composition.

Research Question I: Is there a difference in the team performance of a communication alignment task for different teaming compositions?

> *H01:* There is no difference in the team performance of a communication alignment task for different teaming compositions
>
> *H1:* There is a difference in the team performance of a communication alignment task for different teaming compositions

Research Question II: Is there a difference in the average time of task completion for a communication alignment task for different teaming compositions?

> *H01:* There is no difference in the average time of task completion for a communication alignment task for different teaming compositions
>
> *H1:* There is a difference in the average time of task completion for a communication alignment task for different teaming compositions

Data

The goal of the data collection is to simulate a communication alignment task in which two teammates of varying expertise are tasked with identifying a target from two different perspectives. The experimental design presented in this paper is called the Uncertainty Map Task (UMT) which is a variation of the experiment described in (Griffin et al. 2016). In the UMT, two teammates communicate over a push-to-talk network to describe their respective interfaces, ground their knowledge, and identify a target house. A succession of UMT tasks are presented to the team until the team has received 10 randomly generated target house identification tasks. In the UMT task there are 4 aerial maps and 12 houses for each aerial map (12 x 4 = 48 target houses). Each target house has 4 different perspectives. Figure 1 is an example of Graphical User Interface for one of those perspectives, StreetView_Target_Aerial_ID task where the teammate with the street view describes the target house from the street view perspective and the other teammate is tasked with identifying the target house on the aerial map. Whether a teammate receives the target perspective or the identification perspective is also randomly determined in the experiment. When the teammates agree they have identified the correct house, they press a DONE button and receive feedback on whether they correctly identified the house.

Figure 1: Street View Target and Aerial Identification Task. Teammate I must describe the streetview of the target house (top) and Teammate II must identify the house on the aerial map (bottom)

There are a total of 24 participants in the data collection, 12 experts and 12 novice. There are a total of 126 randomly generated team combinations: expert-expert, expert-novice, and novice-novice with 42 samples per class. The expert-novice and expert-expert have larger sample sizes, but were randomly downsampled because the class size was skewed due to scheduling and availability of subject panel members.

Results

A one-way analysis of variance (ANOVA) was conducted to evaluate the relationship between team composition and team performance. The independent variable, team composition included three levels: expert-expert, expert-novice, and novice-novice. The dependent variable was the percentage correct of identified houses. The ANOVA was significant at the 0.05 level, $F(2,123) = 4.2$, $p = 0.0173$.

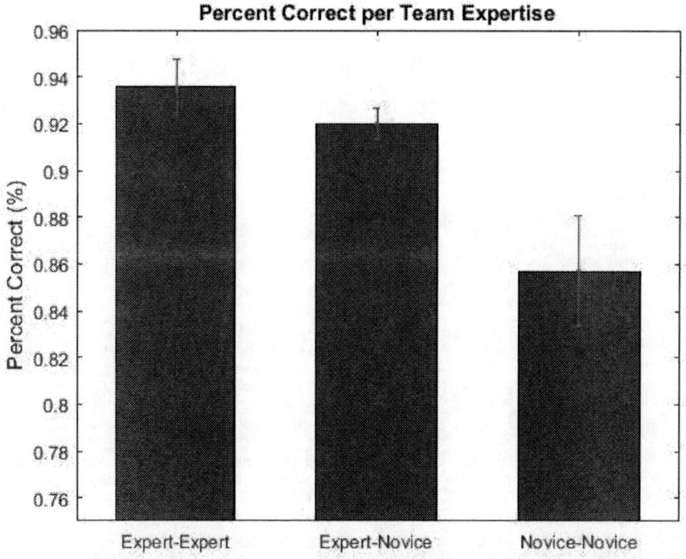

Figure 2: Percent Correct per Team Expertise

Figure 2 illustrates the mean percentage correct for each team composition. These results indicate that there is a significant difference between the means of different team composition illustrating less novice teammates results in degraded teaming performance. These results suggest that we can reject the H01 null hypothesis.

A one-way analysis of variance (ANOVA) was conducted to evaluate the relationship between team composition and average time completion. The independent variable, team composition included: expert-expert, expert-novice, and novice-novice. The dependent variable was the average time of completion to complete a target house identification task (measured in seconds). The ANOVA was significant at the 0.05 level, $F(2,123) = 19.01$, $p = 3.48e08$.

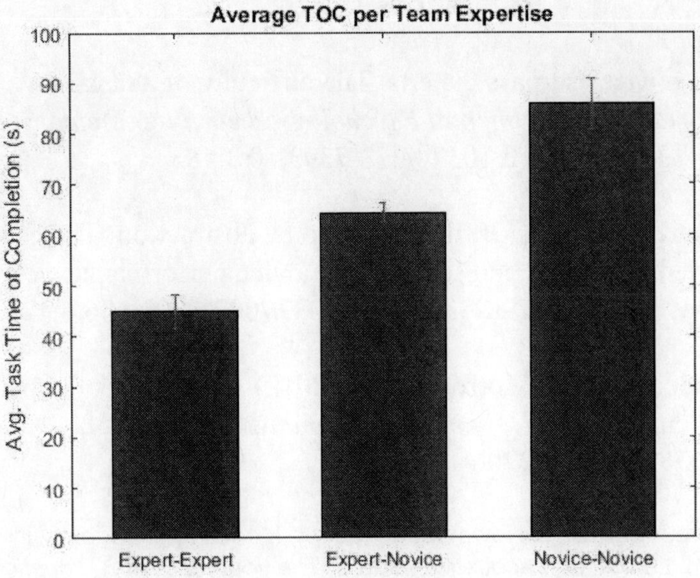

Figure 3: Average Task Time of Completion per Team Expertise

Figure 3 illustrates the mean average time of task completion in seconds for each team composition. These results indicate that there is a significant difference between average time taken to complete a collaborative communication knowledge alignment task for different team composition. teams with more experts complete the task quicker. These results suggest that we can reject the H02 null hypothesis.

Conclusion

The primary focus of evaluating task performance as a function of teaming composition was accomplished through the rejection of both the H01 and H02 null hypotheses, providing evidence that both team performance and average trial completion time of a task are affected by the expertise level of that team composition. Expert-expert teams were found to both perform the highest and have the lowest average trial completion time and novice-novice teams were found to have perfom the poorest and have the highest average trial completion time. This corroborates results found in (Solomon, 1990) from a collaborative communication alignment task. The novice-expert composition performed better and had a lower average trial completion time than the novice-novice team, implying that a novice's performance overall improves when paired with an expert as opposed to a novice. This information can be used in the future to better inform team composition in terms of expertise to produce the best overall performance

References

Adelson, B. (1984). When novices surpass experts: The difficulty of a task may increase with expertise. *Journal of Experimental Psychology: Learning, Memory, and Cognition, 10*(3), 483-495. doi:10.1037/0278-7393.10.3.483

Hinds, P. J., Patterson, M., & Pfeffer, J. (2001). Bothered by abstraction: The effect of expertise on knowledge transfer and subsequent novice performance. *Journal of Applied Psychology, 86*(6), 1232-1243. doi:10.1037//0021-9010.86.6.1232

Nokes-Malach, T. J., Meade, M. L., & Morrow, D. G. (2012). The effect of expertise on collaborative problem solving. *Thinking & Reasoning, 18*(1), 32-58. doi:10.1080/13546783.2011.642206

Romigh, G., Rothwell, C., Greenwell, B., & Newman, M. (2016). Modeling uncertainty in spontaneous speech: Lexical and acoustic features. *The Journal of the Acoustical Society of America*, *140*(4), 3401-3401.

Sheridan, H., & Reingold, E. M. (2014). Expert vs. novice differences in the detection of relevant information during a chess game: Evidence from eye movements. *Frontiers in Psychology, 5*. doi:10.3389/fpsyg.2014.00941

Solomon, G. E. (1990). Psychology of Novice and Expert Wine Talk. *The American Journal of Psychology, 103*(4), 495. doi:10.2307/1423321

LEARNING FROM THE PAST: AIRLINE ACCIDENTS & THE ICAO LANGUAGE PROFICIENCY PROGRAM

Simon Cookson
J. F. Oberlin University
Tokyo, Japan

In response to multiple airline accidents involving pilot-ATC communication breakdowns, ICAO implemented a worldwide language proficiency program in 2011. The official guide to the program, Document 9835, cites four accidents in which insufficient English proficiency of pilots or air traffic controllers was a contributing factor. The purpose of this study is to investigate the relevance of the four accidents to current airline operations. A survey was distributed to UK-based pilots using BALPA online discussion forums. The survey explored: respondents' awareness of the accidents cited by ICAO; sources of information; and the role played by English proficiency in the accidents. This paper presents the results of the survey and identifies areas for further research.

Between 1976 and 2001, a total of 1,460 people died in a series of airline accidents that involved pilot-ATC communication breakdowns. In response to these accidents, the International Civil Aviation Organization (ICAO) initiated a program to improve the language proficiency of pilots and air traffic controllers. This program was developed in the early 2000s, and came into full effect in 2011. The official program guide, ICAO Document 9835, cites four accidents for which "insufficient English language proficiency on the part of the flight crew or a controller" was a contributory factor (ICAO, 2010, p. 1-1). These accidents were:

- 1977 Tenerife runway collision;
- 1990 Cove Neck, New York, fuel exhaustion crash;
- 1995 Cali, Colombia, controlled flight into terrain (CFIT);
- 1996 Charkhi Dadri, New Delhi, mid-air collision.

Some researchers have questioned the assumptions underlying ICAO's language program and called for further empirical research (Estival, Farris & Molesworth, 2016). The aim of this study is to explore the relevance of the accidents cited by ICAO to present-day operations.

Method

Research Questions

Five research questions were drawn up to investigate airline pilots' awareness of, and attitudes towards, the accidents cited in Document 9835. These research questions were used to design an online survey. The questions were as follows:
RQ1. Is studying past airline accidents important for improving airline safety?
RQ2. Which of the accidents cited by ICAO have pilots heard of?
RQ3. What are pilots' sources of information about these accidents?
RQ4. Did insufficient English proficiency of pilots play a contributing role in these accidents?

RQ5. Did insufficient English proficiency of air traffic controllers play a contributing role in these accidents?

Online Survey

A 31-item questionnaire was created using the SurveyMonkey online survey tool. It was designed to be completed in a short time (i.e. 10-15 minutes). To ensure that the question items were rigorous, they were checked by three applied linguists. The questionnaire was distributed to UK-based pilots via two British Airline Pilots Association (BALPA) online discussion forums. One of the forums is for all BALPA members and the other is for British Airways pilots only. These are closed forums where members (ie: airline pilots) discuss professional issues.

Survey Content

The survey has three sections with a mixture of question types: 5-scale Likert items; yes/no closed items; and multiple choice questions. The first section has a single item:

Q1. Studying past airline accidents is important for improving current airline safety. [Strongly agree / Agree / No opinion / Disagree / Strongly disagree]

The second section has five items that are repeated for each accident. Page skip logic is used so that, if a respondent answers "No" to the initial question, they are not asked any more questions about that accident. These are the questions for the first accident (1977 Tenerife):

Q2. On 27th March 1977, there was a runway collision between KLM Flight 4805 and Pan Am Flight 1736 at Los Rodeos Airport on the island of Tenerife. Have you heard of the 1977 Tenerife accident? [Yes / No]

Q3. Where did you hear about the 1977 Tenerife accident? (You can select more than one answer.) [Accident report / Another pilot / Book / Company training / IATA publication / ICAO publication / Internet / Magazine or newspaper article / TV documentary / TV or radio news / Other (please specify)]

Q4. The 1977 Tenerife accident is relevant to current airline operations. [Strongly agree / Agree / No opinion / Disagree / Strongly disagree]

Q5. Insufficient English proficiency of pilots played a contributing role in the 1977 Tenerife accident. [Strongly agree / Agree / No opinion / Disagree / Strongly disagree]

Q6. Insufficient English proficiency of air traffic controllers played a contributing role in the 1977 Tenerife accident. [Strongly agree / Agree / No opinion / Disagree / Strongly disagree]

The final section includes ten questions about demographics: rank, training appointments, flight hours, age, gender, airline, nationality, native language, other languages known and ICAO language proficiency level.

Demographics

There were 92 respondents to this survey, of whom 74 completed all the sections. This a summary of the demographic information:

- 60.8% of respondents were captains and 39.2% were first officers (n=74)

- 23.0% were technical trainers and 2.7% were human factors trainers (n=74)
- 59.5% had 10,000+ flight hours and 18.9% had less than 5,000 flight hours (n=74)
- 47.3% were 46+ years old and 21.6% were 35 or younger (n=74)
- 97.3% were male and 2.7% were female (n=74)
- respondents worked at British Airways (32 pilots), Thomson Airways (17), Virgin Atlantic (5), easyJet (4), Norwegian (2), Air Berlin (1), Qatar (1), flybe (1) and Jet2.com (1) (n=64)
- 87.8% of respondents were British; other nationalities were Dutch (4.1%), Irish (2.7%), Danish (1.4%), German (1.4%), Scottish (1.4%) and Welsh (1.4%) (n=74)
- the native languages of the respondents were English (100% of respondents), Dutch (4.1%), German (1.4%) and Irish (1.4%) (n=74; 3 respondents had 2 native languages and 1 respondent had 3 native languages)

Results

RQ1. Is studying past airline accidents important for improving airline safety?

The first survey item asked whether studying airline accidents is important for improving current airline safety. The respondents think that studying old accidents *is* important:

- 91.3% of respondents strongly agree, and 98.9% agree or strongly agree (n=92)

The respondents were also asked whether each of the four accidents cited by ICAO are relevant to current airline operations. More than 95% think that three accidents are still relevant:

- 1977 Tenerife: 96.6% of respondents agree/strongly agree (n=88)
- 1990 Cove Neck: 96.5% of respondents agree/strongly agree (n=57)
- 1995 Cali: 95.2% of respondents agree/strongly agree (n=63)
- 1996 New Delhi: 79.2% of respondents agree/strongly agree (n=24)

RQ2. Which of the accidents cited by ICAO have pilots heard of?

Respondents were asked if they had heard of the four accidents. All participants know of the 1977 Tenerife collision. A substantial majority have heard of the 1990 Cove Neck and 1995 Cali crashes. However, less than one-third have heard of the 1996 New Delhi accident:

- 100% have heard of 1977 Tenerife (n=92)
- 70.1% have heard of 1990 Cove Neck (n=87)
- 78.1% have heard of 1995 Cali (n=82)
- 32.1% have heard of 1996 New Delhi (n=78)

RQ3. What are pilots' sources of information about these accidents?

The respondents were asked where they had heard about each accident. This a multiple choice question with 11 possible responses including an open "Other (please specify)" comment box. For three accidents, TV documentaries are the most common information source. For the other accident (1995 Cali), company training is the most common and TV documentaries are second. The most common sources are as follows:

- 1977 Tenerife: TV docu. (76.1%), company training (61.4%) & accident report (50.0%) (n=88)

- 1990 Cove Neck: TV docu. (50.9%), accident report (33.3%) & company training (33.3%) (n=57)
- 1995 Cali: company training (66.7%), TV docu. (47.6%) & accident report (46.0%) (n=63)
- 1996 New Delhi: TV docu. (54.2%), accident report (45.8%) & internet (45.8%) (n=24)

Many respondents report multiple sources of information per accident. Aggregating the results for all four accidents, 67.7% of respondents cite two or more sources of information (n=232). Among those who include TV documentaries, the figure rises to 84.9% citing two or more sources (n=139).

RQ4. Did insufficient English proficiency of pilots play a contributing role in these accidents?

Respondents were asked if insufficient English proficiency of pilots played a contributing role in each accident. For three accidents, 58-67% agree with this statement. For one accident (1995 Cali), less than 13% agree, and almost 50% disagree or strongly disagree:

- 1977 Tenerife: 59.1% agree/strongly agree; 19.3% disagree/strongly disagree (n=88)
- 1990 Cove Neck: 66.7% agree/strongly agree; 7.0% disagree/strongly disagree (n=57)
- 1995 Cali: 12.7% agree/strongly agree; 47.6% disagree/strongly disagree (n=63)
- 1996 New Delhi: 58.3% agree/strongly agree; 8.3% disagree/strongly disagree (n=24)

RQ5. Did insufficient English proficiency of air traffic controllers play a contributing role in these accidents?

Finally, the respondents were asked whether insufficient English proficiency of air traffic controllers was a contributing factor in each accident. The strongest agreement is for the 1977 Tenerife accident, with two thirds agreeing with the statement and only 10% disagreeing:

- 1977 Tenerife: 67.0% agree/strongly agree; 10.2% disagree/strongly disagree (n=88)
- 1990 Cove Neck: 29.8% agree/strongly agree; 36.8% disagree/strongly disagree (n=57)
- 1995 Cali: 33.3% agree/strongly agree; 31.7% disagree/strongly disagree (n=63)
- 1996 New Delhi: 50.0% agree/strongly agree; 16.7% disagree/strongly disagree (n=24)

Conclusion

Learning from the Past

Almost 99% of survey respondents agree that studying past airline accidents is important for improving airline safety, with more than 91% expressing strong agreement. Furthermore, over 95% agree that three of the four accidents cited by ICAO are relevant to current airline operations. The exception is the 1996 New Delhi collision, which less than 80% think relevant.

There is considerable variation in awareness of the four accidents. Less than one-third have heard of the 1996 New Delhi accident, compared with 100% for the 1977 Tenerife collision and more than 70% for the other accidents. One reason for the disparity is that the 1977 Tenerife and 1995 Cali accidents (and to a lesser extent 1990 Cove Neck) are often featured in airline non-technical skills training programs.[1] A second reason is that English language publications (including accident reports) are readily available for these accidents, but not for the 1996 New Delhi collision (CAD, 1996; CIAIAC, 1978; NTSB, 1991).

Sources of Information

Strikingly, TV documentaries are the most common information source for three of the accidents. Company training is also important, being the most common source of information for the 1995 Cali accident and significant for two other accidents. In addition, accident reports are prominent, being the second or third most important sources for all four accidents.

TV documentaries are attractive for many reasons: they are visual, aural and dramatic; they have movement, spoken language and sound effects; they may be watched in a short time (40-50 minutes); and they represent accidents in a personal style that foregrounds individuality and personality. However, they have limitations as sources of information about accidents. The limitations include: the selective use of CVR/ATC dialogue, with utterances being re-ordered or re-written; the speech of non-native English speakers being translated into English; and the use of an omniscient narrator who knows an accident will happen although the actual participants did not have this awareness (cf. hindsight bias).

Compared with all respondents, those citing TV documentaries as an information source are more likely to cite two or more sources. It it reassuring that TV documentaries are typically not pilots' only source of information about an accident, but this suggests the need for further research to investigate how they integrate multiple sources of information.

Insufficient English Proficiency

The first finding is that a majority of respondents think the English proficiency of pilots played a contributing role in three of the accidents (1977 Tenerife, 1990 Cove Neck and 1996 New Delhi). The exception is the 1995 Cali crash, for which less than 13% agree. The second finding is that a majority think the English proficiency of air traffic controllers played a role only for the 1977 Tenerife runway collision. For the 1996 New Delhi mid-air collision, 50% of respondents agree, and for the other two accidents the proportion was one third or less. Thus, only in the case of the 1977 Tenerife accident is English proficiency perceived as a problem on the part of both pilots and controllers.

1995 Cali stands out because only a small proportion of respondents (a third or less) think that insufficient English proficiency was a contributory factor in the case of both pilots and controllers. One possible reason is that this has become known as an "automation accident", which illustrates the hazards associated with introducing new technology and unexpected failure modes into the cockpit, rather than as a "language accident".[2] It is noteworthy that the Cali crash is the only one of the four accidents in which all the pilots were native English speakers.

---

[1] This was reported in responses to RQ3 and in online forum comments.

[2] The Colombian accident report cites the flight crew's "uso inadecuado de automatizacion" ("inadequate use of automation") in its listing of the probable cause (CAD, 1996, p. 68).

Limitations and Future Research

The survey scale (n=92) was limited and respondents had a narrow range of backgrounds (100% native English speakers, 97% male and 88% British). Furthermore, to make it simple and quick to use, the questionnaire mainly consisted of closed-ended question types. Despite these limitations, interesting results were generated about the respondents' awareness of, and attitudes towards, the accidents cited by ICAO.

To conclude, there is a critical need for continued research on pilot-ATC communication in the context of the ICAO language proficiency program. This exploratory study has raised a number of questions that warrant further investigation:

(1) How do current airline pilots characterize the accidents cited by ICAO?
(2) How do pilots integrate accident information that comes from multiple sources?
(3) When and where do pilots watch TV documentaries about accidents?
(4) Do they discuss TV documentaries with other pilots? If so, when and where?
(5) What are the limitations of TV documentaries as information sources?

Acknowledgements

I am very grateful to the following colleagues for their input into the design of this study and analysis of the results: Professor John Maher at International Christian University, Tokyo, Associate Professor Paul Joyce at Kindai University, Osaka, and Associate Professor Dan Jackson at Kanda University of International Studies, Chiba.

References

CAD. (1996). *Vuelo controlado hasta caer en tierra, Vuelo 965 American Airlines Boeing 757-223 N651 AA en las cercanias de Cali, Colombia, Diciembre 20, 1995*. Report released by the Civil Aviation Department, Colombia. Bogota: Civil Aviation Department, Colombia. English translation available at: https://skybrary.aero/bookshelf/books/1056.pdf

CIAIAC. (1978). *KLM, B-747, PH-BUF and Pan Am B-747, N736, Collision at Tenerife Airport, Spain, on 27 March 1977*. Report Number CIAIAC A-102/1977. Madrid: Civil Aviation Accidents and Incidents Investigation Commission.

Estival, D., Farris, C., & Molesworth, B. (2016). *Aviation English: A lingua franca for pilots and air traffic controllers*. Abingdon, Oxon: Routledge.

ICAO. (2010). *Manual on the implementation of ICAO language proficiency requirements* (2nd edition). ICAO Doc 9835. Montreal, Canada: International Civil Aviation Organization.

NTSB. (1991). *Aircraft accident report: Avianca, the airline of Columbia, Boeing 707-321 B, HK 2016, fuel exhaustion, Cove Neck, New York, January 25, 1990*. NTSB/AAR-91/04. Washington, D.C.: National Transportation Safety Board.

AVIATION ENGLISH LISTENING AND REPEATING TASK FOR NATIVE ENGLISH SPEAKER AND NON-NATIVE ENGLISH SPEAKER PILOTS

Julia Trippe
Pacific Science and Engineering
San Diego, CA

Aviation English, based on a coded jargon from World War II, is a mandatory form of communication for pilots and controllers in international airspace. The International Civil Aviation Organization also requires proficiency in Conversational English, for use in non-standard communication. However, our past research indicates that Aviation English and Conversational English are distinct varieties of English, suggesting that assumptions about native English speaker proficiency and additive learning for non-native English speakers may be false. To establish how different these language varieties are, we present a study of Aviation English intelligibility for non-native and native English speaking pilots. Results suggest that non-native English speaking pilots exhibit high proficiency in Aviation English without parallel proficiency in Conversational English. Non-native English speaking Aviation English users suffer the unfair burden of having to learn and maintain proficiency in two language varieties. The impact on learning, training and testing of Aviation English is discussed.

A necessary step towards understanding the relationship between Aviation English (AE) and conversational English (CE) is to determine the extent to which these varieties of English are mutually intelligible. After determining that native English speakers (NESs) not versed in AE are scarcely able to understand AE (Trippe & Pederson, ISAP 2017), it remains to be determined if AE-using non-native English speakers (NNESs) can understand CE. The current paper addresses this proposition by examining NNES AE users. If NNES pilots are more proficient in AE than in CE, it may be reasonable to assume that CE proficiency is not necessary for AE proficiency and that initial aviation language training should focus on AE and not on CE, as is the current practice. If this is the case, then dedicated AE training would be a more efficient use of time, energy, and financial resources for pilots and air carriers: enabling students to absorb language lessons in a less stressful environment and reserving valuable aircraft time for flight training rather than language instruction. Additionally, native English speaking AE users cannot be presumed to have proficiency in this variety of English without undergoing testing.

Methods

Participants

In order to establish NNES AE proficiency as compared to their CE proficiency, groups of NNES pilots and NES pilots were given identical oral performance tasks and their results were compared. At a minimum, all participants were Federal Aviation Administration (FAA)-rated private pilots. The NNES pilot group (CP) was made up of 29 (1 female) Chinese flight students, ranging in age from 22 to 26 (M = 23.38, SD = 1.08). The NES pilot (EP) group was made up of 23 (4 female) North American flight students and instructors, ranging in age from 19 to 55 years ($M = 28.30$, $SD = 7.77$). Table 1 summarizes population descriptives: age, total flight time (TT), and instrument flight rules flight time (IFR).

Table 1.
Population Age, Total Flight Time and Instrument Flight Time by Group

	CP ($n = 29$)			EP ($n = 23$)		
	Range	*Mean*	*SD*	*Range*	*Mean*	*SD*
Age	22-26	23.38	1.08	19-55	28.30	7.77
TT	110-200	156.83	25.84	67-7000	1078.30	1767.40
IFR	10-66	37.21	14.79	4-2500	301.65	620.96

Note: CP = Chinese (NNES) Pilots; EP = native English-speaking (NES) Pilots

Procedure

Participants underwent three verbal repetition tasks, starting with a 15-minute verbal working memory (WM) task to establish baseline differences that could affect repetition of verbal elements. This was followed by a five-minute intelligibility task of Standard American English (SAE) to establish CE competency. The final task was a 15-minute AE intelligibility task to determine how well participants perceived AE transmissions. Tasks were administered by computer using Psychopy (Pierce, 2007) software and were self-paced. All participants received the tasks in the same order with no feedback for response accuracy.

Working memory Task. Participants underwent a verbal WM task to determine any correlation between WM and AE abilities. WM was evaluated using the Word Auditory Recognition and Recall Measure (WARRM) (Smith, Pichora-Fuller, & Alexander, 2016) which required participants to repeat Standard English monosyllabic audio stimuli. Possible raw scores ranged from 1.0 to 6.0, depending on number of words consistently remembered after performance of unrelated cognitive tasks. Raw score was multiplied by a factor of 16.67 to make the highest possible score 100.

CE intelligibility task. Participants were asked to repeat ten CE sentences verbatim. The sentences are approximately fifth grade reading level, ranging in length from seven to ten words. Score for the CE task was the percentage of words correctly reproduced of the 83 possible words in the combined ten CE sentences.

AE intelligibility task. Participants were asked to repeat 84 ATCO utterances verbatim. Half of the selected ATCO transmissions consisted of one aviation topic and half contained two topics. Transmissions ranged in length from two to 19 words. Stimuli was organized in eight pseudo-randomized sets in which every dozen utterances included an equal number of one- and two-topic tokens, so that analysis could explore improvement over seven sets of twelve utterances. Score for the AE task was the percentage of words correctly reproduced for each response, each set of twelve responses, and all 84 responses combined for each participant.

Results

Verbal Repetition Task Scores by Group

Native English speaking EPs performed significantly better on all of the tasks than CPs (see Figure 1). As would be expected, EPs scored higher on the CE task and on the WM task, since they are both English repetition tasks. Non-native English speaker CP CE task scores averaged 43.23 (*SD* = 10.45) and EP CE scores averaged 95.55 (*SD* = 3.55). CP WM task scores averaged 50.56 (*SD* = 11.06) and EP WM scores averaged 77.31 (*SD* = 13.60) (see Figure 1). These two task scores confirm expected differences in English proficiency.

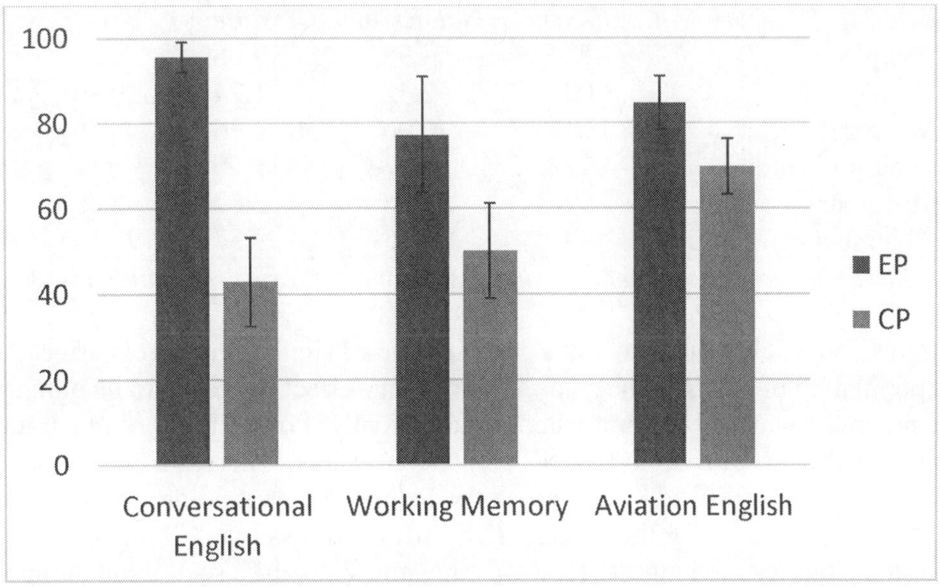

Figure 1. Average Conversational English, Working Memory and Aviation English task scores by Group: Chinese pilots (CP) and native English speaker pilots (EP). Error bars reflect Standard Deviations.

Both groups' results were examined for a learning effect. Using the Bonferroni correction in a pairwise comparison of the seven successive groups of twelve transmissions, it was determined that neither the CP nor EP group showed a learning effect.

Factors Predicting Aviation English Performance

A linear mixed effects regression was performed using nlme package (Pinheiro, Bates, DebRoy, & Sarkar, 2014) in R (R Core Team, 2014) to create the best model fit for predicting AE scores for all responses in the data. The regression included random effects of specific transmission, order presented in the task, and individual participants (subject). The full regression model included fixed effects of pilot group, CE task score, WM task score, age, sex, number of words per transmission, number of topics per transmission, total flight time (TT), Instrument Flight Rules time (IFR), and interaction with group for each of the fixed effects. The final model includes only significant factors that were not correlated (see Table 2).

Examination of correlation of fixed effects using R indicated that TT and IFR were correlated ($r^2 = 0.67$), as were CE and Group ($r^2 = 0.79$), and number of words and number of topics ($r^2 = 0.55$). Accordingly, we examined each of these factors for its contribution to the model. We selected IFR rather than TT, as the measure of AE exposure, because pilots must be in constant contact with ATCOs in the IFR environment. Number of IFR hours was log transformed for inclusion in the regression. We retained both number of words and number of topics as factors, so that possible group differences in language parsing could be discovered.

Table 2.
Linear Mixed Effects Model Summary of Native-English Speaking Pilots (EP) and Chinese Pilots (CP) AE Performance Scores

Predictor	*Coefficient*	*Std. Error*	*t value*	$\chi^2(1)$	*p-value*
Intercept	97.92	4.98	19.66		
EP Group	-3.00	2.42	-1.24	19.80	< .001
WM Score	0.14	0.06	2.36	5.56	0.018
Number of Words	-4.12	0.373	-11.06	91.14	< .001
ln(IFR hours)	1.52	0.58	2.62	6.88	0.009
EP*Number of Words	1.40	0.15	9.57	91.57	< .001

Note: Random effects of Subject, Transmission and Order were included in the model.

Regression results[1] indicate that a combination of significant factors affect AE scores for the entire population of pilots in the study. The primary effect appeared to be number of words in a transmission, which affected group outcomes differently. For every word in a transmission, CPs' AE average scores decreased by 4.12 percentage points, whereas EPs' fell by only 2.72. Model fit was further facilitated by an across-the-board 3-percent point decrement for EPs. These factors combine to indicate, for the shortest (2-word) transmissions, CP and EP scores were almost the same, but for the longest (19-word) transmissions, EPs scored much higher than CPs.

In addition to number of words affecting AE task scores, pilot exposure (as measured by the natural log of their IFR time) was a significant factor. Each unit of ln(IFR) corresponded to a 1.52 percentage point increase in AE scores. This value is greater for the EP group generally, since their mean IFR time is higher (see Table 1). Working Memory task score was also a significant predictor of AE score, probably because this score reflects Standard English proficiency. Every percentage point correct on the WM task, corresponded to a .14 percentage point increase in participants' AE scores. Once again, this effect had a greater benefit for EPs, since their WM scores were higher, on average, than the CP groups' (Figure 1).

CE Task Effect on Non-Native English Speaking Pilots' AE Scores

To determine the possible effect of CE proficiency on non-native English speaker AE scores, we did a separate regression on AE scores for the CP group. The full model for this regression included the above factors in addition to CE task score. The resultant mixed-effects regression model indicates that AE scores for Chinese pilots are significantly predicted by number of words in the transmission and CE score. Similar to the previous regression on both pilot populations, the number of words in a transmission predicted a 4.11 percentage point per word decrease in AE scores for Chinese pilots. Additionally, every percentage point in Chinese

[1] Model fit determination using piecewiseSEM package in R (Lefcheck, 2015), gave a marginal (fixed effects) R^2 value of 0.33 and conditional (including random effects) R^2 value of 0.58.

pilots' CE task score predicted a 0.37-point increase in their AE scores. Chinese pilot AE scores were not predicted by number of topics or WM score.

Discussion and Conclusion

Results of this study indicate that NNES pilots, as represented by a group of Chinese students at a US flight school, exhibit higher proficiency in standard phraseology of Aviation English (AE) than in conversational English (CE). These results suggest that CE ability does not necessarily imply AE ability. Although study results indicate that CE proficiency is correlated with AE proficiency in this population, it is not a consistent predictor. Counterexamples of this relationship are prevalent in the data. Fully 34.5% of CP participants' AE and CE task scores were negatively correlated. Although it requires further study, one possible implication of these findings is that language training specifically focused on AE is likely a more efficient way of increasing AE proficiency than CE training.

Both pilot groups exhibited familiarity with AE as indicated by the fact that there was no adjustment period / learning effect over the brief AE task duration. However, the regression model including all the pilot participants indicates flight experience predicts AE proficiency (see Table XX). This effect appears to be driven by the EP group, since it was not a significant factor in the within-group analysis of Chinese pilots. Results from our previous study on the same EP group suggest that their AE learning curve is initially steep and shallows out with flight experience, reaching asymptote at about 100 hours of IFR time. Although the small number of higher time pilots in the EP population restrains us from generalizing these findings, one conclusion that could be drawn is that, although a brief exposure to AE (during testing) may not be sufficient to increase proficiency, longer exposure does. It is impossible to test this theory on the CP group data, since the Chinese pilots in the current study were all in the early phases of their flight training and had similar, low numbers of IFR hours ($M = 37.21$, $SD = 14.79$), as compared to the native-English speaking pilots in the EP group ($M = 301.65$, $SD = 620.96$).

Regression results indicate that the primary factor in determining difficulty of repetition for both pilot groups was number of words in a transmission, especially for NNES pilots. These findings are consistent with non-native speech studies regarding the cognitive load of translation (Estival & Molesworth, 2016; Farris, 2007). Even comparing pilots with similar flight experience, we would expect NES pilots to have higher AE proficiency than their NNES counterparts, since CE and AE share vocabulary and phonotactics, requiring less translation for NESs.

This study seeks to improve international pilot language training by enhancing the industry's understanding of NNES acquisition of AE standard phraseology. Consistent with our previous study, it appears that CE proficiency does not imply AE proficiency. In the case of the current study, CPs lack of CE proficiency did not limit their AE proficiency. Rather, it appears that the determining factor in their AE abilities was exposure to actual ATCO speech during flight training. Since the rhythm and usage of AE are different from CE (Borowska, 2017; Trippe et al. 2018), language education for professional pilots should be in AE standard phraseology.

Just as for NNES pilots, a short period of AE ground training for NES pilots should enhance AE proficiency similar to the first hundred hours of AE exposure in IFR flight and would serve to prepare pilots for more fluent AE communication. Future research can determine the proper amount of time in listening and repeating actual ATCO transmissions to replicate this initial instrument flight experience. Since flight training is expensive and stressful for pilots, a language-training module for practicing pilot/ATCO communication before and during flight training would be highly beneficial.

AE standard phraseology training should be the basis for all AE communication. Conveyance of more complicated messages could be addressed by expanding AE standard phraseology to include non-routine situations. Emergency and other high-stress situations should not require CE fluency, especially since it is recognized that the cognitive load of speaking in a second language adds to the stress that may accompany such a situation.

References

Borowska, A. (2017c). Avialinguistics. *The Study of Language for Aviation Purposes, Frankfurt am Main: Peter Lang (in print).*

Estival, D., Farris, C., & Molesworth, B. (2016). *Aviation English: A Lingua Franca for Pilots and Air Traffic Controllers*: Routledge.

Farris, C. (2007). *The Effects of Message Length, L2 Proficiency and Cognitive Workload on Performance Accuracy and Speech Production in a Simulated Pilot Navigation Task*: Concordia University (Canada).

Lefcheck, Jonathan S. (2015) piecewise SEM: Piecewise structural equation modeling in R for ecology, evolution, and systematics. *Methods in Ecology and Evolution*, *7*(5), 573-579. doi:10.1111/2041-210X.12512

Pinheiro, J., Bates, D., DebRoy, S., & Sarkar, D. (2014). R Core Team (2014). *nlme: linear and nonlinear mixed effects models*. R package version 3.1–117. Retrieved from https://cran.r-project.org/web/packages/nlme/index.html.

R Core Team (2014). *R: A language and environment for statistical computing*. Vienna, Austria: R Foundation for Statistical Computing. Retrieved from: http://www.R-project.org/.

Smith, S. L., Pichora-Fuller, M. K., & Alexander, G. (2016). Development of the Word Auditory Recognition and Recall Measure: A working memory test for use in rehabilitative audiology. *Ear and Hearing*, *37*(6), e360-e376.

Trippe, J., & Baese-Berk, M. (2019). A prosodic profile of American Aviation English. *English for Specific Purposes*.

Trippe, J., & Pederson, E. (2017). *Aviation English intelligibility*. Proceedings of the 19th International Symposium on Aviation Psychology, Dayton, Ohio.

HELMET MOUNTED DISPLAY FORMAT AND SPATIAL AUDIO CUEING FLIGHT TEST

Thomas Schnell
University of Iowa Operator Performance Laboratory (OPL), Iowa City, IA 52242
Eric Geiselman, Brian Simpson
United States Air Force Research Laboratory (AFRL), 711th HPW, Dayton, OH
Henry Williams, Naval Medical Research Unit Dayton (NAMRU-D), Dayton, OH

Historically, the objective of new technology development has been to enhance pilot performance (such as situation awareness) without causing problems such as Spatial Disorientation (SD). However, when improperly designed or poorly integrated, such technologies may actually reduce performance and increase the likelihood of unintended consequences. SD continues to be a serious problem in the military flight domain and it is critical that both the potential to cause problems as well as support effective defensive mitigation strategies be considered early in the development of new technologies. Past research has shown that new technologies can change operator behaviors. For example, the availability of visual information provided via Helmet-Mounted Displays (HMDs) results in pilots looking farther off-axis for longer durations than when the information is not provided. This paper discusses an ongoing flight test that is conducted on an instrumented L-29 fighter jet trainer that is equipped with a Spatial Audio Horizon Cueing (SAHC) system and an HMD with a conformal referenced (CR) symbology and a forward referenced (FR) symbology. The participants are fighter pilots conducting a series of flight maneuvers while tracking targets with an HMD cueing system. We are investigating the effectiveness of SAHC, in dynamic flight, in conjunction with HMD symbologies, as audio cueing has a high probability to transition into the latest generation fighter aircraft, where seamless integration will be critical to ensure reliable information is consistent with, and complementary to, the existing visual symbology.

Introduction

Spatial Disorientation (SD) is a leading cause of aviation mishap fatalities. A recent review of 601 USAF Class A mishaps from 1993 to 2013 found SD to be causal in 72 (12%) of these mishaps, resulting in 101 fatalities (Poisson & Miller, 2014). The same review compared fatality rates of Non-SD and SD-related mishaps and found that 16.1% of Non-SD mishaps involved a fatality while a staggering 61.1% of the SD-related mishaps were fatal. Similarly, the US Navy cites SD as the number one human factors cause of mishaps (Gibb, Musselman, & Farley, 2012). The military aviation operating environment and the unique technologies utilized there represent a particularly high threat of exposure to SD. Helmet-Mounted Display (HMD) technology is heavily relied upon as a primary pilot/vehicle interface within some 5th generation military aircraft, but it is still mostly unknown how different symbology content and formats may affect SD. One of the main affordances of the HMD is information can be displayed anywhere along the pilot's line-of-sight off the aircraft centerline axis. In terms of SD, significant questions arise regarding what information is needed to support basic orientation while off-axis tasks are being performed. We are undertaking a multi-year effort to determine how that information should be portrayed to balance primary and supportive needs while preventing SD. This applied research project uses an operationally representative testbed aircraft, the OPL L-29 fighter jet trainer, equipped with a fielded-system-representative HMD to compare 3 different off-axis symbology formats, each with and without spatial audio horizon-location cueing, to determine the combination which best mitigates SD and improves performance during operationally representative air-to-air scenarios. One major independent variable manipulation is the comparison of forward referenced aircraft attitude information to symbology that is conformal to the natural horizon both on- and off-axis. A second manipulation is the addition of spatialized audio orientation reference to the outside world to assess its ability to enhance the usefulness of the otherwise exclusively visual symbology. At this stage, the spatial audio cueing is still undergoing laboratory testing at NAMRU-D and AFRL to determine the best spatial arrangement of the localized sound source. We are investigating a modality of the spatial audio system that works similar to a sky pointer or a ground pointer in a Head Up Display (HUD).

Test Objectives

This project will use an operationally representative testbed aircraft, the OPL L-29 fighter jet trainer, equipped with an HMD system to compare 3 different off-axis symbology formats, each with and without spatial audio horizon-location (SAHC) cueing, to determine the combination which best mitigates SD and improves performance during operationally representative air-to-air scenarios. At the time of this writing, we are in the airworthiness approval stages. The specific test objectives are as follows: 1) Compare three different HMD formats, each with and without spatial audio horizon-location cueing, to determine which combination best mitigates SD and improves pilot performance during operationally representative air-to-air scenarios in the actual flight acceleration environment, and 2) Present the same air-to-air scenarios in the DRD and compare the results to actual flight performance for DRD validation purposes.

Experimental Apparatus

A 5th generation fighter aircraft representative Helmet Mounted Display (HMD) was integrated in the OPL L-29 instrumented flight test aircraft and connected to a head-tracked graphics processor that serves as a simulated Distributed Aperture System (DAS) for use in real flight. With this methodology, the Evaluation Pilots (EPs), who are wearing the HMD, can experience a highly realistic DAS environment while operating the L-29 aircraft from the back seat crew station as if they were in a single seat 5th generation HMD fighter environment.

Experimental Procedures and Symbologies

Figure 1 shows the symbologies that will be used in this study. The upper left image is a view of the Virtual Head Up Display (vHUD) symbology that will be seen by the evaluation pilot (EP) in the HMD when he/she is looking forward (bore sight). The test symbologies of interest (Figure 1 upper right, lower left, lower right) will be visible only when viewed off axis or Off Bore-Sight (OBS). Specifically, these symbologies are rendered in the HMD when the pilot's line-of-sight deviates more than 15 degrees laterally or 25 degrees vertically from the aircraft centerline. The upper right image in Figure 1 shows the first OBS test symbology, the Current Display Format (CDF), which was evaluated in the previous test. This symbology is representative of what is available OBS in the baseline aircraft. The CDF displays head-heading, a line-of-sight + symbol (aiming reticle), airspeed, and altitude. There is, however, no aircraft attitude symbology, necessitating cross-check back to the vHUD for attitude reference.

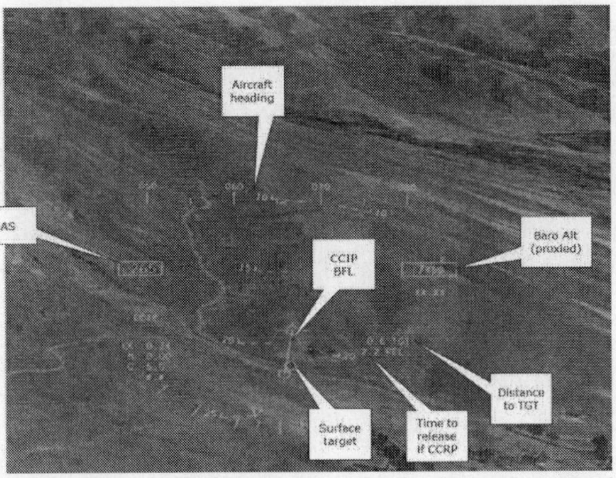

Symbology example of vHUD, looking forward (boresight)

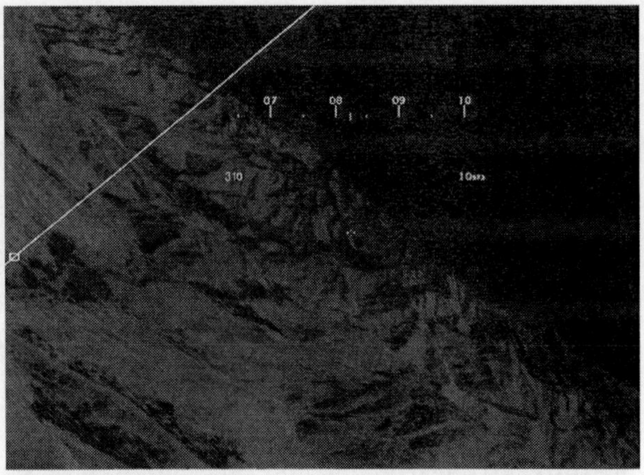

Off-Boresight Symbology #1, Current Display Format (CDF)

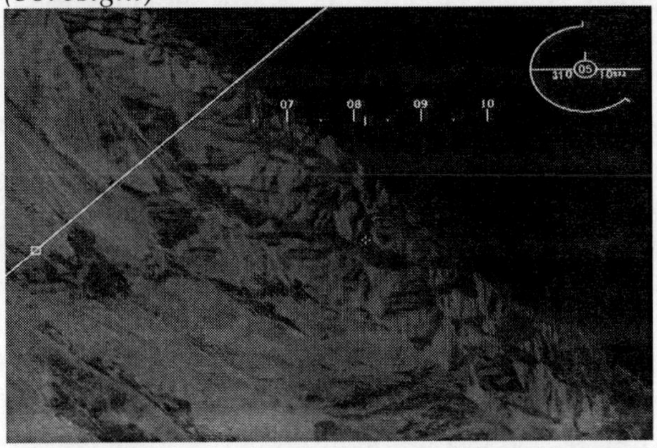

Off-Boresight Symbology #2, Forward Referenced (FR) or Non-Distributed Flightpath Reference (NDFR)

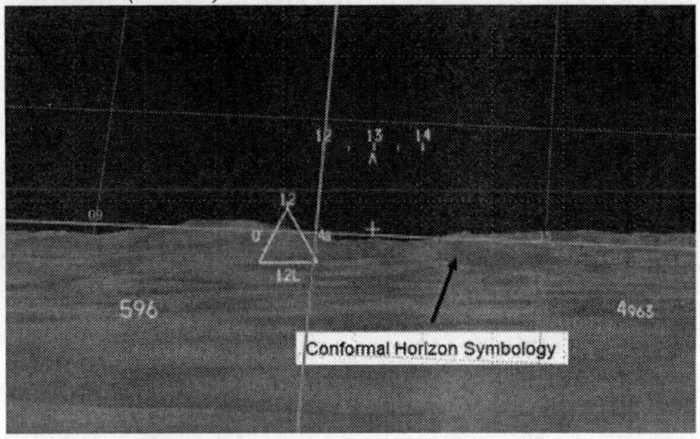

Off-Boresight Symbology #3, Conformal Attitude Reference (CR)

Figure 1. HMD Symbologies used in Phase II

The head-heading indicator correlates to aircraft heading, but directly indicates the heading of the pilot's line of sight. The second test symbology is a Forward Referenced (FR) attitude symbology (lower left image in Figure 1). This symbology is the Non-Distributed Flightpath Reference (NDFR), which was also evaluated in the previous flight test. "Forward Referenced" refers to the fact that the attitude symbology is helmet-stabilized and visible in the same position in the pilot's forward line-of-sight OBS. The NDFR provides attitude, airspeed, aircraft heading, and altitude in one combined symbol. The attitude symbol is a modified arc-segmented attitude reference (ASAR), which changes shape and position with climb/dive and bank of the aircraft, surrounding a fixed aircraft symbol. During straight and level flight, the ASAR is a perfect semicircle below the aircraft symbol with the ends touching the wingtips of the aircraft symbol. As flight path angle decreases, the ASAR gets larger (symbolizing "more earth" in view), approaching a full circle at 90 degrees nose low. Conversely, increases in flight path angle decrease its size. With changes in bank, the ASAR moves around the aircraft symbol, such that the aircraft symbol appears "in the turn." In the example shown, the aircraft is in a slight diving, left-banked turn. The digits "05" in the center of the NDFR (in the left image) represent the rounded heading of the aircraft (around 050 degrees). Altitude and airspeed are 10,573 ft MSL and 310 knots respectively. In the current test, this symbology will be displayed in the lower right-hand corner of the pilot's OBS field of view, rather than the upper right, due to the differing nature of the experimental task. Additionally, due to subjective feedback from EPs in the previous test, the numbers will be slightly larger than shown in Figure 2. The third test symbology is the Conformal Attitude Reference (CR) shown in the lower right image of Figure 1. "Conformal" refers to the fact that the attitude symbology, in this case, is world-stabilized rather than helmet-stabilized. In addition to airspeed, altitude, aircraft heading, and head heading, the CR displays a virtual horizon line which overlays the true horizon as long as the pilot is looking in a direction that keeps the true horizon within the HMD field of view. Additional lines will be drawn above and below the horizon line at 10 degree increments from -90 degrees near nadir to the +90 degrees near the zenith. Vertical lines will be drawn at 30 degree increments of heading. If the true horizon line is above or below the HMD field of view, a decluttered version is rendered where the horizon line becomes dashed and is caged at the proper position (parallel to the actual horizon) on the HMD. A potential issue here is that the pilot may know the direction to the horizon, but may not know the distance to it.

As a possible solution to the issue mentioned above, Spatial Audio Cueing (SAHC) will be integrated into the HMD using 2-channel 3D stereo audio. SAHC will use aircraft state and head tracker data to provide an auditory stimulus dependent on aircraft and head position. Two SAHC conditions will be presented: SAHC On and SAHC Off. The SAHC system works similar to a sky pointer or a ground pointer in a Head Up Display (HUD). The idea is that the sound would be localized on the ground directly beneath the aircraft if the spatial audio cue were to work as a ground pointer, or at the zenith above the aircraft, if it were to function like a sky pointer. Pilot testing is currently underway to determine the best mapping (i.e., to nadir or zenith). In either case, the sound would be lateralized according to the roll angle and longitudinalized based on the flight path angle. The sound is a continuously repeating train of three white noise pulses of 100 ms duration with 100 ms pauses followed by a 500 ms pause. The sound attenuates as a function of angle of bank and climb/dive such that it is at its quietest when the aircraft is level, and loudest when it is at an increasingly unusual attitude in terms of bank and/or flight path angle. The sound is localized in an earth referenced fashion (nadir or zenith) so that its location can serve as a bank-away-from or pitch-away-from cue. Display format (CDF, NDFR, CR) and SAHC condition (without/with) will be combined factorially in a 3x2 repeated-measures design. Each of the three display formats (CDF, NDFR, CR) will be tested with and without SAHC for a total of 6 experimental conditions, or scenarios. Each will be tested once in the simulator and once in live-flight for all EPs. Both display format and presence of SAHC will be within-subjects factors. A "rare event", described in the next section, will occur on the last live-flight run for each EP. Presentation order of the display format and SAHC conditions will be counterbalanced across the 12 EPs to control for learning effects. The simulator and flight portions of the experiment will each consist of 6 flights, one per scenario. For each scenario, the EP will fly a figure-eight pattern while executing a visual tracking task (further explained below). Initial (reference) heading for each run will be a cardinal heading such as due north (360). The EP will begin by setting the bank angle to 60 degrees right wing low (RWL) and initiating a descending turn. He/she will continue for 360 degrees (heading) and descend 1000 ft while maintaining aircraft bank at 60 ±15 degrees. Upon completion of the first circle (crossing reference heading), the EP will reverse course and execute a 60 degree left wing low (LWL) banked turn in the opposite direction to complete the figure-eight. During the second circle, the EP will climb 1000 ft to the original starting altitude. The EP will repeat the figure-eight a second time in the scenario (total of two figure-eights or four circles). Each scenario involves a visual tracking task that requires the pilot to put the HMD center reticle onto a displayed target designation (TD) box. At a random time within 30 seconds after initiation of the first turn of each scenario, a distant object will appear in the EP's OBS visual field-of-regard (above horizon and left or right of vHUD). The object will be superimposed with a TD box rendered on the HMD. The EP will be instructed to place the object in the center of the HMD field-of-view (FOV) and track it (aligned as closely as possible with the aiming reticle) for as long as it remains visible. The object will drift up and down in a sinusoidal pattern. During the final live-flight scenario (last (6th) run), once per EP), the object will depart from the sinusoidal drift pattern in a downward motion into the lower portion of the EP's FOV (below the horizon) and continue on this trajectory. This event is intended to produce an unexpected and potentially disorienting stimulus as a rare event to elicit a measurable flight technical deviation. The EPs will be instructed to adhere to the flight technical parameters discussed in the previous section. Deviations (errors) will be scored. We will also assess spatial orientation performance through the number of significant aileron control reversals (ACR) and elevator control reversals (ECR) exceeding +/- 10 degrees of stick input. Head tracking will be used to determine where pilots were looking, and to infer what information they were seeking and how the HMD impacts this behavior. Some inferences about situation awareness may also be possible when analyzed in conjunction with flight technical performance measures. Specific metrics used in the previous test and planned for this test include duration spent OBS (as percentage of total time), rate of OBS looks or crosscheck frequency (looks per minute), and average duration of each OBS look (seconds). Workload and Situation Awareness will also be assessed. Workload refers to the cost an operator incurs in the performance of a task and may be measured subjectively or objectively (Kramer, 1991). Subjective workload will be assessed by the EP after each run using the Bedford Workload Scale (Roscoe & Ellis, 1990). Electrocardiogram (ECG) will be used to objectively assess workload in real-time using the Cognitive Assessment Tool Set (CATS).

Phase I Results

At this time, we have no data for Phase II of this project as the flight test has not yet started. In lieu, we are showing data from the Phase I flight tests as it illustrates the significant benefits that can be derived from enhancements in OBS symbology. Phase I also used three OBS symbologies. They were 1) Current Display Format (CDF), 2) Distributed Flight Path Reference (DFR), and 3) Non-Distributed Flight Path Reference (NDFR). The scenario involved providing nighttime Close-Air-Support (CAS) to a Joint Terminal Attack Controller (JTAC). During the first scenario of each sortie, the EP checked in with the JTAC, as stated by the briefed simulated air tasking order. The JTAC then provided a situation report, issued an altitude clearance limit, and provided a nine-line brief for the first attack. Following issuance of the nine-line, the JTAC provided a visual Talk-On to the intended first target using visual references that were available on the placemat product/map and which the EP had to identify visually using the HMD DAS. Once the target was identified, the JTAC requested immediate time-on-target for either a show-of-force (SOF) or a bomb-on-target delivery. The full report (Schnell T. et al., 2017) details many different measures of effectiveness. Due to page limitations in this paper, we show one example of a tactically important result, the time it takes to complete the Talk-On. The OBS attitude information available in the DFR reduced the EPs' need to sample the HUD, allowing them to maintain eye-contact with the target area more consistently and for longer durations, thus shortening the overall time required to visually acquire the target. Over numerous measures of effectiveness, including the one highlighted in this paper, the DFR provided the overall best performance during OBS tasks while at the same time being a symbology of minimal clutter. With regard to the duration of the Talk-On, the DFR provided a far shorter duration of around 2 minutes when compared to the duration of 4.4 minutes obtained with the CDF. With a Talk-On duration of 2.7 minutes, the NDFR resulted in a respectable reduction as well. From a flight technical standpoint, the DFR provided the most stable aircraft platform. The DFR and NDFR showed a significant improvement over the CDF, requiring fewer check-looks to the vHUD during the OBS task of the Talk-On phase. In this rubric, the NDFR slightly outperformed the DFR and both test symbologies performed better than the CDF on the tactically significant Talk-On time metric.

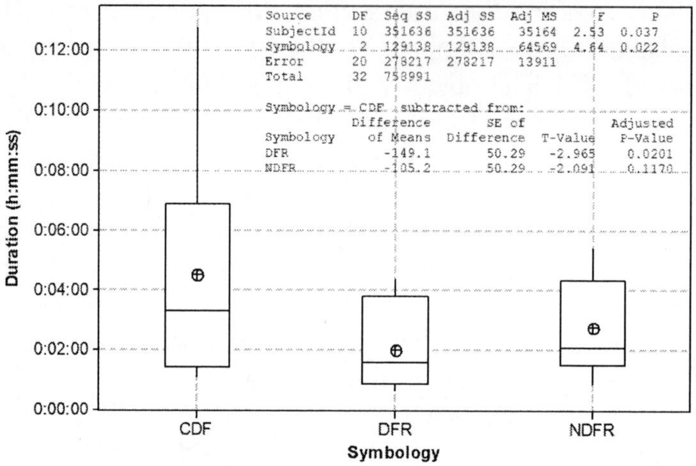

Figure 2. Talk-On Duration by Symbology for N1 Sortie

Acknowledgements

The authors thank the Joint Program Committee-5 Aviation Mishap Prevention Working Group and the Air Force Research Laboratory 711th Human Performance Wing Research, Studies, Analysis and Assessments Committee for funding the projects presented here. Also, we thank the pilots and flight test engineers who volunteered to participate in these activities.

References

Gibb, R., Musselman, B., & Farley, R. (2012). Spatial disorientation… still a problem in 2012. The Combat Edge, July/August: p. 9-10. *The Cutting Edge.* *http://www.acc.af.mil/shared/media/document/AFD-120622-017.pdf*.

Kramer, A. F. (1991). Physiological metrics of mental workload: A review of recent progress. *Multiple-task performance*, 279-328.

Poisson, R. J., & Miller, M. E. (2014). Spatial disorientation mishap trends in the U.S. Air force 1993-2013. *Aviat Space Environ Med, 85*(9), 919-924. doi:10.3357/asem.3971.2014

Roscoe, A. H., & Ellis, G. A. (1990). *A subjective rating scale for assessing pilot workload in flight: A decade of practical use*. Retrieved from

Schnell T, Reichlen C., Geiselman E, Knox J., Williams H., & Ercoline W. (2017, May 8 - May 11, 2017). *A Comparison of Helmet-Mounted Display Symbologies During Live Flight Operational Tasks.* Paper presented at the 19th International Symposium on Aviation Psychology, Dayton, Ohio, USA, May 8 - May 11, 2017.

Schnell T., Reichlen C., & Reuter C. (2017). *Spatial Disorientation Threat Characterization for F-35 Representative Helmet-Mounted Display Use in the Flight Environment, AFRL-SA-WP-TR-2017-0015.* Retrieved from Dayton, OH

EFFECTS OF VISUAL PERCEPTUAL ASYMMETRIES ON PERFORMANCE WHILE USING AN AIRCRAFT ATTITUDE SYMBOLOGY

George A. Reis
Eric E. Geiselman
711 Human Performance Wing, Battlespace Visualization Branch,
Wright-Patterson Air Force Base, United States
Michael E. Miller
Air Force Institute of Technology, Systems Engineering and Management,
Wright-Patterson Air Force Base, United States

In applying the Arc-Segmented Attitude Reference (ASAR) symbology in head-mounted displays (HMDs), it is uncertain if there is an optimal position for the symbology within the display. Vision science literature regarding visual asymmetries suggests that performance may differ depending upon the combination of the location of this symbology within the visual field and whether the user is interpreting the symbology to make *categorical* judgments (e.g., is the aircraft rolling left or right?) or *coordinate* judgments (e.g., what is the aircraft's roll angle). Participants were asked to report aircraft roll and climb/dive angles of briefly presented ASAR symbology within the peripheral visual field on a monitor. There were no performance differences between the left and right ASAR positions in either the coordinate or categorical tasks. There were however trends consistent with horizontal-vertical anisotropy.

Augmented reality (AR) is used to enhance user situation awareness by presenting information in the form of symbols, text, pictures, or video over a real world scene to enhance information transfer. The enhancements can be located in the periphery of the user's visual field to avoid obscuring information with the user's central visual field, particularly when the user views the information within a helmet-mounted display (HMD). In these displays, the observer can either attend to the AR information overtly by making eye fixations on this information or covertly by processing the information without making a fixation. It is proposed that research in perceptual visual asymmetries may advise information design within HMD systems. This premise is derived from experimental research involving a central fixation and an assessment of peripheral visual performance (Hellige, 1993; Bradshaw, 1990; Bourne, 2006). Therefore, it is argued these experimental methods are representative of an observer's performance when covertly attending to peripherally-presented AR information while fixating on real world objects.

Karim & Kojima (2010) categorize perceptual asymmetries in the visual domain as within-field and between-field. In within-field asymmetries, perceptions of stimuli located at a certain location in the field of view (FOV) may differ based on the stimuli's characteristics (e.g., orientation and spatial frequency). In between-field asymmetries, perception of the stimuli differ due to their placement in the user's field of view. Differences in performance have been noted between the lower and upper, as well as between the left and right peripheral regions (Thomas & Elias, 2011; Brederoo, Nieuwenstein, Cornelissen, & Lorist, 2019). Pertinent to the current study, it has been suggested that the left visual field is better than the right in *coordinate* spatial processing and the right is better than the left at *categorical* spatial processing (Kosslyn 1989;

Jager & Postma, 2003). In categorical processing, we determine abstract, prepositional relations between objects (e.g., the cup is to the left of the plate). In coordinate processing, we invoke a measurement process (e.g., the cup is 10 cm away from the plate). The categorical/coordinate asymmetry research typically utilizes simple visual stimuli, thus it is unclear whether similar effects will be observed for more complex, applied visual stimuli. This study explores the effect of interpreting a version of the Arc-Segmented Attitude Reference (ASAR) symbology (Fisher & Fuchs 1992; Geiselman, Havig, & Brewer, 2000; Jenkins, 2008) in categorical and coordinate tasks in various regions of the peripheral FOV. This effort seeks to understand whether performance in categorical or coordinate tasks performed with the ASAR differs with FOV placement.

Method

Participants

Five males participated (mean age: 39 years, range: 22 – 52 years). One participant had experience as a pilot. All reported normal or corrected-to-normal vision.

Apparatus and Stimuli

The ASAR includes a fixed 'ownship' symbol that represents climb/dive angle by its relation to a half-circle arc surrounding the symbol as shown in Figure 1(A). During straight-and-level flight, the upper portion of the circle is not visible and represents the area above the horizon. The visible arc represents the area below the horizon. As the climb angle increases, the visible angle area of the arc narrows in proportion to the climb angle. Conversely, as the dive angle increases, the arc closes towards a circle. During rolling maneuvers, the arc rotates about the ownship symbol. The orientation of the ASAR is 'forward-reference,' depicting climb/dive and roll as if ownship is viewed from behind and the horizon is viewed as if looking forward.

A B C D E

Figure 1. The ASAR representing (A) straight and level flight. (B) 45° climb (C) 45° dive (D) 45° roll left (E) 45° roll right. At a climb/dive of 0°, the half-circle subtended 3.5° of visual angle. The ASAR symbology was presented as white against a gray background.

For application within the visible area of an HMD when the pilot is looking off-axis (away from the aircraft centerline), the ASAR can be small, permitting placement in various locations within the HMD. Thus, this symbology can be placed to reduce clutter and improve operational usability. For example, the ASAR may be placed near the top of the display in air-to-ground mode to move it away from weapon guidance symbology, which is typically presented in the lower portion of the display. For air-to-air, the symbology can be placed at the bottom of the display to de-clutter the upper portion of the display. Placement from one side of the display to the other may be based on separating the ASAR symbology from other forward stabilized symbology. Independent of the ASAR location, its interpretation as an attitude reference is consistent.

It can be argued that roll and climb/dive deviations from straight and level flight can be processed either as categorical or coordinate information, and this dichotomy has practical implications. In interpreting the ASAR, deciding if the symbology indicates a climb or dive, or a roll to the left or right, is a categorization process. This process is important when the pilot's intent is to maintain altitude and heading. On the other hand, determining the magnitude of a climb/dive or roll angle requires coordinate processing. This assessment is important when the pilot decides to change altitude at a specific angle or to maintain a specific roll angle for a certain resultant turn rate.

During the experiment, the participants observed the ASAR on a 23.6-inch ViewPixx display, with its center positioned 57 cm from the participants' eyes. The participants' heads were stabilized with a chin rest. Responses were registered on a SteelSeries XL Bluetooth gamepad controller.

Design and Procedure

Before testing began, participants spent 10 minutes controlling a low fidelity flight simulation during which they received instructions to become acquainted with the dynamic behavior of the ASAR. For each trial, participants fixated on a crosshair centered on the display. They initiated a trial by pressing a button on the right side of the controller. The ASAR was presented briefly for 80 msec. Trials were randomized, without replacement, in each block. A block included a combination of 9 attitude deviations (5° to 85°, at 10° increments) x 2 directions (rolling left and right for roll trials or climbing up and diving down for climb/dive trials) x 8 positions (15° of visual angle from the center of the display, 0° = E, 45° = NE, 90° = N, 135° = NW, 180° = W, 215° = SW, 270° = S, 315° = SE). Participants performed 10 blocks at each of the four combinations of attitude parameter (roll, climb/dive) and spatial processing task (categorical, coordinate), equaling a total of 5,760 trials. After presentation, the ASAR was replaced with a mask of static Gaussian noise to reduce visual persistence.

If the trial was characterized as a roll, categorical condition, participants were instructed to respond with their left thumb on the left or right button of the controller direction pad to match the actual roll direction indicated by the ASAR. Likewise, if the trial was a climb/dive categorical type, the participants responded by pressing the top button (meaning the aircraft was diving) or bottom button (meaning the aircraft was climbing), which map to the mechanization of an aircraft control stick. In both roll and climb/dive categorical trials, response time (RT) and accuracy were recorded. After a response, the Gaussian noise disappeared and the crosshairs appeared again to begin the next trial. In both roll and climb/dive trials, participants were instructed to prioritize accuracy.

If the trial was characterized as a roll, coordinate trial, after the ASAR disappeared in the periphery, the ASAR re-appeared in a straight and level attitude in the middle of the screen (visible against the noise background). The participants then attempted to replicate the roll or climb/dive position by pressing the direction buttons. After the participants obtained the attitude they believed to observe, they pressed a confirmation key and the crosshairs appeared. The Absolute Offset Error (AOE) between the actual ASAR attitude and the participants' response was recorded.

Participants performed either all roll or all climb/dive blocks first. The 10 roll and 10 climb/dive blocks were each divided into two groups (5 blocks categorical, 5 blocks coordinate) and these groups were alternately performed. The sequence of roll and climb/dive blocks and the categorical and coordinate groups were counterbalanced among four participants. The fifth participant received roll blocks first. Before the first time a roll or climb/dive block in combination with a categorical or coordinate tasking was performed, a training block of that combination was completed. Lastly, in every test block, five randomly chosen trials from that set were performed at the beginning to familiarize the participant of the condition being performed.

Results

Results of only the position effects on performance are presented. This statement is not intended to imply that there are no possible interactions between position and angular deviation and direction in roll or climb/dive. Before the data were assessed, values in RT and AOE that were outside of 3.5 standard deviations were identified as outliers and trimmed from the dataset. Accuracy in categorical trials was greater than 95% and there were no apparent tradeoffs between accuracy and RT. The means of RTs and AOEs from the trials in each participant x position cells were chosen to be analyzed. Figures 2 and 3 present AOEs and RTs for climb/dive and roll parameters, respectfully, as a function of position in the FOV.

Table 1 shows the results of the Friedman test carried out to ascertain differences in performance when the ASAR was presented in the eight different positions. Statistically significant differences in performance were observed in the combinations of Climb/Dive-Coordinate, Climb/Dive-Categorical and in Roll-Coordinate blocks. For each of these three combinations, Dunn-Bonferroni tests were performed. The results are shown in the lower portion of Table 1.

Discussion

In general, across all angular deviations and directions in climb/dive and roll, the data suggest no performance differences between the 180° and 0° positions (left vs right) in either the coordinate or categorical taskings. However, there appear to be visual processing differences of the ASAR across the FOV. In particular, the 180° position showed decreased RT and offset error when compared to some other positions. It may be the case that this effect results from pseudoneglect (Jewell & McCourt, 2000) or some other underlying visual attentional asymmetry. Other research (e.g., Corbett & Carrasco, 2011) discusses Horizontal-Vertical Anisotropy (HVA)—where performance on the east-west meridian of the visual field outperforms that of the north-south meridian. The results obtained here trend in line with a HVA as the 0° and 180° positions showed some performance advantages over the 90° and 270° positions. The results from this study indicate that visual asymmetries may exist that influence the design and configuration of HMD symbology.

Climb/Dive Coordinate

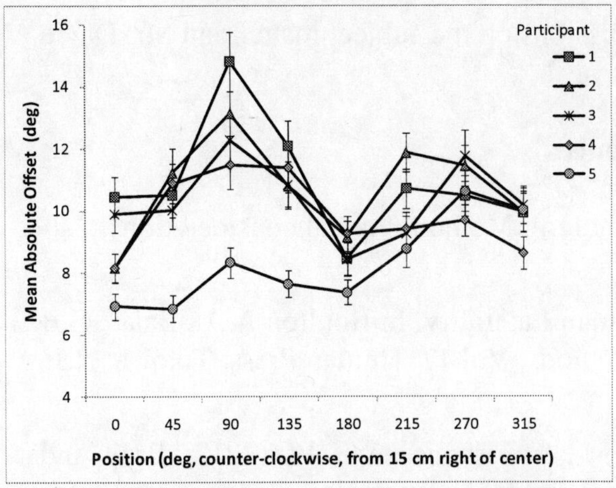

Climb/Dive Categorical

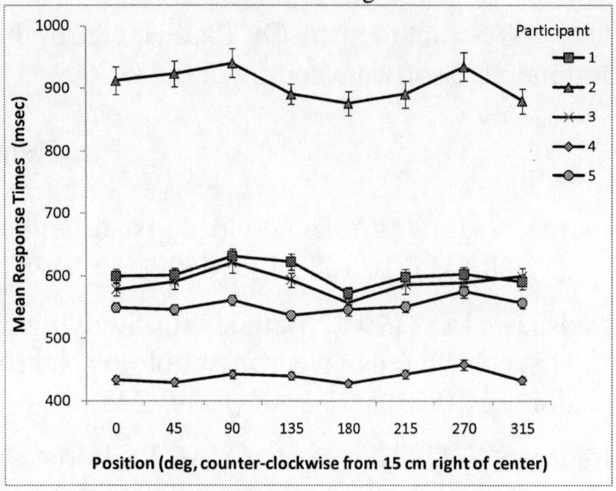

Figure 2. (Left) The mean absolute offsets and (Right) the mean response times across climb/dive directions and angular deviations, at coordinate and categorical trials, respectively, are plotted as a function of the position in the FOV. Errors bars represent ± 1standard error of the mean.

Roll Coordinate

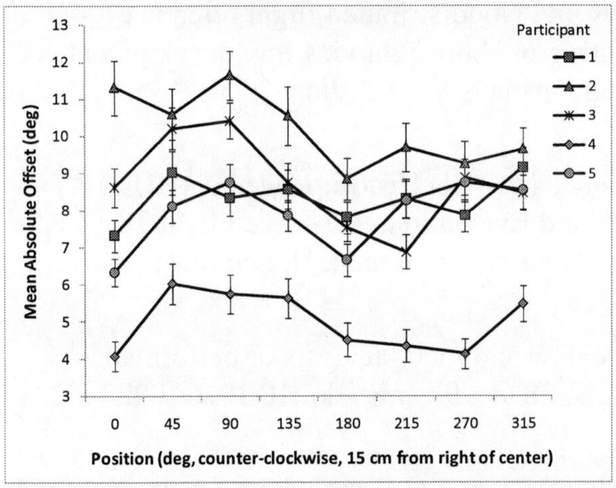

Roll Categorical

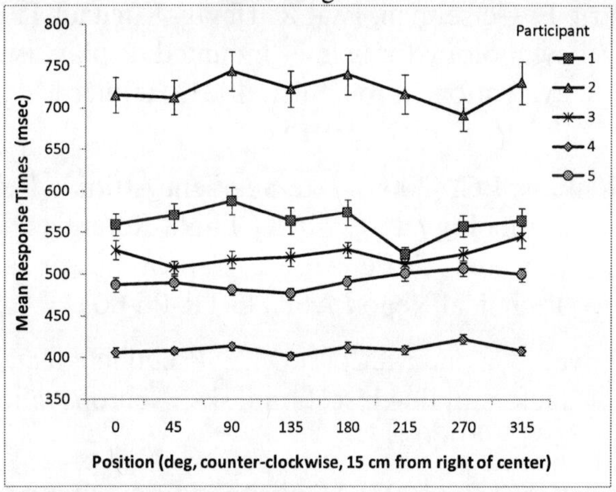

Figure 3. (Left) The mean absolute offsets and (Right) the mean response times across roll directions and angular deviations, at coordinate and categorical trials, respectively, are plotted as a function of the position in the FOV. Errors bars represent ± 1standard error of the mean.

Table1. Results from Friedman and Dunn-Bonferroni post-hoc tests. Assessed at a significance level of $\alpha = .05$. Effect size, *r*, from Wilcoxon signed-rank tests.

Attitude Parameter	Climb/Dive						Roll					
Task	Coordinate			Categorical			Coordinate			Categorical		
$\chi^2(7)$	21.00			20.47			17.533			9.33		
p -value	**.004**			**.005**			**.014**			.230		
Pairwise Tests		*p-value*	*r*		*p-value*	*r*		*p-value*	*r*			
	180 vs 90	**0.014**	0.64	180 vs 90	**0.002**	0.64	180 vs 90	**0.035**	0.64			
	0 vs 90	**0.014**	0.64	180 vs 270	**0.014**	0.64						

Acknowledgements

We want to thank Dr. Paul Havig for discussions on the subject matter and Mr. David Dommett for software coding of the tasks.

References

Bourne, V. J. (2006). The divided visual field paradigm: Methodological considerations. *11*(4), 373-393. doi:10.1080/13576500600633982.

Bradshaw, J.L. (1990). Methods for Studying Human Laterality. In Boulton A.A., Baker G.B., Hiscock M. (eds.) Neuropsychology. Neuromethods, Vol 17. Human Press, Totowa, N.J. doi.org/10.1385/0-89603-133-0:225.

Brederoo, S. G., Nieuwenstein, M. R., Lorist, M. M., & Cornelissen, F. W. (2017). Brain and Cognition Hemispheric specialization for global and local processing: A direct comparison of linguistic and non-linguistic stimuli. *Brain and Cognition, 119*(September), 10-16. doi:10.1016/j.bandc.2017.09.005.

Corbett J.E., Carrasco M. (2011) Visual Performance Fields: Frames of Reference. PLoS ONE 6(9): e24470. doi:10.1371/journal.pone.0024470.

Eric E. Geiselman, Paul R. Havig, Michael T. Brewer, "Nondistributed flight reference symbology for helmet-mounted display use during off-boresight viewing: development and evaluation," Proc. SPIE 4021, Helmet-Mounted Displays V, (23 June 2000); doi:10.1117/12.389157.

Jenkins, J.C. (2008). Arc Segment Attitude Reference (ASAR) Head-up Display (HUD) Symbology as a Primary Flight Reference Test and Evaluation. Air Force Flight Test Center, Edwards Air Force Base, California. Air Force Material Command. United States Air Force. Technical Report AFFTC-TR-06-06.

Jewell, G., and McCourt, M.E. Pseudoneglect: a review and meta-analysis of performance factors in line bisection tasks. Neuropsychologia. 38 (1): 93–110. doi:10.1016/s0028-3932(99)00045-7.

Hellige, J.B. (1993). Hemispheric Asymmetry, What's Right and What's Left. Cambridge, MA./ London England, Harvard University Press.

Jager, G., and Postma, A. (2003). On the hemispheric specialization for categorical and coordinate spatial relations: a review of the current evidence. Neuropsychologia 41, 504–515. doi: 10.1016/S0028-3932(02)00086-6.

Kosslyn, S. M., Koenig, O., Barrett, A., Cave, C. B., Tang, J., & Gabrieli, J. D. E. (1989). Evidence for Two Types of Spatial Representations: Hemispheric Specialization for Categorical and Coordinate Relations. *15*(4), 723-735.

Rezaul, Karim, A. K. M., & Kojima, H. (2010). The what and why of perceptual asymmetries in the visual domain. *Advances in Cognitive Psychology, 6*(6), 103-115. doi:10.2478/v10053-008-0080-6.

Thomas, N. A., & Elias, L. J. (2011). Upper and lower visual field differences in perceptual asymmetries. *Brain Research, 1387*, 108-115. doi:10.1016/j.brainres.2011.02.063.

DISORIENTATION RESEARCH DEVICE TESTING OF SYNTHETIC VISION DISPLAY TECHNOLOGIES

Lawrence Prinzel, Kyle Ellis, Kathryn Ballard, Renee Lake, Stephanie Nicholas, Trey Arthur
NASA Langley Research Center
Hampton, Virginia

Captain Daniel Kiggins
National Institute of Aerospace
Hampton, Virginia

A Commercial Aviation Safety Team (CAST) study of 18 worldwide loss-of-control accidents and incidents determined that the lack of external visual references was associated with a flight crew's loss of attitude awareness or energy state awareness in 17 of these events. CAST recommended development and implementation of virtual day-Visual Meteorological Condition (VMC) display systems, such as synthetic vision systems, to promote flight crew attitude awareness similar to a day-VMC environment. This paper describes the results of a joint NASA/NAMRU-D study that evaluated virtual day-VMC displays and a "background attitude indicator" concept as an aid to pilots in recovery from unusual attitudes. Experimental results and future research directions under this CAST initiative and the NASA "Technologies for Airplane State Awareness" research project are described.

A CAST study of 18 loss-of-control accidents determined that a lack of external visual references (i.e., darkness, instrument meteorological conditions, or both) was associated with a flight crew's loss of attitude awareness or energy state awareness (collectively termed, "airplane state awareness") in 17 of these events. The reports (CAST, 2014a; CAST, 2014b) recommended that, to provide visual cues necessary to prevent loss-of-control (LOC) resulting from a flight crew's spatial disorientation and loss-of-energy state awareness, manufacturers should develop and implement virtual day-VMC display systems, such as synthetic vision systems. In support of this implementation, CAST requested the National Aeronautics and Space Administration (NASA) to conduct research to support definition of minimum requirements for virtual day-VMC displays to accomplish the intended function of improving flight crew awareness of airplane attitude.

Virtual Day-VMC Displays

Virtual day-VMC displays or Synthetic Vision (SV), are intended to provide similar visual cues to the flight crew that are available when outside visibility is unrestricted (i.e., observed under VMC). Their intended function is provide continuous, intuitive attitude, altitude, and terrain awareness, reducing the likelihood of unstable approach, inadvertent entry into an unusual attitude, spatial disorientation, and/or collision with terrain. SV is a computer-generated image of the external scene topography from the perspective of the flight deck, derived from aircraft attitude, high-precision navigation solution, and database of terrain, obstacles and relevant cultural features.

Technologies for Airplane State Awareness

CAST, under Safety Enhancement(SE)-200 defined areas of research needed for design and implementation of virtual day-VMC displays to prevent loss-of-control accidents due to loss of attitude awareness and lack of external visual references. NASA research has been completed that evaluated numerous virtual day-VMC display design characteristics (summarized in Ellis, et al., 2019). To date, the research has been conducted in research facilities including high-fidelity Level D-certified airline training simulators. Although the simulations have yielded valuable data, the facilities are of limited fidelity in inducing vestibular illusions. Hexapod motion-based simulators use Stewart-motion platforms with acceleration onset cueing to deceive the pilot sensory systems into experiencing the motion effects.

This paper describes research that evaluated SV technology in the Disorientation Research Device (DRD), located at the Naval Medical Research Unit (NAMRU-D), Wright-Patterson Air Force Base (NAMRU-D, 2019). The DRD is a simulation facility that has advanced motion capabilities across 6 degrees-of-freedom including sustained planetary accelerations up to 3G in x, y, and z planes. The DRD enabled the extension of the SE-200 research to evaluate the efficacy of display technology and visual dominance through the experimentally controlled induction of spatial disorientation and vestibular illusions.

Experimental Method

Research Pilots

A total of 20 qualified ATP pilots currently flying regional jet aircraft and employed by Part 121 US major or regional airline operators were recruited as subject pilots.

Research Simulator

Designed as a reconfigurable advanced centrifuge device to induce spatial disorientation, the DRD provides simultaneous motion in 6 degrees-of-freedom: (a) roll/pitch/yaw of 360^+ deg rotation; (b) horizontal/vertical travel of 33 ft. and 6 ft., respectively; and, (c) sustained planetary acceleration of 360^+ deg rotation up to 3 G in x, y, and z planes (NAMRU-D, 2019). The DRD was modified to emulate a large commercial transport aircraft including aerodynamic model and the inside of the capsule included transport aircraft controls and displays (Figure 1).

Research Evaluation

Evaluation pilots flew three tasks: (a) Unusual Attitude Recovery (UAR) evaluations; (b) somatogravic illusion take-offs; and, (c) an attitude awareness task, labeled the "plane-following task". The results of the last task are not included in this paper.

Research Displays. Three research display concepts (Figure 2) were evaluated: 1) the baseline was a standard conventional "blue-over-brown" primary flight display; 2) an "SV" PFD display emulated the common display characteristics of commercially available SV systems; and, 3) a "background attitude indicator" concept, termed "SV + BAI". The SV+BAI was presented across the PFD and navigation head-down displays for each pilot, displaying a wide field-of-view attitude horizon to promote ambient vision stimulation (Previc and Ercoline, 2004). The PFD and navigation displays were each presented on 305 x 228 mm LCDs (see Figure 1).

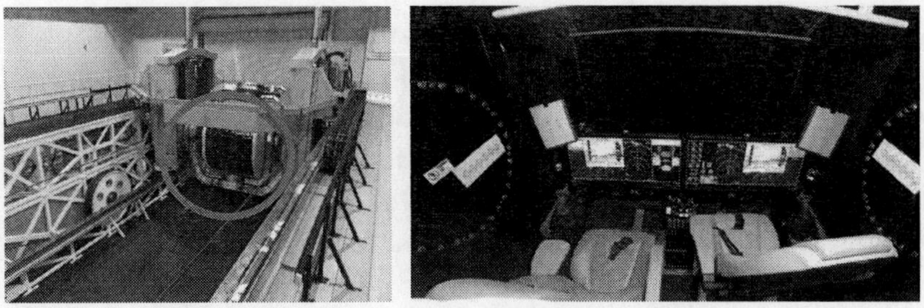

Figure 1. Disorientation Research Device

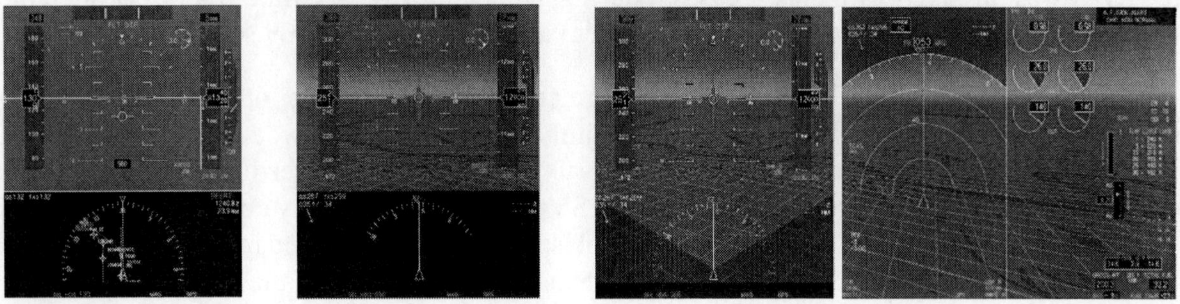

Blue-Over-Brown SV SV + BAI

Figure 2. Research Displays

UAR Scenarios. The UAR scenarios were designed to determine how effectively a pilot could recognize their attitude and execute the proper corrective action.

The initial conditions for the UAR scenarios were: (a) 55° nose-down, 70° left bank; (b) 60° nose-down, 95° right bank; (c) 20° nose-down, 40° left bank; and, (d) 50° nose-down, 45° right bank. The UAR entries were dynamic, meaning the aircraft approached the prescribed UA initial condition with the simulator in motion and the displays blank. The method ensured that pilots would not be able to determine their current attitude using their vestibular sense. At each UA initial condition for the scenario, the displays returned, signaling the EP to recover the aircraft. Twenty-four (24) UAR scenario recoveries were performed by each evaluation pilot, which were fully randomized for each set across participants. Standard recovery training was provided, and pilots participated in practice trials using training set of UAR scenarios.

Somatogravic Illusion. The take-off scenario was designed to induce the "false climb" somatogravic illusion wherein the linear acceleration environment experienced by the pilot's vestibular system may induce the perception of an erroneously high pitch attitude. If induced, the expected pilot response is a pitch down control input. For this scenario, only two display conditions were tested for their efficacy to establish visual dominance in the face of the somatogravic illusion: (a) blue-over-brown, and (b) SV+BAI.

Results

Unusual Attitude Recoveries

The UAR performance was "scored" by the correctness and timeliness of pilot control inputs after the displays were unblanked.

Control Inputs. A generalized linear mixed models (GLMMs) with log link fit revealed that display did not affect the time to make the initial roll input, $\chi^2(2) = 0.574,\ p = 0.574$. There were 18 total cases of incorrect initial roll input. SV+BAI condition had only 2 cases with 0.16 magnitude in the wrong direction (the stick input takes values from -1 to 1) compared to blue-over-brown (7 cases, 0.31) and SV (9 cases, 0.38). Although SV was observed to have 9 incorrect initial roll inputs, the average time-to-correct was only 1.36 sec compared to 2.12 sec and 2.11 sec for blue-over-brown baseline and SV+BAI, respectively. For pitch inputs, the GLMM also did not show a significant effect, $\chi^2\ (2) = 1.54\ ,\ p = 0.463$. There were 39 total cases of incorrect initial pitch input with all displays showing only an average 0.14 magnitude. SV+BAI was observed with only 9 cases requring average of 2.47 sec time-to-correct pitch input compared to blue-over-brown (13 cases, 3.47 sec) and SV (11 cases, 2.09 sec).

Roll Reversals. Roll control reversals are an indicator that the pilot did not fully recognize the aircraft attitude prior to recovery initiation. The data showed only 10 runs (3%) with a roll reversal. However, 7 of 10 roll reversals were in the blue-over-brown baseline display condition but only 1 and 2 roll reversals for the SV+BAI and SV display conditions, respectively. Using a GLMM with a Poisson link showed that these trends are not significant, X^2 (2) = 4.95, p = 0.08. The pilots exhibited excellent performance across the all display types as 10% roll reversals is the generally accepted UAR testing norm (Previc, 2004).

UAR Score. The control recoveries were "scored" by awarding a +1 for the correct initial control input and a -1 for an incorrect initial control input, for both the pitch and roll controllers. The UAR Score is a metric of pilot performance. UAR scores in this test could take 3 values: 2 if the pilot made both pitch and roll inputs correctly, zero ('0') if one of either pitch or roll inputs were incorrect, or -2 if both initial inputs were not correct. The pilots made the correct initial pitch and roll stick inputs 87% of the time. None of the pilots scored a -2 (Figure 3). The SV + BAI display had the highest number of cases where the initial input was correct (93%); however, the test on the UAR score did not reveal a statistical difference between displays. The likelihood ratio test between the two models did not find a significant difference between displays, using an conservative Bonferroni corrected α of 0.0055, X^2 (2) = 7.84, p = 0.02.

Qualitative Results. A significant effect was found for NASA-Task Load Index (TLX) workload ratings on display, F (2, 36) = 9.831, p < 0.05. The Bonferroni-corrected pairwise comparison post-hoc evinced that the blue-over-brown baseline display condition (μ = 25.68, σ = 14.55) was rated higher in workload than both the SV+BAI (μ = 20.34, σ = 12.90) and the SV display (μ = 23.29, σ = 13.27). A significant effect was found for Situation Awareness Rating Technique (SART) ratings on display condition, F (2, 36) = 11.437, p < 0.05. The Bonferroni-corrected pairwise comparison revealed significant difference between blue-over-brown and the two SV display conditions. Pilot participants rated the blue-over-brown display (μ = 106.54, σ = 59.59) significantly lower for SART scores than the SV (μ = 121.90, σ = 52.74) or SV+BAI (μ = 130.05, σ = 53.32). No differences were found between the SV and SV+BAI display for SART scores.

Somatogravic Illusion

The efficacy of the displays to support attitude awareness during the somatogravic illusion was "scored" by the pilot's pitch attitude control where the somatogravic illusion may be more compelling than visual references and lead to erroneous pitch down inputs.The subjects, post-test, were asked about the quality of the illusion. The pilots judged the scenario to be "excellent". Overall, the SV+BAI display was equivalent to the blue-over-brown displays. Out of the 40 runs, no incorrect pitch inputs were made although the pitch attitude went below the target 15 degrees of pitch on some of the blue-over-brown baseline trials (N=3 of 20).

An insignificant effect was found for workload using the NASA-TLX, $F(1, 19) = 3.063$, $p = 0.096$. Pilots tended to rate the SV+BAI display ($\mu = 16.04$, $\sigma = 10.58$) to be lower in mental workload (as measured by the NASA-TLX) compared to the blue-over-brown display condition ($\mu = 20.33$, $\sigma = 16.84$). Pilots were also asked to provide a rating of display efficacy for "visual dominance" during the enhanced motion/somatogravic illusion scenario. The Wilcoxon signed-rank test was conducted comparing blue-over-brown to SV+BAI on the Likert scale rating (1 – 5 scale). The results were that pilots significantly rated the SV+BAI ($\mu = 4.65$, $\sigma = 0.587$; higher for visual dominance compared to the blue-over-brown display ($\mu = 3.25$, $\sigma = 0.851$, condition ($Z = -3.713$, $p < 0.001$).

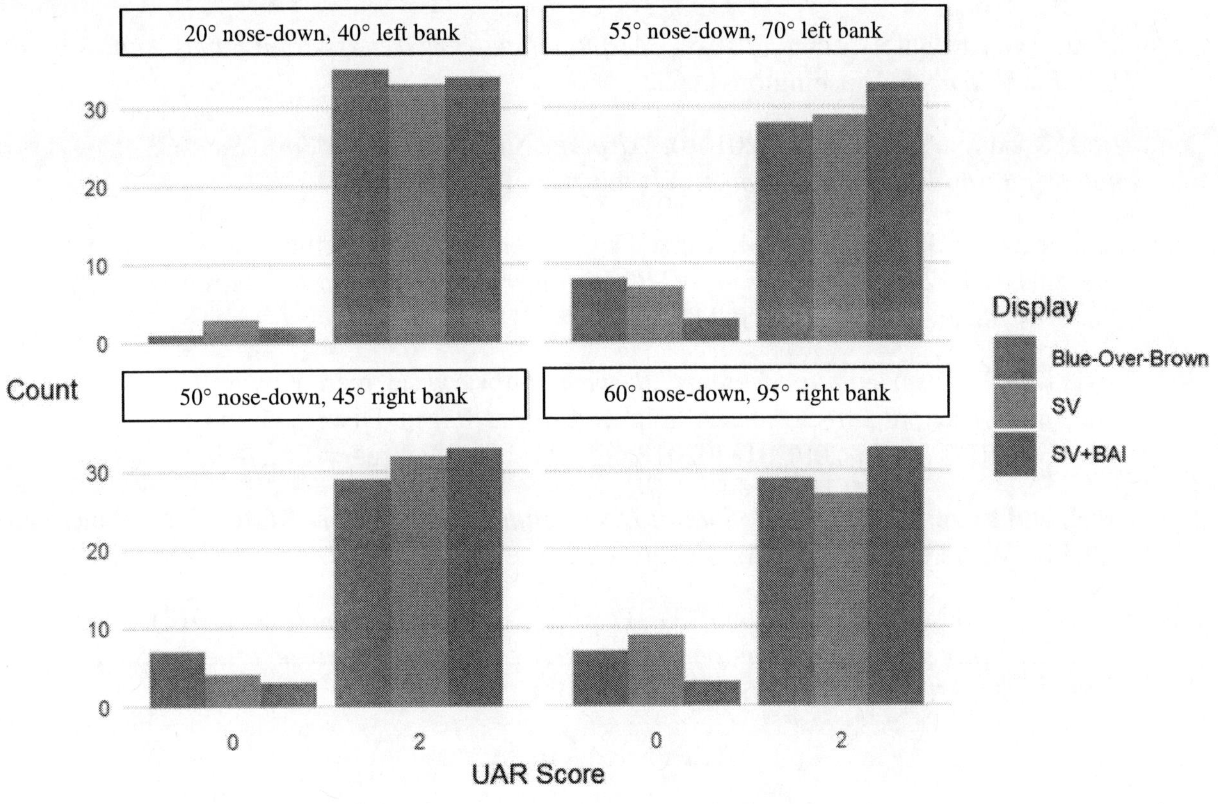

Figure 3. Unusual Attitude Recovery Scores

Conclusions

This research extends the design and development of virtual day-VMC displays for airplane state awareness. The research was a first attempt to evaluate the display visual dominance for efficacy to prevent spatial disorientation induced by experimentally-controlled vestibular illusions. The results confirm previous research that pilots significantly prefer the SV display technology and report much lower workload and higher attitude awareness compared to traditional Blue-Over-Brown PFDs. However, similar to past research and findings in the literature, no statistically significant quantiative performance differeneces between display concepts were found. It is difficult to obtain enough data points to allow for statistical analysis in uncovering if there are performance differences. The types of events tested are naturally rare events, and it is difficult to simulate conditions that would ensure the data would be collected with the necessary number of observations. At minimum, the data suggests that virtual-day VMC displays provide airplane state awareness similar to the blue-over-brown baseline display. The descriptive data validates the need for further spatial disorientation and illusion (both visual and vestibular) test and evaluation. The results also support additional study on the background attitude indicator display concept.

References

Bailey, R.E., Ellis, K.E., and Stephens, C.L. (2013). "Test and Evaluation Metrics of Crew Decision-Making And Aircraft Attitude and Energy State Awareness". AIAA InfoTech Conference, Boston, MA.

Commercial Aviation Safety Team (2014a). *Airplane State Awareness Joint Safety Analysis Team Final Report.* Washington, D.C.: CAST.

Commercial Aviation Safety Team (2014b). *Airplane State Awareness Joint Safety Implementation Team Final Report.* Washington, D.C.: CAST.

Ellis, K.E., Prinzel, L.J., Ballard, K.M., Lake, R.,C., Nicholas, S.N., Arthur, J.J., Bailey, R.E., & Kiggins, D.J. (2019). *Virtual Day-VMC Displays for Attitude State Awareness.* AIAA-2019-0720. San Diego, CA: SCITECH.

NAMRU-D (2019). *Disorientation Research Device.* https://www.med.navy.mil/ sites/nmrc/Dayton/ Directorates/Admin/ SiteAssets/Pages/ Disorientation-Research-Device/DRD%2014%20FEB%202018.pdf. Accessed February 14, 2019.

Previc, F.H. and Erocline, W. (2004). *Spatial Disorientation in Aviation*. Reston, Va. : American Institute of Aeronautics and Astronautics, Inc.

Williams, H.P., Horning, D.S., Lawson, B.D., Powell, C.R., & Patterson, F.R. (2018). *Effects of Various Types of Cockpit Workload on Incidence of Spatial Disorientation in Simulated Flight* (doc NAMRU-D-19-06). Dayton, OH: U.S. Navy.

Acknowledgements

This research was conducted by NASA under the Technologies for Airplane State Awareness (TASA) sub-project, within the System-Wide Safety Project of the Airspace Operations and Safety Program in the Aeronautics Research Mission Directorate. Special thanks to the US Naval AeroMedical Research Unit – Dayton (NAMRU-D) team for providing the facilities, technical expertise, and experimental support required to enable this research. We also thank the individuals Capt. Richard Folga (NAMRU-D), Hank Williams (NAMRU-D), Rick Arnold (NAMRU-D), Bill Ercoline (KBRwyle), and Steve Williams (NASA), for their subject matter expertise, as well as Patrick D. Johnson and Richard Jessop (Metis Technology Solutions) for software implementation expertise.

ERICKSON'S PRACTICE FOR CREWS:
WHAT ABOUT COPING TO THE SITUATION WITH ZEN?

Marielle Plat-Robain
Airbus Operations SAS
Toulouse France

Hypnosis has already been used in very long flights by Bertrand Piccard, one of the two Solar Impulse pilots, to manage fatigue and rest periods. It is also used by navigators like Armel Le Cléac'h during solo races. For this reason we consider it is worth to look at such technics to cope with the constraints of long flights. A study was done in order to explore what kind of benefits hypnosis could bring to cope better with multitask activities constraints like time pressure, good performance demands... We used Multi-Attribute Task Battery II (MATB-II) software to induce different workloads, time pressure and consecutively fatigue sensations. Participants were mostly aeronautical engineers. For each participant, the scenarios were repeated and separated by 3 types of break. Three break conditions of fifteen minutes were then used to differentiate the group of people: "Static rest" or "Exercise" or "Hypnosis" breaks. The hypothesis was that depending on the break condition and the scenario intensity level, variations on performance and participants' feeling could appear. The methodology used task performance index, fatigue and workload subjective scales, eye-tracking data, plus debriefing inputs. The results show that "hypnosis" and "exercise" breaks had interesting effects on fatigue and performance. Unexpected results were underlined by participants' in debriefing about "stressless" effect and calm after using hypnosis to cope with the situation.

Keywords: hypnosis, fatigue, stress.

Crew fatigue is recognized as a safety risk and has been classified by NTSB as a "most wanted" improvement. For the purpose of our work, we will use that the definition of fatigue as "*biological drive for recuperative rest*" (Williamson & al., 2011). Levels of fatigue and levels of confidence impact self-monitoring performances and risk, with a nonlinear relationship between risk and alertness or performance (Folkard & Akerstedt, 2004). The difficulty is mainly due to complex links between objective results and subjective metacognition models that impact the level of one's own activity control loop. Considering the recent example of the Solar Impulse, with pilots using alternative techniques like the Erickson hypnosis techniques for recuperation seems interesting to explore(e.g., solar impulse website). B. Piccard used autohypnosis to manage his resources and lack of sleep during the very long flight and simulation (72 hours). The aircraft does not have an autopilot system to take over part of the tasks, except for wing stabilization (only for periods of twenty minutes). This small window of time allows pilots to take naps or regulate their fatigue. Video testimonies are available; for Bertrand Piccard's 72 hours simulation. During this simulation he slept only 2 or 3 hours out of 24 hours (based on EEG measurements) and he was managing his resources with hypnosis for the rest of the time. He declared that he felt even better when using autohypnosis. The present exploratory work aims at shed light on cognitive resources recovery via hypnosis compared to other short break types.

In related works, Hammer (1954) showed that on highly suggestible students -besides effects on motivation-, hypnosis suggestion improved their performance during a test related to schoolwork. Following this pioneer study, many research teams used the Stroop task; for instance Kaiser & al. (1997) who have shown that the Stroop effect appeared to be enhanced. Other research, like Raz & al. (2002), Landry & al. (2017), Lifshitz & al. (2013), concluded that post-hypnotic suggestion can cancel the Stroop effect by inducing transient alexia. Those studies showed that there are two potential impacts: one coming from the hypnotic state effect itself, and the other from the suggestions that have been induced. The effect of hypnotic suggestion on several well-documented cognitive mechanisms has been studied by Landry & al. (2017). All are considered as automatic processes because they occur outside conscious control and they do not need to be learned. In another direction, sport psychology studied the effect of sport on cognition performance and it appears that effects of physical exercise on cognitive abilities could follow the Yerkes and Dodson's Law of Arousal (Yerkes & Dodson 1908). This means that cognitive performances could be improved following a moderate to high intensity exercise on short to medium duration (less than 1 hour), whereas they decreased for very high intensity or long lasting activity (Tomporowski 2003).

Human Recovery Resource Methodology (HRR 1 & 2)

The exploratory work consist in two successive studies. The first experiment HRR1 was done to investigate if there was any interest in a short break hypnosis session to recover, while experiment HRR2 aimed at replicating the experiment and deepening the analysis to check the rest efficiency with a higher fatigue level. Our assumption, using previous research, was that there will be a better recovery from fatigue using "hypnosis" and "exercise" than "static rest" breaks.

Participants were volunteers' engineers with strong knowledge in Aeronautic tasks. Instructions were: they will assess the effect of a break and that they have to perform as well as they can. Participants had to perform a Multi-Attribute Task Battery (NASA MATBII) mimicking a piloting situation. MATBII was composed of 4 sub-tasks including system monitoring (SYSM), tracking (TRCK), communication (COMM), resources management (RMAN) that are configurable in difficulty, numbers and concomitance. All tasks were used and planned in order to vary the efforts and the consecutive fatigue. The main differences between scenarios were the duration and the level of tasks' demanded: HRR1 repeated scenario was 10mn long with 55 events while HRR2 scenario was double the duration with 30% of tasks' events increase (150 events in 20 minutes: 57 events SYSM, 42 COMM, 31 TRCK and 20 RMAN). The purpose was to vary the workload, resources consumption and consequently the fatigue.

For performing the task, they were seated in front of a computer screen in a windowless room (constant lightening necessary for eye tracker). Thrusmaster T.FlightStick X Joystick was used to perform the tracking task, plus a keyboard or a mouse for other tasks. Adjustments were at operator discretion. Eye activity was registered via a Tobii X1 Light fixed under the screen and the Ogama software. Participants were trained and aware of the performances calculations and constraints. Each sub-task was analyzed with a time spent in abnormal situation (Splawn, 2013) separately and then additionned in global performance index. Difference of performance between scenarios (same scenario) before and after break was always calculated as: Performance after break minus performance before break.

At the end of each MAT-B scenario, a Workload Rating Scale based on the NASA-TLX is prompted on the screen (Hart, S. G., & Staveland, L. E., 1988). NASA-TLX classifies workload in six categories: mental, physical, temporal, performance, effort and frustration. Participants had to rate their global workload on these six scales from "low" to "high" (0/100) except for the performance category "good" to "poor". A subjective global fatigue assessment scale (scored from 1 to 10) was also presented 4 times to participants (Beaulieu-Bonneau, 2012): at arrival (reference), after scenario 1, after break, after scenario 2.

The first experiment -HRR1- involved 19 participants (10 women, 9 men; age range: 23-55 years old) randomly allocated to "Static rest" and "Exercise" breaks, and performed tasks just after having lunch (between 12:30 and 15:00). For the "Hypnosis" break situation, participants were preselected as highly suggestible, using the Standford Modified Scale of Paris Revised Form (Michaud & al., 2007). The second experiment HRR2, involved a different population of 45 volunteers, 15 per group (17 women, 28 men, Mean age = 40 years, 31-49 years, 6 left-handed). Participants were balanced in the three groups regarding their training score for ensuring comparable group performance level. As we increased the workload and the fatigue in the scenario, we judged not necessary to keep the after lunch period as a constraint and extended it at different moment of the day from 08:00 to 18:00 within each group (with a similar distribution between groups). For the HRR2 experiment, an electrooculography (EOG) using 6 electrodes was added to the Tobii (BioSemi). During the test, the hypnosis technic employed was inspired from "Erickson" practice. The script (hypnotherapist verbalizations) was based on 'positive energy' thinking incomes when breath in, 'tensions' outcomes when breath out, plus memory of success reliving associated positive emotions. We used the word -reliving- as it was demonstrated by Faymonville & al (2006) with magnetic resonance images that hypnosis induces a state with a "reliving of the situation" meaning the brain areas active as if the subject was living the situation. Positive emotions related to memory of success were enhanced by hypnotherapist proposed words for related emotions (suggestion): "energy", "joy", "pride", "pleasure", "strength", "enthusiasm" plus "one's own resources mobilization". Those emotions induced a well-known effect on cognition (Fredrickson, 1998), and potentially on motivation, engagement and effort (resources involvement) that participants could involve for their activity (Jonc 2001, Lautrey 2003). A post-hypnotic suggestion was added while getting out the participant from the hypnotic state: "to be here in top form".

Human Resources Recovery Main Results (HRR 1 & HRR 2)

In each experiment, participants increased both their performance and their fatigue rating. In the first experiment, performance increase was slightly better for "Hypnosis" (+7.7%) than for "Exercise" (+6.8%) and "Static" rest (+6.5%). Surprisingly, in the second experiment, participants from the "Static" rest (+7.64%) and the "Hypnosis" (+7.24%) groups were slightly better performers than the "Exercise" group (+5.95%). In both experiments "Hypnosis" showed a small performance increase while Exercise and Static rest exchanges their performance level.

Eye tracker and EOG measurements showed no statistical significant evidence for "static rest" and "hypnosis" groups in arousal increase while it was significant for saccade movements in favor of "exercise" group in the second experiment. Nevertheless, the static rest group showed a higher scoring in blinks than the two others.

Workload scale quotation showed a decrease of general workload (mean group difference) after second scenario in both experiments but more pronounced in HRR1 than in HRR2. HRR1 showed: "Hypnosis" (-18.2), "Exercise"(-13.7), "Rest"(-6.3), while HRR2 showed: "Hypnosis"(-3.9), "Rest"(-2.7), "Exercise"(-1.1).

Fatigue rating showed differences between groups even if not statistically significant in HRR1: Anova, ($F(2,60)=1.723$, $p=0.187$) and group ($F(3,60)=1.076$, $p=0.366$) and for interaction time*group ($F(6,60)=1.034$, $p=0.412$). The absence of significant effect may be explained by the low number (n=6) of participants per group. For that purpose, the HRR2 experiment was performed with increased numbers of participants.

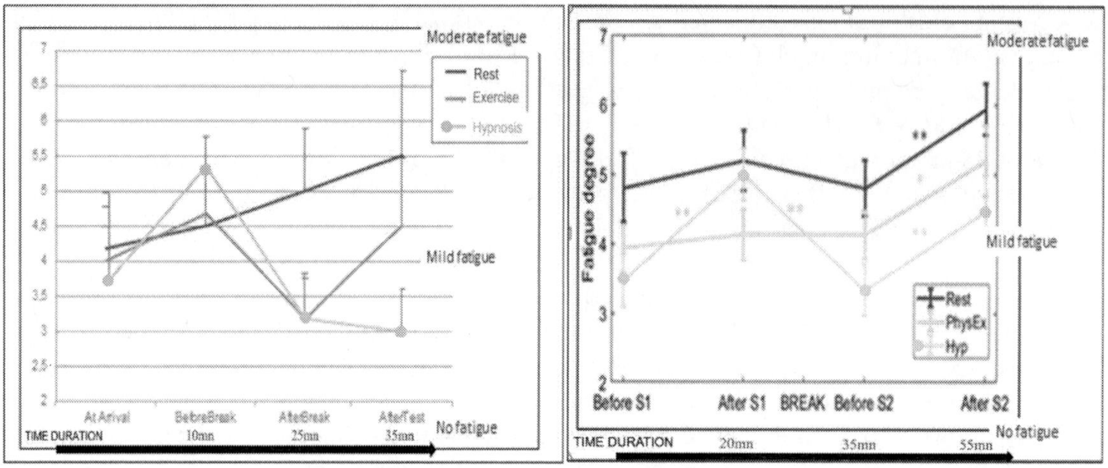

Figure 1&2: *HRR1 & HRR2 self-assessed fatigue*

The most interesting rating in HRR1 is that participants felt less tired at the end of the experiment than at any time before, even at arrival. The second experiment assessment was done in a more demanding situation (mimicking abnormal cockpit workload). In the HRR2 figure, asterisks represent significance to a paired Wilcoxon test intra group after correction for multiple comparisons; * $p<0.05$; ** $p<0.005$. At the end of the second scenario, all participants reported higher fatigue than before the test (rest and hypnosis group $p<0.005$; physical exercise $p<0.05$). The effect of the break is statistically significant, mostly for the "Hypnosis" group (Wilcoxon sign rank paired test, $p=0.0024$, after correction for multiple comparisons, $p=0.0072$). When comparing "after scenario 1" to "before scenario 2" (after break), the Kruskal & Wallis test appears to be statistically significant for all groups (H (2,N=45) =14,41558 $p=0.0007$). When comparing, this time, the groups 2 by 2, only "Exercise" break group and "Hypnosis" break group were significant together and none of them was significant when compared with the "Rest" break group. In both experiments, the "Hypnosis" group rated their fatigue lesser than the other groups at the end of the experiment. They also rated with a higher increase at the end of the first scenario than the two other groups in both studies.

The debriefing showed that the second experiment was judged as requiring a significant effort while performing and induced more fatigue. All participants found strategies to cope with this demanding situation in order to perform better. "Rest" and "Exercise" break groups reported that a break allows strategy adjustment, step back effect and eye relaxation. We did not expect in our assumption the "Hypnosis" break participants' feedback about acting the second scenario without any stress, with a gain in motivation and with less time pressure sensation.

Human Resources Recovery Discussion and Conclusion

In our experiments, fatigue was produces by creating mental workload. The effect of hypnosis we were assessing here was significant on fatigue. Hypnosis was also effective on stress and motivation as debriefed by participants. Our experiments illustrated well the fact that, even with consequent workload and complex multitasks, operators were able to maintain their performance while producing fatigue. The effect on performance was here less marked but could be interesting to deeper study in further experiments as Oakley & Halligan (2009) indicates that hypnosis triggers changes such as motivation, relaxation, and mental absorption. Moreover, Cegarra (2012) states that cognitive resources are used faster when coping with stress. The stressless effect after hypnosis practice, underlined by subjects in debriefings, might help to cope with less resources involved in emotional management and consequently a better performance and less fatigue. This "calm and stressless effect" induced by a hypnotic state was also explained physiologically by subjects' oxytocin hormone increase, DeVries & al.(2003), Uvnäs-Moberg & al.(2004), Varga & al.(2014).

We have to clarify a generic confusion between the hypnotic physiological state that comes naturally (not induced by the subject or a hypnotherapist) and the practice, as they are both commonly called "Hypnosis". Hypnosis as a natural state is described by E. Rossi (1982) as an ultradian cycle of "natural periods of quietness and receptivity" being the natural "common everyday trance", that the body uses naturally to recover. Assuming that hypnotic trance as a natural state exists, hypnosis is the term used by medical and psychology fields for this mechanism but a neurophysiology concept called "mind wandering" seems to correspond to this natural trance state. Mind wandering is defined as "*shifts in attention from outward, stimulus-based processing to inward, introspective cognition*" (Mittner & al. 2014) which is very close to Rossi's definition. This concept provides physiological description about this "common everyday trance" hypnotic state mechanism. This "natural state", which is coming without the will of the subject, is very interesting to investigate as this state has an effect on cognitive activity (Mittner & al. 2014). The practice of self-hypnosis is a way to recognize this "mind wandering" or "common everyday trance" when occurring without the subject's will. The self-practice in aviation could be interesting as it allows the operator to stop or induce the state on demand, at the best moment for him, in order to take the maximum profit. The practice allows to produce different effects such as fatigue recovery (our script example), going faster and deeper into sleep while being in crew rest for instance and much more. Due to the large choice of effects that might be scoped, the practice need to be further investigated with airlines and authorities in order to determine the best integration option in operational conditions.

References

Beaulieu-Bonneau, S. (2012). Somnolence, fatigue et fonctionnement attentionnel suite à un traumatisme craniocérébral.

Caldwell Jr, J. A., Jones, R. W., Caldwell, J. L., Colon, J. A., & Pegues, A. (1997). *The efficacy of hypnotic-induced prophylactic naps for the maintenance of alertness and performance in sustained operations* (No. USAARL-97-10). ARMY AEROMEDICAL RESEARCH LAB FORT RUCKER AL.

Cegarra, J. (2012). De la gestion de la complexité à son assistance: Contributions en psychologie ergonomique. *Unpublished habilitation à diriger des recherches). Université de de Toulouse II-Le Mirail, Toulouse, France.*

DeVries, A. C., Glasper, E. R., & Detillion, C. E. (2003). Social modulation of stress responses. *Physiology & behavior*, *79*(3), 399-407.

Faymonville, M. E., Boly, M., & Laureys, S. (2006). Functional neuroanatomy of the hypnotic state. *Journal of Physiology-Paris*, *99*(4-6), 463-469.

Folkard, S., & Åkerstedt, T. (2004). Trends in the risk of accidents and injuries and their implications for models of fatigue and performance. *Aviation, space, and environmental medicine*, *75*(3), A161-A167.

Fredrickson, B. L. (1998). What good are positive emotions? *Review of general psychology*, *2*(3), 300-319.

Hammer, E. F. (1954). Post-hypnotic suggestion and test performance. *International Journal of Clinical and Experimental Hypnosis*, *2*(3), 178-185.

Hart, S. G., & Staveland, L. E. (1988). Development of NASA-TLX (Task Load Index): Results of empirical and theoretical research. In *Advances in psychology* (Vol. 52, pp. 139-183). North-Holland.

Jong, R. D. (2001). Adult age differences in goal activation and goal maintenance. *European journal of cognitive psychology*, *13*(1-2), 71-89.

Kaiser, J., Barker, R., Haenschel, C., Baldeweg, T., & Gruzelier, J. H. (1997). Hypnosis and event-related potential correlates of error processing in a Stroop-type paradigm: A test of the frontal hypothesis. *International Journal of Psychophysiology*, *27*(3), 215-222.

Landry, M., Lifshitz, M., & Raz, A. (2017). Brain correlates of hypnosis: A systematic review and meta-analytic exploration. *Neuroscience & Biobehavioral Reviews*, *81*, 75-98.

Lautrey, J. (2003). La psychologie différentielle à l'épreuve de la variabilité intra individuelle. *Psychologie différentielle: recherches et réflexions. Rennes: Presses Universitaires de Rennes.*

Lifshitz, M., Bonn, N. A., Fischer, A., Kashem, I. F., & Raz, A. (2013). Using suggestion to modulate automatic processes: From Stroop to McGurk and beyond. *cortex*, *49*(2), 463-473. https://doi.org/10.1016/j.cortex.2012.08.007.

Mulder, G., & Moray, N. (1979). Mental workload: Its theory and measurement.

Michaux, D., Halfon, Y., & Wood, C. (2007). *manuel d'hypnose pour les professions de santé* (pp. 93-97). Paris, France: Maloine.

Mittner, M., Boekel, W., Tucker, A. M., Turner, B. M., Heathcote, A., & Forstmann, B. U. (2014). When the brain takes a break: a model-based analysis of mind wandering. *Journal of Neuroscience*, *34*(49), 16286-16295.

Oakley, D. A., & Halligan, P. W. (2009). Hypnotic suggestion and cognitive neuroscience. *Trends in cognitive sciences*, *13*(6), 264-270.

Raz, A., Shapiro, T., Fan, J., & Posner, M. I. (2002). Hypnotic suggestion and the modulation of Stroop interference. *Archives of General Psychiatry*, *59*(12), 1155-1161.https://doi.org/10.1001/archpsyc.59.12.1155.

Rossi, E. L. (1982). Hypnosis and ultradian cycles: A new state (s) theory of hypnosis?. *American Journal of Clinical Hypnosis*, *25*(1), 21-32. DOI: 10.1080/00029157.1982.10404061.

Santiago-Espada, Y., Myer, R. R., Latorella, K. A., & Comstock Jr, J. R. (2011). The multi-attribute task battery ii (matb-ii) software for human performance and workload research: A user's guide.

Solar impulse 2019: https://www.youtube.com/watch?v=iJRJwoz-yPA https://www.youtube.com/watch?v=xZO35uGAXZM

Tomporowski, P. D. (2003). Effects of acute bouts of exercise on cognition. *Acta psychologica*, *112*(3), 297-324.https://doi.org/10.1016/S0001-6918(02)00134-8.

Uvnäs-Moberg, K., & Petersson, M. (2004). Oxytocin--biochemical link for human relations. Mediator of antistress, well-being, social interaction, growth, healing... *Lakartidningen*, *101*(35), 2634-2639.

Varga, K., & Kekecs2, Z. (2014). Oxytocin and Cortisol in the Hypnotic Interaction1. *International Journal of Clinical and Experimental Hypnosis*, *62*(1), 111-128.

Williamson, A., Lombardi, D. A., Folkard, S., Stutts, J., Courtney, T. K., & Connor, J. L. (2011). The link between fatigue and safety. *Accident Analysis & Prevention*, *43*(2), 498-515.

Yerkes, R. M., & Dodson, J. D. (1908). The relation of strength of stimulus to rapidity of habit-formation. *Journal of comparative neurology and psychology*, *18*(5), 459-482. https://doi.org/10.1002/cne.920180503

A SYSTEM FOR ASSESSING CERVICAL READINESS USING ANALYTICS AND NON-INVASIVE EVALUATION (CRANE)

Zachary A. Kiehl, Deepak Sathyanarayan, and Kent C. Halverson
Aptima, Inc.
Fairborn, OH & Woburn, MA
Michael E. Zabala and Sean Gallagher
Auburn University
Auburn, AL
Brian Farrell
Human Systems Integration
Walpole, MA

Current cervical spine assessment methodologies focus solely on subjective measures, such as pain reports, and range-of-motion (ROM) testing that only measures maximum head excursion and reach (i.e., not dynamic motion). Due to report bias and the potential for negative outcomes of self-reported pain, current clinical assessment methods fail to provide valid, reliable data for medical practitioners to effectively manage long-term cervical health. Furthermore, commercial systems capable of quantitative assessment of cervical spine function are generally sparse and often immature. This paper highlights both the need and a path towards a clinical tool for objective measurement of cervical spine health and functionality. Lastly, a novel solution concept is presented to objectively assess Cervical Readiness using Analytics and Non-invasive Evaluation (CRANE). This solution concept combines cervical spine instrumentation, novel virtual reality (VR) game-based test protocols, robust analytical algorithms, and intuitive presentation of health metrics.

Neck pain and cervical spine injury are known mission degraders across the spectrum of military personnel. Such pain and inury is a undisputed source of increased cost in terms of preventative actions, healthcare treatment, and negative impact on productivity (Hauret, Jones, Bullock, Canham-Chervak, & Canada, 2010). Managing and mitigating long-term health risks associated with Naval aviation is especially problematic. Although the rapid onset of catastrophic injuries to the neck from extreme loading (i.e., high G-forces combined with heavy helmets) continues to be a risk (Harrison, Coffey, Albert, & Fischer, 2015), clinicians often lack the information and tools necessary to properly diagnose, treat, or prevent cervical spine injuries. The inherent negative consequences for aviators that admit to neck injuries or pain is also an exacerbating factor. For example, injury or pain that could lead to performance degradation can result in medical grounding and a "duty not including flying" (DNIF) status. Although some aviators may disclose information during annual clinical visits, others may either seek treatment from private providers or seek no treatment at all. In addition, current clinical tests intended to assess cervical spine health (e.g., range of motion [ROM] tests) have been questioned with respect to their validly, reliability, and overall practicality for clinical use (Jordan, 2000). Unfortunately, most potential solutions (1) neglect the cervical spine by focusing on the lumbar or thoracic spine, (2) lack quantitative data to support physician assessment and patient anamnesis, or (3) are confined to research prototypes that have not undergone clinical trials (Yang, Su, & Guo, 2012).

There are obvious risks with undiagnosed acute and chronic cervical spine injuries. Minor chronic pain can lead to distraction and reduced cognitive abilities, which can impact decision-making during critical phases of operations (Apkarian et al., 2004). Severe injuries that limit ROM or function could thus hinder the aviator's ability to perform effectively. Both situations negatively impact aviator readiness and add unnecessary risk to the mission, and these risks are not isolated to aviators. Ground-based operators and support personnel also experience cervical spine health risks and are exposed to similar point of care testing (POCT) consequences. Fortunately, recent advances in wearable sensor technologies (e.g., inertial measurement units [IMUs]) have made it possible to obtain data for developing motion models that can characterize human behaviors and proclivities. Similarly, advanced predictive analytic methods leveraging machine learning can identify subtle patterns in longitudinal data. This proceedings paper outlines the need and path forward toward a medical material solution that allows clinicians to more accurately diagnose injury and utilize preventative care to maintain healthy cervical function or reduce the likelihood of injury progression.

Approach

With no immediate solution present, this paper will describe a novel method and medical material solution for objectively assessing cervical spine neck function. This approach focuses primarily on a clinical medicine application, though some approaches and research may generalize to continuous assessment of aviator spine health and performance. This overarching objective is supported by three subsidiary topics of discussion: (1) application to the military clinical use case, (2) a review of current sensing components and solutions, (3) an investigation of data modeling techniques and prototype development. Each of these research areas will be discussed in the following section.

Discussion

Military Clinical Use Case

Navy aviators and other military personnel operate in unique, demanding environments in which they are subjected to a wide range of forces that can result in acute injuries and lead to either long-term or permanent disabilities. Forces on the cervical spine are a combination of (1) static axial forces related to head-mounted weight (i.e., wearing a helmet with additional equipment such as night vision goggles [NVGs]), make-shift counter weights, or helmet mounted displays; and (2) dynamic torsion forces related to head-turning (e.g., “check-six” maneuver), often with flexion or extension, to perform in-cockpit tasks or outside scanning across a wide range of g-force situations. Although treatment costs and disability compensation are significant financial issues, a potentially greater issue is the risk of aviators performing missions while experiencing cervical pain or restricted ROM that could either distract or prevent them from successfully completing their mission.

Unfortunately, military clinicians often lack sufficient information to effectively manage aviator cervical health for three reasons: (1) measurement frequency, (2) measurement validity, and (3) measurement reliability. Formal cervical spine assessment typically only occurs during annual flight physicals or when pain is reported. Outside of these events, aviators who experience cervical pain or injury may not report symptoms to medical practitioners (hence foregoing examination or treatment). Many military aviators avoid disclosing medical information that could jeopardize their careers and result in medical grounding (i.e., DNIF). Consequently, some aviators manage their symptoms outside of the military clinic, either through private clinicians or pain medication using over the counter (OTC) medications. This situation exacerbates the military neck health epidemic by figuratively blinding medical practitioners to aviator cervical health status. Considering the nature of the cervical spine issue (i.e., repeated stressors and impact loads over long periods of time), designing a clinical diagnostic system on a single, crude ROM measure seems extremely suboptimal.

The other current limitation to effectively managing aviator cervical spine health is a lack of validity and reliability in the primary means of measurement, which is typically the ROM test. Past research has concluded that this ROM test is limited in that it only measures static range and is not sensitive to the dynamic aspects of cervical spine function, such as linear and rotational acceleration (Youdas, 1991). Currently, no commercial off the shelf (COTS) sensor system exists to objectively measure cervical spine health. Reliance on a relatively crude method such as the ROM test makes it vulnerable to patients "gaming" the test and enduring pain associated with uncomfortable motions. Consequently, medical researchers are limited in the models they can develop using only head excursion data, as opposed to using dynamic kinematic data. Furthermore, the single set of data obtained from the ROM test administered during the annual clinical visit limits the ability to conduct longitudinal assessments of cervical health, which would help medical practitioners diagnose, treat, and ideally prevent neck injury. More importantly, data obtained from the ROM test can be systematically biased due to the strong motivation for aviators to not reveal cervical issues. To address this challenge, an objective sensing solution must be developed to integrate within existing military clinical workflows, which normally consists of period health assessment (PHA) questionnaires, in-person ROM tests, and intermittent health exams as requested by the servicemember. The results of these assessments would need to interface with current and future U.S. military's electronic health record systems, such as the Armed Forces Health Longitudinal Technology Application (ALTHA) and the Military Healthcare System GENESIS.

Review of Current Spine Sensing Solutions

The concept of objective spine health monitoring is relatively new. Despite the relative nascency, COTS sensing systems do exist; however, these systems primarily focus on the thoracic or lumbar sections of the spine (Rodgers, Pai, & Conroy, 2015). Fortunately, the increasing demand for highly functional and data-driven technology has led to increased research on wearable sensors designed for human use. To support the overarching objective of improving aviator cervical spine health through non-invasive evaluation, this article will briefly examine current spine sensing solutions and the requisite components. Most research or commercial sensing systems utilize one or more of the following sensors: accelerometers, gyroscopes, magnetometers, or electromyography sensors (EMG). These Micro-Electro-Mechanical Systems

(MEMS) components can be fused to create inertial or motion-tracking sensors, such as IMUs. Although these sensors are prevalent in numerous electrical technologies (automobiles, aircraft, smartphones), their uses for human monitoring applications are still relativley new. EMG is a technique for recording and evaluating the electrical activity produce by skeletal muscles. This activity can be collected either invasively or non-invasively through the use intramuscular or surface EMG (sEMG), respectively. Paired with various temporal and spectral analytical techniques, EMG can be used to detect muscle fatigue, muscle dysfunction, and muscle activity.

A few of the most notable COTS sensing systems include the following: the Xsens various inertial motion tracking modules (Xsens Technologies B.V., The Netherlands); Delsys's Trigno platfrom (Delsys, Inc., Natick, Massachusetts), which combines the capabilities of EMG and IMU into one unit; dorsaVi's ViSafe, ViPerform, and ViMove inertial tracking platform (dorsaVi Ltd., Australia), which provide solutions for occupational, athletic, and clinical applications; and StrongArm's wearble inertial harness and FUSE platform (StrongArm Technologies, Inc Brooklyn, NY, U.S.). Despite dissimilar wearble form factors and data dashboards, these technologies all leverage inertial measurement through the use of MEMS for healthcare, occupational, and athletic applications. A description of a detailed review of these technologies is not within the scope of this paper; however, such a review reveals that there are few technologies specifically designed for the spine or for clinical environments, and precisely zero mature solutions that are designed specifically for clinical measurement of cervical spine function. Most sensing efforts specifically focused on the cervical spine have been confined to basic research settings, such as academia (Papi, Koh, & McGregor, 2017).

Data Modeling and Protoype Solution

Implementation of medical devices is the clinical setting is often limited or slowed by necessary supporting infrastructure for device operation (Lukas, et al., 2007). Cervical spine measurement devices, such as the CRANE solution, need to provide accurate clinical assessments through a lightweight form factor with little supporting architecture. Thus, effective solutions would leverage MEMS sensors (e.g., IMU, EMG) without the need for room-based motion tracking of the neck for functional assessment. IMUs can be used to provide orientation and 6 degree of freedom (6-DOF) accelerations of the head and torso while sEMG electrodes are placed bilaterally to track muscle activity on key neck extension and neck flexion muscles during clinical assessment tasks. Tasks would include both standard clinical assessments (i.e., ROM) as well as VR-aided functional tasks (e.g. simulated air-to-air combat). Although the quantification of cervical spine disorder using motion measure has not been heavily researched, empirical examinations of other areas spinal regions has been completed (Marras et al., 1999)

Over the course of a POCT assessment, the IMU/EMG instrumentation and VR-system would collect fine-grained head-torso kinematics and muscle activity data. Raw data can be processed to calculate secondary features such as head trajectory, head acceleration, head jerk, and neck muscle activation patterns and characteristics. Over multiple POCT sessions, the CRANE system would allow for longitudinal tracking of cervical spine functionality. Long-term and chronic overuse injuries often emerge after extended periods of untreated sub-symptomatic states, gradually accumulating soft-tissue damage until significant functional deficits are exhibited (Salmon, Harrison & Neary, 2011). Analysis of minute deviations in cervical spine trajectories and fatigue characteristics over time may be indicative of injury development. Clinical insights supported by objective data would aim to supplement current clinical assessments and provide early detection of movement-pain compensation (and potentially injury development) within the POCT environment.

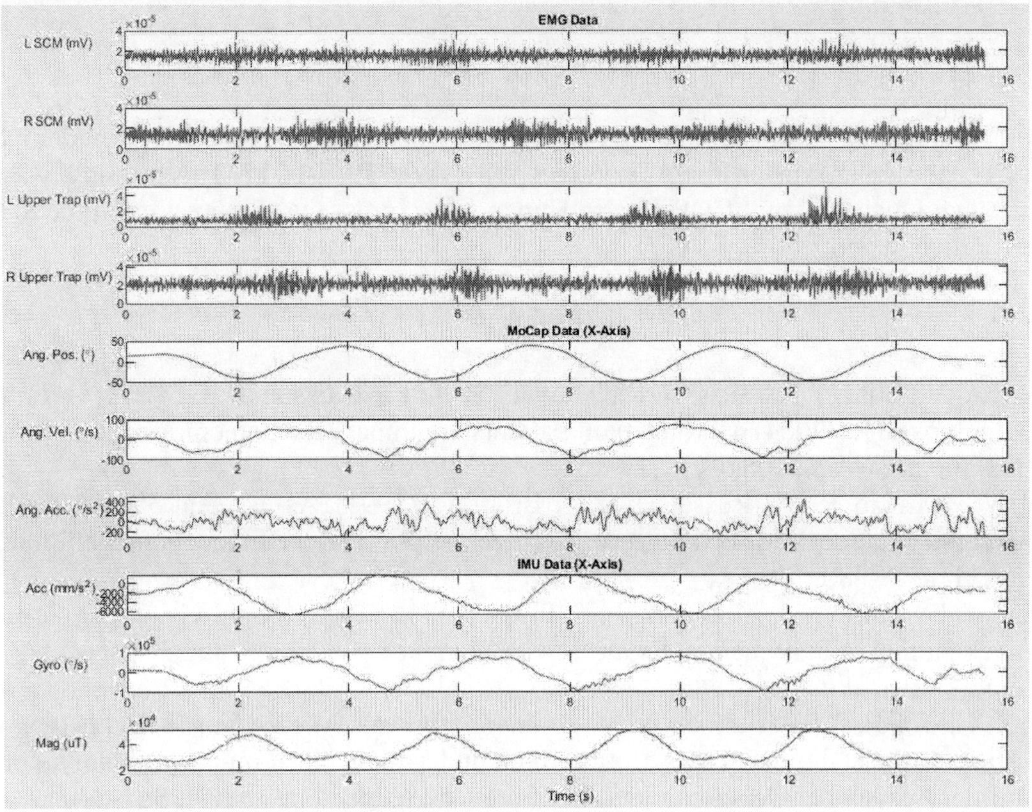

Figure 1. Signal outputs from the CRANE system's IMU and sEMG sensors over cycles of head circumduction. (Top-to-bottom) sEMG sensors tracked left-right sternocleidomastoid (SCM), and left-right upper trapezius. Although not part of the in situ CRANE system, motion-capture provides a reference for IMU-collected head kinematics, including position, velocity, acceleration, and orientation. Though complete 6-DOF motion is captured by the system, only X-axis motion components are depicted for brevity.

Conclusion

Though this research intiative is still in its infancy, early research and development has shown a clear need for a medical material solution capable of objective cervical spine asessement. This article aims to provide a review of existing spine health monitoring capabilites along with a path forward towards cervical spine monitoring in military settings. The initial environment of interest is that of a military clinical environment (e.g., use of the solution during an annual flight physical), with a military clinician serving as the initial user. Outside of clinical use, a long-term objective is to pursue continous and longitudnial spine-health monitoring initiatives in operational settings, such as during long-duration flight missions. A gradual move toward continous monitoring is needed for myriad reasons. Clinical exams are certainly needed to diagnose and treat injury or illness; however, most of this treatment is reactive by nature, as oppossed to proactive. Quite simply, many aviators do not seek treament in the clinic until they are debilitated by often-irreversabile spine pain and damage. Additionally, modern aircraft expose the body to static and dynamic forces that were previously inconceivable. Although research examining short-term exposures is prevalent, the longitudnial effects of such forces and moments is relatively sparse. This article and supporting body of work hope to highlight current research gaps and ameliorate impediments to military spine health monitoring initiatives.

Acknowledgements

This work was funded by the U.S. Army Medical Research Acquisition Activity under a Small Business Technology Transfer Phase I contract W81XWH-19-C-0017. This research was coordinated and managed by LCDR Dustin Huber. We express our sincerest gratitude to the sponsor for their support.

References

Apkarian, A. V., Sosa, Y., Krauss, B. R., Thomas, P. S., Fredrickson, B. E., Levy, R. E., & Chialvo, D. R. (2004). Chronic pain patients are impaired on an emotional decision-making task. Pain, 108(1-2), 129-136.

Hanten, W. P., Lucio, R. M., Russell, J. L., & Brunt, D. (1991). Assessment of total head excursion and resting head posture. Archives of physical medicine and rehabilitation, 72(11), 877-880.

Harrison, M. F., Coffey, B., Albert, W. J., & Fischer, S. L. (2015). Night vision goggle-induced neck pain in military helicopter aircrew: a literature review. Aviation, space, and environmental medicine, 86(1), 46-55.

Hauret, K. G., Jones, B. H., Bullock, S. H., Canham-Chervak, M., & Canada, S. (2010). Musculoskeletal injuries: description of an under-recognized injury problem among military personnel. American journal of preventive medicine, 38(1), S61-S70.

Jordan, K. (2000). Assessment of published reliability studies for cervical spine range-of-motion measurement tools. Journal of manipulative and physiological therapeutics, 23(3), 180-195.

Lukas CV, Holmes SK, Cohen AB, et al. Transformational change in health care systems: an organizational model. Health Care Management Review. 2007;32(4):309-320.

Marras, W. S., Ferguson, S. A., Gupta, P., Bose, S., Parnianpour, M., Kim, J. Y., & Crowell, R. R. (1999). The quantification of low back disorder using motion measures: Methodology and validation. Spine, 24(20), 2091.

Papi, E., Koh, W. S., & McGregor, A. H. (2017). Wearable technology for spine movement assessment: A systematic review. Journal of biomechanics, 64, 186-197.

Rodgers, M. M., Pai, V. M., & Conroy, R. S. (2015). Recent advances in wearable sensors for health monitoring. IEEE Sensors Journal, 15(6), 3119-3126.

Salmon DM, Harrison MF, Neary JP. Neck pain in military helicopter aircrew and the role of exercise therapy. Aviat Space Environ Med 2011; 82:1-10.

Yang, C. C., Su, F. C., & Guo, L. Y. (2012). A new concept for quantifying the complicated kinematics of the cervical spine and its application in evaluating the impairment of clients with mechanical neck disorders. Sensors, 12(12), 17463-17475.

Youdas, J. W., Carey, J. R., & Garrett, T. R. (1991). Reliability of measurements of cervical spine range of motion—comparison of three methods. Physical therapy, 71(2), 98-104.

THE TRAINING OF OPERATORS IN SINGLE PILOT OPERATIONS: AN INITIAL SYSTEM THEORETIC CONSIDERATION

Daniela Schmid
German Aerospace Center (DLR)
Braunschweig, Germany

Neville A. Stanton
University of Southampton
Southamtpon, UK

Single Pilot Operations (SPO) represent a viable concept for commercial aviation in near future. It will require different training regimes to contemporary airliners' pilots because the single-pilot's and remote operator's (including the dispatcher's) responsibilities and accompanying procedures will change, both in air and on ground. This initial system-theoretic analysis of suggestions for training of single-pilot and remote-copilot identified the agreement of job rotation of both. Hence, pilots can still be trained in new single-pilot specific procedures in a special training fleet to includes the apprenticeship-style training in SPO. Advanced new automation tools will be challenging to be included into training. On this basis, skill degradation becomes an issue and must be solved. Nonetheless, the training issue can be early tackled by applying worker competencies' analyses when investigating different concepts in depth early in the design process of SPO.

During the last two decades, research on reducing the crew of an airliner has been hot topic (Lachter, Brandt, Battiste, Matessa, & Johnson, 2017). Different Concept of Operations (ConOps) for commercial Single Pilot Operations (SPO) have been proposed, fragmentarily prototyped and empirically investigated (e.g. Vu, Lachter, Battiste, & Strybel, 2018). This research suggest that SPO seems viable in near to mid-term future. Whilst none of the ConOps was superior to any of the others in any type of human factors research question, system-theoretic analyses excluded, the option of single-pilot without ground-based support. It was shown to be less resilient to safety and security intrusions (Harris, 2018; Revell, Allison, Sears, & Stanton, 2018; Schmid & Korn, 2018; Stanton, Harris, & Starr, 2016). Table 1 (see next page) summarizes the remote operators' most promising and most researched ConOps in SPO, which is most likely to be pursed in future research. In the NASA ConOps, the ground operator is either organized and trained as a hybrid operator, either performing multi-aircraft and dedicated support or as specialist performing one of the high-level functions (Bilimoria, Johnson, & Schutte, 2014). In all other ConOps, no variations in unit structures of the remote copilot have been considered.

The emerging issue of training for transport SPO, and the concomitant changes required, has only been considered marginally in theoretical discussion. It was first discussed regarding certification requirements for such SPO (Harris, 2007; Wilson, Harron, Lyall, Hoffa, & Jones, 2013). Future large transport aircraft employing commercial SPO will be categorized as Part 121 and 125 operations, which currently require a minimum flight crew of two pilots. Hence, as such, SPO are not complient with CFR Part 121 and 125 operations, therefore the regulations would need to be changed or, alternatively, SPO must demonstrate an equivalent level of safety. One challenge is providing new training procedures that fit with the ConOps. The single-pilot as main Pilot Flying (PF) and the remote operater as main Pilot Monitoring (PM), and occasional PF, or even super-dispatcher, would require different and/or additional trainings than is currently the case. Their job would be very different in normal, off-nominal and emergency situations (Wilson et al., 2013). Fallback, or spouse, training has already been considered (see also: Comerford et al., 2013). It refers to non-pilot personnel on-board that receive an abbreviated training sufficient to land the single-piloted aircraft safely in the case of pilot incapacitation. This idea, however, has not been pursued any further.

A technical interchange meeting was coordinated and hosted by NASA to work out the issues and make recommendations for various aspects of SPO that require research (Comerford et al., 2013). New procedures in SPO would require new methods for training because the apprenticeship-style training that is currently employed for current multi-crew operations (MCO) would cease to exist. Pilots could become Captain of a single-piloted aircraft immediately. It is unknown what type and how much training a remote-copilot would need. Skill degradation has to be countered more effectively because of a greater extent of advanced automation tools onboard. Flight length, refering to temporal duration, might become more important and require different training to counter fatigue. Ground personnel also might require different training. For example, an enhanced dispatcher would require additional training to cover their enhanced duties. In summary, the minimum training regulations for each type of SPO-operator would have to be identified, redesigned and certified before they become operational.

Table 1.
The categorization of remote operators in dominant ConOps of SPO.

<table>
<tr><th>Remote operator</th><th>Functions</th><th>ConOps (references)</th></tr>
<tr><td>Super dispatcher</td><td>Multi-aircraft support: (1) monitoring 30 aircraft, contact to airline operations center; (2) Off-nominal: supporting crew decision-making;</td><td rowspan="2">NASA ConOps (Lachter et al., 2017; Matessa, Strybel, Vu, Battiste, & Schnell, 2017; Vu et al., 2018)</td></tr>
<tr><td>Remote copilot</td><td>Dedicated support: off-nominal and emergency support in pilot functions and/or take over of control in case of need</td></tr>
<tr><td>Remote copilot</td><td>Normal: Flight planning and navigation support; support as PM and of communications with ATC/ATM
Off nominal and emergencies: support as PM and/or PF</td><td>Support of one aircraft at a time (Revell et al., 2018; Schmid & Korn, 2018; Schmid & Stanton, 2018; Stanton et al., 2016)</td></tr>
</table>

Notes. For reasons of space, the most representative references of series of studies are given. Peer-reviewed literature is preferred.

The present system-theoretic approach aims to provide a first theoretic (re)view on the issue of training in SPO. No study has properly investigated the training issues for commercial SPO to the best of the authors' and Vu et al.'s (2018) knowledge. In this systematic analysis, possible upcoming issues in training design are identified and considered from a system-theoretic approach. Using the System-Theoretic Accident Model and Process (STAMP; Leveson, 2004) it is possible to identify potential hazards and resulting risks from the training of remote-operators in SPO. This study examplifies how STAMP can be used predictively to assess the single hazards around one component of the hierarchical control structure in system operations (Grant, Salmon, Stevens, Goode, & Read, 2018; Leveson, 2015; Revell et al., 2018). The use of the method in a predicitive context has been less dominant in literature. Nonetheless, its potential should be considered in how it can provide structured hazard analysis early in the design process because concomitant adaptions of regulations to SPO on each level will be required anyways.

Method: System-Theoretic Accident Model and Process (STAMP)

The System-Theoretic Accident Model and Process (STAMP) views as interrelated components that are kept in a state of dynamic balance by feedback loops of information and control (Leveson, 2004). The feedback loops stretch from the top of a control hierarchy of a supranational level via national and organizational level to the lowest levels where the operating process, including its operators, are located. Across this hierarchy, the three primary components of STAMP take effect: the safety constraints, hierarchical safety control structures, and process models (Leveson, 2011, 2017). The safety constraints are design principles, processes, as well as regulatory, legal, insurance, cultural, and other social control. Safe operations emerge from enforcing appropriate constraints on the system's behaviour and from an adaption of them to changes across the time. Incidents and accidents are viewed as an inadequate enforcement of the safety constraints on the behaviour at each level of a sociotechnical system. In this sense, safety is viewed as a control problem rather than a component reliability problem.

The safety constraints are displayed in a safety control structure across each level of a hierarchy from regulators on the top down to the across all levels to the operating process (Leveson, 2017). Such hierarchical control structures can be constructed for both system development and system operations. Both contain the control loops between and across all system components that are currently in place in development and operations. As part of them, actuators implement control actions and sensors provide feedback about the state of the given process. Controllers initiating both can be human or machines. In doing so, they enforce the safety constraints.

These control processes operate between all components that are connected with each other in feedback loops. The third major component of STAMP are these process models. The related hazard analysis of System-Process Process Analysis (STPA) technique predicts how the behavioural safety constraints can be, or were, violated in a incident or accident scenario (Leveson, 2011). Four types of Unsafe Control Actions (UCAs) can occur: (1) A control action is not provided, (2) A UCA is provided, (3) A control action is applied too early/late, and (4) A control action is applied for a too short/long time. Afterwards, the analysists determine how the potenatially hazard control actions could occurr. In our application, STPA is not carried out because we do not investigate a specific incident or accident scenario.

In contrast, we consider the function allocations for remote operators and the single-pilot in SPO that have been proposed in research until now to the best of the authors' knowledge. They have been summarized previously in the introduction. In doing so, this analysis investigates a specific change of the hierarchical safety control structure at a limited scope defined by the research question of the training issue in SPO.

In the present part of a study, we focused on the control loops that are involved in the remote operators' training, which will have to change to adapt to commercial SPO. The whole hierarchical control structure for commercial MCO and SPO is described and presented in Schmid, Vollrath, and Stanton (2018) and Schmid and Stanton (2018). The results were reviewed and validated by a commercial pilot in two half-day sessions (28; male, CPL, frozen ATPL, 1,400 of 1,800 hrs. were undertaken in a B737-800). Based on that, we took the control loops and investigated them further regarding what would have have to change in the remote operators' and single pilot's training when SPO is introduced in future. We also included all considerations of the training issues in SPO that have been raised previously during in the course of research on commercial SPO.

Results: The Control Loops of Training in Single Pilot Operations (SPO)

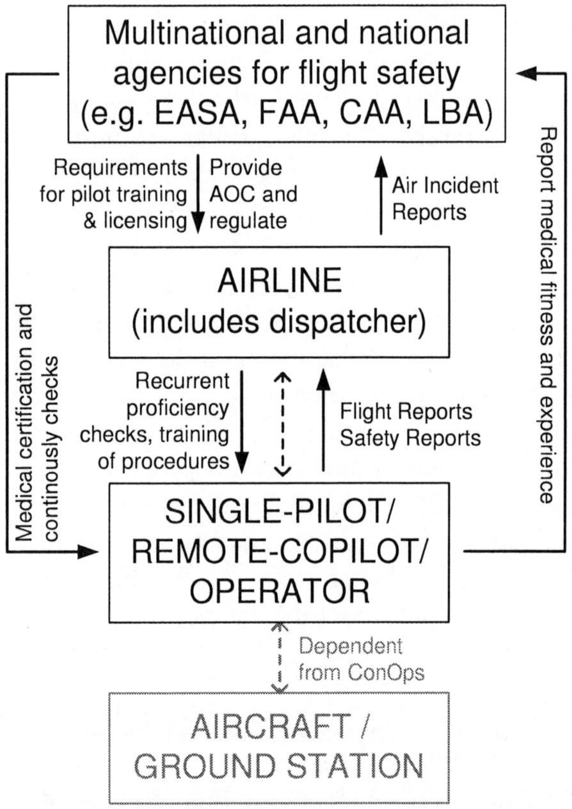

Figure 1. Detail of the hierarchical control structure of SPO for training.

Figure 1 shows the part of the hierarchical control structure of SPO which is relevant for training. The medical certification including checks and the mandatory report of medical fitness will remain. The same accounts for the duty to report any incidents and accidents to the respective upper level controller. In contrast, the procedures, training, and proficiency checks will change to match the requirements for pilot training and licensing that will have to be updated.

First suggestions have already been made how the training of the single-pilot could be approached (Comerford et al., 2013). The apprenticeship-style training could be redesigned in an alternate arrangement of a second pilot observing a single-pilot during operations before he transits to own SPO duties. Operating a special training fleet for SPO pilots complements this idea of the dual trainer concept of military jets (Schmid & Korn, 2017). A fleet would be flown by two pilots that is already equipped with SPO' technology and automation tools. In general, SPO will be introduced after MCO which is why first experience will need to be gained in MCO before switching to SPO. Nonetheless, this concept would serve pure SPO because the Captain of the trainer's fleet aircraft can still be undertaken by an experienced single-pilot.

On ground, the dedicated support or the remote-copilot's functions would essentially require the same skills as those of a conventional pilot (Matessa et al., 2017; Schmid & Korn, 2017). Here, the last quoted ConOps proposes that the single-pilot's and remote-pilot's job can be performed in job rotation to keep skills of each position fresh. Hence, the concepts of training suggested above would include the training of a remote-pilot in depth due to the job rotation in SPO. Finally, the super-dispatcher's job has been already discussed (Matessa et al., 2017). Their job could either performed by a commercial pilot or an enhanced dispatcher. Both of them would require an additional training in performing super-dispatch functions for using the advanced automation tools.

Discussion

In summary, the present paper has identified a dominant concept of training in SPO: the single-pilot's and remote-copilot's job should be performed through job rotations (Matessa et al., 2017) with a trainer fleet concept to tackle the main training issues (Schmid & Korn, 2017). Combining new advanced automation technology with the design of user interfaces gives designers more freedom regarding the whole setup of the single-pilot cockpit and the remote-operator's Ground Station (GS). Here, the future new operating procedures coupled with advanced new automation tools are challenges to be included into training. Nonetheless, the job rotation of both positions ease the training burden by educating each pilot in both jobs to keep the knowledge of the system current. The training will have to be tailored to the new single-piloted aircraft, including the GS, which will differ from contemporary aircraft models. Skill degradation might become an issue, depending on how automation is used. Hence, the result of the present study provides a basis to foster the analysis of worker competencies and start investigating training issues of different ConOps of SPO in place.

Possible changes in training, and later concomitant regulations, are a precondition for system operations that often appear later in the design process, but it is worth considering them much earlier on if SPO is to be successful. The additional costs of training personnel has not been included in any benefits and cost models of SPO until now (Graham, Hopkins, Loeber, & Trivedi, 2014; Malik & Gollnick, 2016; Norman, 2007). New costs, including the infrastructure, can affect the cost-benefit practicability of the SPO concept. Nonetheless, the training issue will become relevant in practice when a distinct ConOps, including advanced automation and the data-link technology, are mature enough. In general, establishing a training regime for SPO operators seems viable as SPO itself.

References

Bilimoria, K. D., Johnson, W. W., & Schutte, P. C. (2014). Conceptual Framework for Single Pilot Operations. In *Proceedings of the International Conference on Human-Computer Interaction in Aerospace*. New York, NY, USA: ACM.

Comerford, D., Brandt, S. L., Lachter, J., Wu, S.-C., Mogford, R., Battiste, V., & Johnson, W. W. (2013). *NASA's Single-Pilot Operations Technical Interchange Meeting: Proceedings and Findings*. Moffett Field, CA, USA: NASA.

Graham, J., Hopkins, C., Loeber, A., & Trivedi, S. (2014). *Design of a Single Pilot Cockpit for Airline Operations*. Paper presented at the Systems and Information Engineering Design Symposium (SIEDS), Charlottesville, VA, USA.

Grant, E., Salmon, P. M., Stevens, N., Goode, N., & Read, G. J. M. (2018). Back to the future: What do accident causation models tell us about accident prediction? *Safety Science, 104*, 99–109. doi:10.1016/j.ssci.2017.12.018

Harris, D. (2007). A human-centred design agenda for the development of single crew operated commercial aircraft. *Aircraft Engineering and Aerospace Technology, 79*(5), 518–526. doi:10.1108/00022660710780650

Harris, D. (2018). Network Re-analysis of Boeing 737 Accident at Kegworth Using Different Potential Crewing Configurations for a Single Pilot Commercial Aircraft. In D. Harris (Ed.), *Engineering Psychology and Cognitive Ergonomics* (pp. 572–582). Cham, Switzerland: Springer.

Lachter, J., Brandt, S. L., Battiste, V., Matessa, M., & Johnson, W. W. (2017). Enhanced ground support: Lessons from work on reduced crew operations. *Cognition, Technology & Work, 19*(2-3), 279–288. doi:10.1007/s10111-017-0422-6

Leveson, N. G. (2004). A new accident model for engineering safer systems. *Safety Science, 42*(4), 237–270. doi:10.1016/s0925-7535(03)00047-x

Leveson, N. G. (2011). *Engineering a Safer World: Systems Thinking Applied to Safety*. Cambridge, MA, USA: MIT Press.

Leveson, N. G. (2015). A systems approach to risk management through leading safety indicators. *Reliability Engineering and System Safety, 136*, 17–34. doi:10.1016/j.ress.2014.10.008

Leveson, N. G. (2017). Rasmussen's legacy: A paradigm change in engineering for safety. *Applied Ergonomics, 59*, 581–591. doi:10.1016/j.apergo.2016.01.015

Malik, A., & Gollnick, V. (2016). Impact of Reduced Crew Operations on Airlines – Operational Challenges and Cost Benefits. In *16th AIAA Aviation Technology, Integration, and Operations Conference* (pp. AIAA 2016-3303). Reston, VA: American Institute of Aeronautics and Astronautics.

Matessa, M., Strybel, T., Vu, K., Battiste, V., & Schnell, T. (2017). *Concept of Operations for RCO/SPO*. Moffett Field, CA, USA: NASA.

Norman, M. (2007). *Economic opportunities and technological challenges for reduced crew operations*. Technical paper prepared for: Crew Vehicle Interface Element, NASA. Hampton, VA, USA.

Revell, K. M., Allison, C., Sears, R., & Stanton, N. A. (2018). Modelling distributed crewing in commercial aircraft with STAMP for a rapid decompression hazard. *Ergonomics, (in press)*. doi:10.1080/00140139.2018.1514467

Schmid, D., & Korn, B. (2017). *A Tripartite Concept of a Remote-Copilot Center for Commercial Single-Pilot Operations*. Paper presented at the AIAA SciTech 2017 Forum, Grapevine, TX, USA.

Schmid, D., & Korn, B. (2018). The Operational Issue of an Airliner's Reduced-Crew Caused by Data-link Break-up to Remote Support. *Proceedings of the Human Factors and Ergonomics Society Annual Meeting, 62*, 71–75. doi:10.1177/1541931218621016

Schmid, D., & Stanton, N. A. (2018). How are laser attacks encountered in commercial aviation? A hazard analysis based on systems theory. *Safety Science, 110*, 178–191. doi:10.1016/j.ssci.2018.08.012

Schmid, D., Vollrath, M., & Stanton, N. A. (2018). The System Theoretic Accident Modelling and Process (STAMP) of medical pilot knock-out events: Pilot incapacitation and homicide-suicide. *Safety Science, 110*, 58–71. doi:10.1016/j.ssci.2018.07.015

Stanton, N. A., Harris, D., & Starr, A. (2016). The future flight deck: Modelling dual, single and distributed crewing options. *Applied Ergonomics, 53*(Part B), 331–342. doi:10.1016/j.apergo.2015.06.019

Vu, K.-P. L., Lachter, J., Battiste, V., & Strybel, T. (2018). Single Pilot Operations in Domestic Commercial Aviation. *Human Factors, 60*(6), 755–762. doi:10.1177/0018720818791372

Wilson, J., Harron, G., Lyall, B., Hoffa, B., & Jones, E. (2013). *Understanding Single Pilot Operations (SPO) Certification Requirements*. Moffet Field, CA, USA: Research Integrations Incorporated prepared for NASA Langely Research Center

OPERATIONAL ALERTING CONCEPT FOR COMMERCIAL SINGLE PILOT OPERATIONS SYSTEMS

J.P.Reitsma, M.M.van Paassen, M. Mulder
Delft University of Technology,
2629 HS Delft, the Netherlands

Reducing high workload levels are a major challenge to enable single pilot operations. Where the pilot is currently supported with many automated systems, the role of mission planner is relatively unsupported, i.e., the flight crew is required to integrate and combine information from various sources to extract the implications on the missions' high-level goals to determine if the mission can still be completed safely and successfully. An operational alerting display is developed to provide the pilot with a clear overview of the current and future operational flight constraints. This enables the pilot to determine if the initial plan is valid under the existing conditions. The display is not limited to system malfunctions, but combines the full spectrum of operational constraints, e.g., weather and airport operations. The display concept was tested on usability with a commercial pilot to provide a preliminary performance indication on the effectiveness of the concept.

Reduced pilot operations (RPO) for commercial flights is predicted to reduce direct operation costs, Bilimoria (2014). Other costs reductions can be found on both the operating and manufacturing side. Furthermore, RPO could reduce the issue of the projected pilot shortage. These factors make a strong business case, especially for short haul operations. It is therefore not surprising that there is much interest from industry for RPO. However, major challenges have to be resolved in order to enable such operations. One of the challenges is to handle high workload situations. It is likely that automated systems will relieve the pilot in command from many tasks, Harris (2007). Especially, the system management task will likely to be, at least partially, being taken over by automated systems. The pilot would be in command of all the automation resources, but what remains crucial is that the pilot is aware of the implications of the systems for the safe and expeditious conduct of the flight. The role of the pilot will be that of a flight planner (both on a strategic and tactical level); a communicator with ATM facilities, and a surveillance operative, Harris (2007).

Currently, obtaining information to determine if the flight plan is affected is a rather taxing and time-consuming task. Integrating information from the various sources and converting them to operational constraints is becoming more supported for certain applications. However, it is still much relying on pilots' experience and expertise, Bailey (2017). In this paper, a flight plan evaluation tool is designed with the objective to support the operator more effectively, i.e. reducing workload of determining the implications of events introduced by system and the environment. In-flight, the pilot is mainly evaluating the flight plan constraints and modifying the plan if necessary. Therefore, enabling effective evaluation is an essential step to enable effective flight planning.

The flight plan evaluation support has been developed with Applied Cognitive Work Analysis methodology introduced by Elm (2003). It is a pragmatic framework to determine in a stepwise and traceable manner information and presentation requirements. The requirements are based on an expert knowledge model, the Functional Abstraction Network (FAN). This model is built with operating manuals and interviews with subject matter experts. From the model, cognitive work requirements are derived, from which information requirements are derived. That ultimately feed into presentation requirements. This set of requirements on both information and presentation objectively guide the visual form in terms of content. The concept itself can take many visual forms, which can be adjusted to optimize usability and interaction.

Method

The initial concepts are static, not interactive pictures, displaying a normal and non-normal scenario. The concept was tested on usability for two non-normal conditions, i.e. a hydraulic failure and a generator drive failure similar to recent experiments performed by Bailey (2017), with an experienced 737-800 pilot. The objective was to get initial feedback on the concept and test if the pilot could extract the implications of the failure events.

The pilot was presented with a display of the overhead panel, glare shield and pedestal, that represented the flight deck effects in a static manner. The conventional flight deck and the concept were presented on large LCD displays positioned on a desk. A paper version QRH and dispatch information was presented digitally. First, the hydraulic failure was presented on which the pilot had to act, determine the implications and decide the following actions in a conventional way. After, he was presented with the concept and asked if the consequences were as he initially had considered. This was repeated with the generation drive failure. After each scenario, features of the display were discussed.

The scenario was a flight from London Stansted to Innsbruck. The events were injected at the location of 15 minutes prior to top of descent, west of Zurich. Alternate airport was planned to be Salzburg. In scenario 1, conditions of runway 08 at LOWI had a braking action of medium forcing to divert.

Result

The FAN captures the high-level goals of commercial flight operation. In order to successfully and sustainably create value as a company with flight operation, the company needs to guarantee safety, compliance, expected comfort levels and operate according to flight schedule within the estimated budgeted costs. These five factors are the high-level goals that need to be satisfied to make the operation to a success. The operation itself can be classified into three types of movement, namely the movement of the vehicle in air, the movement of the vehicle on the ground and the movement of the passengers/payload. All these types of movement introduce specific functions, e.g. braking with landing gear is only applicable on the ground. This first concept, for the moment, includes only flight and ground operations of the vehicle.

Two types of abstract functional blocks, namely path and space, where found that can be used to determine if a flight plan is still satisfying the higher-level goals, i.e. safety, compliance, schedule, cost and comfort. One can define paths and spaces for each of these high-level goals, for example compliance space (defined by intentional constraints) is generally more restricted than the safe space (defined by causal constraints).

The functions introduced by the system and environment can restrict the path horizontally, vertically, in velocity or in time. The path needs to be within the operating space in order to be unrestricted. The restrictions of the path and space are the result of lower level functions for example, the maneuvering capability, resources, navigation, surveillance, communication, ability to protect the payload and planes health and so on. These lower level functions are enabled by the power generation functions, i.e. electrical, mechanical, hydraulic, pneumatic. If these systems don't function properly, the complete system is not able to perform certain paths any longer. Multiple flight plans can be analyzed by applying the path and space principle and from interviews pilots it was found that pilots are actually assessing multiple flight plans mentally, i.e. the normal flight plan to the planned destination but also the contingency plans, e.g. to an alternate airport and/or nearest airport in case of diversion.

The basic principles of path and space are applied to the safety and compliant goals, for simplification purposes at the moment. The initial concept is presented in Figure 1, shows how the space is restricted horizontal and vertically, how the path is restricted horizontally (top view display in the center), vertically, in time and velocity (with time/distance display) related to safety and compliance. Furthermore, it shows which systems (bottom right) and environmental effects (bottom center) imply these restrictions.

Normal condition

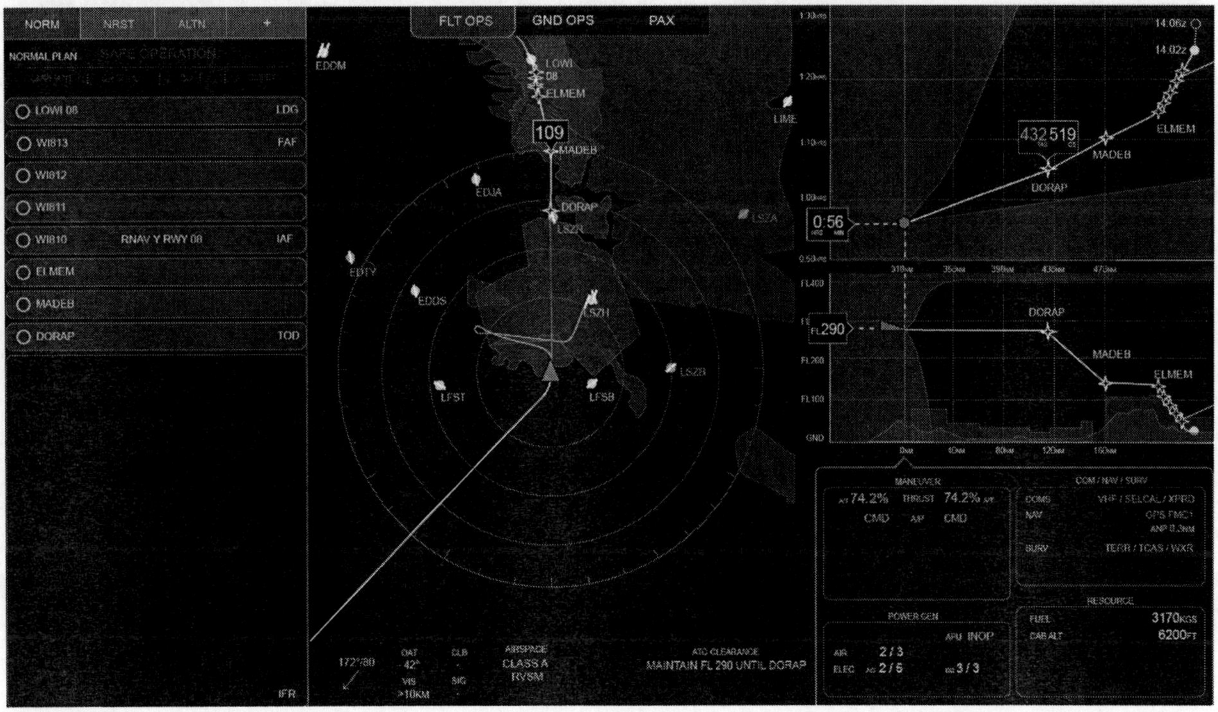

Figure 1.

Flight plan evaluation concept in the normal condition.

The flight plans under current evaluation are presented in on the left-hand side in Figure 1, the landing phase is shown at the top and the current leg or 'now' effects are presented on the bottom. It shows all the waypoints. The cause of the restrictions is presented in the list at the leg where it will be affected.

The airports in the center display are color-coded based on the conditions and ability to land safely and within the standard operating procedures. Further, the contingency plans are displayed in cyan. Restricted space is presented as grey, where black is unrestricted space. The paths are color-coded with amber or red in case any restrictions are broken. The cause of the restriction can be obtained from the list.

Non-Normal condition

In figure 2, the effects on the path and space are shown for a hydraulic system A failure. The landing distance is increased and the alternate plan (LOWI, go/around, divert to alternate), is not safe any longer due to increased fuel consumption due to the inability to retract the gear once lowered, which reduced the final reserve fuel under 30 minutes at LOWS. On the button right, the hydraulic power generation sources are reduced to one out of two and the hydraulic fluid is low. Furthermore, Autopilot A is not available any longer, which has the result that the minimum usable height is increased to 158 ft, which has the consequence that CAT II ILS at LOWS RWY 15 is not authorized.

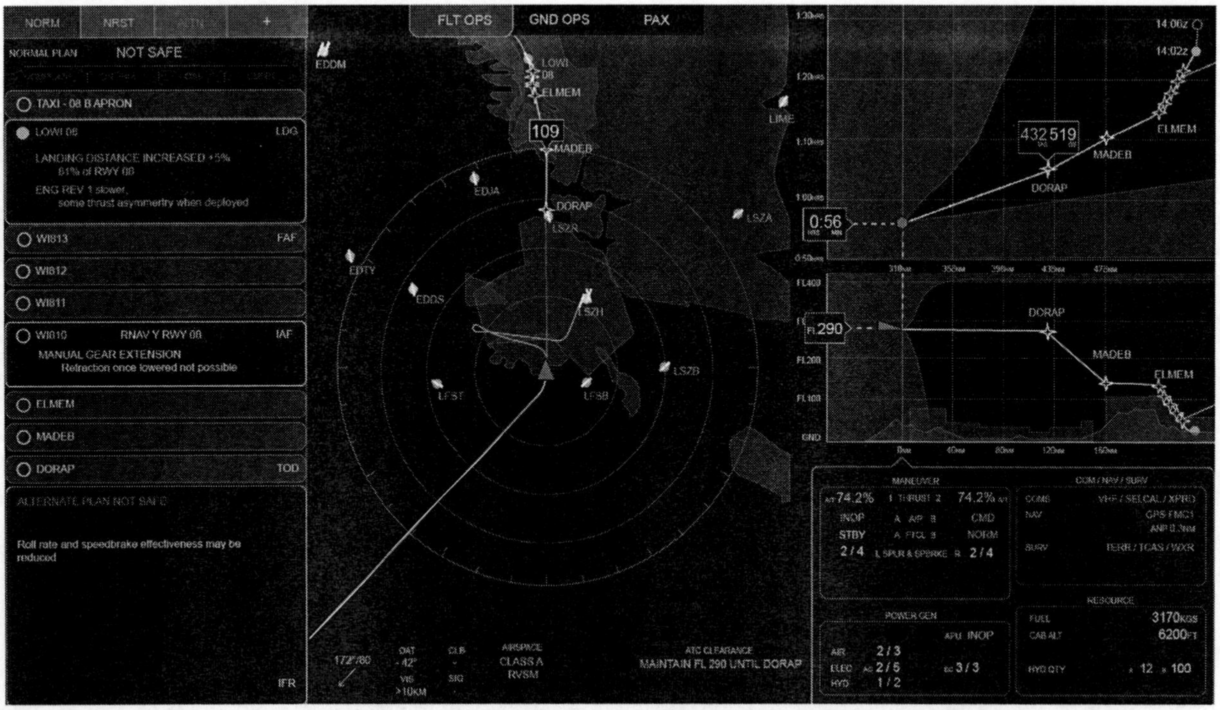

Figure 2.
Flight plan evaluation concept in the non-normal condition of a hydraulic failure.

First impressions

The first impression was that the majority of the attention was focused on the list with the operational effects. The system status and the time / distance display were not used much. The pilot had a clear picture, from the conventional method (QRH and performance tables), that LOWI with braking action less than good was not a safe option, therefore the fact that the item about landing distance increased in the list was not a surprise. However, that implied that trying an approach and lowering the gear was not considered to be realistic. This caused a mismatch with the thinking process. The question that remained was where to divert to. For that case the support is, in its current form, too leading, i.e. only a divert to Zurich is shown. Where, other fields were also a feasible option(s) worth reviewing, e.g. direct to LOWS or another closer airport. Preferred is full option space to choose the suitable diversion airport and not a single indicated path.

Besides this, the link between the system failure to the actual consequence was not always understood. Determining what operational effects were caused by what system/environment effect was not always clear. For example, separating between the effects of an inoperative autopilot and the result of the load shedding was not easy. Also, the feeling existed that some steps were skipped in restoring the systems and the consequences were shown too pessimistic. This included that the cabin altitude pressurization auto mode was inoperative in the second scenario. The feeling existed that it could have be restored by switching it to alternate mode and not fully in manual operating mode.

Furthermore, the list of legs and affected items is presented future-up and felt counterintuitive since the waypoints listed in the FMS are presented the other way around. Another comment was that the time/distance display, especially for ground operations display was not very usable, since it is very difficult to accurately predict taxi time and merely an indication would be sufficient.

Discussions

These initial findings show that it is challenging to match the presented information to the operator's thinking process. If effects or plans do not completely align, a mismatch occurs and the concept becomes a burden rather than a support Westin 2015. Wording and conventions need to be taken further into consideration. The workflow of the concept should be reconsidered to further streamline the process and support the pilot step by step from event to consequence. Furthermore, the concept in its current form is found to be too leading. The pilot should be given more flexibility to evaluate nearest alternates.

The majority of the consequences were extracted in advance using the conventional manner. Some consequences were not extracted at first glance, especially which approaches are authorized, e.g. CATII and RNAV. The reason for this is could be that weather was not limiting ILS operations, and the note 'land at nearest suitable airport' was making the RNAV approach in proximity of high terrain not favorable, when compared to a closer large international airport with multiple runways and approach types.

Linking the consequences and system effects was not completely clear. This might be due to the physical distance between the system status display and the flight plan list. The time / distance display was not found to be effective yet. However, the plan that was affected by time and speed constraints was a plan that was found to be not a realistic option for the pilot. Currently, expected arrival time and fuel are presented by the FMS by two lines, which is less difficult to comprehend than the time/distance display. Scenarios in which time/speed restrictions play a crucial role should be tested to determine if the potential of this display. This initial test was performed statically and the pilot was given as much time as he required. This doesn't represent the real environment, in which interruptions occur, and time is limited. Furthermore, the display was not interactive yet.

Conclusions

The designed concept to evaluate flight plan constraints was designed and tested on usability. This initial feedback provided great insight on how important it is to align the display to the metal workflow of the pilot. Guiding the pilot through the complete process, i.e. from event to consequence, would make it easier to understand the cause-effect relations. The principle of the path and space restrictions is still very useable, but the pilot should be given more flexibility to obtain information. Here, the concept will profit from interaction to extract relevant information in the right context. The concept will probably proof its value better, with some initial training and human-in-the-loop experiments under real, dynamic scenarios, which will be future steps.

Acknowledgements

Special thanks to the group of commercial pilots for sharing their expertise through interviews, providing constructive feedback from the user perspective and making this usability study reality. We also want to thank Lars Fucke, and Randy Mumaw for providing valuable input that lead to the design of the concept.

References

Bailey, R.E., Kramer L.J., Kennedy K.D., Stephens C.L., and Etherington, T.J. (2017). An assessment of reduced crew and single pilot operations in commercial transport aircraft operations.*AIAA/IEEE Digital Avionics Systems Conference - Proceedings, volume 2017.*

Bilimoria, K. D., Johnson, W. W., and Schutte, P. C. (2014). Conceptual framework for single pilot operations. In Proceedings of the International Conference on Human-Computer Interaction in Aerospace (p. 4). New York, NY: AC

Elm, W, Potter, S, Gualtieri, J, Roth, E, Easter, J.R. (2003). Applied Cognitive Work Analysis: A pragmatic Methodology for Designing Revolutionary Cognitive Affordances. *Handbook of Cognitive Task Design. 357-382.*

Harris, D. (2007). A human-centered design agenda for the development of single crew operated commercial aircraft. *Aircraft Engineering and Aerospace Technology. 79. 518-526. 10.1108/00022660710780650.*

Westin, C, Borst, C, Hilburn, B. (2015). Strategic Conformance: Overcoming Acceptance Issues of Decision Aiding Automation?. *IEEE Transactions on Human-Machine Systems. 46. 41-52.* 10.1109/THMS.2015.2482480.

FROM CREWED TO SINGLE-PILOT OPERATIONS: PILOT PERFORMANCE AND WORKLOAD MANAGEMENT

Anja K. Faulhaber
TU Braunschweig, Institute of Flight Guidance
Braunschweig, Germany

Higher levels of automation have come to replace human roles in the cockpit. Therefore, a further reduction of the crew size from two pilots to one has become an option. Such single-pilot operations (SPO) need to provide at least the same safety standards as today's two-crew operations (TCO). The present study aims at identifying potential issues in pilot performance and workload during SPO as opposed to TCO. Fourteen pilots flew short ILS approach and landing scenarios in a fixed-base A320 flight simulator. A 2x3 factorial design was used with the factors crew configuration (TCO and SPO) and scenario (baseline, turbulence and abnormal). Performance data and subjective workload ratings were collected. The results suggest that workload might be problematic mostly during abnormal situations. The design of adequate support solutions for such situations will be a major challenge for the implementation of SPO.

Commercial aircraft are commonly operated by two pilots – the pilot flying (PF) and the pilot monitoring (PM). This crew configuration could change in the future, considering the current discussion about a possible reduction of the crew size to one pilot. Economic factors are the main drivers for the transition toward these so-called reduced-crew or single-pilot operations (SPO). Airlines want to save costs, gain more operational flexibility and prepare for an expected pilot shortage due to the growing demand for commercial aviation (Bilimoria, Johnson, & Schutte, 2014; Comerford et al., 2013). The reduction of crew size has, in fact, a historical background in commercial aviation. During the past decades, cockpit crews have gradually been reduced from initially five crew members to today's two-crew operations (TCO). So far this 'de-crewing' has not led to any safety issues when it was accompanied by adequate technological support (Harris, 2007). In light of this historical trend and taking into account the ongoing technological progress, a transition to SPO seems like the logical next step.

However, the implementation of SPO will be more complex than the previous transitions from five to two crew members. Eliminating the second pilot means eliminating a part of the redundancy in the cockpit which has been a foundation for safe operations in aviation. Additional support through automation might not be enough anymore to ensure safe flight conditions either (Bilimoria et al., 2014). It has even been proposed that we need a revolutionary approach entailing a complete rethinking of the pilot's role and hence of the allocation of tasks between human and machine (Boy, 2014; Sprengart, Neis, & Schiefele, 2018). Further research is required in this context to form a profound basis for a possible reconfiguration of the flight deck for SPO.

From a human-centered perspective, one of the major challenges in the introduction of SPO is workload (Koltz et al., 2015). Especially during abnormal scenarios, workload can reach critical levels in SPO (Bailey, Kramer, Kennedy, Stephens, & Etherington, 2017; Etherington, Kramer, Bailey, Kennedy, & Stephens, 2016). For normal scenarios, results from previous studies are ambiguous. Bailey et al. (2017) investigated workload during normal TCO and SPO conditions and found that workload ratings for TCO were higher than expected, almost at the same level with the SPO workload ratings. An additional post-test questionnaire did, however, reveal a significant result for the effect of crew configuration on workload ratings. The authors made limitations in their study design responsible for biased workload ratings, leaving open questions regarding the general validity of the study results. As human workload does play an important role in the conceptualization of SPO, these open issues require further investigation.

The present study aims at tackling these open issues and providing a better understanding of workload and performance in SPO. Therefore, a flight simulator study was conducted. The study design was loosely based on Etherington et al. (2016) and Bailey et al. (2017) but a focus on the approach and landing phases of flight was chosen. These are particularly demanding phases for pilots and can be expected to reach critical levels in SPO (Koltz et al., 2015). Additionally, a within-subject design was used to avoid effects of individual differences. The complete design of the study will be explained in more detail subsequently.

Material and Methods

Participants

Fourteen pilots (1 female) participated in the study. They were aged between 26 and 56 years (M = 41.14, SD = 9.44) and their flying experience ranged from 300 to 22000 flight hours (M = 6204, SD = 6271). Five of them were captains and first officers each and one of them was senior first officer. The remaining three participants didn't report a rank because they were not working for an airline at the time. Participation was voluntary and unpaid. The study was performed according to institutional and national standards for the protection of human subjects.

Experiment Design

The study was conducted in a fixed-base A320 flight simulator. A 2x3 factorial within-subject design was chosen. The factors were crew configuration (TCO and SPO) and scenario (baseline, turbulence and abnormal). Participants flew one trial per condition resulting in a total of six experimental trials. The task for each trial was to manually fly short ILS approach and landing scenarios at Frankfurt Airport, runway 25 left. Each trial lasted about 2.5 minutes. The initial situation was always the same: The scenario started 8 nm from the runway at an altitude of about 2600 ft. Airspeed was set to 180 kt, the landing gear was still retracted and the flaps were already extended to 15° (indication 2). The view was clear and there were no clouds. The wind was calm; only in the turbulence scenario moderate turbulence was simulated. In the abnormal scenario, an engine fire was induced when the participants reached an altitude of 1800 ft.

The NASA Task Load Index (TLX) (Hart & Staveland, 1988) was used to assess subjective workload ratings directly after each scenario. It consists of six workload subscales – mental demand, physical demand, temporal demand, performance, effort, and frustration – which are all rated on a scale from 0-100. Qualitative data in the form of video and audio recordings from each session as well as debriefing interviews were collected. Additionally, eye tracking data and simulator parameter were recorded. Only data of the PF sitting in the left seat were collected. The present paper will focus on the analysis of the TLX scores and observed behavior patterns related to performance and workload management.

Procedure

Pilots participated in teams of two. Upon arrival, they were briefed on the experiment, received the material (checklists, charts and a Quick Reference Handbook) and gave informed consent. Afterward, each participant was allowed one or two training trials as PF depending on their prior experience with the A320. If they felt confident after the first training trial, the second one was skipped. The experiment started with the first participant as PF in the SPO condition. The second participant was waiting in the briefing room. After finishing the three scenarios in the SPO condition, the second pilot joined in as PM and together they flew the same scenarios in the TCO condition. Then participants changed roles and seats – the first participant became PM and the second one PF. Now the second pilot flew all three scenarios first in the TCO condition with the PM and afterward alone in the SPO condition. Hence, half the participants started with the SPO condition while the other half started with the TCO condition. After each scenario, the PF completed the NASA TLX. The order of the scenarios was balanced; each participant was assigned a different order. When the experimental trials were completed, a short debriefing interview with both participants concluded the session. The total duration was about two hours.

Results

In order to investigate the effects of the factors crew configuration and scenario on the perceived workload, a two-way repeated measures ANOVA was performed on the NASA TLX data. The level of significance was $p \leq 0.05$. Matlab was used for all analyses.

Workload

The results showed that workload was at the same level for SPO and TCO baseline conditions but trended higher for the turbulence and abnormal conditions in SPO (Figure 1). The baseline condition yielded nearly the same mean values for TCO (M = 37.74, SD = 20.74) and SPO (M = 37.62, SD = 15.11). In the turbulence condition, there was a small difference with means of 37.68 (SD = 13.81) for TCO and 43.15 (SD = 13.39) for SPO. As expected, the abnormal condition received the highest workload scores and the most prominent difference in ratings with mean scores of 50.3 (SD = 14.36) for TCO and 56.9 (SD = 17.88) for SPO.

However, the results from the ANOVA revealed that the effect of the factor crew configuration did not reach significance ($F_{1,13} = 2.54$, $p = 0.135$, $\eta^2_p = 0.163$) and neither did the interaction effect ($F_{2,26} = 1.15$, $p = 0.331$, $\eta^2_p = 0.082$). A significant main effect was found though for the factor scenario ($F_{2,26} = 8.02$, $p = 0.002$, $\eta^2_p = 0.382$). A bonferroni post-hoc test showed that this effect applied only to the abnormal scenario compared to both the baseline ($p = 0.021$) and turbulence scenarios ($p = 0.026$). The post-hoc comparison of baseline and turbulence scenarios did not reach significance.

The subscales of the NASA TLX were additionally analyzed separately to understand which of them were affected most by the factor crew configuration and which contributed most to the overall workload rating. A look at the unweighted mean scores showed that the subscales for mental demand and effort received the highest mean workload scores in general (Figure 2). With the exception of the performance subscale, scores were consistently higher in SPO as opposed to TCO conditions, even though the difference remains relatively small.

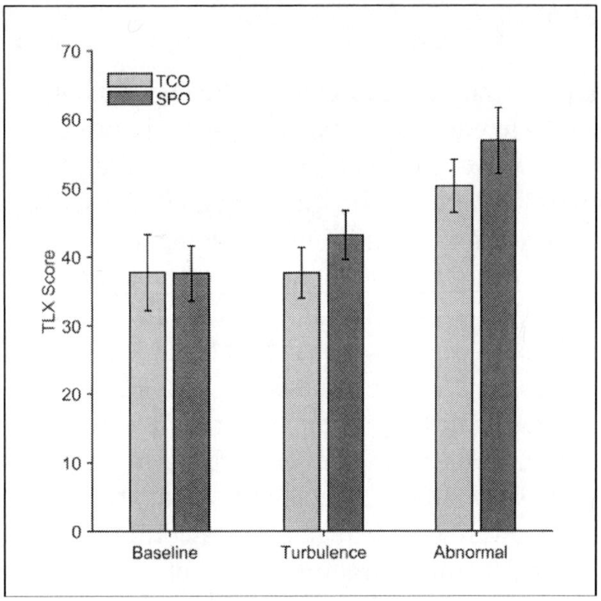

Figure 1. Bar graphs showing NASA TLX unweighted mean composite workload scores representative of the 2x3 factorial design. Error bars show standard errors of the mean.

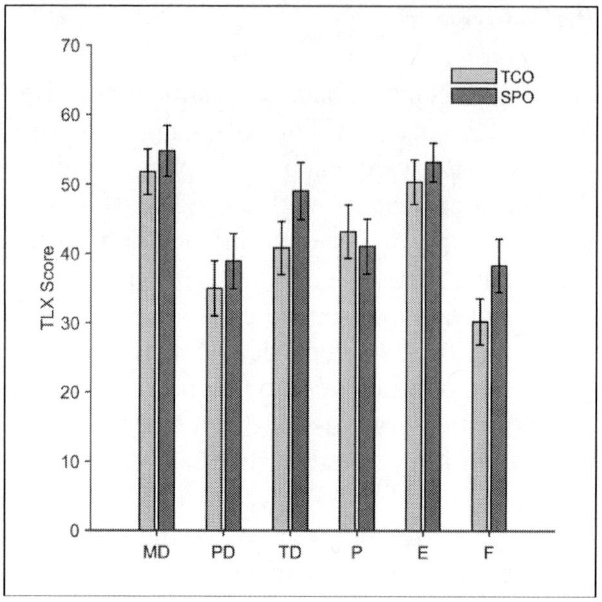

Figure 2. Bar graphs showing the NASA TLX unweighted mean workload scores for each subscale for the factor crew configuration. Error bars show standard errors of the mean.

The NASA TLX mean subscale workload scores for the factor crew configuration are summarized with the respective ANOVA results in Table 1. The subscales for temporal demand and frustration show the highest difference in mean scores. These were also the only subscales for which a significant effect of the factor crew configuration was found. In conclusion, temporal demand and frustration seem to be the subscales affected most by the crew configuration while mental demand and effort contribute most to the overall workload ratings.

Table 1.
Statistics for the NASA TLX unweighted workload scores for the factor crew configuration.

	Mean (SD)		ANOVA		η^2_p
	TCO	SPO	$F_{1,13}$	p	
Mental	51.79 (21.24)	54.76 (23.84)	0.62	0.445	0.046
Physical	35 (25.71)	38.93 (25.72)	0.99	0.337	0.071
Temporal	40.83 (24.86)	49.05 (26.65	8.17	**0.013**	0.386
Performance	43.21 (24.96)	41.07 (25.7)	0.22	0.650	0.016
Effort	50.36 (20.76)	53.21 (18)	0.40	0.536	0.030
Frustration	30.24 (21.75)	38.33 (25.15)	7.23	**0.019**	0.358
Composite	41.91 (17.27)	45.89 (17.26)	2.54	0.135	0.163

Note. Significant effects are highlighted in boldface.

Performance

Qualitative analysis of pilot's behavior patterns during the experiment revealed that participants developed different strategies to manage workload in the SPO condition. The majority of participants (9 out of 14) talked to themselves or called out each step while following the landing checklist. Some of them even made exactly the same calls they were used to from TCO. Thinking aloud was however never mentioned nor asked for during the briefing session and this could hence be interpreted as a way to handle workload. Further analyses of pilot performance showed that checklist usage was more consistent in TCO. When distracted by other tasks, the PF was generally more prone to forget the completion of the landing checklist. An interesting case of this type happened during the abnormal scenario in the SPO condition and led to a crashed landing because the landing gear had not been extended. The participant confused the warning sound indicating that the gear was still retracted with the alarm triggered by the engine fire and became aware of this mistake shortly before touchdown when it was too late. There were similar situations in the TCO condition, where the PF did not actively demand to check the status of the landing checklist because he or she was distracted by other tasks. In these cases, the PM reminded the PF of the checklist and suggested further steps if necessary.

In general, there was no consensus on whether the abnormal procedure for the engine fire should be performed at all during the approach and landing phases of flight. In fact, only four participants performed the procedure in both TCO and SPO abnormal conditions consistently. Six of the participants decided that it would always be best to concentrate on the landing and to disregard the warning completely. They only informed ATC about the situation and cleared the warning. Interestingly, the remaining four participants performed the procedure thoroughly in the TCO condition. In the SPO condition, however, they either decided to disregard the warning or they started the abnormal procedure checklist and aborted before they could complete it. One of the participants even commented that it would be risky to perform the procedure without someone else watching over it.

Discussion and Conclusions

The present study aimed at investigating workload and performance in SPO compared to TCO conditions. Results revealed that workload was not perceived as higher in baseline SPO conditions but only in scenarios involving turbulence or abnormal procedures. This is to a certain extent in line with the results from previous studies (Bailey et al., 2017; Etherington et al., 2016). However, the differences in workload ratings between SPO and TCO were small and several participants even reported after the experiment, that they did not perceive workload as a major issue for SPO. Comparing the NASA TLX ratings to results from other studies shows, though, that the highest mean of 56.9 from the SPO abnormal condition is higher than 75% of all TLX scores obtained from aircraft piloting tasks (Grier, 2016). It can, therefore, be considered as relatively high. In particular the temporal and frustration dimensions of workload were found to be affected most by the SPO condition. Observation of pilot performance also indicated that higher workload did lead to more errors and less accuracy in the completion of tasks, especially during the abnormal SPO condition. Further challenges for the implementation of SPO are therefore to design adequate support for such high workload situations. Several concepts have already been proposed such as a ground operator (Lachter et al., 2014) or a harbor pilot (Koltz et al., 2015). However, further research in this area is required.

Acknowledgements

This research was conducted within the graduate program "Gendered configurations of humans and machines. Interdisciplinary analyses of technology" (KoMMa.G) funded by the federal state of Lower Saxony, Germany. The author would like to thank Prof. Dr. Peter Hecker and Dr. Thomas Feuerle. Special thanks go to the Institute of Flight Guidance at the German Aerospace Center (DLR) Braunschweig, in particular to Tatjana Kapol and Maik Friedrich.

References

Bailey, R. E., Kramer, L. J., Kennedy, K. D., Stephens, C. L., & Etherington, T. J. (2017). An assessment of reduced crew and single pilot operations in commercial transport aircraft operation. *IEEE/AIAA 36th Digital Avionics Systems Conference (DASC), St. Petersburg, FL, USA*. https://doi.org/10.1109/DASC.2017.8101988

Bilimoria, K. D., Johnson, W. W., & Schutte, P. C. (2014). Conceptual framework for single pilot operations. *International Conference on Human-Computer Interaction in Aerospace (HCI-Aero 2014), Santa Clara, CA*. https://doi.org/10.1145/2669592.2669647

Boy, G. A. (2014). Requirements for single pilot operations in commercial aviation: A first high-level cognitive function analysis. *Complex Systems Design and Management International Conference, Paris, France*.

Comerford, D., Brandt, S. L., Lachter, J., Wu, S.-C., Mogford, R., Battiste, V., & Johnson, W. W. (2013). *NASA's single-pilot operations technical interchange meeting: Proceedings and findings*. Moffett Field, CA.

Etherington, T. J., Kramer, L. J., Bailey, R. E., Kennedy, K. D., & Stephens, C. L. (2016). Quantifying pilot contribution to flight safety for normal and non-normal airline operations. *IEEE/AIAA 35th Digital Avionics Systems Conference, Sacramento, CA*. https://doi.org/10.1109/DASC.2016.7778094

Grier, R. A. (2016). How high is high? A meta-analysis of NASA-TLX global workload scores. *Proceedings of the Human Factors and Ergonomics Society Annual Meeting, 59*(1), 1727–1731. https://doi.org/10.1177/1541931215591373

Harris, D. (2007). A human-centred design agenda for the development of single crew operated commercial aircraft. *Aircraft Engineering and Aerospace Technology, 79*(5), 518–526. https://doi.org/10.1108/00022660710780650

Hart, S. G., & Staveland, L. E. (1988). Development of NASA-TLX (Task Load Index): Results of empirical and theoretical research. In P. A. Hancock & N. Meshkati (Eds.), *Human Mental Workload* (pp. 139–183). Amsterdam: North Holland Press.

Koltz, M. T., Roberts, Z. S., Sweet, J., Battiste, H., Cunningham, J., Battiste, V., et al. (2015). An investigation of the harbor pilot concept for single pilot operations. *Procedia Manufacturing, 3*, 2937–2944. https://doi.org/10.1016/j.promfg.2015.07.948

Lachter, J., Brandt, S. L., Battiste, V., Ligda, S. V., Matessa, M., & Johnson, W. W. (2014). Toward single pilot operations: Developing a ground station. *International Conference on Human-Computer Interaction in Aerospace (HCI-Aero 2014), Santa Clara, CA*. https://doi.org/10.1145/2669592.2669685

Sprengart, S. M., Neis, S. M., & Schiefele, J. (2018). Role of the human operator in future commercial reduced crew operations. *2018 IEEE/AIAA 37th Digital Avionics Systems Conference, London*. https://doi.org/10.1109/DASC.2018.8569803

AN EMPIRICAL TEST OF AN ENHANCED AIRSPEED INDICATOR

Julia Trippe
Pacific Science and Engineering
San Diego, California

Robert Mauro
Decision Research & University of Oregon
Eugene, Oregon

Analysis of airliner accidents and incidents identified a class of events in which structurally, mechanically, and electronically sound aircraft decelerated through the minimum safe operating speed to the stick-shaker activation speed. For a subset of these events the automation was *no longer actively controlling to the airspeed target*, a condition which the Primary Flight Display does not explicitly indicate. Increasing the salience of critical automation information may enhance the ability of the flight crew to detect, recognize, and diagnose when an aircraft will inappropriately decelerate, prior to a speed deviation. In the current study, we designed and tested a modification of the airspeed tape on the Primary Flight Display to explicitly annunciate the absence of active speed control. Our experiment showed that professional pilots were faster at recognizing an airspeed anomaly when using the Enhanced Airspeed Indicator as compared to the traditional air speed tape. No speed/accuracy trade-off was observed.

Modern properly functioning fully-automated airliners should not inadvertently stall. But they do. Sherry & Mauro (2014) examined 19 incidents in which mechanically and electrically sound airliners decelerated through minimum safe operating speed to stick shaker activation (5 knots above stall speed). Analysis of these events revealed that the automation was not actively controlling airspeed. In some cases, inadvertent auto-throttle (A/T) deactivation caused the airspeed target to be neglected. In several other cases, circumstances caused the automation to transition into an unexpected A/T "dormant" mode in which airspeed was not controlled. In all of these cases, the crews failed to respond in a timely manner. Typically, the events occurred while the aircraft was properly decelerating, thereby masking the inappropriate behavior of the autoflight system. These incidents of "controlled flight into stall" demonstrate two problems in the flightdeck "human-machine system." First, there is no clear indication on the flightdeck that the automation is or is not controlling airspeed. Second, pilots frequently do not fully understand the operation of their autoflight systems (Sarter & Woods, 1995). In this paper, we focus on the task of increasing the salience of automation control of airspeed (or lack thereof).

Airspeed is typically displayed on the Primary Flight Display (PFD) on nearly all modern airliners. The PFD collocates, on a single screen, indicators that were previously separate, thus allowing the pilot to obtain data about the orientation and status of the aircraft with very little eye movement. However, the PFD symbology in common use can be misleading. Airspeed is displayed on a moving "tape" on the left-hand side of the PFD (see Figure 1). The target airspeed is displayed above the tape and marked by a "bug" on the tape when that speed is within the range displayed on the tape. Current speed is displayed in a window superimposed on the tape.

This display does not indicate whether the automation is attempting to achieve the airspeed target. It only indicates that the target has been set.

To determine the state of the automation, the pilot must read and interpret the indications on the Flight Mode Annunciator (FMA) displayed at the top of the PFD (see Figure 1). Automation states are indicated by cryptic abbreviations. Changes in mode are indicated by blinking text for a few seconds. In some incidents and accidents, pilots have evidently misinterpreted the FMA abbreviations (e.g., Asiana 214, TA 1951). FMA labels are often ambiguous or overloaded (Feary, McCrobie, Alkin, Sherry, Polson, Palmer, & McQuinn, 1998) and automation control mode changes are not explicitly annunciated (Sherry, Mauro, & Trippe, 2019). Studies indicate that flight crews do not use the FMA as intended (Feary, et al., 1998; Norman, 1990, Degani, Shafto, & Kirlik, 1999). One eye-tracking study found that 62% of the pilots observed did not rely on the FMA during anticipated mode changes and 45% of the pilots did not refer to the FMA after an unanticipated change (Mumaw, Sarter, & Wickens, 2001).

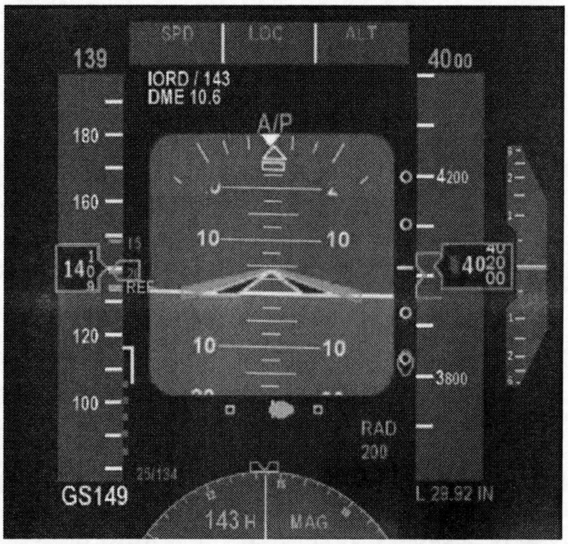

Figure 1. Boeing 777 Primary Flight Display.

In this study, we examined whether pilots' abilities to determine what aspect of the autoflight system (if any) was in control of the airspeed could be improved by making small changes in the color coding and text used on the display.

Method

Enhanced PFD design

Our goal was to design simple, relatively low-cost modifications to the existing PFD design that would substantially improve pilots' ability to determine whether the automation was in control of the aircraft's airspeed. We defined certain parameters based on previous cognitive research. Our design required explicitly indicating, in a highly salient and unambiguous manner, the current state of airspeed control. At minimum, this design needed to indicate who/what was controlling the airspeed (i.e. AFS or manual control) and what component of the system was setting the target (i.e. Mode Control Panel (MCP), Flight Management System (FMS), or none).

The design had to be easily interpretable without necessitating access to memorized rules. We also determined that it was more important for pilots to understand the need for intervention than to understand why. As noted by Vicente and Rasmussen (1988), relying on skill-based behavior rather than having to access rule-based or knowledge-based behavior could save valuable cognitive resources and time.

In addition, given the "limited real estate" of the already densely populated PFD, any modifications would have to be crafted so as not to interfere with other flight information. We determined that the most effective solution would be to locate our changes in the same place that pilots were already looking for airspeed target and control information: the middle third of the airspeed tape, using target source colors consistent with all flight deck displays (magenta = FMS, green = MCP, white = manual). Collocating this new layer of information with familiar indications would also serve to alert crews to the fact that it could not be interpreted as usual. We chose to visually block inactive target indications with Xs so that they would not be relied upon automatically (see Figure 2). Target source is indicated by the color of the target number (at top of tape) and target "bug" (on the tape). Control is indicated by color and Xs. If targets are magenta with no Xs, the AFS is controlling the airspeed to FMS targets. If the targets are green with no Xs, the AFS is controlling the airspeed to MCP targets. If the targets are X'd out, then the AFS is not controlling airspeed. Appropriately color-coded targets are visible under Xs to indicate what they would be if automation were reengaged.

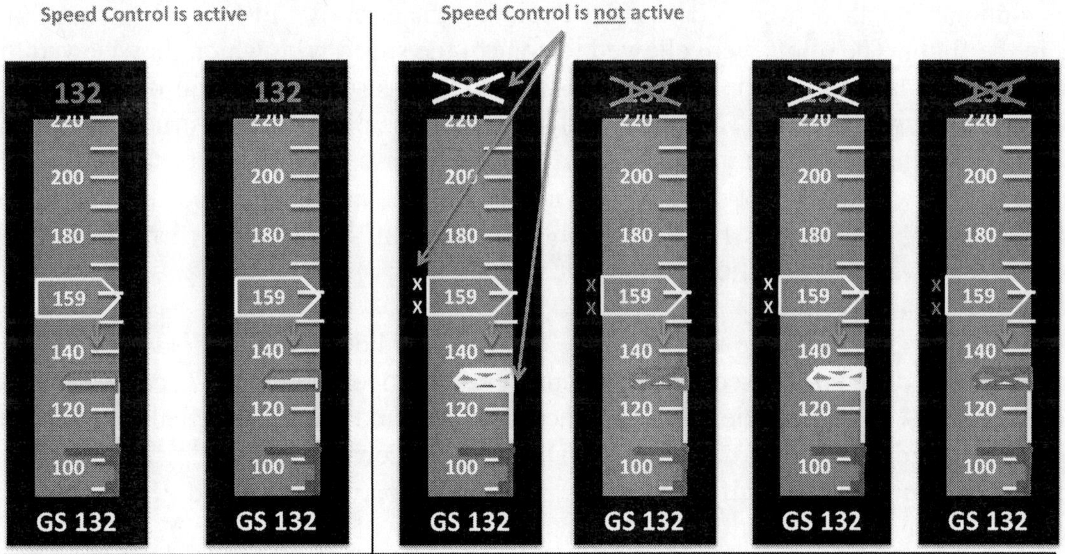

Speed Control Active		Speed Control **not** Active			
A/P and A/T Engaged		A/P or A/T **not** Engaged	A/P or A/T Engaged	A/P or A/T **not** Engaged	A/P or A/T Engaged
VNAV not Engaged	VNAV Engaged	VNAV not Engaged		VNAV Engaged	
Mode controlling to Airspeed		-	Mode **not** controlling to Airspeed	-	Mode **not** controlling to Airspeed

Figure 2. Enhanced PFD Airspeed tape with explicit indication of the absence of airspeed due to engagement status or control mode. Targets and bugs X'd out and appropriately source colored.

Study design

To test the proposed Enhanced PFD, we created an online study depicting the above enhancements overlaid on a traditional Boeing 777 PFD. We then recruited 31 Boeing pilots, currently working in the 777 (12 participants), 737 (12 participants), or 767, 757 or 747 (total of 7 participants) with an average flight time of 11,733 hours (SD: 4,740).

Participating pilots were shown a series of 24 flight scenarios. Each scenario was composed of a set of 3-4 still frame "slides" depicting chronological "snap shots" of either a traditional PFD or an enhanced PFD as it would have looked at sequential time periods during a particular flight maneuver (e.g., departure, descent, localizer intercept, approach). All of the flight scenarios began with the auto-throttle in control of the airspeed. For each flight scenario, they were first shown a brief written description of the scenario along with ATC-like instructions in italics that described what would happen in the upcoming slides. For example:

VS Descent to GS (3 slides)
Inbound on LOC to KORD RWY 14R
Maintain 4000
Cleared ILS 14R

If the scenario placed the pilot's aircraft on an approach, the participants were also shown an approach plate with the approximate vertical and lateral position of the aircraft indicated. After the description slide, there were 3 or 4 slides depicting the aircraft's PFD at advancing time points in the flight. The pilots were allowed to control the speed with which they viewed these slides. Before the last slide in the scenario was shown, an instruction slide appeared indicating that the next slide would be the final slide and reminding the pilot of the instructions which were: "As soon as the final PFD slide in the scenario appears, please tell us as fast as possible without making mistakes whether or not there is a deviation from the intended airspeed and/or flight path. Press the "y" key for "yes, there is a problem" or the "n" key for "no, there is no problem. This slide will advance automatically after your selection."

The pilot's reaction time from the time that the last slide was presented until a key was pressed was recorded. Twelve scenarios depicted a normal operation and 12 scenarios depicted a problem. Two sets of 24 scenarios were produced, set "A" and set "B." For half of the pilots, set A scenarios were depicted on a traditional PFD and set B scenarios were depicted on an enhanced PFD. For the other half of the pilots, the relation between sets and PFD type were reversed. The order of presentation was counterbalanced by PFD type.

Results

Thirty-one pilots participated in the experiment. As expected given the design and the experience level of the pilots, the pilots made very few errors. The proportion of incorrect responses did not differ by type of PFD (Traditional PFD: 10.8%, Enhanced PFD: 8.9%; X2(1)=0.703, n.s.). For 17 scenarios there was no difference in speed of responses due to type of PFD. One scenario was incorrectly presented. For 6 scenarios there was a statistically significant difference. In all of these cases, pilots were faster to respond when the scenario was presented using the enhanced PFD (see Table 1). The mean reaction time difference for these scenarios was 10.0 seconds less for the enhanced PFD, with a mean standard deviation of 6.04 seconds.

Table 1.
Significant Effect of PFD Type on Mean Log(Speed) of Correct Responses by Scenario

		Mean Reaction Time (milliseconds)		**Mean Ln(Speed)**				
		PFD Type		**PFD Type**				
Scn #	**Problem**	**Traditional**	**Enhanced**	**Traditional**	**Enhanced**	**t**	**df**	**p**
13,14	N	9416	6259	.16	.61	2.277	25	.032
15,16	N	10963	7595	-.02	.48	2.327	26	.028
17,18	N	10642	6936	.04	.51	2.404	29	.023
23,24	N	12348	7901	-.10	.47	2.263	22	.034
41,42	Y	6744	3443	.60	1.33	2.549	28	.017
45,46	Y	9965	4131	.16	1.06	3.877	26	.001

Note: Reaction times are in milliseconds. To stabilize distributions for analysis, reaction times were converted into Ln(Speed)=Ln((1/RT)*10000). Higher ln(speed) corresponds with lower reaction times. t=2-sided independent sample t-test value; df=degrees of freedom, p=probability.

Discussion

In this study we did not attempt to load the pilots or to distract them with mechanical failures or operational issues of the sorts that populate accident reports. All that the pilots needed to do was to view the PFD and determine whether the display indicated that there was a deviation from the expected path in altitude or airspeed. As would be expected, the participants, all of whom were experienced pilots, made very few errors reading their PFDs. Had many errors been observed, we would have questioned the fidelity of the study. Experienced pilots do not make many errors on such routine tasks.

To determine whether the modifications to the PFD would have any effects, we used reaction time, a measure that is much more sensitive than error rates. On most scenarios the participants were equally fast at responding to the scenarios regardless of PFD type. However, whenever there was a statistically reliable difference in speed of response, pilots were faster to respond correctly when they were using the enhanced PFD whether or not there was a problem. This pattern of results indicates that the modifications caused no problems and caused a measurable increase in performance.

The PFD modifications used in this study were designed to rely solely on relatively minor software changes to existing PFDs. Although it may be possible to achieve increased performance from other more elaborate modifications, our aim was to devise simple modifications that could be retrofitted to existing equipment at relatively minor development, certification, and implementation costs. Using this simple, easily implemented PFD enhancement that could be adapted for any platform, there was a measurable increase in pilot performance responding correctly to unexpected automation behavior. The scenarios used in this study reflected real world incidents that could easily end in catastrophes. The clear indications of lack of automated airspeed control allowed a fast skill-level response. In a time sensitive, safety critical environment, a few seconds could mean the difference between landing safely and crashing.

To be confident in the effectiveness of the modifications that we developed, the enhanced PFD used here should be compared to traditional PFDs in a flight simulator using pilots flying scenarios that tax the pilots' abilities as they are frequently taxed in actual operations and abnormal situations. If the enhanced PFD continues to demonstrate superior performance compared to traditional PFDs under these conditions, we would have greater confidence in the wisdom of making the investments required to develop, certify, and modify the PFD in existing airliners.

References

Degani, A., Shafto, M., & Kirlik, A. (1999). Modes in human-machine Systems: Constructs, representation, and classification. *International Journal of Aviation Psychology*, *9,* 125-138. Vertical Guidance Understanding. NASA/TMm1998-112217

Mumaw, R.J., Sarter, N.B., and Wickens, C.D. (2001): Analysis of Pilots´ Monitoring and Performance on an Automated Flight Deck. 11th International Symposium in Aviation Psychology. Columbus: Ohio State University.

Norman, D. A. (1990). The 'problem' with automation: inappropriate feedback and interaction, not 'over-automation.' *Philosophical Transactions of the Royal Society B: Biological Sciences*, *327*(1241), 585-593.

Sarter, N. B., & Woods, D. D. (1995). How in the world did I ever get into that mode? Mode error and awareness in supervisory control. *Human Factors: The Journal of the Human Factors and Ergonomics Society*, *37*(1), 5-19.

Sherry, L. R. Mauro (2014) Controlled Flight into Stall (CFIS): Functional Complexity Failures and Automation Surprises. 2014 Integrated Communications Navigation and Surveillance (ICNS) Conference, April 8-10, 2014

Sherry, L., Mauro, R., & Trippe, J. (2019). Design of a primary flight display (PFD) to avoid controlled flight into stall. Aviation Psychology and Applied Human Factors.

Vicente, K. J., & Rasmussen, J. (1988). On Applying the Skills, Rules, Knowledge Framework to Interface Design. *Proceedings of the Human Factors Society Annual Meeting*, *32*(5), 254–258.

Acknowledgements

This work was funded by NASA NRA NNX12AP14A. Special thanks to Lance Sherry and Immanuel Barshi for technical suggestions.

OPERATIONAL ALERTING ON MODERN COMMERCIAL FLIGHT DECKS

J.P.Reitsma*, M.M.van Paassen*, M. Mulder*
* Delft University of Technology,
2629 HS Delft, the Netherlands

The glass cockpit and EFB enable new ways of information presentation and interaction on the flight deck of modern commercial jets. This information supports crews in flight plan management, which essentially entails evaluating the plan against (ever-changing) flight constraints and, if necessary, modifying it. Flight constraints emerge from the interaction between the system and its operational environment. Understanding the constraints, and checking the flight plan against these constraints, requires selection and combination of information from many sources. Operational alerting can support this process, by prioritizing and formatting information to match the operational context. A number of modern flight deck systems are evaluated on how they support alerting in an operational ready format. From the comparison we can conclude, that there is a trend towards operational alerting, especially on a tactical level.

The glass cockpit has been around for a while now, and with the introduction of the Electronic Flight Bag (EFB) the paperless cockpit became reality. Flight crews have faster access to more information than ever before. But the question that remains is how does all this information support the flight crew in performing their task? The main responsibility of the flight crew is the success of the operation and the well-being for the passengers and plane. To ensure this, a well-considered flight plan is crucial to achieve a successful operation. Flight planning starts before the flight, McGuire et. al., (1991) and includes determining the flight constraints, i.e. departure and arrival times, reviewing weather, aircraft range, reviewing terminal constraints, optimizing the horizontal profile and vertical path, planning for contingencies. The result is a flight plan that is tailored to achieve the operational goals with most up-to-date information. If the plane and its systems are found to be fit, the flight plan can be executed. While executing the plan, the flight crew is constantly monitoring the world and aircraft systems to see if any of the original flight constraints are changed and/or broken, a likely event in real world scenarios. If they detect any deviation, they are required to act, either by continuing with the initial plan or modifying the plan, by re-routing, reconfiguring or communicating. This whole process is flight plan management.

Flight plan management is therefore an information problem in which the flight crew has to combine information, from various sources, about the operating environment and the airplane systems. Sequentially, they have to transform this information to determine how the previously assumed constraints are affected and what the consequences are for their operation. The interaction between the plane and environment is crucial since the threats to safe operations are not only system malfunctions, but the majority of the threats are caused by effects from the world, like weather, traffic, terrain, ATC and airport conditions (Thomas, 2003). These sources are highly dynamic, due to the chaotic nature of the world and are difficult to accurately predict in advance. Continuously updated information is needed to determine the implications on the flight plan.

Evaluating the flight constraints and implications on the flight plan is essential, (Harris 2007), but can be a taxing task especially under highly dynamic circumstances. During recent experiments, Bailey et. al., (2017) found that pilots under severe workload are unable to fully comprehend the limitations after certain failures. Further, understanding the limitations of the plane heavily relies on the pilot's expertise and experience (Mumaw, 2017). This leaves us with the question, is the flight crew with all its resources on the modern flight deck sufficiently supported for the flight plan management task? In particular, how are the modern flight decks presenting information to the flight crew to evaluate what can and cannot be done operationally?

Method

Five modern flight decks families are evaluated on how they represent information of the operating environment and system status with respect to the intended plan. First, information available on the flight deck from the operating environment is reviewed. This is done separately since various airplane types have similar means to obtain this kind of information. Second, information about the system status is reviewed for various flight decks, including the Boeing 737 NG, Boeing 717/MD11, Airbus A320/A330/A340, Boeing 777/787 and Airbus A380/A350. A hydraulic reservoir failure on a single system will be used as a case study to show the differences in presenting system implications. Obviously, the impact of the failure on the airplane status will be different on the various airplanes, but our interest is how and if operational implications are presented.

Results

Terminal / Route Information - Information about the airport facilities, standard procedures and routes, are published in the Aeronautical Information Publication (AIP) or airport facility directory (AFD). Besides these publications, the crew has also an Operations Manual provided by the operator in which operational information can be found that the operator may deem necessary for the proper conduct of flight operations. This include for example, preferred routes, SOPs, operating minima, escape routes, and minimum flight altitudes. Day-to-day information about the current conditions are communicated by Notice to Airmen (NOTAM), while short notice information is provided by ATC or the operator. Despite the recent change to electronic format of these manuals and NOTAMs, the content of information is similar to the paper version.

One step towards integrating and transforming the content to a more operational format is done on Boeing's Airport Moving Map (AMM) and Airbus's On-Board Airport Navigation System (OANS) and shows the location of runways, taxiways, and other airport features in relation to the airplane position. Additionally, the status of the runways and taxiways is shown, e.g. closed taxiways and active runways. The crew can now clearly see if the flight plan is crossing any constraints on the ground.

Terrain Information - Besides the charts and procedure published in the AIP and OP, modern Flightdeck are equipped with, real-time terrain information provided by (E)GPWS. These systems present terrain and alerts if the predicted path is colliding with terrain in the near future. The terrain information is integrated on the flight deck by the Navigation Display (ND) and / or on a Vertical Situation Display (VSD). Where, Synthetic Vision Systems (SVS) present terrain constraints integrated on the Primary Flight Display (PFD).

Weather Information – Weather is dynamic and can be difficult to predict accurately during the planning phase. Current weather conditions are distributed by ATIS or D-ATIS, which is just the digital version of ATIS. TAFs, SIGMETS, AIRMETS, PIREPs, forecasts, prognostic charts, wind/temp charts at different flight levels, are provided through the dispatcher and are often available in digital format. Furthermore, planes are equipped with weather radar that can detect real time precipitation and turbulence. This information is readily available and integrated on the ND. Weather radar furthermore alerts for wind shears on the PFD and presents its location on the ND. Whereas the forecast needs to be requested and then processed mentally.

Traffic Information – Traffic is highly dynamic and to avoid collisions the Traffic Collision Avoidance System was developed. Predicted collisions are alerted on the PFD and ND. TCAS also provides a solution to avoid traffic. However, only planes equipped with a transponder can be detected and avoided. With the introduction of ADS-B traffic positions are made available and allow for airborne and ground traffic situation awareness, either displayed on the ND or EFB.

ATC Clearances / Requests – Obtaining information once airborne is only possible due to communications. Communication is mostly done by voice (either through VHF, HF, or satellite), however with the introduction of datalink it became possible to send and receive information in an electronic format. Clearances and requests can be sent digitally with controller pilot data link communication (CPDLC). Which allows even to upload clearances to the FMC and therefore integrate it into the flight plan. Information can either be provided by ATC, e.g. for clearances, or by the Company, e.g. gate information. Clearances are integrated by OANS on the Airbus A350, by color coding a cleared and requested path, showing intentional constraints.

System Implications: Boeing 737 NG - The Boeing 737 continuously evolved since it was introduced back in 1967. However, information presentation regarding the status of the engines and systems on the Boeing 737NG is however very similar to the classic 737. Dials are replicated in an electronic format, together with alerting block lights. The main alerting method relies on annunciator lights in front of the pilot together with corresponding lights on the overhead / pedestal panel. The crew has to scan the flight deck to determine what systems are causing the malfunction. Once the lights are identified, the crew will consult the QRH, either a paper or digital version, and look-up the corresponding alerting light. This will guide him/her through a non-normal checklist, which assists in reconfiguring the system to prevent and minimizing further deterioration of the plane systems. Once the failure is stabilized, the QRH provides the implications for the remainder of the flight. For the hydraulic system failure, various alert light across the flight deck will illuminate. After the reconfiguration of the systems, the crew is left with instructions and notes that are useful for the remainder or the flight, see Figure 1 for an example of this.

3. Check the Non–Normal Configuration Landing Distance table in the Advisory Information section of the Performance Inflight chapter.

...

5. Plan for manual gear extension.

Note: When the gear has been lowered manually, it cannot be retracted. The drag penalty with gear extended may make it impossible to reach an alternate field.

Note inoperative Items:

Autopilot A inop

Autopilot B is available.

....

Figure 1. Example of checklist items that have implications on a flight plan.

This is not only a lot of information to interpret, but it is also not straightforward to determine what the exact effect is. The first step for the crew is to determine when an affected system will be used. Next, one has to determine if and how it impacts the intended operation. The landing distance for example needs to be checked with tables, which require additional information about the weather and runway conditions. As an example, take the note about the manual extension, shown above. It requires considerable effort to figure out if a go-around and reaching the alternate field after lowering the gear is still possible. This is already a challenging task on the ground, needless to say that this is a difficult task once airborne.

System Implications: Boeing 717 / MD 11 - The Boeing 717 and MD11 share a similar flight deck, termed *the advanced flight deck.* The main system display is the Engine Alerting Display (EAD) this shows the engine status with an overview of all systems alerts. The B717 has also synoptic displays, which show the status of a particular system with alert related to the applicable system. Even though alert messages are presented in a centralized alphanumerical manner, flight crews are still depended on the assistance of the QRH, which is similar to the B737. An addition to the B737 is that the B717/MD11 also include a consequence page in which the alerts and consequence are summarized, Morgan (1992). In the case of the hydraulic failure, the consequence page would include: *"SPOILER INBD FAIL | REDUCED ROLL RATE AVIALABLE".* This is very similar to the notes from the QRH, only now in an electronic format.

System Implications: Airbus A320 / A330 / A340 - Airbus provide system alerting through the electronic centralized aircraft monitor (ECAM) system. The status of the plane is automatically sensed and the appropriate actions to reconfigure the systems to prevent any further damage are shown. These actions are sensed and marked green when the system is in the correct state. After all actions are performed, a page appears with the limitations and inoperative systems, see Figure 2. This can be compared with deferred items, notes, inoperative systems and consequence page. The A320/A330/A340 are also equipped with synoptic displays.

STATUS	INOP SYS
- L/G........................GRVTY EXTN	GREEN HYD
LDG DIST PROC..................APPLY	SPLR 1+5
	CAT 3
SLATS/FLAPS SLOW	NW STEER
CAT 2 ONLY	AUTO BRK

Figure 2. Example ECAM items that have implications on a flight plan.

These indications tell the crew to check the landing distance and that they can do a CAT II approach with Autoland. It's still up to the pilot to look up the landing distance in the tables and determine if an Autoland with Cat II will be sufficient for their operation. Although this overview is quite clear, some items, e.g. the capability to retract the gear once lowered, is not provided and has to come from pilot's experience and system knowledge. If this implication is not considered by the flight crew, the consequences can have a major impact on the flight plan.

System Implications: Boeing 777/787 - EICAS, which was first introduced on the Boeing 757, is Boeing's main system status and alerting display. It shows alert messages in a centralized alphanumerical manner with an indication if a dedicated checklist exists. This checklist will appear in the electronic checklist (ECL), which is similar to the QRH but has the functionality to sense if systems are in the correct position, like the ECAM system. The notes will be stored on a dedicated page. So, finding the checklists and storing the notes are easier with ECL. However, integration with the flight plan needs to be done by the flight crew.

System Implications: Airbus A380 / A350 - On the A350 and A380, the ECAM system is provided with more real estate due to the larger displays. The inoperative systems are split-up into two categories, namely 'All phases' and 'Approach & Landing". This makes it easier for the flight crew to determine in what flight phase the effects will limit the operation. However, much of the actual impact needs to come from the crew themselves.

System and environment: ROPS & RAAS - The Airbus' Runway Overrun Prevention System (ROPS), (Airbus, 2011) and the Honeywell Runway Awareness and Advisory System (RAAS) (Clark, 2011) are systems that integrate the airplane configuration and status with the operational environment, e.g. runway conditions, weather. They calculate the stopping distance required on a specific runway under various conditions. This is done in real time and considers changing conditions like wind. It will alert if the runway is too short. This system off-loads the crew from making the calculation of the landing distance for the current configuration. The system makes the calculation eight times per second, faster that the crew can ever do. The system is providing the crew with essential information if a landing is possible yes or no. The brake-to-vacate function is another operational focused function, which can determine how to apply and configure the brakes to vacate the runway at an optimal taxiway.

Discussion

From this case study, it can be observed that the accessibility of the information is largely improved by the introduction of electronic presentation. ECAM and ECL made it easier to obtain the required checklist. However, much of the content is similar to the paper version and the crew still has to combine all the limitations to determine when and what the effects are on the flight plan. This requires time, effort and continuous attention, which are scarce in flight and during non-normal events. Secondly, there is a trend in integrating information and provide operational alerting, e.g. TCAS, EPGWS, weather radar, airport map and ROPS/RAAS. These systems provide alerts in case collisions, or overruns are predicted. Alerts like, RUNWAY TOO SHORT, or NO TAKE OFF are clear in terms what operation cannot be performed. However, the support from these alerts and systems are limited to the tactical level, a cause of this limitation is due to restrictions in the available data, e.g. traffic can only accurately be observed in a short time span. Finally, checklists provide guidance after a malfunction with notes, limitations and deferred items, but considerable effort need to be spent by the crew to determine how events affects the operation. Therefore, system-wise the crew is relatively unsupported to fully comprehend and predict the repercussions of a change in system status.

Conclusion

Comparing the various flight decks, we identified that more recently introduced flight decks and systems are integrating and transforming information in a more operational format. However, currently operational alerting is limited to support on a tactical level, but this could be expanded to combine more information for the entire flight plan, supporting the flight crew also on a strategical level. This will make it easier for pilots to obtain an overview what operations can and can't be done, which is beneficial during high-workload, complex and time-critical events. Systems that can assess the intended plan(s) based on up-to-date information have the potential to off-load the pilot, improve the quality of the assessment, reducing unconsidered effects and reducing the dependency on pilot's experience and expertise, which is favorable with reduced flight crew experience with non-normal events.

References

AIRBUS SAS - Flight Operations Support and Services (2011), A350-900 Flight Deck and Systems - Briefing for Pilots – *Issue 02 – September 2011*

Bailey, R.E., Kramer L.J., Kennedy K.D., Stephens C.L., and Etherington, T.J. (2017). An assessment of reduced crew and single pilot operations in commercial transport aircraft operations. *In AIAA/IEEE Digital Avionics Systems Conference - Proceedings, volume 2017.*

Clark, S, Trampus, G (2011), Improving Runway Safety with Flight Deck Enhancements. *In AERO magazine QTR 01/11, The Boeing Company.*

Harris, D. (2007). A human-centered design agenda for the development of single crew operated commercial aircraft. *Aircraft Engineering and Aerospace Technology. 79. 518-526. 10.1108/00022660710780650.*

McGuire, J. C., Zich, J. A., Goins, R. T., Erickson, J. B., Dwyer, J. P., Cody, W. J., & Rouse, W. B. (1991). An Exploration of Function Analysis and Function Allocation in the Commercial Flight Domain *(NASA CR-4374). Long Beach, CA: McDonnell Douglas Aircraft Company.*

Morgan, J, Miller J (1992), MD-11 Electronic Instrument System, *Proceedings IEEE/AIAA 11th Digital Avionics Systems Conference, Seattle, WA, USA, 1992, pp. 248-253.doi: 10.1109/DASC.1992.282148*

Mumaw, R. and Feary, M., (2017). Aircraft Capability Management. *Presentation at the NextGen Flight Deck and Human-Autonomy Teaming Symposium, NASA Langley Research Center*

Thomas, M.J.W. (2003), Improving organisational safety through the integrated evaluation of operational and training performance: an adaptation of the line operations safety audit (LOSA) methodology, *Human Factors and Aerospace Safety, Vol. 3 No. 1, pp. 25-45.*

PILOT INFORMATION NEEDS FOR ELECTRONIC DATA-DRIVEN CHARTS

Joseph M. Jaworski
Cherokee CRC, LLC
Oklahoma City, OK

Michelle Yeh
Federal Aviation Administration
Washington, DC

Cathy Swider
Federal Aviation Administration
Washington, DC

Electronic charting technology is evolving from "fixed" raster-based charts to data-driven charts, in which information elements shown on the chart can be re-configured during flight. Specifically, we were interested in indentifying a set of minimum information requirements for a concept in which pilots brief with a fixed chart showing *all* information elements but then fly with an electronic chart, which may or may not include all the information elements that were briefed. Two hundred twenty-nine pilots rated the importance of information elements shown on four different types of aeronautical charts. We analyzed the data using one-way chi-square tests to identify a criticality "level" for each information element. This information was then used to identify a "minimum set." This paper presents an overview of the findings.

Aeronautical charting has evolved with changes in display technology, expanded use of global position systems (GPS), and increased processing capabilities. With each evolution, the usability of the aeronautical chart needs to be considered. For example, early research in the design of aeronautical charts focused on the usability of paper Instrument Approach Procedure (IAP) charts, which provide a visual representation of the information pilots need to fly an approach. Pilots indicated that these charts were cluttered - yet sometimes excluded needed information, and were difficult to read to the extent that pilots could not find information (Cox and Connor, 1987; Ashworth, McBain, Bassett, Moran, Soderlind & Buck, 1975). Additionally, the presentation of information (e.g., the layout, font, symbology) differed across chart providers. To address these concerns, the Volpe Center conducted a series of studies in the 1990s to improve information search on IAP charts. The results of this research led to the introduction of the "briefing strip" format which had the following properties:

- A briefing strip at the top of the chart to promote briefing as a critical component of flying an approach, and to present the required information in a logical order in one place.
- A boxed layout for heading and frequency information (see Multer et al., 1991).
- Graphical icons to depict missed approach information (see Osborne & Huntley, 1992).

As aeronautical chart information moved from paper to electronic mediums, research examined how to organize and "layer" information elements, so that the information could be added or removed. Pilot surveys were conducted to identify critical information elements for instrument approach charts (Hansman and Mykityshyn, 1995a) and surface moving maps (Yeh and Chandra, 2005). Additionally, Schvaneveldt, Beringer and Lamonica (2001) conducted a survey to identify critical information elements for flying in general. Collectively, the results showed that "critical" information elements differed depending on the phase of flight. Hansman and Mykityshyn reported that pilots were interested in the ability to declutter information but were concerned about the ability to retrieve the suppressed information when needed.

As electronic charts become integrated into flight decks, the design of the chart may diverge further depending on the manufacturer's design philosophy. The simplest electronic chart is a raster image that is an electronic version of a paper chart. A symbol identifying own-aircraft position may be added if the raster chart is geo-referenced. More complex are vector- and data-driven charts, which provide more capabilites to the end user than raster charts by encoding information about each information element, so that the chart can be re-rendered and re-scaled when the pilot zooms in (or out), allowing the size of the symbols and text to resize in a corresponding way. Users can also add or remove layers of information or select symbols to see more information about that symbol. Thus, the information on the electronic chart can become more specific to the task at hand, the pilot can use manual or automatic decluttering to customize the information, and the chart can be integrated with other map information.

The Federal Aviation Administration (FAA) was interested in understanding whether a minimum set of information elements could be defined for these customizable electronic charts. A couple of attempts have been made so far to characterize the information requirements. For example, SAE ARP 5621 provides a categorization of information elements based on subject matter expert opinion for electronic charts intended to be used as a replacement for paper charts. The SAE Committee decomposed nine chart types into the information elements shown on the charts and evaluated the criticality of each information element for presentation on a fixed chart for briefing or a moving map format for flying the procedure. Each information element was rated as a criticality based on the following:

- Level 1: information elements that can not be removed
- Level 2: information elements that should be shown initially but could be removed by pilot action
- Level 3: information elements that do not need to be presented initially and can be manually selected (or deselected)

Due to the number of information elements, we refer the reader to SAE ARP 5621 for the full classification. These levels, based on subject matter expert opinion, provide an initial framework for organizing information elements.

Pepitone, et al. (2014) provided data for a preliminary validation when they examined the criticality of information elements for integrating instrument flight rules (IFR) procedural chart information onto a forward flight deck display (e.g., a primary flight display (PFD) or multi-function display (MFD)). Twenty Honeywell pilots participated in a card-sorting task in which they rated the criticality of the information elements for flying a procedure using three levels, similar to the ones identified in SAE ARP 5621. The results provided some validation of the SAE framework, but the study was limited in that the data reflected the opinions of corporate pilots only and no statistical analyses were reported.

We wanted to further examine the criticality ratings provided in SAE ARP 5621 and Pepitone et al. Our focus was to identify a set of minimum information elements for a display concept in which pilots brief with a fixed chart that shows *all* information but then fly with a configurable electronic chart, which may or may not include all the information briefed. Our study addressed four different chart types (IAP, Instrument Flight Rules (IFR) Enroute, Standard Terminal Arrival Routes (STARs), and Standard Instrument Departures (SIDs)).

Method

Participants

Participants were recruited in two ways. First, 600 pilots, randomly selected from the Civil Aerospace Medical Institute (CAMI) Aeromedical Pilot Database, were invited to participate in the survey via email. Additionally, 600 invitations were sent via US postal mail to those pilots. To participate in the survey, pilots needed to have flown IFR in the previous 6 months and be a user of FAA, Jeppesen, or U.S. Government (military) charts. These pilots were characterized by pilot type (air transport, corporate, military, general aviation) based on information fields on pilot licenses recorded in the database. Due to a low response rate from the first sample, a second random sample of 600 pilots was selected and invitations were sent for participation. In this first effort, 258 pilots responded (a 21.5% response rate), but only 186 met the criteria for inclusion.

The participants recruited from the Aeromedical Pilot Database were primarily air transport and corporate pilots, so we conducted a second recruiting effort with local universities, military bases, and flying clubs to recruit general aviation and military pilots. We sent emails to 151 pilots, of which, 43 met the criteria for inclusion (28%).

In total 1,351 pilots were invited to participate; 326 responded (a 24% response rate). Of these, only 267 met the inclusion criteria. 229 pilots completed the survey.

Surveys

The purpose of this research was to gather pilot opinions of the importance of information elements shown on four types of charts: IAP, IFR, STARs, and SIDs. Due to the number of information elements on each chart, we created two surveys: one that included information elements on IAP/IFR charts (221 information elements), and the other with information elements from SID/STAR charts (206 information elements). Participants were randomly assigned to a survey. 114 pilots responded to the IAP/IFR survey, and 115 to the SID/STAR survey. The number of participants by pilot type are shown in the table below.

Table1. *Distribution of Participants by Pilot Type*

Pilot Type	IAP/IFR Participants	SID/STAR Participants
Air Transport	30	30
Commercial	30	30
General Aviation	30	30
Military	24	25

The median time to complete the IAP/IFR survey was 38.5 minutes; the median time to complete the SIDs/STARs was 28.9 minutes.

Participants completed a background questionnaire first before being presented with the information element survey. In the survey, participants were instructed to rate the importance of aeronautical information elements for a new charting concept using customizable electronic charts that are interactive and customized to display only information elements needed to execute the procedure. In particular, we emphasized that the customizable electronic chart would show only the information relevant to the procedure being flown. Category definitions were modified from the SAE ARP 5621 for the specific intended function. Pilots were asked to rate each information element individually with respect to aircraft operation when executing the procedure and not on the frequency of use. Ratings were made along four levels of importance. A fifth level was included if participants did not know the information element.

- 1 = Required to be displayed continuously for the safe and successful execution of the instrument flight procedure.
- 2 = Displayed initially, but can be removed and recalled for reference, as needed.
- 3 = Not displayed initially, but can be displayed manually for reference, as needed.
- 4 = Not required to execute the procedure.
- Don't know/Unsure

Pilots were presented with charts that depicted as many of the information elements being rated as possible. A sample is shown in Figure 1.

Standard Terminal Arrival Routes

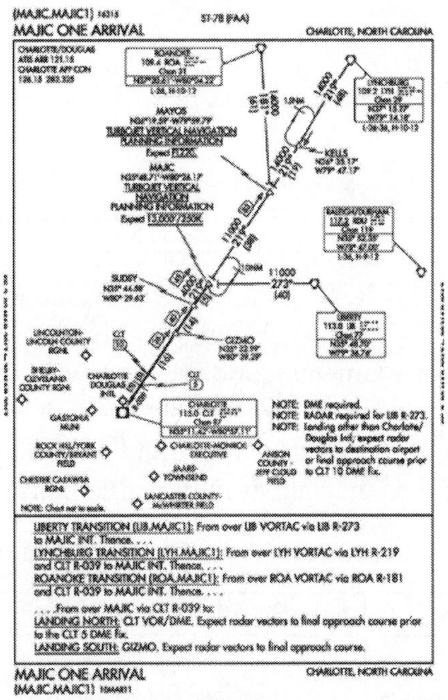

STARs1. **Focusing on the chart as an example of STARs Charts**, use the scale below to rate the ***importance*** of display elements when using a *customizable electronic chart* to *execute* a procedure.

	Required continuously (Level 1)	Displayed initially - can be removed/recalled (Level 2)	Not displayed initially - can be displayed manually (Level 3)	Not required to execute procedure (Level 4)	Don't know/Unsure
Identification					
Revision Date	○	○	○	○	○
Chart Index Number/Page Number	○	○	○	○	○
Effective Date	○	○	○	○	○
City/Location Name	○	○	○	○	○
Airport Name	○	○	○	○	○
Airport ICAO Identifier*	○	○	○	○	○
Procedure Name (e.g., Canoga Eight)	○	○	○	○	○
Procedure Identifier (e.g., CNOG8.VNY)	○	○	○	○	○
Changes*	○	○	○	○	○

* *Denotes display elements that may not be represented on the example chart.*

Click the box below to review *importance* definitions.
❑ Review definitions

Figure 1. Example of STAR survey with response options.

Approximately 32% of the information elements on the IAP/IFR survey and 53% of the information elements on the STAR/SID survey were not depicted. An asterisk denoted this.

Results

The frequency of responses for each level for each information element were calculated and analyzed using a series of chi-square tests. We developed the following framework with which to analyze the data:

1. Did pilots feel that the information element was required to be displayed to successfully execute the procedure? (Levels 1, 2, and 3 vs. Level 4)
2. If yes to 1, did pilots feel that the information element was required to be displayed at all times to successfully execute the procedure? (Level 1 vs Level 2, or Level 1 vs Level 3)
3. If the information element was not required at all times (Levels 2 and 3), did pilots feel that the information element should be displayed initially (Level 2 vs. 3)?

We compared the results of our analysis to the subject matter expert assessments captured in SAE ARP 5621 and the data provided by Pepitone, et al. (2014) as an intial validation. A subset of the critical (Level 1) information elements for each chart (post-comparison) are shown in Table 2 below. **This is not a complete list.** For a full list, the reader is referred to the technical report (in preparation).

Table 2. Sample of Level 1 (highest criticality) elements by chart type.

IAP Airport Elevation Airport Identifier All appropriate Navaid Symbols Communications Tower Frequency FAF (Maltese Cross) FAF Crossing Altitude (MSL) (HAT) Fix Altitude Fix Information Fix Name/Identifier Fix Symbol GS Intercept Altitude (MSL) Holding Pattern – Holding Pattern Depiction Landing Minimums CAT 1 Decision Altitude (DA) Landing Minimums – Minimum Descent Altitude (MDA) Landing Runway Number …	**IFR** Airway Designator Airway Magnetic Course Airway Symbol (center line) Area Minimum Altitudes – OROCA Sector Altitudes Indication of compulsory reporting Intersection, Waypoint, or Fix Name Intersection, Waypoint, or Fix Symbol Minimul Crossing Altitude (MCA) Navaid Frequency Navaid Identifier Navaid Name Navaid Symbol Segment Minimun Cruising Level or MEA Indication of MET Report Required Air Defense Identification Zones (ADIZ) …
SID Airport Elevation Airport Identifier Airport Name Course Definition – Heading Course Definition – Radial Course Definition – Track Instrument Procedure Course/Tracks – Identifier Instrument Procedure Course/Tracks – Symbol Intersection /Fixes on Procedures – Identifier Navaids for Fixes – Identifier Navaids for Fixes – Symbol Navaids for Legs – Symbol Procedure Name Transition Course Depiction Transition Name …	**STAR** Airport Identifier Airport Name Course Definition – Heading Course Definition – Radial Course Definition – Segment Mileages Course Definition – Track Holding Pattern Depiction Instrument Procedure Courses/Tracks – Identifier Instrument Procedure Courses/Tracks – Symbol Intersection/Fixes on Procedures Identifier Navaids for Fixes – Symbol Primary Airport Elevation Primary Airport Runway Layout Procedure Name Textual Information Crossing Altitude Restrictions …

Conclusions

This study presents a first step in identifying critical information elements for configurable electronic charts. The results shown here reflect pilot opinions of the importance of each information element for a new charting concept in which pilots brief with a fixed chart and fly with a reconfigurable electronic chart. Our next step is to ensure that the relationships between information elements is reflected appropriately (e.g., that related items that need to be shown at the same time are categorized the same way). Validation, potentially through simulation testing, is also needed to ensure that the prototype charts can be used during flight.

Acknowledgements

The research was conducted under the Flight Deck Program Directive/Level of Effort Agreement between the Federal Aviation Administration NextGen Human Factors Division (ANG-C1) and the Aerospace Human Factors Research Division (AAM-500) of the FAA Civil Aerospace Medical Institute (CAMI). The authors would like to thank our program manager, Katrina Avers; Dan Jack, Shijing Liu, and Suzanne Thomas of Cherokee CRC, LLC; and Stephanie Chase and Danielle Hiltunen of the US DOT Volpe Center for their contributions. Finally, we would like to thank all the pilots who participated.

References

Ashworth, McBain, Bassett, Moran, Soderlind & Buck. (1975). Special Air Safety Advisory Group Report to the Federal Aviation Administration. Washington, D.C., US Department of Transportation.

Cox, W. J., & Connor, C. W. (1987). *Report of Safety Survey: Human Integration of Approach Charts*. MIDWEST SYSTEMS RESEARCH INC DAYTON OH.

Hansman, R., Mykityshyn, M. (1995a). *An Exploratory Survey of Information Requirements for Instrument Approach Charts*. DOT/FAA/AAR-95/2. March 1995.

Multer, J., Warner, M., Disario, R., & Huntley, M. (1991). *Design Considerations for IAP Charts: Approach Course Track and Communication Frequencies* (DOT/FAA/RD-91/19). Washington, DC: US DOT Federal Aviation Administration.

Osborne, D. W., and Huntley, M. S. (1992). *Design of Instrument Approach Procedure Charts: Comprehension Speed of Missed Approach Instructions Coded in Text or Icons* (DOT/FAA/RD-92/3). Washington, DC: US DOT Federal Aviation Administration.

Pepitone, D., Ball, J., & Letsu-Dake, E. (2014). *NextGen Flight Deck information Displays: Recommendations for Managing Display Complexity with Electronic Chart Information* Version 2. Deer Valley, AZ: Honeywell.

SAE ARP 5621, *Electronic Display of Aeronautical Information*, January 17, 2011.

Schvaneveldt, R. W., Beringer, D. B., & Lamonica, J. A. (2001). Priority and organization of information accessed by pilots in various phases of flight. *The International Journal of Aviation Psychology, 11*(3), 253-280.

Yeh, M., and Chandra, D. C. (2005). *Designing and Evaluating Symbols for Electronic Displays of Navigation Information: Symbol Stereotypes and Symbol-Feature Rules* (DOT/FAA/AR-05/48, DOT-VNTSC-FAA-05-16). Washington, DC: US DOT Federal Aviation Administration.

DEPICTION OF VERTICAL FLIGHT PATHS FOR NEXTGEN ARRIVAL AND DEPARTURE INSTRUMENT FLIGHT PROCEDURES

Divya C. Chandra
USDOT Volpe Center
Cambridge, MA

Charts for instrument approach procedures have a vertical profile view that pilots can refer to during the final stages of the approach to the runway. However, there is no similar view of vertical flight path on charts for arrival and departure procedures. We studied whether a depiction of vertical flight path is feasible for arrivals and departures and whether it could help pilots manage procedures such as Optimized Profile Descents (OPDs), which are becoming common in the Next Generation Air Transportation System (NextGen). We identified sample procedures with challenging features including multiple flight path transitions, multiple constraints, course reversals, and many waypoints. Then we developed low-fidelity static prototypes and gathered informal feedback from airline pilots and chart developers. Based on their feedback and prior research, we developed a list of design considerations for both static and dynamic displays of vertical flight path for arrivals and departures.

The Next Generation Air Transportation System (NextGen) relies upon Area Navigation (RNAV) and Required Navigation Performance (RNP), which allow aircraft to fly more precise lateral routes using satellite navigation and/or other aircraft navigation systems. These form the basis of Performance-Based Navigation (PBN), an important component of NextGen. NextGen leverages PBN for the design of new RNAV Standard Instrument Departures (SIDs) and RNAV Standard Terminal Arrival Routes (STARs). A SID defines a path from a specific runway to the enroute airspace, and a STAR defines a path from the enroute airspace to a termination point from which the aircraft can join an Instrument Approach Procedure (IAP) to land at a specific runway. RNAV SIDs and STARs are common at major airports in the United States, such as Denver, Dallas-Fort Worth, and Atlanta. SIDs and STARs can extend over long distances (well over 50 miles and even beyond 200 miles in some cases) and they extend well into high altitude airspace (above 18000 ft). They are developed for routes that Air Traffic Control (ATC) uses to manage flights into and out of airports. They are presented to pilots on published aeronautical charts and encoded into the navigation database of the aircraft's Flight Management System (FMS) so that the autoflight systems can follow the route as programmed.

This paper describes an effort to identify design considerations for the depiction of vertical flight paths for SIDs and STARs. We are interested in this issue because charts for IAPs have a vertical profile view that pilots find useful, but there is no similar view of vertical flight path on charts for arrival and departure procedures. Our questions were, what is an adequate depiction of a SID/STAR vertical flight path and is this depiction feasible to implement? Could this depiction help pilots to manage compliance with vertical flight path constraints on speed and altitude? To explore these issues, we first partitioned the design problem. Our primary focus was the design of vertical flight path depictions for *static* (pre-composed) aeronautical charts (paper or Portable Document Format, PDF). Secondarily, we considered data-driven (dynamic) views of the vertical flight path for SIDs and STARs.

We begin this paper with background on SIDs and STARs and related findings from other displays of vertical flight path. Then we present the method for this effort, including the assumptions and scope. Though we prototyped several options, we present only the two that received the most favorable reviews by airline pilots and chart developers. Finally, we list design considerations for depiction of vertical flight path on arrivals and departures.

Background

RNAV SIDs and STARs have more turns and more altitude and speed constraints than older SIDs and STARs (which use conventional, line-of-sight, ground-based navigation aids). Optimized Profile Descents (OPDs) are a specific type of RNAV STAR. They have many vertical constraints (on speed and/or altitude) that allow the aircraft to descend continuously for fuel efficiency. The vertical constraints on OPDs also allow ATC to anticipate the aircraft vertical flight path, within bounds, which releases airspace for other traffic flows.

To fly RNAV SIDs and STARs, including OPDs, pilots are more dependent upon the FMS and, in particular, its Lateral Navigation (LNAV) and Vertical Navigation (VNAV) modes. Use of automated systems, such as the FMS, to manage flight path can be very effective, but introduces its own vulnerabilities (PARC/CAST, 2013). Managing automated systems can also increase pilot monitoring workload (Flight Safety Foundation, 2014).

Butchibabu, Midkiff, Kendra, Hansman, and Chandra (2010) found that problematic STARS, identified from the Aviation Safety Reporting System (ASRS), had more altitude constraints and more waypoints than baseline STARs. In other words, pilots have more difficulty managing the vertical flight path constraints on RNAV STARs with many vertical constraints, such as OPDs. Based on conversations with pilots, we found that this is more likely if the VNAV algorithms are less sophisticated or if the aircraft is not equipped with VNAV (Chandra and Markunas, 2017). For SIDs, Butchibabu et al. (2010) found that lateral deviations were more common than vertical flight path deviations. However, in discussions with industry pilots, we learned that vertical flight path constraints that do appear on SIDs are very difficult for pilots to manage, and that is why they are less common. Recent discussions with industry pilots confirmed that managing vertical flight path constraints on OPDs remains a top concern.

Vertical profile views on IAP charts graphically depict altitudes that the aircraft must meet during the last stages of the approach. It is logical to anticipate that a similar display might help pilots fly SIDs and STARs accurately, so we examined the design and use of the IAP vertical profile view. From Chandra and Markunas (2017), we learned that pilots go back and forth between the plan view on the IAP chart and the profile view; they use both. The vertical profile and the plan view are aligned in specific ways (names of waypoints, for example), and some data are duplicated (e.g., altitudes are shown in both places). The vertical profile view is less cluttered than the plan view; it simply shows less information for a shorter portion of the approach. Pilots can use the profile view for reference in flight during the final stages of the approach. However, pilots with more advanced avionics may focus on the primary flight displays and monitor other flight deck systems, such as VNAV, so they may not look at the chart's profile view as much.

Vertical Situation Displays (VSDs) are electronic flight deck displays of vertical flight path. SAE ARP 5430 (2013) contains detailed design guidance for VSDs. Although there are many potential uses for the VSD, Boeing designed theirs primarily to mitigate controlled flight into terrain during approach (Boeing, 2002). The VSD is a strategic, real-time display, whereas pilots use the static vertical profile on an IAP for planning and reference. Also, the VSD shows ownship position; in order for pilots to correlate the view on the VSD with the view on the static IAP vertical profile, they would need to know where their own aircraft is on the static depiction.

Electronic data-driven charts are also in development and could be used to show vertical flight paths. Such charts allow pilots to access all the information from a static chart (much of which is not available on primary flight deck electronic displays) in an electronic, customizable format. For an introduction to different types of electronic charts, including data-driven charts, see Chandra, Yeh, Riley, and Mangold (2003). Additional information on data-driven charts can be found in SAE ARP 5621 (2004), and Larson (2011).

Method

Before developing our prototypes, we scoped the effort by making several assumptions to simplify the work. For example, we decided to prototype only static concepts, not data-driven electronic concepts, which would have required software development and many design choices. We also decided not to constrain the size of the depictions. Instead, we wanted to identify the best possible design, even if it did not fit on existing charts. Existing charts have very little free space, and finding any at all will be a significant problem. We also chose to make only black and white depictions. Current SID/STAR charts from the United States government are not printed in color, although that could change in the future. We did not consider how notes would be shown, the scale of the vertical axis, or symbology for speed constraints (for which we kept today's standard format, lines above and below the speed). Lastly, we did not evaluate how far along the route the vertical flight path should be depicted– whether the depiction should extend to the last waypoint with a constraint, or to the end of the whole route, up to the last waypoint.

Another key decision we made was about what pilot task(s) the display would support. To make this decision, we first brainstormed about how the depiction might be used and came up with four possibilities. After discussing these options with airline pilots and chart developers, we concluded that the most likely use for the display would be to assist with route verification during set up in FMS. This is in agreement with a finding from Chandra and Markunas (2017) that pilots do not use static charts to monitor the flight path in real time; they use static charts primarily to review and brief the route during flight preparation. The other possible uses we considered were for the depiction to (a) highlight segments with potentially steep climb or descent gradients, (b) help pilots build an internal mental representation that they could reference later in flight, and (c) help pilots understand any route amendments from ATC. Of these, only the first (highlighting steep climbs/descents) was considered to be valuable.

We also had to select specific arrival and departure procedures for our prototypes. To do this, we first created a list of challenging features. These were multiple transitions (i.e., branches in the flight path), multiple altitude or speed constraints, multiple waypoints, and course reversals. Many RNAV STARs have multiple transitions because the paths go in different directions for the different landing directions and runways at the airport. SIDs have transitions to merge paths from different runways and transitions to merge onto different airways in the enroute airspace. Although transitions can overlap to some extent, each typically has a unique vertical flight path. So, any airport with multiple runways will have more transitions (for both SIDs and STARs), and there will be more distinct vertical flight paths for those routes.

We considered five different RNAV instrument flight procedures for the prototypes and eventually used three: the FRDMM THREE STAR (an OPD) into Washington, DC (KDCA), the EDETH FIVE SID out of Salt Lake City, UT (KSLC), and the LEETZ TWO SID also from Salt Lake City. (The two Salt Lake City SIDs are no longer in use and the FRDMM has been updated.) We used these procedures to illustrate different prototype options including: four display formats, four types of altitude-constraint symbols, two types of flight path representation (line segments vs. smooth), two types of horizontal-axis scaling (equal intervals vs. notional), and different amounts of data duplication from the plan view. The four display formats include two different table views, two different 2-D graphical formats, and one 3-D format. The four different altitude constraint symbols included the standard chart convention (lines above/below the altitude), shading, the FMS convention (letters), and new triangle symbols. We did not create prototypes for all combinations of these features, just some illustrative examples for discussion and feedback.

Findings

Our discussions with pilots and chart developers provided useful feedback on the design of the prototypes views. Some key points are summarized below.

Altitude-constraint symbols. The group recommendation was that we should match the symbol for altitude constraints with existing chart symbols, rather than develop a new symbol. Matching the FMS notation was not useful.

Horizontal axis. The group recommendation was that we should include a numerical value for the distance between waypoints (even if the scaling is notional). Pilots will use this along with the altitude requirements to estimate if there will be a steep descent or climb gradient. Equal intervals between waypoints along the horizontal axis were misleading, but exact scaling was unnecessary.

Display formats. The 3-D format was not effective. It produced a depiction that was difficult to interpret; turns and descents were confusable.

The two prototypes that received the most favorable feedback are shown in Figure 1, along with an IAP vertical profile view (a) for comparison. One is a graphical 2-D view (b) and the other is a table view (c), both for an RNAV SID at Salt Lake City, Utah. In Figure 1(b), the upward path (a climb) indicates either altitude increase or turns in an intuitive integrated view. This view may help pilots to identify steep or level vertical gradients and sharp turns. It uses the standard symbols for altitude constraints. However, this graphical view still requires a separate image for each transition with a unique vertical path. The pros of the table view in Figure 1(c) are that it can be used to verify the constraints quickly, its format corresponds to a familiar layout of the FMS Control and Display Unit (CDU), and it uses the standard symbols for altitude constraints. The drawbacks of the table in Figure 1(c) are that each route transition requires its own table, and the table column format will not match all CDUs.

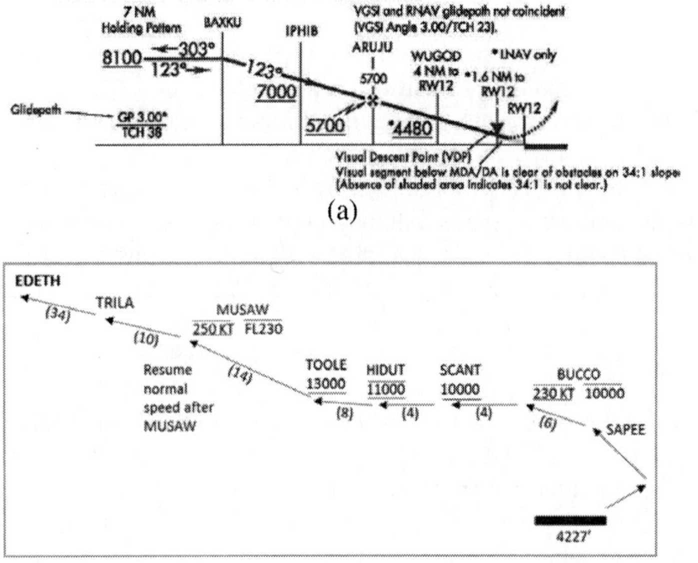

(c)

Depart runway heading		
At 4727' (500' AGL), LEFT turn		
Direct **SAPEE**. 5.6 NM		
BUCCO 3.7 NM	230 KT	10000
SCANT 3.9 NM		10000
HIDUT 7.6 NM		11000
TOOLE 13.9 NM		13000
MUSAW (Resume normal speed after MUSAW) 10.4 NM	250 KT	FL230
TRILA 34.2 NM		
EDETH		

Figure 1. Example vertical profile view from Federal Aviation Administration (FAA) chart legend (a), sample graphical format for Salt Lake City, Utah EDETH FIVE RNAV SID (b), and sample table format (c) for the same procedure, which has multiple transitions, multiple constraints, and a course reversal. The depictions in (b) and (c) end before the enroute transitions begin, after the EDETH waypoint.

In addition to receiving specific feedback on the prototypes, another important result of this effort was that, while developing the prototypes, we identified design considerations for static depictions of vertical flight path. Some of the design considerations were in-scope, in that we considered them in our prototypes. These include the display format, the symbology for altitude constraints, type of flight path representation, horizontal-axis scaling, and duplication of data between the vertical flight path depiction and the plan view. Some design considerations were not considered in our prototypes. These were space limitations, use of color, notes, how far the vertical profile should extend along the route, symbology for speed constraints, and the vertical-axis scale.

Data-driven charts have additional design considerations. These charts are presented on an electronic display and managed through a graphical user interface. As such, there are general software considerations such as user interaction with display and data elements, the physical display size, display configuration (automatic or user-controlled), and sources of data for the display. More specific considerations for data-driven vertical flight path depictions are:

- When should pilot use this depiction and why?
- What is the priority of this information relative to other available data?
- Will ownship be available?
- How is an electronic data-driven vertical profile view different from a VSD?

Summary and Conclusions

We explored design considerations related to creating vertical flight path depictions for SIDs and STARs with difficult features such as course reversals, multiple constraints, transitions, and waypoints. These procedures, which include OPDs, are being developed as part of NextGen. We selected some candidate procedures to prototype in static formats, then gathered feedback on these prototypes informally from airline pilots and chart developers. We identified several design considerations for depictions of vertical flight path for both static and electronic data-driven formats.

The most likely reason for a pilot to use a static vertical flight path depiction is to check constraints along the route during a briefing in preparation to fly the procedure. The depiction could provide an efficient way to verify the route in the FMS or help draw attention to potential challenges in segment gradients. However, static depictions of vertical flight path are not useful for in-flight monitoring because they do not provide the pilot with the necessary information at the right time.

Another problem for these depictions, which we did not address, is that they take a lot of space, even for just one transition. And, in fact, most major airports have multiple runway and enroute transitions, each with a unique vertical flight path. It would take a lot of space to show every possible transition and flight path. The operational utility of any depiction will depend on how it is integrated with other chart data. We know that depicting the vertical flight path for a SID/STAR with many flight path transitions, waypoints, and constraints is not likely to fit within the currently available space on static (paper or PDF) charts, even if we minimize the size of depictions and list only waypoints with constraints. Incorporating these depictions on static charts may be costly and of limited utility.

Even separating a SID or STAR with multiple transitions across multiple chart images (as discussed in Chandra and Markunas, 2013), may not yield sufficient free space on a chart. Vertical flight path depictions that require multiple pages will be unwieldy and are unlikely to produce sufficient benefits to make them worthwhile, given the extent of costs and resources required to update so many charts. For these reasons, we do not expect that including depictions of vertical flight path on static SID/STAR charts would provide an operational benefit. There is some hope that data-driven depictions of vertical flight paths will be helpful, but there are also many unanswered questions about how these depictions would work in practice.

Acknowledgements

This document was prepared for the FAA with funding from the Human Factors Division (ANG-C1). This work supports the NextGen operational improvements for the Use of Optimized Profile Descent (OI 104124) and for RNAV SIDs, STARs, and Approaches (OI 107103). This paper was prepared for the FA6Y Interagency Agreement (Human Factors of Flight Operations and Technologies). It was funded via the FY17 RE&D System Development – NextGen Air/Ground Integration PLA, Task 12.

Thanks to Alison Bisch for help with this project. Thanks also to the pilots and chart developers who contributed input on the prototypes, to our FAA technical sponsors, Kathy Abbott and Jeff Kerr, and to our program manager, Sabreena Azam. The views expressed herein are those of the authors and do not necessarily reflect the views of the Volpe National Transportation Systems Center or the United States Department of Transportation.

References

Boeing. (October 2002). *Vertical Situation Display for improved flight safety and reduced operating costs.* Retrieved from http://www.boeing.com/commercial/aeromagazine/aero_20/vsd.pdf.

Butchibabu, A., Midkiff, A., Kendra, A., Hansman, R.J., & Chandra, D.C. (2010). Analysis of safety reports involving area navigation and required navigation performance procedures. *Proceedings of the International Conference on Human-Computer Interaction in Aeronautics* (HCI-Aero 2010). 3 5 November, Cape Canaveral, FL. Retrieved from https://www.volpe.dot.gov.

Chandra, D.C., & Markunas, R. (2017). *Line Pilot Perspectives on Complexity of Terminal Instrument Flight Procedures*, (DOT-VNTSC-FAA-17-06). Cambridge, MA, U.S.DOT Volpe National Transportations Systems Center. Retrieved from https://www.volpe.dot.gov.

Chandra, D.C. and Grayhem, R. (2013). *Evaluation of a Technique to Simplify Area Navigation and Required Navigation Performance Charts*, DOT-VNTSC-FAA-13-02, Cambridge, MA, U.S. DOT Volpe National Transportation Systems Center. Retrieved from https://www.volpe.dot.gov.

Chandra, Divya C., Yeh, M., Riley, V., & Mangold, S.J. (2003). *Human Factors Considerations in the Design and Evaluation of Electronic Flight Bags (EFBs), Version 2.* DOT-VNTSC-FAA-03-07. USDOT Volpe Center: Cambridge, MA. Retrieved from https://www.volpe.dot.gov.

Flight Safety Foundation. (2014, November). *A practical guide for improving flight path monitoring.* Retrieved from https://flightsafety.org/toolkits-resources/flight-path-monitoring/.

Larson, G.C. (2011). Transformers: Jeppesen Unfolds a New Map. *Business and Commercial Aviation* pp.48-52.

Performance-based operations Aviation Rulemaking Committee (PARC)/Commercial Aviation Safety Team (CAST) Flight Deck Automation Working Group. (2013). *Operational use of flight path management systems.* Retrieved from www.faa.gov.

SAE International. (2013). *Human Interface Criteria for Vertical Situation Awareness Displays.* ARP 5430. Retrieved from http://standards.sae.org/arp5430/.

SAE International. (2004). *Electronic Display of Aeronautical Information (Charts).* ARP 5621. Retrieved from http://standards.sae.org/arp5621.

MISSION-READY ADAPTIVE DECISION AIDS: THE NEW PERFORMANCE MODEL FOR CREW STATION DESIGN

Captain Kevin M. Smith U. S. Navy (Ret.)
Mesquite, Nevada

Discussed are three theories: 1) Baysian probability theory, 2) signal detection theory, and 3) operational decision theory. To combine the three is to come to an understanding of how one can operate effectively in complex environments. Complex environments present unique challenges from a human performance perspective. They require applying uncommon skill sets to allow for optimization of performance. Applying analytic methods to clarify and respond to mission critical events is most urgent. The analytics of uncertainty is presented. Three mission critical decisions are discussed: to continue or abandon the mission, to perform the approach go-around maneuver, and to determine the takeoff go-no go. These are at the heart of optimizing mission level performance.

As a general class of phenomena, complex environments contain complex situations and systems. Complex environments are one of the most challenging to consider, in large measure because of our inability to understand and predict; they can be fraught with uncertainty. If one is planning to operate in a complex environment by employing large-scale dynamic systems, conventional reasoning cannot be used. Complex entities are non-deterministic by nature because complexity theory informs us that complex systems exhibit novel behavior and emergent properties, rendering these entities and phenomena to a class by themselves residing outside of conventional wisdom.

Tackling the decision problem for large scale dynamic systems utilized in the field of aviation is of immediate importance yet is arguably the most difficult. This is because very little is understood with respect to optimizing the performance of such systems, and previous attempts have not considered the levels of uncertainty associated with such systems.

The Overall Mission Continuation Decision

Operational decisions are singular among all other classes of decisions and represent the most important command activity. Operational decision theory supports operational decision making (ODM) and helps identify and optimize operational decisions. ODM provides for the broad situation awareness needed to identify risk and the structure to manage a rising risk profile.

An advanced qualification design team defined observable mission related activities. They realized various mission tasks were *not* performed in a linear sequence. High performing crews did something unexpected—they prioritized their tasks.

What is critical for mission success is how well flight crews and the captain solve problems in a complex environment. Non-linear problem solving took center stage. This breakthrough came with the following insights: all air carrier mission activities are highly planned, often using sophisticated planning tools. While all activities are planned, excellent pilots do not plan real-time activities. Some are discarded altogether. These pilots prioritize and select tasks using some kind of decision-making process to optimize mission outcome.

This decision making process gained definition after Keeny and Raiffa invented a branch of mathematics that dealt with the numerical weighting of multiple attributes. This defined operational decisions, identified key decisions and specified triggers that activate certain decision pathways. High performing pilots were selecting optimum pathways, but this had yet to be understood.

An operational decision for pilots is now defined by Smith and Hastie (1992) as containing three unique components:

1. It must often be performed using incomplete information.
2. Once airborne, it is always performed under increased time compression.
3. Consequences of poor decisions are often catastrophic, placing the aircraft, crew, passengers, and the corporation in jeopardy.

Determining Risk

The operational decision for the air transport mission conforms to the following rules.

1. If the risk to the completion of the mission is low, then continue with the original mission plan.
2. If the risk to the mission is moderate, then modify the mission to either reduce or stop the risk from rising.
3. If the risk is high, abandon the mission plan and activate available alternatives. This we refer to this as "divert—reject—abandon."

The nature of risk means we must deal with it or it can get worse—it will be a rising risk. In aviation systems, unless decisive action is taken during critical events, risk will continue to rise to a point where one experiences a catastrophic mission failure. This point is called the critical event horizon.

Rising risk can be explained by using the risk continuum. The risk continuum is organized into three zones. When risk rises it passes through zone one (low risk), to zone two (moderate risk). If the encountered event is critical enough, or if risk has not been mitigated, it will likely become high risk, zone three. Each zone has a certain action. For low risk, continue with the mission plan. For moderate risk, modify the mission plan to arrest the rise or lower the risk. For high risk, where catastrophic failure is probable, abandon the mission plan and immediately implement survival measures.

Operational Envelope

Four hostile agents create boundary conditions. These boundaries define when it acceptable to operate and when it is not. They are the edges of an operational envelope. See Figure 1. Low risk resides within the envelope. Risk factors that impact the mission but do not place the aircraft outside the boundaries should be considered as moderate. The operational strategy will be to modify but not abandon the mission.

Hostile agents come from four general directions. These are:

1. Any adverse condition, such as adverse wind, freezing precipitation, and so forth.
2. Restricted visibility. This can often limit the ability to land at a particular airport, causing great concern if insufficient fuel remains to proceed to an alternate airport.
3. Mission critical alerts and warnings. This could be such things as terrain alert, traffic alert, or thunderstorm detection.
4. Human and system limitations. System limitations could be speed or altitude, where human limitations could be fatigue, task overload, or inexperience.

Figure 1. Operational envelope created by boundary conditions.

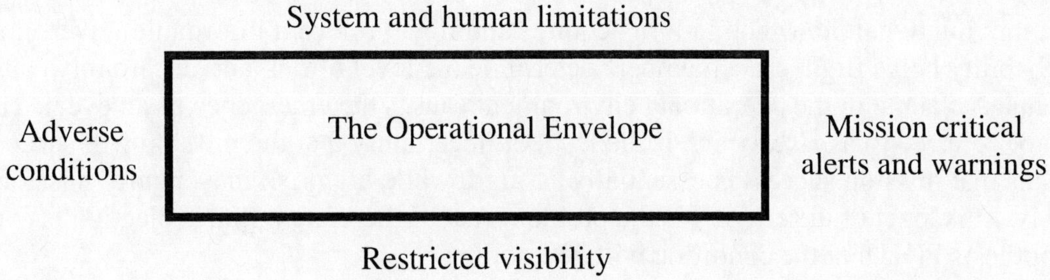

Some agents are more dangerous, have more energy, or can bind with other agents, causing a dangerous "cumulative effect" if undetected. Figure 2 shows the combined vector is the resolved hypotenuse of the triangle formed. The combined vector travels to the corner of the mission space; this is a rising risk situation that must be addressed immediately.

Figure 2. The cumulative effect, when two sides of the operational envelope and bonded together to create a high risk situation.

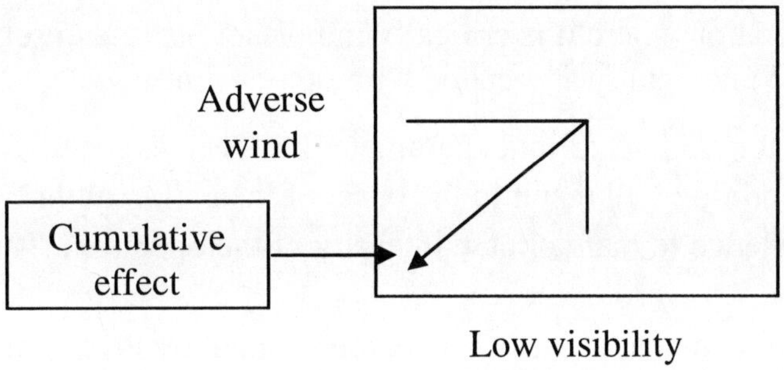

Let us look at an example of risk, the operational envelope, and the cumulative effect. On December 12, 2005, flight 1248 attempting to land at the Chicago-Midway airport crashed (National Transportation Safety Board, 2007). At the time a significant Midwest snowstorm made the weather exceptionally poor. The flight crew had to deal with four hostile agents that had entered the mission space:

1. Braking action advisories in effect with fair to poor braking action reported.
2. Short runway with no overrun.
3. Adverse wind, with an 8 knot tailwind reported.
4. Low visibility and approaching landing minimums.

Using the operational envelope, we can see that the mission should be immediately abandoned. To attempt the approach and landing leaves the rising risk unchecked and it will pass beyond the critical event horizon, resulting in catastrophic mission failure. In this case this is precisely what happened.

Mission performance is optimized by first understanding the prevailing risk and then knowing what to do about it. When risk begins to rise, the flight crew must prioritize or discard activities to manage a rising risk profile. If risk mitigation measures are not effective, and the risk is high or projected to go to high, then the mission plan must be abandoned, and survival measures must be taken.

The Unstable Missed Approach Decision

Baysian Probability

Bayesian probability is based on the concept that the likelihood of an event can be understood in terms of a moving dynamic, which in turn acquires additional relevant information over time. Bayesian probability helps flight crew members determine the level of risk and uncertainty caused by certain events. A change in the operational environment causes the emergence of an event. These impedimenta can be referred to as *risk*, which is the level of uncertainty that the mission will succeed. Low risk means that mission success is essentially assured, while high risk may signify mission failure is most likely. This level of uncertainty is the problem space where the initial projection is expressed as a hypothesis P(H) that the condition will deteriorate.

1. P(H) is the probability of encountering a mission critical event and its impact on the mission.
2. P(E) is additional evidence that has been encountered expressed in probabilistic terms.
3. P(E/H) is an updated figure of merit for E given that H is true.
4. P(H/E) is the unknown that we wish to determine. It is the updated level of uncertainty.

Problem Solving Under Conditions of Uncertainty

Our case study involves an operation where it is critically important that the onset of instability be determined notwithstanding the uncertainty associated with other events.

- At point A on the trajectory, an operational parameter has been exceeded. The likelihood that this out-of-tolerance condition will result in the onset of instability at the approaching point D, is key operational knowledge to maintain the integrity of the operation. This is represented by P(H). See Figure 3.
- Additional evidence is obtained at point B, and is represented by P(E). This evidence may be germane to the operation, and it could influence the determination of the onset of instability.
- At the conditional point along the trajectory, identified as point C, P(E) is assessed. Given that P(H) is true, P(E) is updated and given a value commensurate with this updated evidence. This probability is represented by P(E/H).

It is necessary to determine at point D on the trajectory whether the onset of instability will occur. If it is highly probable that instability will occur, then best practices dictate that the mission should be abandoned prior to reaching this limit. This probability is the determined value of P(H/E). This is the unknown that we wish to discover, or, in another way of presenting it, this is the unknown in the mathematical equation that represents all three previous points.

The mathematical equation is represented below.

$$P(H/E) = \frac{P(H) * P(E/H)}{P(E)}$$

Figure 3. Case study of problem solving under conditions of uncertainty.

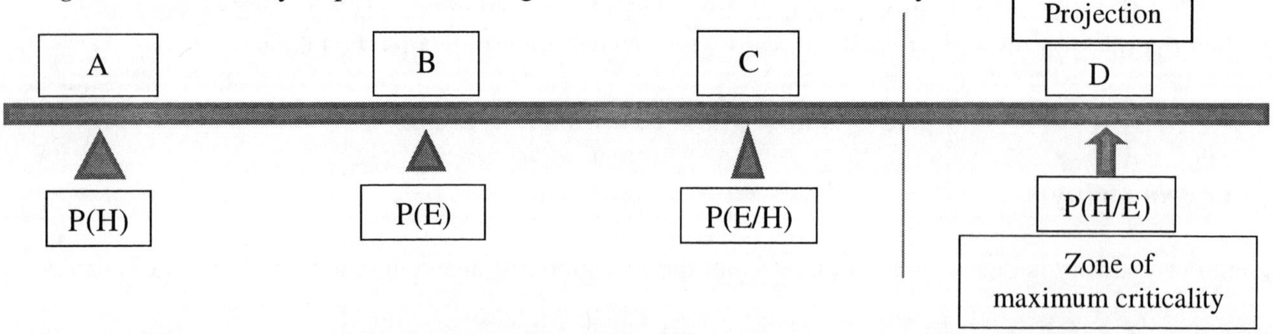

The Takeoff "Go/No-Go" Decision

In takeoff operations in large transport aircraft, the crew is constantly monitoring the environment and aircraft to assess the emergence of any operational risk. When the danger is either high or low, the decision is relatively straightforward. However, when the value is in the midrange, the decision becomes difficult. This region is called the zone of ambiguity.

Unexpected Operational Difficulties

Xc labels the decision criteria that is used to make the optimum decision. This situation is more complex than first realized when the area surrounding Xc is examined more closely. This is shown in Figure 4.

Figure 4. Zone of ambiguity is where the two areas overlap, showing where decisions become difficult.

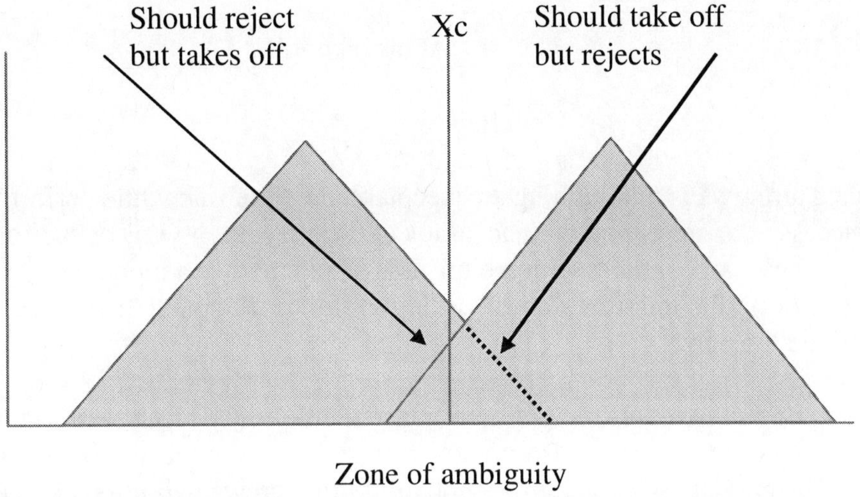

Optimizing the Decision Function

The analytical approach focuses on improving the discrimination between the two states (danger—should reject; low risk—can continue). This should be the primary goal of improvements in technology. But improvements in "discriminatory training" are also necessary. This corresponds to changing the relationship between the distributions by moving the distributions farther apart.

Operational Analysis of the Takeoff Decision

We will use the takeoff operation to show the decision analytic structure and its properties. The decision structure is represented in Figure 5. The key choice points are depicted. They involve the choice of continuing takeoff as planned, continuing takeoff with modifications to the operational plan, or aborting takeoff due to significant danger. In this example, the key choice points involve the primary choice to continue the takeoff as planned or reevaluate the takeoff plan. The secondary choice is contingent on the first—either to modify the operation to accommodate a rising risk or to abort the takeoff due to excessive risk. The key activities associated with each choice point are the mechanism by which the decision is executed. It is important to realize that an optimum decision selection criterion, called an alternative, is the ability to select the most accurate decision path with respect to the prevailing risk at the time. Choice Point A represents the primary binary decision. Notice that this entails the evaluation of risk. This is important for many reasons. Risk analysis is critical in selecting the correct path. While execution of the proscribed maneuver is important, it is at Choice Point B after the risk is examined. Many studies as well as documented operational experience have suggested that the takeoff accident rate is excessive. But while this insight is important they focus mostly on the maneuver execution phase rather than first examining the higher-order skill requirements involving the optimization of the operational decision.

Figure 5. Decision analytic structure with key choice points.

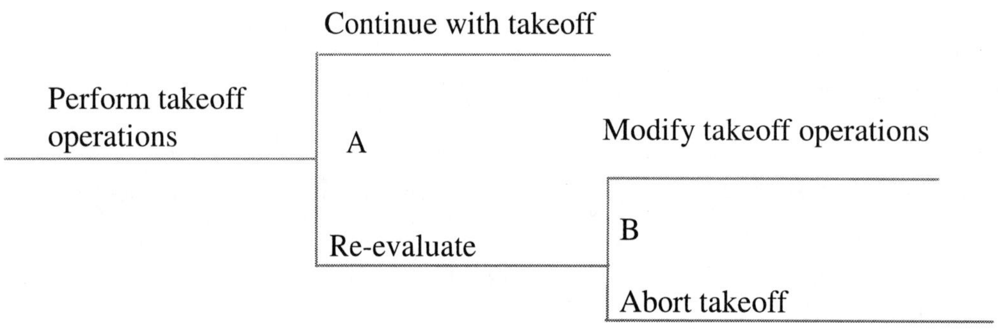

Conclusion

A serious challenge facing aircrews is maintaining an acceptable level of risk while performing a mission. Key to their success is to determine with accuracy and clarity if a low risk situation prevails or is anticipated. If so, then crews can continue with the mission as planned. If a moderate risk posture is evident, then crews must modify the mission plan accordingly. If the risk posture is judged to be high, then crews must discontinue the current plan.

The effective management of risk involves the optimum placement of the decision criteria, which we have labeled Xc, along the risk dimension. It also involves the reduction of the ambiguity zone through discrimination methods. Current data shows that the probability that crews will not make the correct abort decision with respect to accurate assessment is 54 percent while 46% were correct. Flight crews are incorrectly assessing risk and aborting takeoffs at an alarming rate.

Among the many solutions that have been proposed to reduce takeoff accidents, several proposals have involved moving the decision criteria, Xc. However, such an administrative adjustment of the decision criteria should not be undertaken without a careful analytical study of improving the discrimination capabilities of prevailing risk.

References

Smith, K.M & Larrieu, S. (2015) Mission Adaptive Display Technologies and Operational Decision Making in Aviation. Hershey, IGI-Global.

National Transportation Safety Board. (2007, February 10). Aviation Accident Report AAR-07/06. Retrieved January 3, 2015, from http://www.ntsb.gov/investigations/AccidentReports/Reports/AAR0706.pdf

Smith, K. M., & Hastie, R. (1992). Airworthiness as a design strategy. In *Proceedings of the FSI Air Safety Symposium*. San Diego, CA.

IF WE'D ONLY LISTEN! WHAT RESEARCH CAN TELL US ABOUT AIRCREW FATIGUE.

Christina (Missy) Rudin-Brown
Ari Rosberg
Danielle Krukowski
Human Factors and Macro Analysis Division,
Transportation Safety Board of Canada (TSB)
Gatineau, QC, Canada

Performance decrements associated with fatigue are significant risk factors of occupational, motor vehicle, and aviation accidents. The substantial number of recent aviation occurrences involving aircrew fatigue and the slow progress of related rulemaking prompted the TSB to include fatigue management on its 2018 Watchlist of key safety issues. At the same time, a finding of aircrew fatigue in a 2017 NTSB investigation into a near-taxiway landing prompted some journalists to argue that there are few, if any, research studies showing how fatigue affects flying ability, and that current efforts in fatigue management may not be effective. This paper explores research in psychology where effects of fatigue on human performance were identified, and describes correlative changes in pilot performance, with a focus on the approach and landing phases of flight. Examples from recent TSB air transportation safety investigations are used to illustrate.

Around midnight on July 7, 2017, Air Canada flight 759, on approach to San Francisco International airport, almost landed on a taxiway where four commercial aircraft holding hundreds of passengers and crew were lined up waiting to be cleared for departure. The taxiway was adjacent and parallel to the active runway. In September 2018, the U.S. National Transportation Safety Board (NTSB) released its report into the incident, which represented one of the closest calls of a potentially disastrous accident in history. The NTSB determined that the flight crew had misidentified the taxiway as the landing runway. Amongst other findings, aircrew fatigue due to circadian disruption and the length of continued wakefulness was cited as a contributing factor to the crew's misidentification of the intended landing surface, their ongoing expectation bias, and their delayed decision to initiate a go-around. Although both crew met Canadian flight duty time and rest requirements, at the time of the incident the pilot flying had been awake for more than 19 hours, and would have been permitted by the regulations to remain on duty for a further 9 hours. The NTSB concluded that Canadian aviation regulations in effect at the time of the incident did not always allow for sufficient rest for reserve pilots, a situation that could result in these pilots flying in a fatigued state during their window of circadian low.

As a consequence to this incident, some journalists (e.g., Nunes, 2018) argued that there were few, if any, research studies showing how fatigue affects a pilot's flying ability, and that current efforts in fatigue management in air transportation may not be effective. While anything that increases the demand for more research into the effects of fatigue on aviation safety is appreciated, the journalist's conclusion is unfounded. Knowledge and understanding of recent psychological research in fatigue and human performance does, in fact, tell a great deal about how fatigue affects a pilot's performance.

While there are typically between 1000 and 1100 aviation occurrences reported to the TSB each year under mandatory reporting requirements, practical considerations dictate that only about 2% are fully investigated by the TSB. Therefore, it is challenging to estimate statistically the prevalence of fatigue-related aviation accidents. Nevertheless, review of those investigations where fatigue was concluded to have played a role can increase our understanding of the issue. Between 1991 and 2017, there were 43 TSB air investigations that made formal findings related to fatigue; 22 made findings about fatigue as a cause or contributory factor, and 26 made findings about fatigue as a risk. Thirty-four investigations made findings regarding fatigue on the part of air crew and, of these, 12 (35%) involved the approach and/or landing phases of flight.

The objective of this paper is to review some of what is currently known in psychology about the effects of fatigue on human performance that would predict corresponding changes in pilot performance. In the interest of brevity, the focus is on the approach and landing phases of flight, with examples from TSB aviation investigations used to illustrate.

Sleep-related fatigue and flying

Although there are various other types of fatigue that can affect human performance (e.g., physical, mental, and lethargic fatigue), fatigue that is related to the amount and quality of sleep obtained is biological in nature. Consequently, it cannot be prevented by, for example, characteristics of personality, intelligence, education, training, skill, compensation, motivation, physical size, strength or practice. In this context, "fatigue" is conceptualized as a continuum between being asleep and being fully awake. Sleep-related fatigue can be caused by acute or chronic sleep disruptions, extended periods of wakefulness, circadian (daily) rhythm effects, medical and psychological conditions, and sleep disorders. Performance impairments associated with fatigue are significant risk factors and predictors of occupational accidents and injuries (Dawson et al., 2011), motor vehicle accidents (TIRF, 2016), and aviation occurrences.

Flying an aircraft is a complex activity. It requires a human operator – the pilot – to perceive, process and integrate many different sources of information in a high workload environment, to perform many sub-tasks concurrently, and to make and change plans on short notice. To make effective decisions in this environment, pilots need to have an accurate understanding of their goals, decisions, and information requirements, as well as the state of the aircraft, other crew, and any passengers. Situation awareness – the perception of elements in the environment within a volume of time and space, comprehension of their meaning, and projection of their status in the near future (Endsley, 1995) – is paramount.

The approach and landing phases of flight represent periods of high workload even for highly skilled flight crew, and comprise a number of psychological elements, including:

- attending to, and perceiving, instruments, controls, and outside environmental conditions;
- making control inputs in a timely manner;
- remembering information and inputs; interacting effectively with other crew and ATC; and
- assessing and understanding the changing situation and associated risks; making effective decisions and solving problems – to either commit to landing or to perform a go-around.

The following sections present some of the primary psychological constructs that underpin pilot performance during the approach and landing phases of flight. The findings of research examining the effect(s) of fatigue on these constructs is briefly summarised and, for each construct, the corresponding effects on actual pilot performance that would be predicted are described and illustrated using occurrences investigated by the TSB.

Attention

Human attention and the capacity to process information are limited. Because human information processing takes place constantly, and because there is so much information available in the environment, it is necessary for pilots to cope with this flow by filtering out less important information to attend to the important information. A pilot's ability to attend to critical stimuli within their environment will be impaired if they are distracted or inattentive, and will result in impaired situation awareness (Endsley, 1995). In one study (Sanders & Reitsma, 1982), university students performed watchkeeping sessions during which either a centrally or a peripherally located signal was presented intermittently. One night's sleep loss slowed responses and led to missed stimuli, and impaired reactions to peripherally presented targets more significantly than centrallly presented ones. Similarly, vigilance, or the ability to sustain attention on a task for a given period of time, is impaired by fatigue (Dinges et al., 1997), which also increases distractibility. Fatigued participants performing the psychomotor vigilance task (PVT), a simple visual-manual reaction time task, with an attractive distraction task show a significant increase in both head turns towards the distracting stimulus and lapses on the PVT compared to when they are not tired (Anderson & Horne, 2006).

Fatigue's limiting effects on attention will impair a pilot's ability to attend to and detect stimuli in the environment. The TSB has investigated air transportation accidents where fatigue has had negative effects on pilot attention during the approach and landing phases of flight. For example, on 22 December 2012, a Fairchild SA227-AC Metro III charter flight departed Winnipeg, Manitoba at 1939 Coordinated Universal Time as a charter flight to Sanikiluaq, Nunavut (TSB investigation A12Q0216). Following several unsuccessful visual approaches, visual contact with the runway environment was made after passing the missed approach point. Following a steep descent, a rejected landing was initiated at 20 to 50 feet above the runway; however, it was too late - the aircraft struck the ground approximately 525 feet beyond the departure end of the runway. The 2 flight crew and 1 passenger sustained serious injuries, 5 passengers sustained minor injuries, but 1 lap-held infant was fatally injured. The investigation determined that frustration, fatigue resulting from a flight delay and shortened sleep the night before the flight, and an increase in workload and stress resulted in crew attentional narrowing and a shift away from well-learned, highly practised procedures, which contributed to the accident.

Information processing

Fatigue reduces the rate of information processing, which can affect the speed with which a person can identify important information, process and react to it. One often-cited research finding is that 17 hours of wakefulness produces slowing in psychomotor functioning equivalent to a blood alcohol concentration (BAC) of 0.05% (Dawson & Reid, 1997). Belenky et al. (2003) chronically sleep-restricted participants for one week and compared them to non-sleep-restricted controls. For those who were restricted to 3 or 5 hours of sleep per night, PVT responding slowed and lapses (non-reactions) increased steadily over the 7 days.

Slowed or inaccurate information processing would be expected to impair pilot performance during approach and landing, when a pilot must perceive and react to constantly changing information and stimuli. The TSB has investigated air occurrences where impairment in information processing from fatigue has played a causal role. For example, at 2145 eastern daylight time on 02 August 2001, a Cessna 182 aircraft departed on an instrument flight rules flight from Kuujjuaq to La Grande-Rivière, Quebec (TSB investigation A01O0210). Thirty nautical miles north of destination, the pilot established radio contact with the Flight Service Station (FSS), who informed the pilot that the ceiling was 300 feet overcast with a visibility of 8 statute miles in haze. The pilot was planning to conduct a global positioning system approach; however, minutes later at 0223 local time, the aircraft struck the ground. The pilot, who was the sole occupant, was fatally injured. Counting ground and air times, the pilot had likely been on duty 8 to 10 hours per day for the two days preceding the accident and, on the accident day, flew a distance of over 1900 nautical miles, for a flight and duty day of over 20 hours. The flight was conducted single-pilot, crossed six time zones over relatively featureless terrain, in instrument meteorological conditions (IMC), at night. Based on these factors, it was concluded that fatigue likely affected the pilot's performance and contributed to the accident.

Memory

Short-term memory allows the temporary storage of information. Working memory temporarily stores information while it is being manipulated for tasks such as reasoning. Both types of memory are impaired by fatigue. Babkoff et al. (1988) tested participants who were deprived of sleep for up to 72 hours. Participants responded to a visual memory and search task where randomly chosen target letters were selected from background non-target letters. As the target letters were not presented together with the letter matrix, participants had to remember the target letters during the task. Response speed and accuracy decreased over time depending on the level of sleep restriction.

A pilot must be able to recall a large body of knowledge from memory, and must be able to apply this information using working memory to a constantly changing operational environment, especially during approach and landing. The TSB has investigated accidents where fatigue's effects on memory have had causal or contributory effects. For example, in June 1994, a Swearingen Merlin II was returning to Thompson, Manitoba after having completed a MEDEVAC flight (TSB investigation A94C0088). After being cleared for approach, the aircraft descended below the minimum beacon-crossing altitude, struck the non-directional beacon tower in a wings-level attitude, and crashed. Both crewmembers were fatally injured, and the flight nurse was seriously injured. The TSB determined that the flight crew lost altitude awareness during the localizer back course approach and allowed the aircraft to descend below a mandatory level-off altitude. Contributing factors to the occurrence were the crew's deviation from a published approach procedure, ineffective in-flight monitoring of the approach, rapidly developing localized fog conditions, and pilot fatigue. The accident occurred just after midnight after both pilots had been awake for about 17 hours and on duty for about 9 1/2 hours. The investigation concluded that the decision to deviate from the published approach profile – in essence, taking a shortcut – was consistent with the effects of fatigue, as were the failure to reset navigational instruments and altimeter and accurately monitor the approach.

Problem solving / decision-making

As time awake increases, general cognitive performance deteriorates. Problem solving, vigilance and communication tasks show a 30% decrement after 18 hours of wakefulness and a 60% decrement after 48 hours (Angus, Pigeau, & Heslegrave, 1992). Duty requirements that extend a crewmember's time awake beyond 18 hours can be expected to result in marked impairment in information processing and problem solving abilities. Fatigue will also reduce a person's flexibility in their problem-solving approach to a situation that is perceived to be different from the routine, so that they perseverate and repeat previously ineffective responses (Horne, 1988). Perseveration increases the likelihood that the normal routine will be maintained, leading to a failure to revise the original plan and make it difficult for a fatigued person to devise and try a novel solution. This has obvious repercussions for pilots during approach and landing.

In general, performance on more complicated tasks, such as decision-making, is degraded to a higher degree than performance on easier tasks such as simple counting when people are fatigued (Lamond & Dawson, 1999). In a simulator study using military pilots (Previc et al., 2009), 34-hour sleep deprivation was associated with degraded flying precision, but no changes in instrument scanning, suggesting that the fatigue impaired information processing and decision-making rather than eye-scanning behaviour. In another military simulator experiment (Caldwell et al., 2004), pilots showed significant decrements in mood, cognition, central nervous system activation, and flight skills in the predawn hours during a night without sleep. Control errors sometimes doubled and pilots experienced central nervous system alterations, which degraded information-processing capacity and reaction time. Information processing was impaired, and a wide array of basic capabilities were degraded as a result of compromised vigilance, poor situation awareness, and slow reaction time. If an emotional component is involved, as is often the case in emergency situations, decision-making is significantly impaired after 23 hours of wakefulness, even if large amounts of stimulants like caffeine are consumed (Killgore, Grugle, & Balkin, 2012).

The TSB has investigated accidents where aircrew problem-solving and decision-making impairments caused by fatigue were a causal factor. For example, in January 2012, an Enerjet Boeing 737-700 was operating from Fort St. John to Fort Nelson, British Columbia (TSB investigation A12W0004). At 1117 Mountain Standard Time, during the landing rollout, ENJ401 overran the runway end by about 230 feet. There were no injuries to the 112 passengers or 6 crew members and no damage to the aircraft. The investigation determined that the captain did not attain appreciable sleep in the 24 hours preceding the flight and was fatigued, which led to the captain continuing the approach when the aircraft was not in a stabilized configuration, consistent with what is known with "fatigue-induced reduction in forward planning and a focus of attention towards salvaging the flight".

Conclusions

Psychological constructs and related skills that are sensitive to the effects of fatigue are often similar to those skills that are required in an emergency, or high workload, situation such as during the approach and landing phases of flight. Based on the research reviewed in this paper, the contention of some journalists (e.g., Nunes, 2018) that there are few, if any, research studies showing how fatigue affects a pilot's flying ability, is obviously unfounded.

Acknowledgements

The views expressed are those of the authors and not necessarily of the TSB. We would like to gratefully acknowledge the contributions of the many TSB investigators and other staff who worked on the investigations reported herein.

References

Anderson, C., & Horne, J.A. (2006). Sleepiness enhances distraction during a monotonous task. Sleep, 29(4), 573-576.

Angus, R., Pigeau, R., & Heslegrave, R. (1992). Sustained operation studies: From the field to the laboratory. In C. Stampi (Ed.) *Why We Nap: Evolution, Chronobiology, and Functions of Polyphasic and Ultrashort Sleep*. Boston: Birkhäuser.

Babkoff, H., Mikulincer, M., Caspy, T., Kempinski, D., & Sing, H. (1988). The topology of performance curves during 72 hours of sleep loss: A memory and search task. *Quarterly Journal of Experimental Psychology*, *40*, 737-756.

Belenky, G., Wesensten, N., Thorne, D., Thomas M., Sing, H., Redmond, D., Russo, M., & Balkin, T. (2003). Patterns of performance degradation and restoration during sleep restriction and subsequent recovery: A sleep dose-response study. *Journal of Sleep Research*, *12*, 1-12.

Caldwell, J., Caldwell, J., Brown, D., & Smith, J. (2004). The effects of 37 hours of continuous wakefulness on the physiological arousal, cognitive performance, self-reported mood, and simulator flight performance of F-117A pilots. *Military Psychology*, *16*, 163-181.

Dawson, D., Noy, Y.I., Härmä, M., Akerstedt, T., and Belenky, G. (2011). Modelling fatigue and the use of fatigue models in work settings. *Accident Analysis & Prevention*, *43*, 549-564.

Dawson, D., & Reid, K. (1997). Fatigue, alcohol and performance impairment. *Nature*, *388*, 235.

Dinges, D., Pack, F., Williams, K., Gillen K., Powell, J., Ott, G., Aptowicz, C., Pack A. (1997). Cumulative sleepiness, mood disturbance, and psychomotor vigilance performance decrements during a week of sleep restricted to 4-5 h per night. *Sleep*, *20*, 267-277.

Endsley, M. (1995). Toward a theory of situation awareness in dynamic systems, *Human Factors, 37*, 32–64.

Horne, J. (1988). Sleep deprivation and divergent thinking ability. *Sleep*, *11*, 528-536.

Killgore, W., Grugle, N., & Balkin, T. (2012). Gambling when sleep deprived: Don't bet on stimulants. *Chronobiology International*, *29*, 43-54.

Lamond, N. & Dawson, D. (1999). Quantifying the performance impairment associated with fatigue. *Journal of Sleep Research*, *8*, 255-262.

Nunes, A. (2018). Ottawa must diverge from its wrong path to air safety. *The Globe & Mail, Opinion*, October 7. Retrieved from: https://www.theglobeandmail.com/opinion/article-ottawa-must-diverge-from-its-wrong-path-to-air-safety/).

Previc, F., Lopez, N., Ercoline, W., Daluz, C., Workman, A., Evans R., & Dillon, N. (2009). The effects of sleep deprivation on flight performance, instrument scanning, and physiological arousal in pilots. *International Journal of Aviation Psychology*, *19*, 326-326.

Sanders, A. & Reitsma, W. (1982). Lack of sleep and covert orienting of attention. *Acta Psychologica*, *52*, 137-145.

TIRF (Traffic Injury Research Foundation). (2016). Fatigue-related fatal collisions in Canada, 2000-2013, October 19. Retrieved from: www.tirf.ca.

TESTING THE APPLICABILITY OF A CHECKLIST-BASED STARTLE MANAGEMENT METHOD IN THE SIMULATOR

Annemarie Landman, Delft University of Technology / TNO, The Netherlands
Sophie H. van Middelaar, Delft University of Technology, The Netherlands
Eric L. Groen, TNO, The Netherlands
M. M. (René) van Paassen, Delft University of Technology, The Netherlands
Adelbert W. Bronkhorst, TNO, The Netherlands
Max Mulder, Delft University of Technology, The Netherlands

Several checklist-based methods have been proposed to help pilots manage startle in unexpected situations. In the current experiment, we tested how pilots reacted to using such a method, which featured the mnemonic *COOL*: Calm down – Observe – Outline – Lead. Using a motion-based simulator outfitted with a non-linear aerodynamic model of a small twin-propeller aircraft, twelve pilots practiced using the *COOL* method before performing four test scenarios involving startling events. Application of the full method in the test scenarios was high (90-100%), and pilots rated the method on average as useful (4 on a 1-5 point Likert scale). The first two steps of the method were seen as the "core" of the method. However, pilots also displayed difficulty with prioritizing dealing with immediate threats over executing the method. The results are promising, but they also warn us to be cautious when introducing a startle management method.

Recently, there is an increase in focus on training pilots to manage the startle effect. The term "startle" is often used to designate a combination of a true startle response (a reflexive increase in stress) and a surprise (a mismatch of information with one's mental model; Rivera et al., 2014). A surprise requires one to adjust the mental model to the situation, which can be very difficult under high stress. The inability to solve this can result in confusion, loss of overview and panic (Landman et al., 2017). It has been proposed that the training of piloting skills in a more unpredictable and variable manner makes performance more robust in surprising situations in operational practice (Casner, Geven & Williams, 2013; Landman et al., 2018). A different, more generalized approach to the problem is to teach pilots a startle management method based on a checklist. Three examples of such methods are "Unload-Roll-Power" (of a "mental upset"; Field et al., 2018), "Reset-Observe-Confirm" (ROC; Boland, 2018), and "Breathe-Analyze-Decide" (BAD; Martin, 2016). These methods can supplement existing decision-making aides for pilots and are mainly aimed to guide pilots through the first moments of being startled and surprised.

The current study was performed to obtain data regarding pilot evaluation of the usefulness of such a startle management method, and their ability to prioritize immediate threats over applying the method. To our knowledge, no such data has been published yet, although data suggest that pilots generally liked the ROC and URP methods (Boland, 2018; Field et al., 2018). If pilots find a startle management method useful and easy to apply in startling situations in the simulator, this would be a first step towards its validation and its implementation in training practice.

Method

Participants

Twelve Dutch, currently employed commercial airline pilots participated in the experiment. The pilots came from five different companies. Due to their initial pilot training, all pilots had some flying experience (i.e., circa 25 hours) in a small, multi-engine propeller (MEP) aircraft, similar to the one that was featured in the experiment. One pilot had more experience (i.e., 100 hours).

Table 1.
Characteristics of the participants.

	Mean (SD)
Age (years)	37.4 (12.7)
Experience large aircraft (hrs)	7172 (5549)
Experience small SEP/MEP (hrs)	265 (107)
Employed (years)	13.5 (10.8)

Table 2.
Characteristics of the participants (cont.).

	N
Aerobatics experience	2
Glider rating	4
Instructor	4
Rank: Captain	4
Rank: First officer	6
Rank: Second officer	2
Gender: male	12

Apparatus

The practice and testing took place in the SIMONA research simulator at the Delft University of Technology. This is a six-degrees-of-freedom full-motion simulator with a hydraulic hexapod motion system. The simulator has a collimated 180 degrees horizontal by 40 degrees vertical field of view for outside vision rendered with FlightGear. Sound effects were played over a 5.1 surround sound system.

The non-linear aerodynamic model of a Piper PA-34 Seneca III, a light twin-engine propeller aircraft was used. The flight deck was modeled after a generic multi-crew cockpit. The flight controls and instruments include a control column and pedals with force feedback, pitch trim on the column, throttle, gear, and flap lever with three flap settings: 0°, 25° and 40°. The (digital) instruments included a Primary Flight Display (PFD), a gear- and flap indicator, Exhaust Gas Temperature (EGT) display, RPM and torque indicators, fuel quantity and oil temperature/pressure displays.

Training intervention

The startle management method tested in the experiment was based on existing methods and it was taught using the mnemonic *COOL*:

- C - Calm down. Take a deep breath, sit upright, relax arms and shoulders and become aware of applied control forces.
- O - Observe. Take a step back and observe the situation. Call out the basic instrument readings: pitch, speed, bank angle, altitude and vertical speed. Call out what the aircraft seems to be doing (e.g., "continuously yawing to the right") as well as other unusual perceptions such as noise. Check secondary instruments and configuration if relevant.
- O - Outline. Following the observations, zoom in on the problem and formulate a hypothesis on the cause.
- L - Lead. Formulate a plan for immediate and/or future actions.

It was emphasized that immediate actions needed to fly the airplane took precedence, and that the method did not need to take up much time. Pilots were told that the purpose of the experiment was to test the usefulness of the method, and they were encouraged to apply it in the scenarios.

Tasks

The tasks were performed as single-pilot crew. For familiarization, pilots first flew four left-handed traffic patterns from takeoff to landing (see, Figure 1). Required settings (as displayed in Figure 1) were available on a checklist in the cockpit, and the stall alarm was demonstrated in the last pattern. Pilots then came out of the simulator to receive information on startle and surprise and instructions on the *COOL* method. They went back into the simulator and performed four practice scenarios. First, they flew a standard pattern in which they were asked to execute the *COOL* method approximately six times. They then performed an approach and landing with strong crosswind and a malfunctioning rudder. The third scenario consisted of a standard pattern with an RPM indicator failure on the left engine. In the fourth scenario, an engine failure occurred shortly after rotation.

Next, pilots were informed that they would perform four test scenarios. The scenarios were designed to offer a variety of instrument-related and controllability-related issues, most of which were familiar to the pilots. Each scenario consisted of flying a pattern (Figure 1) during which one of the following issues occurred. FLAP: when selecting flaps 25, the left flap malfunctioned and remained up, which caused a roll and a yaw moment. MASS: a heavy piece of cargo broke loose after rotate during takeoff, and shifted with a scraping noise towards the tail, causing a violent pitch-up moment. STALL: before leveling off, a bird struck the angle of attack vane, creating an impact sound and causing a continuous (false) stickshaker and (false) stall audio alarm. To provide enough space for a recovery, pilots were tasked to climb to 2000 ft in this scenario, after which they descended back to 1000 ft at downwind. UAS: an instrument malfunction caused the indicated airspeed to diverge from the actual airspeed by -1 kt every second, starting at rotate.

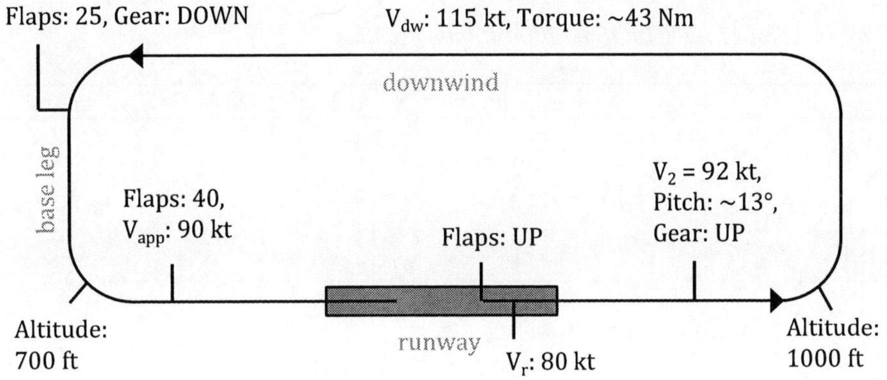

Figure 1.
The standard traffic pattern flown in the experiment.

Dependent measures

Following each test scenario, pilots filled in a questionnaire. They reported if they had applied the *COOL* method, and if so, which steps. An audio recording was used to confirm whether pilots called out the instrument readings (*Observe*). If applied, pilots rated the perceived usefulness of the method in the scenario on a 1-5 scale labeled: very little – little – moderate – much – very much. Pilots rated their perceived startle and surprise on a 0-10 point scale with the labels 'not at all' and 'extremely' at the endpoints. Pilots rated perceived anxiety on a similar visual-analogue scale (Houtman & Bakker, 1989), and mental effort on the Rating Scale for Mental Effort (RSME; Zijlstra & van Doorn, 1985). An open interview was performed at the end of the experiment, to collect pilot clarifications of ratings, their impressions of the method and suggestions for improvement (if any).

The audio recordings were also used to investigate if pilots inappropriately executed the method while there were immediate issues to attent to. For this, it was checked if pilots started *Obseve* before recovering the upset in MASS (bringing pitch angle back below 20 degrees).

Data analysis

The median startle and surprise scores are reported for each scenario as a manipulation check.
To evaluate usefulness of the *COOL* method, the number of pilots applying the steps in each scenario, as well as an overview of all usefulness scores, are reported for each scenario. Low ratings will be analyzed and discussed independently.

Results

Application of the *COOL* method

The self-reported application of the method is shown in Table 4. Even though application was encouraged, pilots sometimes did not apply the whole method. Averaged over the four scenarios, pilots applied the whole method in 89.6% of the cases. Application of *Observe* was reported by all pilots in all scenarios. This was confirmed in all audio recordings except one in UAS, however, two recordings were lost. As reasons for not applying the whole method, pilots named time-criticallity, distraction or not finding the method applicable.

In STALL, of the eight pilots who unloaded and of whom audio was available, one performed *Observe* before unloading. In MASS, of the eight pilots who exerienced a pitch angle exceeding 20 degrees, five executed *Observe* before recovering.

Table 4.
Self-reported application of the COOL method items.

	FLAP	STALL	MASS	UAS
Calm down (n)	11	12	10	12
Observe (n)	12	12	12	12
Outline (n)	12	12	10	12
Lead (n)	12	11	11	12
Full method (n)	11	11	9	12

Example of the *COOL* method application

Table 3 shows the audio transcript of a pilot executing the *COOL* method in STALL. As can be seen, the four steps seem to follow each other naturally. *Observe* started with the bigger picture (speed, attitude), and then zoomed in on the problem (vibrations, engine parameters, stick shaker, pitch-power). The pilot rated the method as useful in this scenario (4 out of 5), and reported moderate startle and surprise (respectively 6 and 7 out of 10).

Table 3.
An audio script showing an example of the COOL method being applied in STALL. Author comments are in [brackets].

Category	Pilot comments
Calm down	*Wow! I feel something. COOL!* [Pilot breathes]
Observe	*Observe! Speed 97. Pitch 12. 2000* [ft] *coming up. Heading 185. Gear up, flaps up. I feel vibrations? Engine torque normal. Quantities, temperatures... I have a shaker. Speed, pitch, power is okay.*
Outline	*I'm not stalling. No clue what it is. My first impression: a false stall warning.*
Lead	*Okay, heading 008. I'm on downwind. Descending to 1000* [ft].

Perceived usefulness of the *COOL* method

Pilots rated the method generally as useful in the test scenarios (see Figure 2), with medians of 4 in STALL, FLAP and UAS, and one median of 3 in MASS. All low individual scores (i.e., below 3) except for one score in MASS, were due to pilots finding the scenarios not difficult or startling enough. The low score in MASS was due to the scenario being *"too time-critical for the method to be applicable"* according to the pilot. Other critique or suggestions for improvement were: *"The method can be extended."* (someone suggested adding 'Options'), *"It slows you down / interferes with thinking."*, *"It is a bit too long."*, *"It may be too distracting in a more complex cockpit."*, *"Observe and Outline can be combined"*.

Pilots who gave high ratings remarked the following: *"The workload was okay."*, *" It is especially applicable when highly startled."*, *"It has a natural flow."*, *"Calm down and Observe are important and are the core of the method."*, *"It forces you to look around."* and *"It prevents tunnelvision"*.

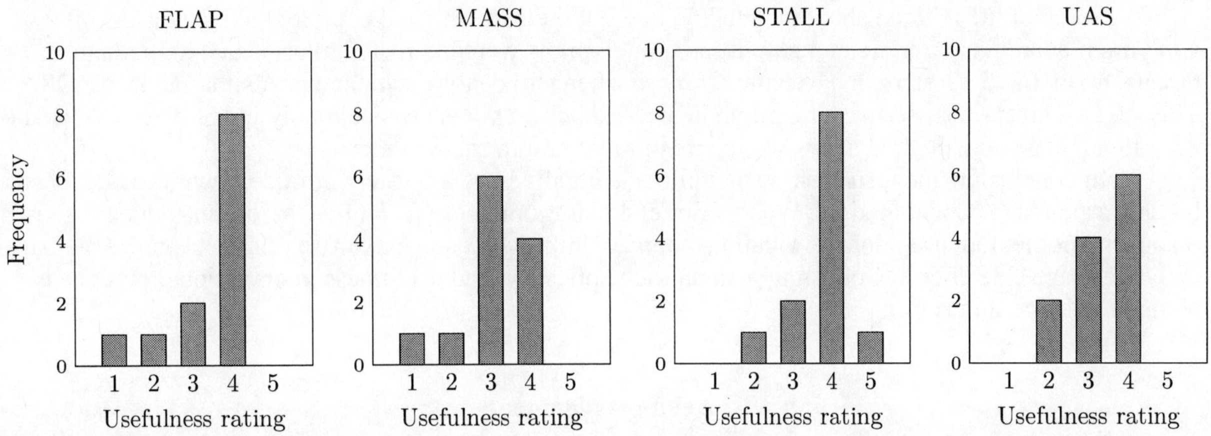

Figure 2.
Perceived usefulness of the COOL method in the four test scenarios.

Manipulation check

The pilots found the scenarios generally challenging, considering the scores in Table 4. Startle and anxiety were rated above the midpoint of the scale in all scenarios except UAS. All scenarios were rated very surprising (7-8), even though pilots knew that malfunctions would occur. Mental effort (RSME) was scored around 60, which is between "rather much effort" and "considerable effort" on the scale. MASS seemed to be the most challenging, possibly due to it being an unfamiliar problem, which caused considerable controllability issues.

Table 4.
Median ratings of startle, surprise, mental effort and anxiety in each scenario.

	FLAP	MASS	STALL	UAS
Startle (0-10)	6	6-7	6-7	4-5
Surprise (0-10)	7	8	7-8	7-8
RSME (0-150)	62	77	56	60
Anxiety (0-10)	5.25	6.40	5.15	4.30

Discussion

The results of the current experiment are promising for the applicability of checklist-based startle management methods. Following a practice session with the *COOL* method, and after encouragement to try the method, all pilots applied it in the test scenarios. Still, the steps: *Calm down*, *Outline* or *Lead* were skipped in some (circa 5 %) of the cases. This suggests that accurately applying the method in real situations might be difficult, because the stress level would likely be higher and there would likely be several months between practicing the method and applying it.

The method was rated highly useful in the test scenarios (4 out of 5). Pilots also had several suggestions for improvements. Summarized, the method could benefit from being simplified, especially as it is to be applied in more complex situations in operational practice. The last two steps (*Outline* and *Lead*) seemed not very necessary and could be left out. *Observe* could be simplified by reducing the number of parameters to check, and by focusing on their meaning (e.g. "airspeed is okay") instead of on their absolute values (e.g. "airspeed is 100 knots"). Pilots suggested that in a two-pilot crew, the first step (*Calm down*) could be applied by both, and the pilot monitoring could then perform *Observe*.

Some of the critique about the method being too distracting and complex may be due to pilots applying it at an inappropriate moment. Even though pilots were instructed to prioritize immediate threats, many (62.5 %) started to execute *Observe* when still dealing with an upset situation in MASS. This suggests that extensive training might be needed before pilots can accurately judge when to execute a startle management method and when to attend to more important matters.

In conclusion, the results show that pilots generally have a positive attitude towards a checklist-based startle management method. We recommend that training organizations, before introducing a startle management method, use pilots' evaluations to make improvements, maximize pilot acceptance, and also avoid negative side effects. Collecting data on the application and usefulness in operational practice is recommended as a next step.

Acknowledgements

The paper is based on a larger study on the effects of a checklist-based startle management method on performance measures, and it is planned to be part of a doctoral dissertation entitled: Managing Startle and Surprise on the Flight Deck.

References

Boland, E., Managing Startle & Surprise, Presented at PACDEFF 2016, 2016. Retrieved from http://pacdeff.com/ wp- content/uploads/2016/11/PACDEFF- Startle- Surprise- Management.pdf.

Casner, S. M., Geven, R. W., & Williams, K. T. (2013). The effectiveness of airline pilot training for abnormal events. *Human factors*, *55*(3), 477–485.

Field, J. N., Boland, E. J., Van Rooij, J. M., Mohrmann, J. F. W., Smeltink, J. W. (2018). Startle Effect Management. (report nr. NLR-CR-2018-242), European Aviation Safety Agency.

Houtman, I. L. D., & Bakker, F. C. (1989). The anxiety thermometer: a validation study. *Journal of personality assessment*, *53*(3), 575–582.

Landman, A., Groen, E. L., Van Paassen, M. M., Bronkhorst, A. W., & Mulder, M. (2017). Dealing with unexpected events on the flight deck: a conceptual model of startle and surprise. *Human factors*, *59*(8), 1161–1172.

Landman, A., van Oorschot, P., van Paassen, M. M., Groen, E. L., Bronkhorst, A. W., & Mulder, M. (2018). Training Pilots for Unexpected Events: A Simulator Study on the Advantage of Unpredictable and Variable Scenarios. *Human factors, 60*(6), 793–805.

Martin, W.L., (2016). Developing Startle and Surprise Training Interventions for Airline Training Programs, Presented at PACDEFF 2016. Retrieved from http://pacdeff.com/wp-content/uploads/2017/08/ PACDEFF- FC- Forum- Presentation- on- Startle.pdf

Rivera, J., Talone, A. B., Boesser, C. T., Jentsch, F., & Yeh, M. (2014, September). Startle and surprise on the flight deck: Similarities, differences, and prevalence. In *Proceedings of the Human Factors and Ergonomics Society Annual Meeting*, *58*(1), 1047–1051. Sage CA: Los Angeles, CA: SAGE Publications.

SCD: A 3 STATES STARTLE COPYING DISPLAY TO MANAGE DELETERIOUS EFFECTS OF EXTREME EMERGENCY SITUATION.

BEY Christophe
Akiani SAS
Bordeaux, France
HOURLIER Sylvain
Thales/ENSC, Human Engineering for Aerospace Laboratory (HEAL)
Bordeaux, France.
ANDRE Jean-Marc
University of Bordeaux, ENSC-BdxINP, IMS UMR CNRS 5218
Bordeaux, France

The management of cognitive resources are central in the case of a decision-making process by pilots. We undertake a study involving Airbus 400M pilots and allowing to understand these mechanisms and to propose recommendations for the design of a tool to assist in the management of their cognitive resources. We find that in the most critical cases and under strong temporal pressure, the maintenance of control of the situation corresponds to a survival type behavior which alone can allow a return to the metarules (back to basics).

Our display management proposal allows the pilot to maintain control of the situation regardless of his capabilities. It allows a phase of stabilization by a reduction of the stress, then a phase of "soft" recovery of the control and the management of the mission on larger spatiotemporal dimensions. Three modes of entry in the HMI are envisaged: spontaneous, proposed and on demand.

In a previous work (Bey 2016), we described our research project on the management of cognitive resources in the case of a constrained decision process. We have found that in critical situation and under strong temporal pressure, the pilot panel that we analyzed (14 crews of the French Air Force) is divided into two categories: those who implement adaptive strategies ensuring systematically the success of the mission; and those who do not implement this type of strategy and for whom the result is much more random. The adaptive strategies put in place are based on a change of objective. We noticed that successful ones focus on (i) maintaining the aircraft in flight without attempting to resolve the breakdown at first and, because they are aware of the seriousness of the situation, (ii) a survival strategy.

The consequence of this strategy is the refocusing on very short deadlines of around 20 seconds (limited to the return to manual flying, compensation and control of the aircraft). Then, once the control of the plane is assured, they gradually increase the temporal span of their control to longer sequences (of about one minute: joining the axis in safety, passage in manual ILS, parameters of the final). Finally, they allow themselves the control of a much larger time span (of several minutes) on which they will be able to devote themselves to the management of the breakdown and its consequences with a view to the landing (consultation of the ECAM with strong hierarchy in the choice of data retained). They then make their landing in nominal conditions. These conclusions from the experimental approach, show that in the most critical cases, the maintenance of control the situation corresponds to a survival-type behavior ("Survival Skills") which alone can allow a return to metarules ("back to basics"). "). The conditions of entry into survival behavior are: 1) A metacognitive management of the detection of exceptional situations without immediately applicable solution: "Understanding that we do not understand".

Metacognition then allows us to fall back on survival behaviors, in order to re-credit cognitive resources by disengaging them from too high levels, based on rules or knowledge. 2) An acceptance of the transgression as a solution to an exceptional situation. This is conditioned by the pre-exposure (training) to the "backup" transgression. 3) A call for aeronautical "survival skills" grounded FNCM type (Fly, Nav, Com, Manage). As part of this study, our goal is to propose a new management model based on two essential queries: 1) Ensuring immediate survival; 2) Ensuring survival, by limiting temporal swings. The means implemented can be based on the re-crediting of cognitive resources by: 1) lowering the level of stress: 2) lowering the level of information processed (no parallelization, reducing the scope of future possibilities). On the basis of these results, our desire is now to move the HMIs towards help with the recovery of situational awareness and decision-making. These new HMIs should provide help on some or all of the following: 1) To credit time to the pilot (delegation to himself in time, reordering of his priorities); 2) Increase resources by delegation (incentive to delegate to co-pilot or artificial intelligence); 3) Reduce the requirements of the situation (limit constraints, e.g., be content with rough speed management, but sufficient to ensure safety); 4) Change goals (and act on the requirements / resource relationships they involve).

The SCD model: A 3 state Startle Copying Display

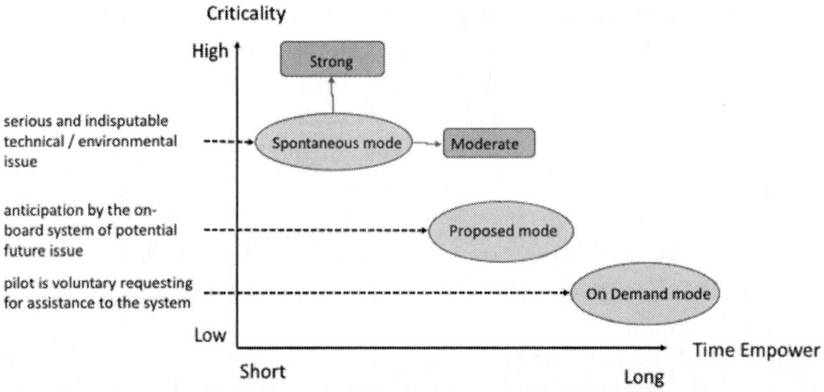

Figure 1. The 3 input modes of the HMI according to the criticality and urgency of the situation.

Our display management proposal must allow the pilot to maintain control of the situation regardless of his capabilities. It allows a phase of stabilization by a reduction of the stress, then a phase of "soft" recovery of the control and the control of the mission on larger spatiotemporal dimensions. To achieve this objective, it deals hierarchically with the following 3 points: 1) Accompany the reduction of temporal and spatial spans; 2) Prioritize actions and reinforce the return to basics and the application of "airmanship"; 3) Accompany the re-extension of temporal and spatial control spans (allow the progressive extension of the temporal span of anticipation -cf. Lini, 2015-). However, it must be considered that the widening of temporal empanels increases the field of possibilities and consequently the cognitive cost of anticipation. To manage the harmful effects of an extreme emergency, we propose a HMI system based on three modes of entry according to criticality and urgency: on demand, spontaneous and proposed (figure 1).

Presentation of the proposed new HMIs

These screens allow pilots to have aggregated and contextualized information regarding the objectives of their mission. They are the result of interviews conducted with participating crews about the level of integration and abstraction of the information they need to decide on the solution to be implemented in the face of the stakes and risks encountered.

The screen in Figure 2a shows the status of the main systems: ENGINES, FUEL, SYSTEM, COMMS, NAVIGATION, FPL, TIMING. For each of them a color label defines the state of the system (i.e. nominal, degraded or broken down).

The screen in Figure 2b shows the situation point in the form of a radar diagram. This visualization presents the environmental dimensions, systems and critical oil, related to the success of the mission. In the example proposed, a system state (typically a plane equipment failure) combined with a limitation of communications, loss of a backup, and this in the context of an unforeseen weather degradation, presents a critical risk on the success of the mission, especially during the landing phase.

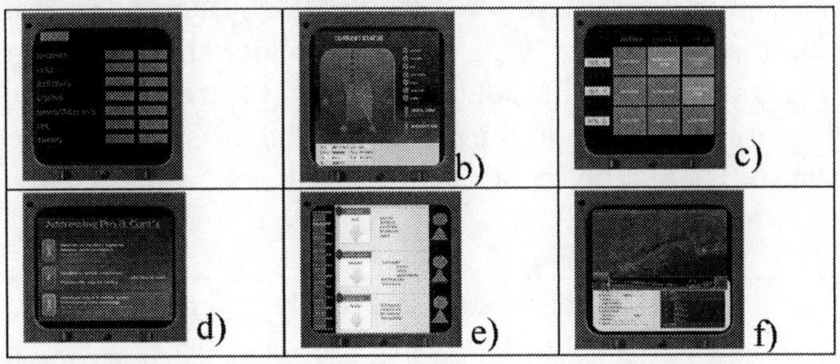

Figure 2. New proposed HMIs. 2a) presentation of the state of the systems, 2b) of the situation point in the form of a radar diagram. 2c) possible solutions with indication of their criticality according to 3 parameters (Human, Environment, System), 2d) advantages, disadvantages and limits of the selected solution, 2nd) Do list "with associated timeline. 2f) Briefing, presentation of the activation of the selected solution.

The screen of Figure 2c presents possible solutions with an indication of their criticality. 3 parameters are considered: Human, Environment, System. This representation makes it possible to assess the difficulty of setting up the chosen solution considering the airplane state, the weather environment in particular (TAF, METAR), but also the reception infrastructures (NOTAM) and the Crew state in terms of airmanship and situational awareness. The color codes used are: Green for Safety, Blue for Low Risk, Orange for Moderate Risk, and Red for Critical.

The screen in Figure 2d shows the advantages, disadvantages and limitations of the selected solution. For each solution, the list of advantages and disadvantages is addressed, as well as the limits of the scope of application. This visualization makes it possible to choose the most suitable solution for the dynamic environment and the airplane and crew state.

The screen in Figure 2e shows a "Do list" with an associated timeline. On the left vertical strip, the time scale is represented in a chronological way, if possible in a coherent and coherent way of the temporality of the realization of the different actions to carry out. On the vertical banner on the right the difficulty to realize the step of the do-list is materialized by the disks, while the criticality is represented by the triangles. Color codes are always applied according to the same display philosophy. Red for Critical, Orange for Severe, Yellow for Moderate, and Green for Easy.

Finally, the screen of Figure 2f presents the briefing. It is the activation of the selected solution. A cartographic background makes it possible to visualize the trajectory and the possible alternatives (clearance and diversions). A horizontal strip, located just below, represents the vertical section of this trajectory. On the left side, a time scale with associated schedules is shown. In the lower part of the screen, on the left side in the synthetic version, a briefing is presented to remind the action plan. In the right-hand part, the remarkable points of the flight plan and the criticality associated with these markers are displayed in tabular form.

The different modes available

On-demand mode

It corresponds to a request from the pilot who asks for assistance from the system. The pilot feeling overwhelmed (feeling "behind the plane") or absent can request the activation of the HMI "on demand" to facilitate his understanding of the current state and the resumption of his involvement in the control of the aircraft. In this case, we assume that there is no critical situation or committed alarm. The system's response will be proposed in the form of successive briefing steps (the pilot validating the progress from one to the other). The proposed IMH sequence (Figure 3) therefore dynamically illustrates the presentation of futures according to the context of the mission and potential vulnerabilities ("what-if"): situation report, matrix solutions, solution example, checklist, briefing, exit.

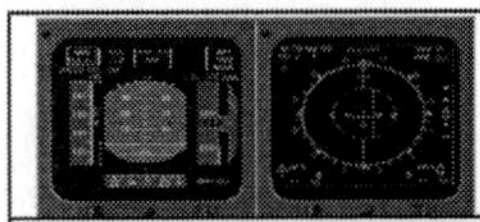	Step 1: Nominal status of displays (PFD and ND).
	Step 2: Entering priority control mode (simplified PFD and neutralized ND).
	Step 3: Entering System Status mode (simplified PFD and ND on the Aggregate System Status page).
	Step 4 : Entry into Mission Constraint Projection mode (simplified PFD and ND on Situation Point page).
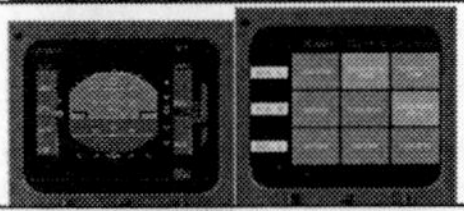	Step 5: Entering Solution Selection Mode (Simplified PFD and ND in Risk Matrix)
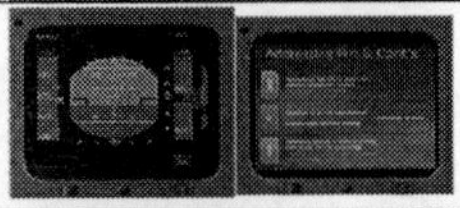	Step 6: Entry in advantages and disadvantages mode of the solution (simplified PFD and ND on advantages / disadvantages page)
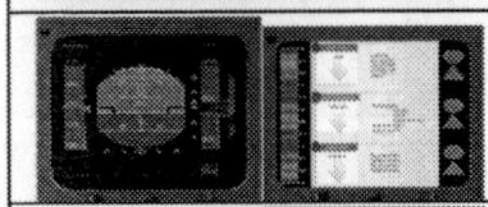	Step 7: Entry in page application mode of the selected solution (simplified PFD and ND on page Do-List)
	Step 8: Entering Briefing and Realization mode (simplified PFD and ND on Briefing page)
	Step 9: Return to standard mode, output after event resolution (PFD and ND classic)

Figure 3. Proposed HMI sequence for on-demand management.

The spontaneous mode.

It corresponds to an instantaneous display without intervention or acceptance by the crew. It occurs following a serious and indisputable technical / environmental problem (e.g. loss of both engines) or in the event of detection by the aircraft's monitoring system of a condition "incompatible" with the operation of the aircraft (considering the flight data, the system deduces that the pilots' response is inadequate without prejudice to the reasons for this behavior). In this case, the dispensed information is only the data essential to flight control (speed, altitude, variometer, artificial horizon, available energy - such as engine revolutions). In this input mode, corresponding to a critical case, these basic data will override all others. Two display options are possible: strong simplification and weak simplification.

Spontaneous with strong simplification. In this case, only the main screen is kept with the simplified display (Figure 4). The other screens are off (Second screen, parameter screen, FMS and ECAM interfaces). In this input mode, the risk of "startle effect" is high. One way to minimize this risk will be to gradually manage the system's alarm escalation, or even modify its procedures (for example, limit audible alarms).

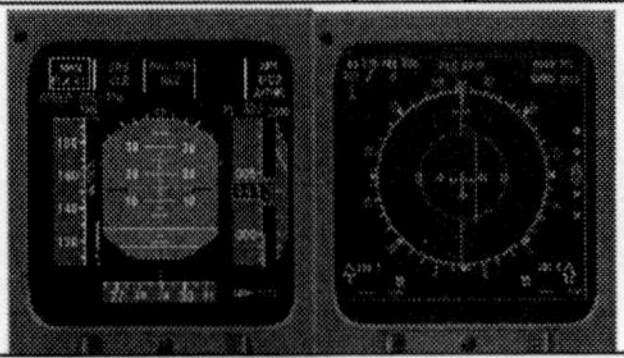	a): Nominal status of displays (PFD and ND).
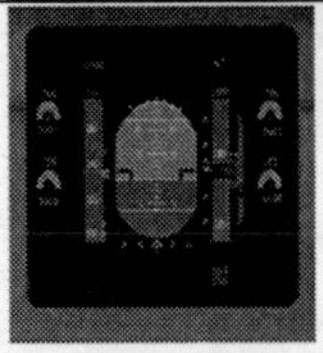	b): Main screen proposal in simplified display mode. The screen information is condensed from the original PFD and increased by the representation of the energy management (effective thrust) This allows a real-time management of the dynamic monitoring of the aircraft situation.

Figure 4. Representation of PFD and ND displays. a): Nominal status of displays (PFD and ND); b) Representation of the main screen proposal in simplified display mode

The sequence of the appearance of the different screens in spontaneous mode in Strong Simplification Option (Figure 5) would be as follows : 1) Simplified PFD appears on the classic PFD screen after the event has been detected (Figure 5, step 1); 2) The Simplified Navigation page appears automatically, once the system detects that the aircraft is under control or at the request of the pilot (Figure 5, step 2); 3) The System States appear on the second screen, once the trajectory has been secured or at the request of the pilot, in the form of a list associated with a performance index (Figure 5, step 3); the PFD is always in a simplified version. The pilot identifies the event and chooses to deal with the incident; 4) Switch to problem solving mode with the Do-List on the left page (Figure 5, step 4); 5) Return to nominal display once the incident is resolved or at the request of the pilot (Figure 5, step 5).

Spontaneous with own simplification option. In this case, the simplified screen (Figure 4b) is displayed on the main screen (PFD). The other screens are maintained as is.

The proposed mode

It corresponds to a projection of the system on a more distant term. It proposes to anticipate a potential future criticality and leaves the pilot the prerogative to choose or not an assistance within the framework of an alternative strategy. In this case, the "pilot in the loop" flight management approach is preserved. This anticipation may occur if the aircraft's monitoring system detects a lack of performance in the operation of the aircraft. Although not used in this context, flight data are already available, such as the monitoring of pilot actions on the interactors (detection of the rate of production of errors or non-actions symptomatic of a state of fatigue or advanced stress). They will be considered in the design of the new HMI and associated AI. In other cases, the aircraft's monitoring system analyses the flight data in real time and projects them into a future adapted to the mission context (oil, meteorology, infrastructure, flight plan, etc.). It proposes a refocusing on immediate objectives (maintaining flight control), but above all it complements it with adaptive strategic proposals (including what-if management). In this mode, the management of the pages viewed can be inspired, either by the spontaneous mode in high or low version or by the on-demand mode, depending on the criticality and origin of the incident envisaged by the system.

	Step 1: Entering "piloting" mode first when the critical incident occurs (simplified PFD and neutralized ND). Realization of FLY.
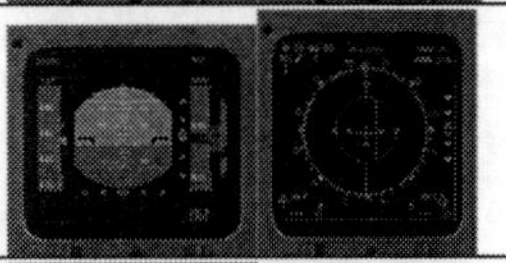	Step 2: Entry into the "navigation" mode, once the piloting has been carried out (simplified PFD and ND scale adapted to time). Realization of NAVIGATE
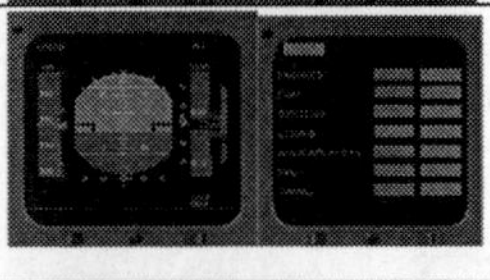	Step 3: Entering the "System Status" mode, once the trajectory has been secured (simplified PFD and ND on the Aggregate System Status page). MANAGE realization.
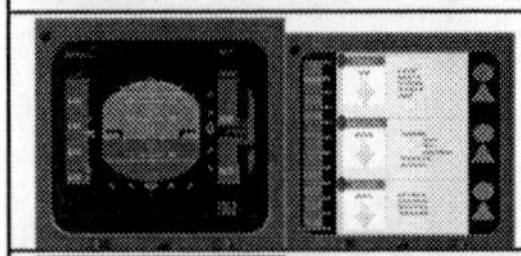	Step 4: Entering the application page mode of the solution (simplified PFD and ND on Do-List page). Realization of MANAGE & COM.
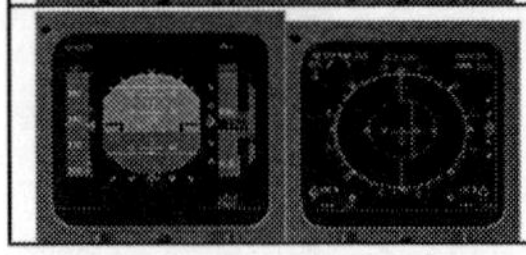	Step 5 : Return to standard mode, after the event has been resolved (PFD and ND classic).

Figure 5. Description of the sequence of the appearance of the different screens in spontaneous mode in option strong simplification

Conclusion

Beyond the modification of interactions with the cockpit (HMI) it is clear that pilot selection and training are key factors in performance and risk management in mission execution. The conclusions of our research show the relevance of airmanship training and the interest in also developing the pilots' ability to adapt their time projection to the criticality of the event (time span). The solutions proposed in this work are based on these findings. In crisis situations in constrained times, the 3 levels of assistance (Startle Copying Display) allow pilots to be guided back to the basics of aircraft control. 1) Accompany the reduction of temporal and spatial spans. 2) Prioritize the actions to be taken and reinforce the return to basics and the application of "airmanship". 3) Accompany the re-extension of temporal and spatial control spans (allow the progressive extension of the temporal span of anticipation).

References

BEY, C., (2016). Gestion des ressources cognitives et stratégies d'adaptation court terme chez les pilotes d'aéronefs, Doctoral dissertation, Bordeaux University, France.

LINI, S., FAVIER P.A., ANDRE J.M., HOURLIER S., BEY C., VALLESPIR B., BARACAT B. (2015). Influence de la profondeur temporelle d'anticipation sur la charge cognitive en environnement aéronautique. Le Travail Humain, 78, (3), p239–256.

PILOT EVALUATIONS OF A NON-VERBAL STARTLE AND SURPRISE MANAGEMENT METHOD, TESTED DURING AIRLINE RECURRENT SIMULATOR TRAINING

Annemarie Landman,
TNO, Delft University of Technology, The Netherlands
Eric L. Groen
TNO, The Netherlands
Marc Frank
Luxair, Luxembourg Airlines
Gunnar Steinhardt
Cargolux International Airlines, Luxembourg
M. M. (René) van Paassen
Delft University of Technology, The Netherlands
Adelbert W. Bronkhorst
TNO, The Netherlands
Max Mulder
Delft University of Technology, The Netherlands

Aviation safety organizations have recommended that airline pilots are trained for startle and surprise. However, little information is available on useful training interventions. Therefore, a training intervention trial was executed during airline recurrent simulator training. The method consisted of a slow visual scan from the side-window, over the instruments, ending with facing the other pilot. Following a recorded video instruction, 38 airline pilots in two-pilot crews performed a training scenario in which they could apply the method. Data on application and evaluation of the method were obtained from each pilot. Few pilots actually applied the method (18.4%), and many gave low ratings to applicability of the method in the scenario, as well as in operational practice. Results show that a startle management method, as well as manner in which it is trained, should be carefully evaluated before being implemented in training practice.

Aviation safety organizations have recently recommended that pilot training should include specific means to deal with startle and surprise. Although startle is commonly used to designate both startle and surprise, strictly seen, startle refers to a reflexive stress response, whereas a surprise occurs when information is encountered that does not fit within one's mental model of the situation (Rivera et al., 2014). If both are experienced simultaneously, there needs to be an adjustment of the mental model under high stress, which can be very difficult (Landman et al., 2017). This may result in panic, cognitive lockup and total confusion. Training interventions that have been proposed include teaching pilots certain actions to "break out" of this state. An example of this would be a checklist specifically focused on relaxation, problem analysis and decision-making. The application rate of such a method was high in an experimental setting and pilots generally appreciated it, however, some also indicated that it was too distracting or complex (Landman et al., 2019).

The current study tested a simpler startle management method, consisting of a slow scanning motion of the head from the side window, over the instrument panel, ending with facing one the other pilot. The reasoning behind this method were as follows. First, it may help one consider the overall situation, including the other pilot's state, instead of immediately zooming in on the problem. Looking out the side window, which is also used in initial training and aerobatics, can be used to obtain natural sense of the aircraft's attitude. Second, the method buys time and prevents intuitive reactions to a problem that is not fully understood yet. In a similar manner, standard procedures to recover from spatial disorientation include a first step of recognizing and confirming the spatial disorientation, before attempting to recover (e.g., Previc & Ercoline, 1999). Third, performing a slow, conscious motion may instill a sense of control and stimulate goal-directed processing, as high stress is known to shift attentional control towards being more stimulus-driven (Eysenck, Derakhshan, Santos & Calvo, 2007). Potential advantages that this method may have compared to a checklist, are that it is faster, simpler, more active and more specific (compared to e.g. the command to "Observe"). The current paper describes an early-stage trial of this method, to obtain data on its application and perceived usefulness in a representative sample of airline pilots in a standard training setting.

Method

Participants

Data were collected of 38 B737 pilots (18 captains and 18 first officers) and 18 Bombardier Q400 pilots (9 captains and 9 first officers). One dataset of a captain was excluded as this person was involved in the intervention method design. For privacy reasons, no other personal data was collected. The experience level of the B737 pilots was generally higher than that of the Q400 pilots, with circa 2,500-25,000 hours compared to 600-12,000 hours. Pilots were informed that their data would be processed anonymously. They were also free to refrain from filling in the questionnaire, but there were no refusals.

Training intervention

The experiment took place during a recurrent simulator training session at Luxair, Luxembourg Airlines. The training intervention consisted of an 8-minute instructional video, in which a type rating instructor gave information about startle and surprise, and outlined the intervention method: 1) Turn your head to the outside shoulder, look out of the side window. 2) Turn your head back in a continuous movement, check your flight instruments. 3) Continue turning and see your colleague's flight instruments. 4) Continue turning and have a look at your colleague. 5) Now turn back and evaluate the situation. The total duration of the method can be under 10 seconds. The video demonstrated execution of the method from a first-person view in the cockpit.

Scenarios

The B737 training session was a Line Oriented Flight Training (LOFT), which consisted of a complete flight from Tenerife (TFS) to Luxembourg (LUX). In cruise, the crew received warnings from ATC about an explosive device being on board. Sharing workload with the first officer, the commander would need to order a search and prepare the cabin for descent. During descend, the device would trigger, causing an elevator runaway. Since the explosion in the B737 scenario was most startling, this scenario was expected to be the most suitable for applying the startle management method.

The Q400 training session consisted of practicing several flight situations. The scenario that was used for the experiment involved a double engine malfunction, one after the other. The standard procedure in this case would not be adequate, as it would cause both engines to be shut down simultaneously. The inadequacy of standard procedures was expected to be surprising and stressful.

Dependent measures

During the debriefing of the training session, the pilots filled in questionnaires, which were collected in sealed envelopes. As a manipulation check of the scenarios, the following variables were rated on a 1-5 scale, ranging from *very little* (1) to *very much* (5): Surprise by the ATC warning (B737) or engine malfunctions (Q400) and Startle by the device explosion (B737) or engine malfunctions (Q400). Anxiety following the events was rated on a 10 cm horizontal visual-analogue scale ranging from *none at all* to *maximum* (Houtman & Bakker, 1983). Mental demand and perceived time pressure following the ATC message (B737) or engine malfunction (Q400) were rated on the NASA-TLX mental demand and temporal demand subscales (Hart & Staveland, 1988). Finally, pilots also indicated whether they were informed by colleagues about the events in the scenario.

Next, pilots were asked if they applied the training intervention during the scenario. If confirmed, they were asked at which moments they applied it, and to what extent they felt that it helped them, as rated from *very little* (1) to *very much* (5). On a similar scale, all pilots rated how useful the method would be in operational practice. If pilots did not apply it, they indicated if this was mainly because they forgot, because they didn't find it applicable to the situation, or because they used a different method to manage their startle.

Results

Manipulation check

The manipulation check shows that pilots found the scenarios moderately surprising and stressful, scoring on average around the midpoint on the scales (Table 1 and 2). It is interesting that startle and surprise scores spread from the lowest to the highest endpoints, indicating that pilots may experience the same scenario very differently. Anxiety levels are similar between the groups, while the Q400 group reported somewhat higher surprise and the B737 group more startle. In the B737 group, 45 % (17) of the pilots were informed about the scenario, whereas 54 % (20) were not, and one skipped the question. The Q400 pilots all reported not being informed.

Table 1.
Pilots' subjective experience of the B737 bomb threat scenario.

	Mean (*SD*)	Minimum	Maximum
Startle (explosion) (1-5)	3.05 (1.21)	1	5
Surprise (message) (1-5)	3.11 (1.13)	1	5
Mental demand (message) (5-100)	51.7 (16.1)	15	75
Time pressure (message) (5-100)	57.2 (20.0)	20	95
Anxiety (message) (0-10)	4.5 (2.3)	0.0	7.5
Anxiety (explosion) (0-10)	5.1 (2.3)	0.0	10.0

Table 2.
Pilots' subjective experience of the Q400 double engine malfunction scenario.

	Mean (*SD*)	**Minimum**	**Maximum**
Startle (1-5)	2.61 (.92)	1	4
Surprise (1-5)	3.33 (.91)	1	5
Mental demand (5-100)	58.6 (17.2)	35	100
Time pressure (5-100)	54.4 (17.2)	25	90
Anxiety (0-10)	5.1 (1.9)	1.7	8.0

Application of the startle management method

In the B737 group, 9 out of 38 pilots (24 %) applied the method in the scenario. Eight when the explosion occurred, and one as an extra scan to check for issues. Of those not applying the method, most indicated that they forgot (37 %), or found it not applicable (37 %). Others reported they used a different method to manage startle (26 %).

In the Q400 group, 4 out of 18 pilots (22 %) applied the method in the scenario. Most pilots did not find in applicable in the scenario (56 %), some forgot (16.7 %) and one used a different method (5.6 %). All in all, the application rate of the method was low and it was similar in the different scenarios.

Perceived usefulness of the startle management method

The perceived usefulness of the method in the scenarios is shown in Figure 1. As can be seen in the figure, there were many in the B737 group who rated the method of very little use, whereas those in the Q400 group rated it little to moderately useful.

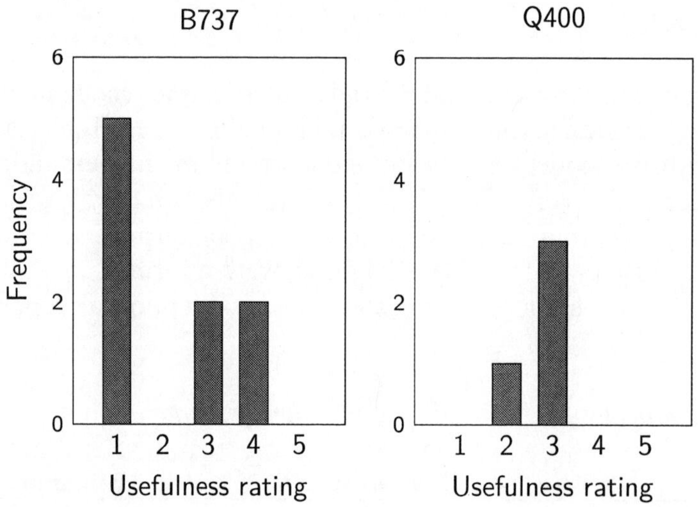

Figure 1.
Pilots' perceived usefulness of the startle management method in the scenario. Only those who applied the method are included.

The perceived usefulness of the method in operational practice is shown in Figure 2. It was similar to the ratings of usefulness in the scenario. Although the Q400 group seemed a little more positive towards the method, both groups included a relatively large proportion of pilots who rated the method of "very little" or "little" use in operational practice.

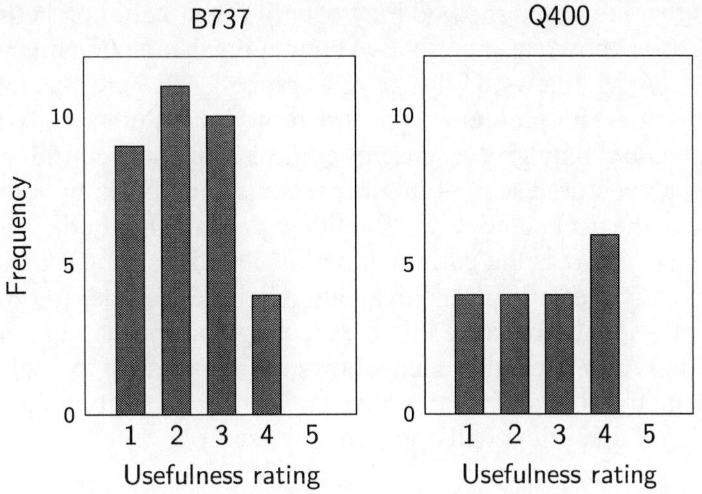

Figure 2.
Pilots' perceived usefulness of the startle management method in operational practice.

Discussion

This experiment shows that pilots found the tested startle management method generally of little to moderate use in the scenarios and in operational practice. It is most notable that the intervention had a very low application rate (22-24 %), and a relatively large proportion of pilots (22-24 %) rated the method to be of very little use in practice. This is in contrast to a checklist-based startle management method, which was applied by all, and rated, on average, to be of very little use by 8 % of the participating pilots (Landman et al., 2019). There are some methodological aspects of the current study that may have caused lower ratings and application compared to the other study. First, there was very little time available in the experimental training session to explain the reasoning behind the method. With more time, the reasoning behind the method can be explained and there would be more room for discussing the method beforehand, which may improve the pilots' openness towards it. Second, the tested startle management method was introduced during a mandatory training session. On the one hand, this mandatory setting could make the pilots more resistant to accepting the method. On the other hand, the current sample group is more representative of the general pilot population, compared to a group who participated in an experimental study based on invitation (Landman et al., 2019). Third, many pilots, especially those in the Q400 engine malfunction group indicated that the scenario was not startling enough for the method to be applicable. One remarked that it might be more useful "in cruise, when not mentally prepared for a malfunction, as we are in the simulator."

Besides the manner in which the instructions of the method were given, there are some aspects of the method itself that can be adapted to improve it. First, most pilots who applied the method in the experiment, applied it together with their fellow pilot, indicating that if one pilot takes the initiative to execute the method, the other pilot is likely to join. The application rate might thus be improved by adding a callout at the start (e.g., "Let's do a scan"). Second, pilots indicated that they particularly experienced looking out of the side window as unhelpful. Some remarked that they thought it would be disorienting in-flight; that it seemed senseless; and that it caused them to lose time. Some of these objections can potentially be tacked with an explanation of the purpose behind the "senseless" and counter-intuitive actions. However, these objections may also indicate that the tested method may benefit from including actions that are more task-focused. Task-focus is known as an effective coping mechanism against performance stress (see e.g., Baumeister 1984, Matthews, Hillyard, & Campbell, 1999). Consciously working on (part of) a solution to the stressful problem, even if that means systematically gathering information or simplifying the situation, may give a sense of control and instill confidence. Perhaps placing more emphasis on a structured scan of the instruments and checking verbally with one's fellow pilot would improve the acceptance and effectiveness of the method.

In conclusion, whereas the current experiment had a strong practical approach, this made it difficult to accurately measure pilots' evaluations of the method. In order to obtain a more accurate picture, pilots could be tasked with executing and evaluating a method in a more experimental setting. Also, the experiment shows the importance to reserve time and resources for the development, training and testing of a startle management method, so that the end product is an effective method that pilots will apply in practice.

References

Baumeister, R. F. (1984). Choking under pressure: self-consciousness and paradoxical effects of incentives on skillful performance. *Journal of personality and social psychology*, *46*(3), 610-620.

Eysenck, M. W., Derakhshan, N., Santos, R., & Calvo, M. G. (2007). Anxiety and cognitive performance: attentional control theory. *Emotion*, *7*(2), 336-353.

Hart, S. G. & Staveland, L. E. (1988). Development of NASA-TLX (Task Load Index): Results of empirical and theoretical research. *In P. A. Hancock and N. Meshkati (Eds.) Human Mental Workload. Amsterdam: North Holland Press.*

Landman, A., Groen, E. L., Van Paassen, M. M., Bronkhorst, A. W., & Mulder, M. (2017). Dealing with unexpected events on the flight deck: a conceptual model of startle and surprise. *Human factors*, *59*(8), 1161-1172.

Landman, A., Van Middellaar, S. H., Groen, E. L., Van Paassen, M. M., Bronkhorst, A. W., & Mulder, M. (2019). Testing the applicability of a checklist-based startle management method in the simulator. *In Proceedings of the 20th International Symposium on Aviation Psychology*, Dayton, OH.

Matthews, G., Hillyard, E. J., & Campbell, S. E. (1999). Metacognition and maladaptive coping as components of test anxiety. *Clinical Psychology & Psychotherapy: An International Journal of Theory & Practice*, *6*(2), 111-125.

Previc, F. H., and Ercoline, W. R. (Eds.), (2004). *Spatial Disorientation in flight.* Reston, VA: American Institute of Aeronautics and Astronautics.

Rivera, J., Talone, A. B., Boesser, C. T., Jentsch, F., & Yeh, M. (2014). Startle and surprise on the flight deck: Similarities, differences, and prevalence. In *Proceedings of the Human Factors and Ergonomics Society Annual Meeting* (Vol. 58, No. 1, pp. 1047-1051). Sage CA: Los Angeles, CA: SAGE Publications.

HUMAN FACTORS ELECTRONIC KNEEBOARD DESIGN GUIDELINES FOR MILITARY TACTICAL AVIATION

Robert Bridgman[1], Kelly J. Neville[1], Lauren Massey[1], Curtis Krauskopf[1], Ali Mizan[1], John Mooney[2], and Dylan Schmorrow[1]

[1] Soar Technology, Inc.
Orlando, Florida

[2] BGI, LLC
Dallas, TX

Great strides have been made in reducing the reams of paper-based materials that pilots were once required to bring into the cockpit. Much of that paper-based information is now available to pilots on electronic devices known as electronic kneeboards (EKBs). The main goal of this paper is to describe a design strategy we are using that integrates interdisciplinary perspectives and engages users in the design process. We describe the use of this design strategy to specify and design EKB applications (i.e., *apps*) that are uniquely supportive of the work demands faced by tactical pilots. As a result of the work described herein, we will be integrating multiple apps in support of high level goals, i.e., developing *super apps*. Future work will focus on developing these super apps so that they are responsive to situational changes. Future work additionally includes addressing key challenges associated with navigating within the EKB information space.

The advent of handheld computing devices dramatically changed the aircraft cockpit. In addition to changing the way pilots obtain information, it opened the door to a great deal more information and functionality. Originally used to give pilots easier access to the reams of documentation they brought into their cockpits, these electronic devices have continued to grow in both popularity and capability. They now feature a wide and still-growing variety of applications designed to give pilots easy access to information that previously required extensive search in their manuals or provide additional information that they may not have even had accessible.

Although the goal of using electronic knee boards (EKBs) sounds straight forward; give pilots easy access to useful information when they need it, achieving it is quite challenging. In particular, designing a device that supports dynamic and complex cognitive work, involves multiple parallel threads of activity, which is not directly observable. In addition, the design of the EKB's pilot interface matters greatly. Not only do EKBs bring easily accessible informtion into the cockpit but they also introduce the risk of increased workload, distraction, confusion, and head-down time. The outcome achieved depends heavily on the design used to implement the capabilities.

Designing to support complex, cognitive work relies heavily on deep user involvement and an iterative design process that involves frequent user testing in domain-relevant contexts. Notably, the design team was aware that there would be resistance to adopting these digital technologies as replacements for what many believe are "perfectly functional" physical tools. In order to address this potential barrier, we ensured that the design team had deep and relevant experience in the target-use domain of military tactical aviation. Additionally, evaluations with targeted end users were conducted in a way that focused on how cognitive work is performed versus on how a proposed EKB design would fit into their work. This helped to mitigate potential resistance to the proposed digital replacement technologies.

The goal of this paper is to describe a design strategy that engages users in the design process to both achieve a user-centered design and minimize the common and challenging obstacle of user resistance to change. This strategy includes the composition of our team, our design methods, and techniques we used to obtain user inputs and feedback.

Methods

Design Team and Philosophy

The design team consists of members representing multiple disciplines and perspectives, as advocated by Woods and his colleagues for design in complex technology-rich work domains (e.g., Roesler, Woods, & Feil, 2005; Woods, Tittle, Feil, & Roesler, 2004). Our team consists of three pilots (one licensed private pilot, one U.S. Navy Reserves F/A-18 pilot, and one former U.S. Navy MH-60 pilot), two software engineers, and two human factors psychology professionals. Two team members have experience in more than one key area: one pilot is also a software engineer and another has a graduate degree in human systems integration.

Design negotiations among the team members facilitate the interdisciplinary sharing of a wide range of design possibilities, constraints, and decision criteria, all of which contribute to effective design. According to Roesler et al. (2005):

> Balancing across the perspectives on design from the point of view of practitioners, innovators, and technologists presents a rich structure of relationships that can encourage innovation that results in more useful products (p. 211).

In our team, these design negotiations have tended to take the form of brainstorming sessions led by the two military aviators. As the two discuss their ideas about where and how in their work different EKB functions, or *apps*, could be useful, the other team members ask questions and take notes. Discussions about high-level work-support needs periodically segue into discussions about specific app designs, at which time the rest of the team participates more fully. The human factors and software engineering professionals at this point contribute ideas about how a given design might be implemented and scoped and raise questions about design interactions with work demands and general usability. Following each brainstorming session, individual team members develop design artifacts or prototypes based on the discussions, and these typically are presented and used as a point of discussion in the next brainstorming session.

The iterative, multi-disciplinary approach meant that designs are continually evaluated from three critical perspectives. These are the same perspectives advocated by Woods et al.: a technologist (our software engineer), a cognitive engineer (our human factors professionals), and a reflective practitioner as problem holder (our team's pilots). There may be an infinite number of ways to design an EKB or EKB app that supports fighter pilots but many will not integrate well into the work and work environment. The joint participation of these three types of perspective-holders allows us to develop and evolve designs that are more likely than others to succeed; more likely because the multidisciplinary team is able to pre-emptively identify and negotiate many of the competing priorities, constraints, and affordances that a successful design needs to address.

Subject Matter Expert (SME) Evaluation

Design evaluation sessions were conducted with five former military pilot volunteers who were paid a consultant's rate for their time. The sessions focused on the usefulness and usability of EKB designs in the context of single-seat tactical aircraft piloting. Two research team members, a Navy Reserves F/A-18 pilot and a human factors professional, met with each pilot for approximately one hour. Each pilot was asked to participate in two of three evaluation activities (time did not permit participation in all three). These activities were designed to obtain:

- Feedback on icons used to represent an initial set of identified apps,
- Insight into how and when pilots would use the apps (and therefore would expect the apps and their information and functionality to be readily available to them), and
- Feedback on the design of a *grading app* to be used by instructors in the grading of training flights.

The two activities in which each pilot participated were chosen on the basis of the pilot's expertise (e.g., one pilot was an instructor and so was shown the grading app) and on the activities conducted in preceding sessions to ensure that data from each activity were obtained from at least three pilots. Because of the qualitative, semi-structured nature of each evaluation activity, different amounts of time were spent by each pilot on each activity.

The remainder of this paper focuses on the evaluation and design of EKB app icons and functionality, i.e., the first two of the above list of bullets. (Information about the grading app design and evaluation can be obtained by contacting the authors.)

Icon feedback. In this evaluation activity, the pilot was told he would perform a series of app icon searches. For each trial, the pilot was shown an app icon. After studying the icon for roughly 3 to 5 seconds, a research team member removed the cover sheet over a matrix of icons positioned on the pilot's right thigh, at which time the pilot was to search the matrix and announce when he found the target icon. Icons in the matrix included distractor icons in addition to the full set of actual app icons. A different matrix of randomly positioned icons was used for each trial. An example of this matrix configuration is shown in Figure 1.

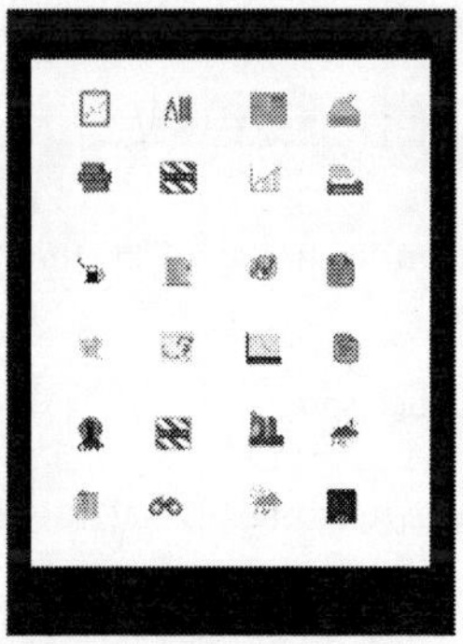

Figure 1. An example of the matrix of icons presented to SMEs.

We did not collect search performance data as our focus was on pilots' design feedback and recommendations. We asked each pilot about his subjective experience of searching for the icon. Specifically, we asked the pilot to rate on a 5-point scale if the icon was easy or hard to find relative to other icons and to think about and discuss characteristics of the icon contributing to the assessed difficulty. Pilot responses were recorded by one of the research team members using pen and paper.

Insight into when and how pilots would use the apps. For this evaluation activity, pilots were given a description of the purpose of each proposed app while being shown a laminated cut-out version of the app's icon. Each app's name was included on the icon cut-out. Pilots were then given the full set of 15 icons and asked to organize them into groups of apps they considered similar (we did not define 'similar' and instead left the choice of what counts as similar to the individual pilot). Each pilot was given multiple copies of each icon cut-out and a stack of blank paper rectangles and a pen. They were told they could use the copies to place icons in more than one grouping and that they could use the pen and paper to add additional apps to the set. Each pilot explained his groupings to the research team members while creating them and elaborated further after completing them. One research team member took hand written notes as they spoke.

Results

Application Groupings

Each SME was given an opportunity to group app icons into self-defined categories (see, e.g., Figure 2). These groupings provided insight into how and when the pilots anticipate using each app and suggested additional apps that the development team hadn't considered.

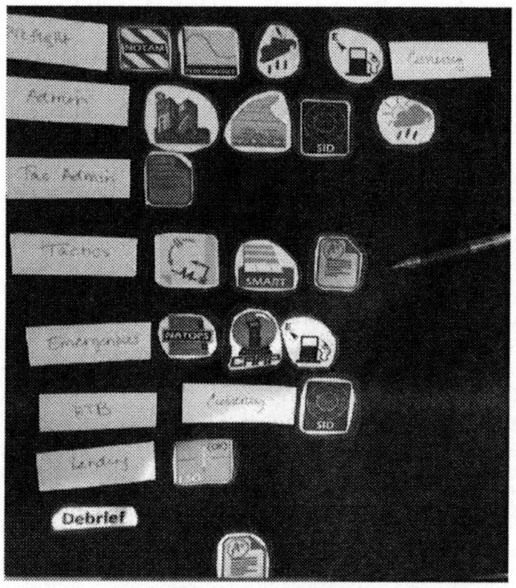

Figure 2. An example icon grouping.

Four of the SMEs created groups of apps that centered around three main types of pilot activity: the tactical mission, navigation and administrative work, and non-normal or emergency conditions. Table 1 displays a specific grouping created by the fifth SME, who introduced the idea of first and second tier applications within a major activity category. This adds another layer of organization on top of the situation-specific groupings in which apps are organized soley by pilot goal or activity set.

Table 1.
Example Application Grouping

Preflight	Inflight: First Tier	Inflight: Second Tier
Standard Instrument Departure (SID) app	Changes in relevant data, including: -Notifications about degraded weather conditions -New NOTAMS related to pilot's flight path	Performance charts/calculator
Weather app	Bingo support (e.g., nearby runways and fuel required to reach them)	NATOPS
Notice to Airmen (NOTAM) app	Pilot's scratchpad	Navigation charts
Aircraft Discrepancy Book (ADB)	Yellow Brick Road, i.e., course rules, app	Smart Pack
Geo-Scratchpad	Annotatable briefing and air combat maneuver (ACM) rules (i.e., frequently used reference documents)	Grading apps
	Standard arrival (STAR) charts	Flight log

Thus, a primary finding of this research is the recommendation of *super apps* that integrate numerous functions in service of a particular aircrew goal. A given function may be integrated into more than one super app; however, active function features presented to the user will be specific to the overarching goal and thus will vary by super app. Super-app concepts that emerged from the data include the following:

- *Emergency super app*. This app would provide aircrew with an integrated flow of resources for responding to emergencies. It would present relevant checklists, weather information, airfield and fuel information, aircrew performance charts, navigation charts, and other navigation tools and would do so in a way that minimizes extra work by the aircrew.
- *Mission super app*. This app would support inflight tactical planning and performance and may also support the debriefing of the tactical portion of a mission. It would feature an integrated suite of apps that aircrew would use to sketch out and view the tactical game plan; access weapons delivery profiles; view tanker locations and engagement zones; and record shot data, merge data, and more.
- *Inflight Guide super app*. The Smart Pack app would be an app that assists aircrew with navigation and administrative aspects of a mission. It would draw from flight guidance sources to present or highlight relevant route details, weather, NOTAMS, communications frequencies, patterns, routine checklists, and more along an aircrew's route.

A second major outcome of the research is the recommendation that the aircrew kneeboard card, i.e., mission line-up card, be continuously available at a central point of reference. SMEs recommended its use as a kneeboard 'homepage' in conjunction with an electronic scratchpad that the aircrew member could switch to via a sideways swipe. The kneeboard card lists the key basic elements of the mission at hand to help aircrew keep track of, for example, who is doing what when and what communications on which frequencies will mark their progress.

Conclusions

In this research, multiple SMEs interacted with our initial design elements to help us determine how to improve their fit to the demands of their work environment. SMEs responded positively and encouraged the integration of the various functions and features into super apps. Accordingly, in the next phase of this project, we will focus on integrating multiple functions to produce super apps, starting with the Emergency super app.

'Super apps' consist of more features, information, and functionality than individual apps and each new element represents many new possibilities for interaction with existing elements and aircrew. Further, although a super app's functions are all used in support of the same high-level objective, the conditions surrounding their use may change dramatically over the course of a mission. Consequently, we expect adapting super app designs to the complexities and variability of the flight environment to be an extensive process. To support this process, we are seeking flexible design, engineering, and evaluation methods that facilitate the development of complex integrated, interactive, and context-responsive capabilities.

Our near-term work also must address design challenges associated with supporting aircrew in interacting with and viewing the contents of the real estate-limited kneeboard apps.

Major challenges to be addressed have surfaced in this study and been reiterated in literature the team has reviewed. In particular, panning and zooming have, to date, not served as effective means for helping pilots view details on charts, checklists, and other information sources that are much larger than the mini iPad display (e.g., Chandra & Kendra, 2009; Sweet et al., 2017). We need to either develop alternative mechanisms for improving content visibility or add mechanisms that compensate for problems encountered when panning and zooming on navigation charts especially, but also on other resource material. Currently, we are evaluating one such alternative—the use of touch-based magnification bubble overlays—and need to continue working to identify other possibilities.

Similarly, navigation among pages and functions is far from straightforward given the quantity of content, the reasons any given content element might be relevant to the pilot, and the multiple ways different elements relate to one another. Research suggests current navigation schemes may pose safety hazards (e.g., Evans et a., 2013). Thus, it is imperative that we design improved techniques and tools for helping pilots keep track of what it is they are looking at, what they have recently seen and how to get back to it, and what's available that they might want to see or use next and how to get to it.

Acknowledgements

This research was sponsored by the Naval Aviation Training Systems Program Office (PMA-205). The views expressed in this paper are those of the authors and do not represent the official views of the organizations with which they are affiliated.

References

Chandra, D. C., & Kendra, A. (2009). Review of safety reports involving electronic flight bags. *Proceedings of the 15th International Symposium on Aviation Psychology*. Dayton, OH: Wright State University.

Evans, E. T., Young, S. D., Daniels, T. S., & Myer, R. R. (2013). Usability of EFBs for viewing NOTAMs and AIS/MET data link messages. In *Digital Avionics Systems Conference (DASC), 2013 IEEE/AIAA 32nd* (pp. 2A2-1). IEEE.

Roesler, A., Woods, D. D. & Feil, M. (2005). Inventing the Future of Cognitive Work. In W. Jonas, R. Chow, N. Verhaag (eds.), *Design-System-Evolution: Application of systemic and evolutionary approaches to design theory, design practice, design research and design education*. Bremen, Germany: European Academy of Design.

Sweet, J., Vu, K. P. L., Battiste, V., & Strybel, T. Z. (2017). The Comparative Benefits and Hazards of EFBs and Paper Documents in the Cockpit. In *Proceedings of the Human Factors and Ergonomics Society Annual Meeting* (Vol. 61, No. 1, pp. 99-100). Sage CA: Los Angeles, CA: Sage.

Woods, D., Tittle, J., Feil, M., & Roesler, A. (2004). Envisioning human-robot coordination in future operations, *IEEE Transactions on Systems, Man, and Cybernetics: Part C, 34*(2), 210-218.

Effects of Verbal vs Graphical Weather Information on a Pilot's Decision Making during Preflight

Warren Pittorie
Florida Institute of Technology
Melbourne, Florida
Meredith Carroll
Florida Institute of Technology
Melbourne, Florida
Deborah Carstens
Florida Institute of Technology
Melbourne, Florida

Advancements in technology has made obtaining a graphical/textual preflight weather briefing easier than a traditional verbal briefing. This study compared weather briefings delivered in a verbal format (i.e., written narrative) to those delivered in a graphical format (i,e., radar map plus textual data) in a within-subjects study that altered the order in which participants received each format. Thirty-six pilot participants read and responded to weather briefings for two flight scenarios that when put together, created a simulated multi-leg flight. Each pilot's decision making and confidence in their decision was captured via Likert-scale and open-ended questions following each scenario. Decision making response was measured based on whether participants made a "go" or "no-go" decision, along with ratings of decision confidence. This paper will present the study methods and results, as well as a discussion on weather briefing design and delivery.

According to Fultz & Ashley (2016), adverse metrological conditions have contributed to 35% of general aviation (GA) accidents with 60% of these accidents occurring in instrument metrological conditions (IMC). Analyzing current and forecasted weather conditions along the intended flight path, prior to takeoff, is crucial to the safety of any flight. Due to advancing technology in the GA industry, pilots are now able to access weather information in various formats that can aide or hinder their decision-making abilities. The goal of the weather briefing prior to takeoff is to increase a pilot's situational awareness of all of the weather-related hazards along the intended flight route. Situational awareness is defined as the perception of critical elements in the environment, comprehension or their meaning, and the projection of their status into the future (Wickens, Hollands, Parasuraman, & Banbury, 2012). Situational awareness is critical within the aviation environment because many responsibilities that fall on a pilot are time-critical and occur within a dynamic environment. A study by Topçu (2017) also supported that individuals must monitor a dynamic environment as this could alter their goals and desired outcomes, therefore affecting decision-making. Despite the number of weather-related GA accidents that have occurred, Fultz and Ashley (2016) suggest that relatively little is known about the overall characteristics of these types of accidents. This research gap in weather-related accidents prompted this study into pilot decision-making and confidence under dynamic environmental conditions.

This study focused on evaluating the impact of preflight weather briefing format on pilot decision making. Two formats that are currently available to pilots were selected for analysis: a verbal weather briefing and a graphical/textual weather briefing. A verbal weather briefing is typically obtained when a pilot makes a phone call to a professional weather briefer. Weather information is shared verbally over the course of this phone call, and the pilot is able to take notes during the briefer's narrative. A graphical/textual briefing is typically obtained through the use of a computer or tablet. Pilots are able to view weather charts and diagrams as well as textual data presented in a Meteorological Aerodrome Report (METAR) format. A METAR is a compilation of the observed weather elements at an individual ground station, such as an airport (FAA, 2016). These two formats were analyzed for their effects on a pilot's decision-making ability and decision confidence.

Methods

This study utilized a survey method which presented pilots with a paper-based scenario and associated weather briefings. Each scenario was comprised of two legs, one to Sebring, FL and one to Okeechobee, Fl, each with their own hazardous weather. The study incorporated a within-subjects independent variable (IV) of weather briefing format, so each participant received both a verbal and a graphical/textual weather briefing. There were two between-subjects IV's in this study: the order of the two flight legs and the order in which each briefing format was received. Both flight order and format order were counterbalanced, creating four orders in which the scenarios could be presented to the participant.

Participants

Participants were selected through convenience and snowball sampling strategies, with all participants being certified Private Pilots or higher. The study sample was comprised of flight students and flight instructors from a flight school in central Florida. Flight students and flight instructors were recruited via email, direct messaging or word of mouth. Once a participant was selected for the study, they were assigned a number in chronological order from 1-36. One of the four surveys, labeled A, B, C, and D respectively, were assigned to each participant. The order of the distribution of these surveys was randomized via a dice roll. The number "3" was the result of the first dice roll, and this corresponded with survey C being paired with Participant 1. The survey order was then paired in alphabetical order with each subsequent participant.

Questionnaire Procedure

The survey was shared with each participant in a digital format though a Google Form. Results were then exported from the Google Form responses into an Excel spreadsheet for data analysis. Participants began each survey by providing demographic data, which included total flight hours, simulated and actual instrument hours, and level of certification. Next, participants proceeded to the scenario portion of the survey. Each survey was split into two halves, with each half representing one portion of the multi-leg flight. The verbal weather briefing information was presented as a written narrative taken from a briefing obtained over a phone call to an official weather briefer. The graphical/textual weather briefing information was presented as images of a radar depiction chart and a line of METAR code. Both scenarios were followed by Likert-scale questions to capture likelihood to make a "go" or "no-go" decision and the confidence each pilot had in this decision. The anchors for decision making ranged from

"Extremely Unlikely" to "Extremely Likely" and anchors for confidence ranged from "Not Very Confident" to "Very Confident." After the Likert-scale questions, several open-ended questions were posed to each participant to answer the "why" behind a pilot's indicated level of likelihood to fly and the confidence in their decision.

Results

Participant Demographics

Thirty-six pilots participated in this study, including seven Private Pilots, twelve Instrument-Rated pilots, eight Commercial Pilots, four Certified Flight Instructors, three Certified Flight Instructors – Instrument, and two Airline Transport Pilots. Of the 36 pilots, 32 were male and four were female. Pilot age was collected, with a mean across participants of 21.2 years (SD = 2.07). Pilots also reported their total flight hours and total instrument hours (simulated and actual). For total flight hours, the mean was calculated as 399.0 hours (SD = 491.5 hours). For total instrument hours, the mean was calculated as 71.1 hours (SD = 32.0 hours).

Effect of Weather Briefing Format and Scenario Order

To account for order effects, the order in which the scenarios were presented and the order in which the formats were presented were coded into two between-subjects variables (scenario order: Okeechobee first, Sebring first; format order: graphical first, verbal first). A repeated measures MANOVA was run with one within-subjects independent variable, weather briefing format, and two between-subjects independent variables, the order of the weather and the order of the two scenarios." The two dependent variables in the study were decision (based on likelihood to "go" of "no-go") and confidence in the decision. No main effect for the format of the weather briefing on either of the dependent variables was found; however, the interaction between the format of the weather briefing and the order of the scenario presented for decision approached significance ($F(1, 32) = 3.956$, $p = .055$, partial $\eta 2 = .110$). When Sebring was presented first, the pilot was more likely to "go" for the verbal format than the graphical format. When Okeechobee was presented first, the pilot was more likely to "go" for the graphical scenario than the verbal. The results are illustrated in Figure 1, which represents the group based on scenario and format order, where A = SV, B = SG, C = OV, D = OG, with S = Sebring Airport, O = Okeechobee Airport, V = Verbal Format, and G = Graphical/Textual Format).

Effect of Weather Briefing Format and Scenario Order

There was also a statistically significant interaction between the format of the weather briefing, the order that the formats were received, and the order that the scenarios were received for decision ($F(1, 32) = 4.833$, $p = .035$, partial $\eta 2 = .131$), and for confidence ($F(1, 32) = 6.201$, $p = .018$, partial $\eta 2 = .162$). Pilots were more likely to "go" in trial one than they were in trial two, regardless of which scenario or briefing format was presented first. The only exception to this was when the Okeechobee scenario l was presented first with a verbal weather briefing. Pilots were more confident when presented with either Okeechobee scenarios first, with confidence reducing in the later Sebring scenarios. Pilots who received Sebring in graphical format first gained confidence after they received Okeechobee in verbal format second. In trial one, pilots were most confident when they received Okeechobee in verbal format and least confident when receiving Sebring in graphical format. In trial two, pilots were most confident when receiving Okeechobee in graphical format and least confident when receiving Sebring in verbal format (see Figure 1 below).

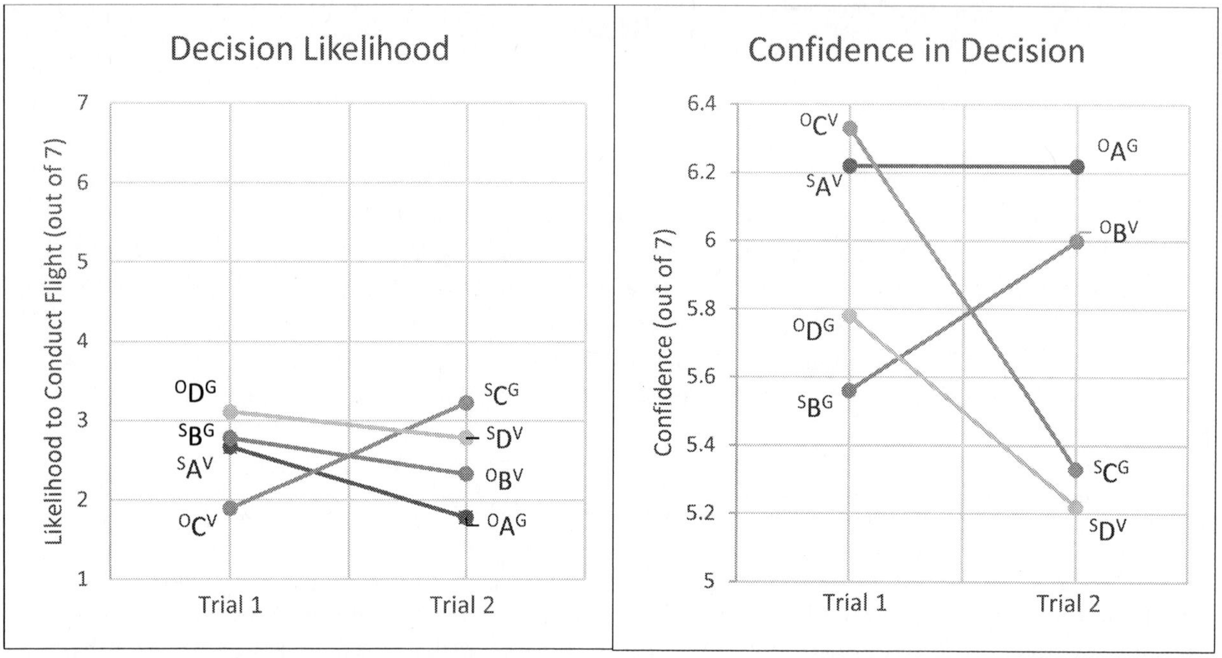

Figure 1. Decision Likelihood and Confidence Between Trials 1 and 2

Discussion

Overall for the decision to "go" or "no-go," participants in trial one were more likely to "go" when they received a graphical briefing first. However, these same pilots were less confident in this decision. This might have been because pilots could visually see where hazardous weather (in this case precipitation) is located in relation to their origin and destination. However, these pilots were not confident in this decision which may be due to pilots having to weave in and out of hazardous weather while enroute; a behavior that does not promote a safe flight. Pilots who received the verbal briefing first were less likely to fly but were more confident in this decision. This may be due to pilots not being able to see precipitation or low cloud ceilings when the format was verbal. Because pilots were provided information only in a verbal briefing, they were confident in terminating the flight due to the lack of multiple sources of information, specifically missing visual information. For the second trial for decision and confidence, pilots who received the Sebring graphical scenario were the most likely to fly but were not confident in their decision. The Sebring flight contained very low cloud ceilings, low visibility, but no precipitation. In the graphical/textual briefing, the text stated that there were low ceilings, but this could not be depicted in the visual aid, which only showed precipitation. For this scenario, no precipitation was present. The textual data listed that fog was present at Sebring while the visual aid did not show any hazard near this airport. The pilots who received this scenario were likely to fly but were not confident in their decision. This may have been due to the graphical/textual data not supporting one another and being unable to portray a common threat between them. The difference between the textual and graphical data was intended as the precipitation radar pictured on each survey is able to show the location and intensity of rain, but not fog. The information between the textual and graphical sources did not directly contradict one another, but it instead required the pilots to more carefully analyze the radar image after reading the line of raw textual weather data. Pilots could have fully understood the hazards that both the precipitation and low clouds posed if the graphical/textual data showed the same information. Likewise, the Okeechobee graphical scenario, when presented second, made pilots the least likely to fly but the most confident in this decision. This may be due to the precipitation being considered a greater threat when presented after the low ceilings from the Sebring scenario. This result shows an order effect when the pilots were presented with the scenario involving low cloud ceilings. Although these low ceilings were first considered hazardous, the pilots were even less likely to "go" but more confident in this decision during the second trial perhaps due to the precipitation being considered a comparatively greater threat.

The most interesting interaction for decision was the difference between trials one and two when the Okeechobee verbal scenario was presented first. This was the only group of pilots whose likelihood to make a "go" decision sharply rose while their confidence in this decision sharply fell when they received the Sebring graphical scenario second. This may have been due to the fact that the precipitation was only being depicted verbally, preventing pilots from visually seeing the location and intensity of the precipitation. This resulted in pilots that were unlikely to fly and were confident in that decision. Upon receiving the graphical/textual Sebring scenario, the pilots saw a visual aid that displayed very low cloud ceilings and no precipitation. The weather diagram showed no precipitation over Sebring, as also depicted in the line of textual weather data. In the textual weather data, Sebring was reporting very low ceilings. Because the weather diagram was not able to depict cloud ceilings, there was no conflict between the textual weather and the diagram either. The pilots were likely satisfied with receiving a more cohesive briefing that did not clearly present hazardous weather and this resulted in the higher likelihood to fly but less confidence due to the high chance of low cloud ceilings.

Conclusion

This study focused on how verbal and graphical/textual weather briefings affected a pilot's decision-making and confidence during preflight. Overall, it appears that graphical/textual weather briefings will result in a pilot being more likely to conduct a flight under deteriorating weather conditions. This is potentially due to the images of the current weather conditions clearly showing the location and intensity of poor weather. Pilots may be less confident when deciding to "go" on a flight after receiving a graphical/textual briefing because even though they deem the flight safe, they area aware that they may be required to navigate around potentially hazardous weather while enroute. Pilots utilizing a verbal briefing may be less likely to conduct a flight due to the lack of a visual aid accompanying the narrative of hazardous weather. Pilots may be more confident in this "no-go" decision because the termination of their flight means that they will not have to worry about dodging hazardous conditions while enroute.

This study has a direct impact on the safety of GA flights and how pilots obtain and analyze preflight weather briefings. This study mimicked a multi-leg flight, each with slightly different hazardous weather conditions. Results suggest that the type of briefing received for each leg should by tailored to the type of weather condition: graphical/textual briefings should be used for precipitation and verbal briefings should be used for areas of low visibility. Reversing the type of briefing might result in a more likely "no-go" decision due to the lack of information in the briefing that would fail to support the weather phenomena. The practical significance of this study for GA flights is that pilots need to be aware of the limitations of each type of briefing. Since pilots do not have control over the type of weather they might encounter, they need to ensure they are utilizing the correct briefing format during preflight. Ideally, a pilot could use both types of briefings, thus eliminating the limitations of each format. However, this is largely dependent on the equipment available to the pilot during preflight.

References

Federal Aviation Administration. (2016). *Pilot's Handbook of Aeronautical Knowledge* (FAA-H-8083-25B). Oklahoma City, OK: United States Department of Transportation.

Fultz, A. J., & Ashley, W. S. (2015). Fatal weather-related general aviation accidents in the united states. *Physical Geography, 37*(5), 291-312. doi:10.1080/02723646.2016.1211854

Topçu, O. (2017). USING SITUATIONAL AWARENESS FOR ADAPTIVE DECISION \ MAKING IN AGENT- BASED SIMULATION. *Middle East Technical University Northern Cyprus Campus*. Retrieved March 6, 2019.

Wickens, C. D., Hollands, J. G., Parasuraman, R., & Banbury, S. (2012). *Engineering Psychology and Human Performance* (4th ed.). New York, NY: Routledge.

A NEW HMI EVALUATION METHOD (MERIA) BASED ON PILOT'S MENTAL REPRESENTATIONS

LETOUZE Théodore
Bordeaux University, BdxINP - ENSC, Lab IMS Cognitique UMR 5218
Talence, France
CRENO Lisa
Centre Aquitain des Technologies de l'Informations et Electroniques
Talence, France
DIAZ Jaime
Centre Aquitain des Technologies de l'Informations et Electroniques
Talence, France
HOURLIER Sylvain
Thales/ENSC, Human Engineering for Aerospace Laboratory (HEAL)
Talence, France
ANDRE Jean-Marc
Bordeaux University, BdxINP - ENSC, Lab IMS Cognitique UMR 5218
Talence, France

Continuous evolution of HMIs is necessary to keep operators in an optimal situation. In this context, we consider mental representations (MR) mobilized by operators as key elements for decisionmaking. Capturing and analysing these representations is not easy with existing tools. We propose a specific method (i.e. "MERIA" for Mental Representation Impact Analysis). Our case study focuses on a group of first officer (Airbus A320) in a dynamic situation with high time pressure. We are interested in cases where the HMI generates MRs that are inconsistent with the situation, resulting in a discrepancy between the prescribed activity and the actual activity. The goal is to identify the link between erroneous MR and the interface that created them. Our modelling structure allows us to create this link and place it in a proper temporal context. We observe that the constitution of the MR is different from one subject to another. However, invariants in the appearance of some erroneous MR make it possible to attribute the causality to an interface element well-defined in space and time. Thus, this analysis allows us to offer recommendations for HMI design to improve decision making. Our results show that the improvement does not lie in a drastic modification of the interfaces. Rather is allows a synchronization of the data coming from the cockpit with the pilot's MR of those data.

Introduction

We only observed the co-pilot activity. They intervene in a dynamic, uncertain, risky situation and they must make multiple decisions under the pressure of real time to achieve their performance objectives (Graziani et al., 2016). In the context of the cockpit of an Airbus A320, our objective is to determine which interfaces allow the co-pilot to build a good mental representation of the situation and which ones do not. In complex environments, HMIs and co-pilot cognitive activities can be evaluated in multiple ways. The methodologies we are interested in are those that aim to jointly evaluate the efficiency of the interface and it's use by the operator. There are various categories of methods (Stanton, 2013): ETS (Annett, 2004), ACRC (Vicente, 1999), SAGAT questionnaire, SPAM method, Situational Awareness Requirements Analysis (Endsley, 1995, 2001, Selcon and Taylor, 1990), etc. None of these methodologies creates a causal link between the user's cognitive process, his mental representation and the HMI used. This is the reason why we have proposed the MERIA (Mental Representation Impact Analysis) method, specifically adapted to this problem (see Letouzé & al., 2019).

Method

The methodological approach of MERIA is based on a triangulation of methods. It has been developed to design interfaces allowing operators to be more resilient in problem solving situations. It combines qualitative and quantitative methods that show, in a detailed and contrasted way, the observed activity (Altrichter, 2008). This method allows us to collect the general activity of the co-pilot (subject of this study) through three points of view: i) the experts describe the sequence of action expected in the scenario (prescribed task); ii), the experience of the co-pilot is collected by an interview (task performed); iii) we collect the general characteristics of the co-pilot (experience, personal data, etc.). The aim of this approach is to improve "the richness and sophistication of our analysis" (Guilbert & Lancry, 2007) and to get as close as possible to the "true value of the information collected" (De Battisti, Salini, & Crescentini, 2006), by crossing the three types of points of view.

Application of the method

The methodology is applied to a population of experts when performing a scripted and constrained activity in terms of progression and duration. The method is applied according to the following process: (1) A scenario representative of the co-pilot's activity is defined precisely. It is also verified that this scenario is reproducible under realistic conditions. The expected performances at each stage are defined by a collective of experts in pilot operations, aeronautics and cognitive sciences. Each performance element is associated with a mental representation. This is why cognitive science experts need to be involved in this phase. (2) A homogeneous panel of co-pilots is recruited. These co-pilots do not know the scenario. (3) The scenario is performed by the co-pilots of the panel in a cockpit of the current A320. During the scenario, cognitive science experts observe the activity and identify key events. A pilot expert comments on the co-pilot's actions to make the activity more explicit. The experts (cognition and aeronautics) are not in the copilot's environment. (4) Immediately after the end of the scenario, the cognition expert conducts a self-confrontation interview. During the interview, the co-pilot is "put back in the situation", the expert making him relive the scenario step by step. This expert identifies the RM associated with each step. (5) This interview shows the evolution of the RM over time using the MERIA grid (e.g. Figure 1). This grid is completed by observation of the experiment (3). We collect one grid per co-pilot. An inter-judge measurement method (Cohen kappa) makes the coding process of the grid more reliable. (6) From these grids, cognition experts identify the problematic interfaces and those that produce the expected effect.

The graphical representation: MERIA grid

The tool is constructed as follows: The prescribed scenarios are represented by white squares placed in the "NODES" column. In the "INPUT" column, squares indicate the different sources of information addressed to the co-pilot. In the "MENTAL REPRESENTATION" column, we coded the elements relating to the co-pilot's actual mental representation. It varies and evolves during the scenario depending on what the co-pilot perceives, understands and anticipates. In the "IMPACT" column, we indicated the consequences of the actions implemented as soon as the mental load, the choice of the airport/runway (Bremen or alternatives) and/or the landing limitations were affected.

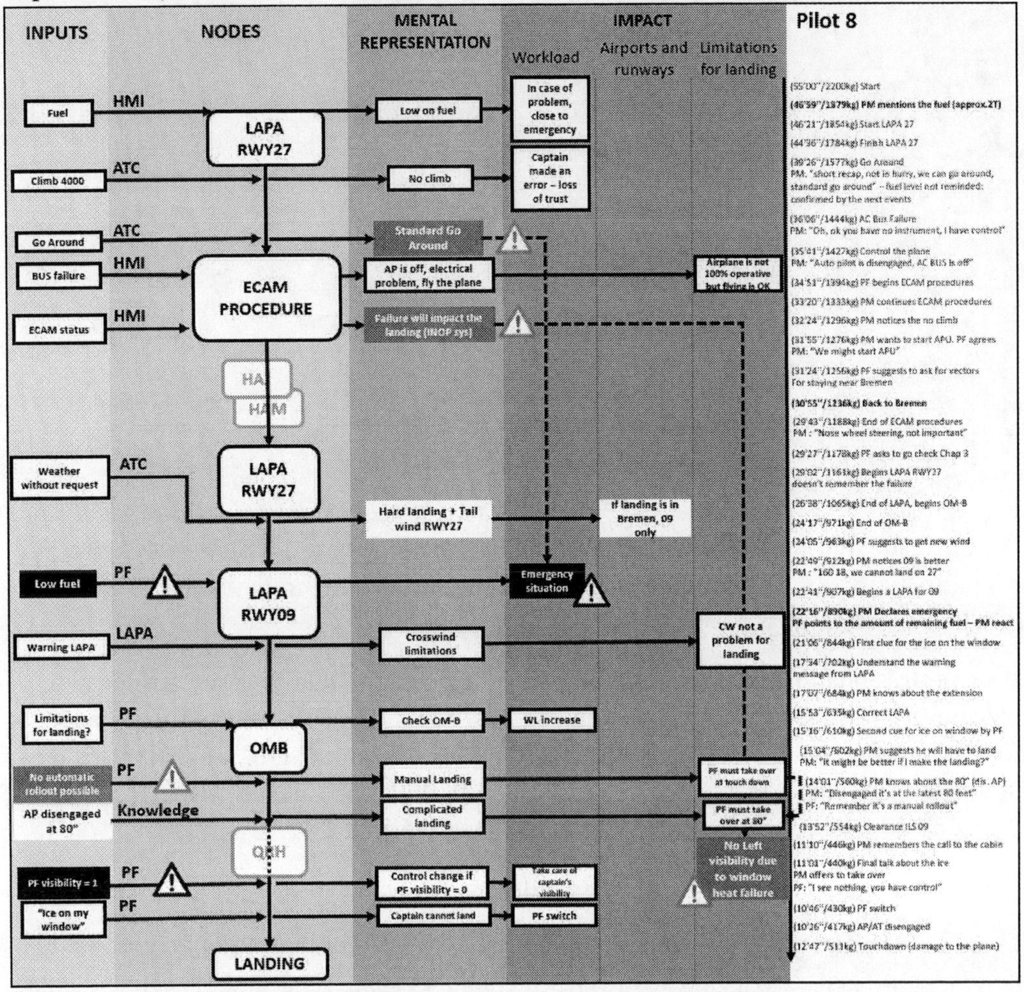

Figure 1. MERIA Model of Pilot # 8. (White without outline: actual performance > prescribed, White with black outline: real = prescribed, Gray: real < prescribed, without being critical, Black: real << prescribed, critical state, Triangles allow to quickly identify the elements that interfere with activity).

Scenario

The context of the critical scenario observed is unique since it begins in the middle of the flight (in a phase just before the approach and landing) so that the co-pilots are not aware of the amount of fuel remaining (fuel on board) to reach the end of the flight. The scenario has 4 phases of unequal duration (see Figure 2). Different "nodes" (or key elements of the scenario) structure these phases.

Description du panel

We have access to a panel of 10 co-pilots from Lufthansa Airlines, trained at Bremen Airport (Germany). They know all the particularities of this airport: runway length, nearby airports, unofficial runway extension, etc. These voluntary and paid participants are "experts" of the task, which offers a particular interest for the analysis of decision-making, the knowledge of the situation and the mechanisms of mental representation studied.

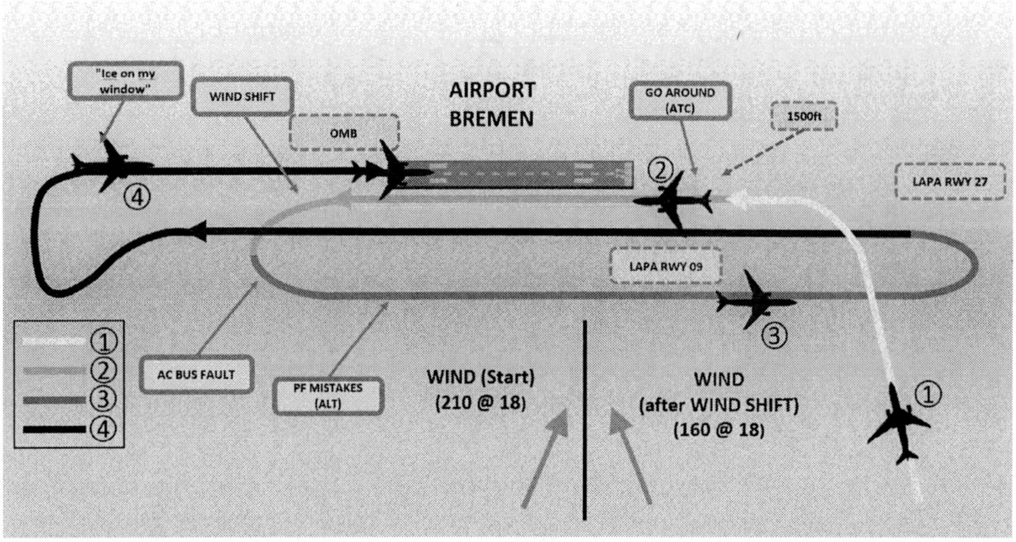

Figure 2. Schematic representation of the prescribed scenario (Phase 1 - Start of the go-around scenario, Phase 2 - Until failure, Phase 3 - Until the decision to land on runway 09, Phase 4 - Up to on landing).

As a professional, none of the pilots recruited are captain, that is to say, responsible for the plane or its passengers. Co-pilots averaged 30.9 years (min: 28, max: 36, standard deviation (SD): 3.28), a total of 4045 flying hours on average (min: 2250, max: 7000, SD: 1569), of which 3125 hours on average on Airbus A320 (min: 250, max: 6000, SD: 1557) and 667.78 hours on average over 12 months (min: 600, max: 750, SD: 42.78).

Results

From the MERIA grids we constructed from the 10 self-confrontation interviews, we were able to identify gaps between the expected mental representations and the actual mental representations. The different phases of the scenario (inputs and associated mental representations), allow us to identify needs for co-pilot. We can also identify services that the system could render to the co-pilot. Our analysis is among us to identify 10 services not rendered by the system that penalizes the mental representations of pilots. These services would be required for the completion of the requested task but the existing system is not designed to respond to it. In some cases, it is the training or the expertise / experience of the pilot that addresses this deficiency.

A total of 56 unreturned services are recorded for all 10 pilots as we can see on Table 1. These services can be categorized (some occurrences can belong to more than one category). It is observed that 9 occurrences can be identified as feedback defects from the system. That is to say a lack of visibility of the signal, or even its absence. 21 occurrences can be identified as a lack of data synthesis (cross-referencing of several types of information) and a lack of explanation of their consequences. 26 can be likened to a lack of spatial clarification of constraints and possibilities. Finally, 33 occurrences are similar to a failure to represent temporal constraints and the evolution of the system over time. The domains of spatial and temporal representations overlapping to a certain extent.

Table 1.
Number of services not provided by the system for all 10 pilots over the entire scenario. During the chosen scenario, the system must provide 10 services.

Service	Number of times the service has not been delivered (/10)	Services provided to the co-pilot 8
Warning about fuel level	9	Yes
Feedback : actions done	2	Yes
Projection of the flight action field compared to the needs to land	10	no
Combining failures to explain consequences and keep it in PM mind	7	no
Projection of the flight action field (fuel + wind + speed) compared to the needs to land	5	Yes
Combine the weather with the needs to land and explain options	7	no
Combine the aircraft state with the needs to land and explain limitations	4	Yes
Combining failures and information from documentation to explain consequences and keep it in PM mind	7	Yes
Explain what should be done to follow procedures	3	Yes
Combining failures and aircraft state to explain who should take control at each time	2	Yes
Total	56/100	3 services non delivered

Through this analysis of the results, and by combining this information with self-confrontation interviews, we observe that the representations provided by the system were not consistent with the representations expected by the co-pilots. The co-pilots evaluate the field of possible temporally (available flight time) and/or spatially (attainable distances). For the fuel on board the aircraft, for example, the system produces an indication in kilograms while the co-pilots convert it into minutes or nautical miles. This inconsistency is found for other information the aircraft provides: the inoperative systems, the co-pilot reflects the type of failure and consequences, the weather is transmitted in code form (METAR) and the data are relative to the ground, the co-pilots reflect in terms of cloud layers and relatively to the aircraft. Such difference between the information provided by the system and the cognitive functioning of the co-pilot creates a blow of information conversions that reduces performance and can potentially lead to conflicts in human-system collaboration or errors.

Conclusion

The use of the MERIA methodology highlights activities for which pilots are not properly assisted by the system. From this point of view, the results are similar to those obtained with SA measurement methodologies. The added value of our approach comes from the fact that the MERIA method highlights the discrepancies between the expected and actual mental representations of the co-pilots. It makes it possible to identify exactly where and when the source or sources of the offset on the manipulated HMI are. These results open the door to new studies on system design and evaluation. In this case of application, the methodology allowed us to identify areas for improvement of the A320 cockpit system. This study highlights that improved co-pilot performance can be achieved through conceptual system changes and improved communication between operators and systems. In conclusion, the MERIA tool provides a solution to the evaluation and improvement of Man Machine Teaming.

Acknowledgement

The results presented here are based on the work carried out under the Future Sky Safety program, funded by the Horizon 2020 research and innovation program of the European Union under Grant Agreement No 640597.

References

Altrichter, H. (2008). *Teachers investigate their work: an introduction to the methods of action research* (2nd ed). London; New York: Routledge.

Annett, J. (2004). Hierarchical Task Analysis. In Neville Anthony Stanton, A. Hedge, K. Brookhuis, E. Salas, & H. W. Hendrick (Eds.), *Handbook of human factors and ergonomics methods* (pp. 329–337). Boca Raton, FL: CRC press.

De Battisti, F., Salini, S., & Crescentini, A. (2006). Statistical Calibration of Psychometric Test. *Statistica & Applicazioni*, *4*(2), 1–25.

Endsley, M. R. (1995). Measurement of situation awareness in dynamic systems. *Human Factors*, *37*(1), 65–84.

Endsley, M. R. (2001). Designing for situation awareness in complex systems. In *Proceedings of the Second International Workshop on symbiosis of humans, artifacts and environment* (pp. 1–14).

Graziani, I., Berberian, B., Kirwan, B., Le Blaye, P., Napoletano, L., Rognin, L., & Silvagni, S. (2016). Development of the Human Performance Envelope Concept for Cockpit HMI Design. In *HCI-Aero 2016 International Conference on Human-Computer Interaction in Aerospace*. PARIS, France. Retrieved from https://hal.archives-ouvertes.fr/hal-01409075

Guilbert, L., & Lancry, A. (2007). L'analyse des activités des cadres : l'intérêt de la triangulation des méthodes. *Le travail humain*, *70*(4), 313–342. https://doi.org/10/fm327w

Selcon, S. J., & Taylor, R. M. (1990). Evaluation of the Situational Awareness Rating Technique(SART) as a tool for aircrew systems design. *AGARD, Situational Awareness in Aerospace Operations 8 p (SEE N 90-28972 23-53).*

Stanton, Neville A. (2013). *Human factors methods: a practical guide for engineering and design*. Burlington, VT: Ashgate Publishing Company.

Vicente, K. J. (1999). *Cognitive work analysis: Toward safe, productive, and healthy computer-based work*. CRC Press.

EFFECTS OF DECISION TYPE AND AID ACCURACY ON USER PERFORMANCE

Lori Mahoney
Joseph Houpt
Wright State University
Dayton, OH

Automated aids provide users additional information for making decisions. The way the aid presents the information requires the user to either make the same decision as unaided or to agree or disagree with the aid's recommendation. In this study, we measured response times and accuracy without an aid and with an aid where either: 1) the subject makes the same decision as the unaided condition, or 2) the subject agrees or disagrees with the automated aid's decision. Results show subjects were more accurate with direct selection decisions, more accurate aids, and easier tasks, with an interaction between decision type and aid accuracy. Subjects were faster with direct selection decisions and more accurate aids, with an interaction between decision type and aid accuracy. Using a cognitive model we found information accumulation rates and caution varied across conditions.

The addition of an automated aid for speeded choice tasks gives the user additional information to make their decision; a correct aid response leads to a faster and more accurate human response (Rovira, McGarry & Parasuraman, 2007; Wickens, Clegg, Vieane & Sebok, 2015), but a number of different factors, including trust, workload, and automation reliability influence automation use, disuse (e.g. underutilization or neglect) and misuse (e.g. overreliance or complacency) (Parasuraman & Riley, 1997). Multiple studies have shown that decreased aid reliability decreases human performance (Rovira, McGarry & Parasuraman, 2007; Rovira, Cross, Leitch & Bonaceto, 2014; Wickens, Clegg, Vieane & Sebok, 2015). Rovira, et al (2014) showed there was little variability in response accuracy for low task demand; however, for high task demand accuracy improved with reliable automation and did not degrade below manual performance with imperfect automation.

Another factor to consider is that the way an automated aid presents the information requires the user to either make the same decision as they would unaided (i.e. select the correct signal) or to agree or disagree with the aid's recommendation. Parasuraman, Sheridan, & Wickens (2000) proposed a 10-level model for automation of decision and action selection functions, such as the output of an automated target recognition system or an aircraft-ground collision avoidance system. The model distinguishes between level 4 where the automation suggests one recommendation, and level 5, where the automation executes the one recommedation if the human user approves. In both cases the user is presented with one option to make a decision, but how the information is presented differs and requires the user to make a different decision. We hypothesize that the agree/disagree (level 5) decision decreases user performance and that there is an interaction with aid accuracy. In the agree/disagree condition an inaccurate or less accurate aid requires the user to determine if the aid provided a correct recommendation by comparing it to their own decision (i.e. make at least two decisions); when this is done in series the response time (RT) is slower for this condition. For the higher accuracy aid the user can select the aid's recommendation without knowing if it agrees with theirs because they know the aid is almost always correct; users will not need to compare the aid's recommendation to theirs for every signal, so their perfomance is improved. The direct selection (level 4) decision does not require the user to evaluate the aid against their recommendation so the user is making one less decision in this condition regardless of the aid's accuracy, therefore we expect the RT to be faster for this condition. We expect this difference in decision type to also affect accuracy. We predict the level 5 decision accuracy will be lower, especially for the less accurate aid, since the user is basing their decision on their own recommendation. Additionally, since the agree/disagree decision is asking a different question of the user than the decision to select the correct signal, either with or without an aid, we expect that the mean drift rate, the response boundary, or both should vary across the test conditions for a linear ballistic accumulator (LBA) model, in which results are accumulated linearly and independently for all responses. The response boundary should be larger for conditions with more difficult tasks and/or less accurate aids since the user will have less certainity in the correct decision and be more cautious in making their decision. The mean drift rate should be lower for the agree/disagree decision condition with the less accurate aid since users are making more decisions and are less efficient. The mean drift rate should be higher for the more accurate aid since users are more efficient with an aid that is almost always correct.

Method

Forty-seven students (32 females, 15 males) from Wright State University participated in this study. All subjects gave informed consent to participate and were given course credit as compensation for their time. Ages ranged from 18 to 36 (M = 19.9). Data from 3 participants were removed due an overall accuracy of less than 70%. For the remaining 44 participants, trials were removed that had response times of zero, where the subject responded faster than the time to present the stimulus, and that had response times slower than 4.16 seconds (99th percentile).

Subjects were presented with long and short vertical rectangular bars while signal uncertainty, automation accuracy, and decision type are manipulated in the context of a manufacturing quality assurance task. Subjects were instructed that short bars were desired and should be selected whereas a long bar should be rejected. We measured RT and accuracy over three decision type conditions for all subjects: the subject decided without automation, with automation (level 4 decision type), or chose to either agree or disagree with the automation's recommendation (level 5 decision type). There were two signal uncertainty conditions determined by the standard deviation of the bar lengths: easy (.15 SD) and difficult (.3 SD), and two automation accuracy conditions: high (95%) and low (80%). The automation accuracy conditions did not apply for the decision type condition of deciding without automation. The order of these ten conditions were counterbalanced across subjects.

Results

Table 1 summarizes the median RT and mean accuracies across all test conditions for all responses (correct and incorrect). For the unaided baseline and level 4 decision type, the hard conditions have longer RT than the easy conditions, for each level of aid accuracy. Additionally, the low aid accuracy conditions for the level 5 decision type have longer RT than all the other conditions. Figure 1 shows the RT distributions by test condition. The level 5 decision type condition appears to have a bimodal distribution for the correct responses in the high accuracy aid condition. This did not occur for the level 4 decisions. By looking at the distribution of RT for individual subjects, and separately for agree and disagree responses (Figures 2a and 2b), it is evident that some subjects respond very quickly to agree with the automated aid, regardless of its recommendation, while other subjects respond with times similar to when directly selecting the signal (i.e. level 4 decision). In trials with the high accuracy aid this will result in correct responses 95% of the time, which creates two distributions of RT – one quicker for those that always quickly agree and one slower for those that evaluate the stimulus and then decide. The trend is not visually obvious for the low accuracy aid; further analysis is needed to determine if the trend occurs for the low accuracy aid as well.

Table 1. *Summary of Response Times and Accuracies Across Test Conditions*

Condition (# trials)	Median Human RT, ms (95% CI)	Mean Accuracy, % (95% CI)
1 = select signal w/o aid, easy (5251)	639 (610 – 669)	84.8 (81.7 – 87.7)
2 = select signal w/o aid, hard (5255)	652 (617 – 688)	75.8 (73.9 – 77.6)
3 = select signal w/high accuracy aid, easy (5266)	669 (641 – 698)	91.4 (89.7 – 93.1)
4 = select signal w/high accuracy aid, hard (5260)	693 (658 – 727)	86.5 (84.6 – 88.4)
5 = select signal w/low accuracy aid, easy (5253)	657 (621 – 694)	85.3 (83.1 – 87.4)
6 = select signal w/low accuracy aid, hard (5243)	692 (647 – 738)	78.7 (76.7 – 80.6)
7 = agree/disagree w/high accuracy aid, easy (5183)	657 (584 – 730)	93.0 (91.2 – 94.9)
8 = agree/disagree w/high accuracy aid, hard (5125)	657 (576 – 738)	89.3 (86.7 – 92.0)
9 = agree/disagree w/low accuracy aid, easy (5193)	811 (741 – 880)	82.7 (79.3 – 86.0)
10 = agree/disagree w/low accuracy aid, hard (5164)	834 (744 – 924)	77.5 (74.7 – 80.3)

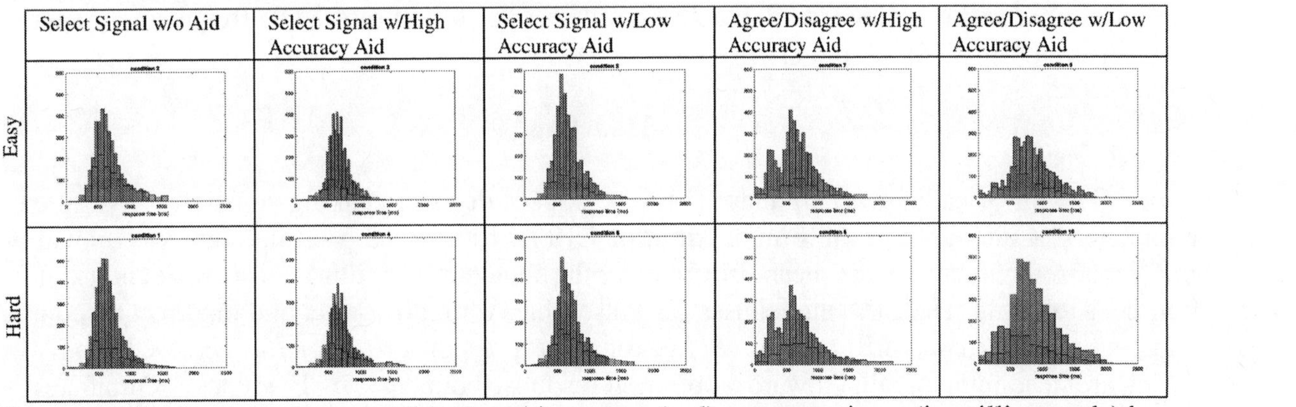

Figure 1. Distribution of correct (blue) and incorrect (red) response times (in milliseconds) by condition. The response times (x-axis) range from 0 to 2500 milliseconds for all plots. The frequency (y-axis) ranges from 0 to 600 for all plots.

A repeated measures ANOVA on median correct RTs in aided conditions indicated decision type, aid accuracy, and the interaction between decision type and aid accuracy were significant ($F(1,301) = 5.6$, $p = .02$, $\eta = .09$; $F(1,301) = 31.7$, $p < .001$, $\eta = .22$; $F(1,301) = 31.7$, $p < .001$, $\eta = .22$). A linear mixed-effects regression model on RT indicated decision type and the interaction between decision type and aid accuracy were significant predictors ($B = 1.26$, $t = 24.1$, $p < .001$; $B = -1.37$, $t = -23.1$, $p < .001$) of RT. Subjects were faster in the select decision type and decision type as a moderator of aid accuracy on RT is stronger for agree/disagree decisions than for direct selection decisions. A logistic mixed-effects regression analysis of accuracy in aided conditions, indicated that decision type, aid accuracy, difficulty, and the interaction between decision type and aid accuracy were significant predictors ($B = -0.13$, $z = -3.6$, $p < .001$; $B = 0.58$, $z = 14.4$, $p < .001$; $B = -0.43$, $z = 15.1$, $p < .001$; $B = 0.38$, $z = 6.4$, $p < .001$) of accuracy. Subjects were more accurate in the select decision type, more accurate with more accurate aids, and more accurate with easier tasks.

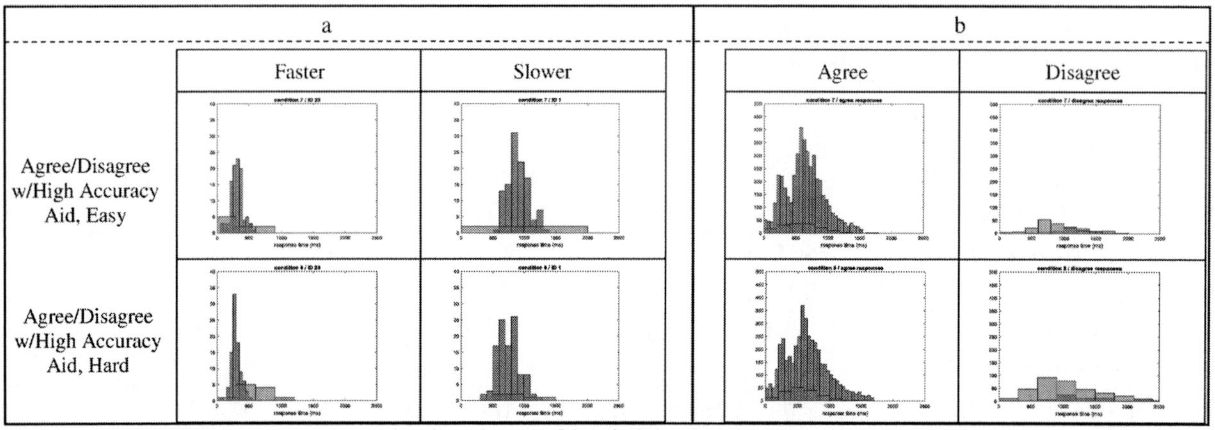

Figure 2. (a). Representative distributions of individual subjects with faster and slower correct (blue) and incorrect (red) response times (in milliseconds) for level 5 decision type in the high accuracy aid condition. The response times (x-axis) range from 0 to 2500 milliseconds for all plots. The frequency (y-axis) ranges from 0 to 40 for all plots.
(b) Distribution of correct (blue) and incorrect (red) response times (in milliseconds) for level 5 decision type in the high accuracy aid condition separated by agree and disagree decision. The response times (x-axis) range from 0 to 2500 milliseconds for all plots. The frequency (y-axis) ranges from 0 to 500 for all plots.

Modeling

The linear ballistic accumulator (LBA) model of decision making consists of five fitted parameters: A, the range of uniform distribution U[0,A] from which starting point k is drawn; b, the response boundary; v, the mean drift rate; s_v, the standard deviation of drift rate; and t_0, the non-decision time. The LBA model uses response times for both correct and incorrect responses and assumes a different drift rate for each (v for correct response and 1-v for incorrect), both of which are heading in parallel toward a common response boundary, b. Evidence accumulates linearly at the mean drift rate, v, for both responses until one reaches the response boundary; this is the model's response and once reached the evidence for the alternative response is discarded.

We compared the fit of four different LBA models using the data from this study: Model 1 fixes all parameters between conditions; Model 2 allows mean drift rate to vary between the 10 test conditions and fixes all other parameters; Model 3 allows response boundary to vary between test conditions and fixes all other parameters; and Model 4 allows response boundary and mean drift rate to vary and fixes all other parameters. The optimal set of fitted parameters for each model was found by maximizing the log-likelihood with a modified version of Steve Fleming's MATLAB code for fitting the LBA model (available at https://github.com/smfleming/LBA). All participants were assumed to be equal for determining the model parameters.

Initially, each model with one parameter varied (i.e. Model 2 and Model 3) was compared to the most restricted, specific model (i.e. Model 1); each had a log-likelihood value that is larger (i.e. less negative) than the log-likelihood for Model 1, indicating a better goodness-of-fit. A χ^2 comparison of each general model to the specific model results in a χ^2 that exceeds the critical χ^2 (16.9) for 9 degrees of freedom given α=.05 for both models (χ^2_{1v2} = 2164; χ^2_{1v3} = 2768), indicating that the increased log-likelihood for the more complex models is merited by the additional parameters in each. Next the more complex, general model that allowed both v and b to vary across all test conditions (i.e. Model 4) was compared to Model 2 and Model 3 separately. The additional complexity in Model 4 provides a higher log-likelihood and that additional complexity results in a χ^2 that exceeds the critical χ^2 (28.9) for 18 degrees of freedom given α=.05 (χ^2_{2v4} = 2090; χ^2_{3v4} = 1487). Table 2 lists the optimal parameters for Model 4.

Table 2. *Optimal parameters for Model 4*

Constants	*Parameters that vary by condition*									
A = 488	*Unaided*		*Level 4 Decision*				*Level 5 Decision*			
t_0 = 1.02 x 10^{-3}	*Easy*	*Hard*	*Easy*	*Hard*	*Easy*	*Hard*	*Easy*	*Hard*	*Easy*	*Hard*
s_v = 0.225			*+ Accuracy*		*– Accuracy*		*+ Accuracy*		*– Accuracy*	
v	0.70	0.64	0.79	0.72	0.71	0.66	0.76	0.73	0.66	0.63
b	739	726	803	784	755	756	689	676	799	786

Note: "+ Accuracy" is the more accurate aid. "– Accuracy" is the less accurate aid.

Discussion

The results from this study show that the type of decision made by a user has an effect on their RT and accuracy and the effect interacts with the accuracy of the automated aid. Additionally, aid accuracy and task difficulty also have an effect on accuracy. Looking only at user accuracy, the results agree with Rovira (2014) for difficult (i.e. high demand) tasks where accuracy improved with reliable automation and did not degrade below unaided performance with less accurate automation. For the easy (i.e. low demand) tasks, we differ from their results in that we find response accuracy improved with more accurate automation and actually degraded below the unaided performance with less accurate automation for the agree/disagree condition. The tasks presented in this study were more abstract than the task in Rovira's study so it is possible that users rely on the aid more for more abstract tasks; this needs further research.

The variations in mean drift rate and response boundary in the LBA model describe varying behavior in subjects across the test conditions. We expected lower response boundaries for easier tasks and/or more accurate aids, but the results were the opposite. The mean drift rate (v) and response boundary (b) are higher for the easy conditions, across all aid accuracies and decision types, indicating that subjects are more efficient, but also more cautious for the easy conditions; the opposite is true for difficult tasks. For the more accurate aid, across all difficulties for level 4 decisions, the v and b are higher, again indicating that subjects are more efficient in their choice, but also more cautious. For level 4 decisions with the less accurate aid, the mean drift rates approximately equal the values for the baseline unaided condition, while the b are higher, indicating that subjects are equally efficient in both conditions, but more cautious with a less accurate aid. The level 5 decision values are interesting because they behave oppositely; *v* increases and *b* decreases for the more accurate aid and *v* decreases and *b* increases for the less accurate aid. Subjects are more cautious and least efficient with the less accurate aid and least cautious and more efficient with the more accurate aid for agree/disagree decisions.

We found that when a lower accuracy aid is presented, users respond faster and more accurately making a direct selection, but when a higher accuracy aid (e.g. ≥ 92% for the aid used in this study) is available, users respond faster and more accurately by agreeing or disagreeing with the aid's recommendation. Users are least cautious and more efficient when agreeing or disagreeing with the more accurate aid. When using a less accurate aid users are less cautious and more efficient making a direct selection. This is the difference between level 4 and level 5 interaction in the model proposed by Parasuraman, Sheridan, & Wickens. When designing and implementing an automated aid the goal is generally to create one that has a high accuracy, but if that is not possible then there is a benefit to implementing the automation where it suggests one alternative, but the human still has authority to execute that alternative or choose a different one.

Acknowledgements

The authors appreciate the contributions of Jane Hwang in assisting with conducting the study. The tasks used in this study were based on previous studies by Zinn, Yamani, & McCarley.

References

Parasuraman, R., & Riley, V. (1997). Humans and automation: Use, misuse, disuse, abuse. *Human Factors, 39*(2), 230–253. Retrieved from https://journals-sagepub-com.ezproxy.libraries.wright.edu/doi/

Parasuraman, R., Sheridan, T. & Wickens, C. (2000). A Model for Types and Levels of Human Interaction with Automation. *IEEE Transactions on Systems, Man, and Cybernetics – Part A: Systems and Humans, 30*(3), 286-297.

Rovira, E., Cross, A., Leitch, E., Bonaceto, C. (2014). Displaying Contextual Information Reduces the Costs of Imperfect Decision Automation in Rapid Retasking of ISR Assets. *Human Factors, 56*(6), 1036-1049. Retrieved from https://journals-sagepub-com.ezproxy.libraries.wright.edu/doi/full/

Rovira, E., McGarry, K., Parasuraman, R. (2007). Effects of Imperfect Automation on Decision Making in a Simulated Command and Control Task. *Human Factors, 49*(1), 76-87. Retrieved from https://journals-sagepub-com.ezproxy.libraries.wright.edu/doi/pdf/

Wickens, C.D., Clegg, B.A., Vieane, A.Z., & Sebok, A.L. (2015). Complacency and automation bias in the use of imperfect automation. *Human Factors, 57*(5), 728-739. Retrieved from https://journals-sagepub-com.ezproxy.libraries.wright.edu/doi/full/

IDENTIFYING A POSSIBLE FUNCTION FOR ARTIFICIAL AGENT ADAPTATION IN VARIABLE TASK RATE ENVIRONMENTS

David J. Canzonetta and Michael E. Miller
Air Force Institute of Technology
Wright-Patterson, AFB, OH

The current research sought to identify a method to calculate agent response time (ART) as a function of inter-arrival time (IAT), which balances human-agent team performance, human engagement, and human workload. A human-in-the-loop experiment evaluated human-agent team performance, as measured by team score, human engagement, as measured by the number of manually performed tasks, and workload, as measured through a subjective questionnaire, as a function of IAT and ART combination. Results demonstrated that task IAT strongly correlated with performance, engagement, and workload, while ART strongly related to engagement. Optimization was applied to the resulting data to determine ARTs which maximized performance while sustaining desirable levels of human engagement and workload. The optimization produced an ART function for application in future research to judge the effectiveness of adapting ART to boost human-agent team performance.

Humans and artificial agents can be teamed together to complete intricate and vital tasks. Successful task completion relies on the balance of human engagement and workload within these teams. For example, an unengaged human operator experiencing underload can face decreased alertness (Parasuraman, 2008). Dynamic function allocation is a common adaptive automation method for maintaining proper workload balance (Schneider, Bragg, Henderson, & Miller, 2018). However, this type of function allocation can force the human to maintain awareness of their present tasks within the current allocation, effectively increasing mental workload (Kaber, Riley, Tan, & Endsley, 2001).

Previous research conveyed that agent responsiveness within a human-agent team can affect human engagement (Goodman, Miller, Rusnock, & Bindewald, 2017). This discovery suggests that a well-timed agent response could provide an alternative approach to achieving the proper balance between human engagement and human workload in systems employing adaptive automation. For situations where environmentally-imposed inter-arrival time (IAT) heavily influences operator workload, calculation of optimal agent response time (ART) as a function of IAT becomes a possible method for task load sharing. The current study varied IAT and ART, measuring their effects on human-agent team performance, human engagement, and human workload. The data collected from this study produced a function for desired ART as a function of IAT to support future research.

Method

Participants

The experiment involved 14 participants (9 male and 5 female). Two participants were left-handed. Mean participant age was 25.4 and ranged from 20 to 31. One participant had previous Space Navigator experience. All but one participant exhibited normal color vision using the Ishihara Color Deficiency Charts (Ishihara, 2012). The participant with apparently irregular color vision obtained the third highest recorded score, indicating their ability to successfully identify the items in the game. Therefore, the analysis included their data. Participants self-reported spending an average of 48.7 hours per week using a computer or similar machine.

Apparatus and Environment

The experiment used a touch-screen tablet application titled "Space Navigator." Space Navigator closely resembles commercially-available air-traffic-control games. In this game, a human and agent work together as peers to achieve the highest score possible. The object of Space Navigator is to navigate red, blue, yellow, or green ships that spawn onto the screen to planets of their corresponding color, while obtaining randomly-appearing bonuses during their routes. The human-agent team receives 100 points upon successful navigation of ships to their corresponding planet. Ships are removed from the screen when they arrive at their appropriate planet. Additionally, the human-agent team receives 50 points for navigating ship paths through bonuses that appear on the screen. A bonus appears at a random on-screen location once every 10 seconds and remains on-screen until collected by a ship. The team loses 200 points when two ships collide. The human can physically draw a ship path with their finger, but if the human does not draw a path within a specified time window, the artificial agent presents a straight-line path from the ship to its appropriate planet. However, this agent path does not account for any bonuses or the paths of any other ships on screen. The human can draw or redraw a route at any time. The agent cannot overwrite a human-drawn route. Participants played all games on a Microsoft Surface Pro 4 in a quiet and secluded location.

Experimental Design and Procedure

The input variables to this study were agent response time (ART) and inter-arrival time (IAT). ART is the time an agent waits to draw a route for a new ship. IAT is the number of seconds between the times that two subsequent ships appear. Previous research narrowed and tested a range of IAT and ART values from 2s to 4s and 2.6s to 8.6s, respectively (Schneider et al., 2018). This research analyzed how the ratio of ART to IAT, referred to as the Adaptation Coefficient (AC), affects score, engagement, and workload (Schneider et al., 2018).

Decreasing IATs result in more ships appearing within a given time. This has the apparent and desired effect of increasing task load by requiring the human-agent team to provide more routes within a given time interval. Since these ships remain in the environment for a period of time to transit to their destination planet, the density of ships in the environment increases, increasing the probability of collisions, and reducing the number of possible collision free routes within the environment. This effect further increases task load as the human must draw or redraw longer and more complex routes.

Figure 1 displays the IAT and ART points used in this experiment, illustrated by points with markers "x" and "o", respectively. Past studies narrowed the sampling area to boundaries and points featured in Figure 1 by demonstrating team performance in the experiment environment remained similar for IAT values greater than 3.4s (Goodman et al., 2017; Schneider et al., 2018). The dashed lines that create the top and bottom boundaries represent AC of 2.0 and 0.5, respectively. These AC were chosen because they represent locations of manageable human workload in the Space Navigator environment, as discovered in previous research (Schneider et al., 2018), although human-agent team performance varied within this range. When IAT is significantly less than 2.6s, the human will struggle to keep up with new tasks, thereby experiencing overload. When IAT is significantly greater than 2.6s, the human will experience large breaks between new tasks, thereby experiencing underload. As ART decreases, the human typically draws routes slower than the agent, which could prevent the human from drawing and thereby decrease human engagement. Conversely, as ART increases, the human can draw routes faster than the agent, so one might assume that human engagement increases.

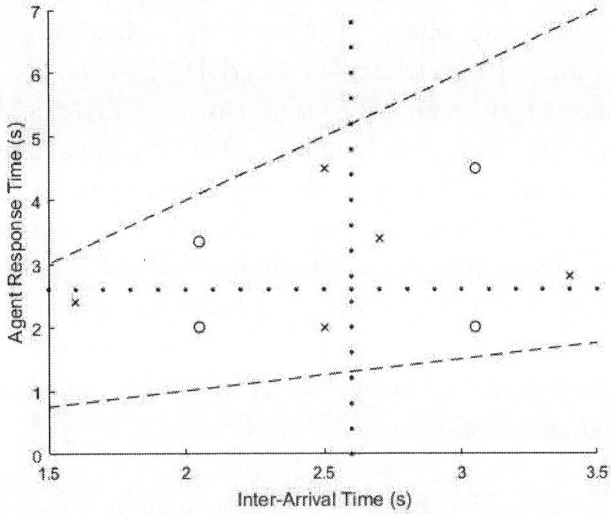

Figure 1. Depiction of Inter-Arrival Time (IAT) and Agent Response Time (ART) points sampled during the current experiment (shown as x's and o's). The vertical and horizontal dotted lines indicate the average human draw time of 2.6 s. The sloped dashed lines indicate a range of values useful for human-machine teaming based on previous research. Points marked with an "o" in Figure 1 represent the centroid of each region within the boundaries provided by the dashed and dotted lines. Points marked with an "x" were selected to be near the boundary extremes to provide insight into human performance near these transition regions.

For each experimental session, the research administrator provided a demonstration of Space Navigator to participants from a narrated script. The participants then played three, 2.5 minute practice rounds, each with an agent teammate, to become familiar with the Space Navigator environment. Practice rounds contained slower than average IAT and ART values to give participants time to understand the mechanics of the game and touchscreen response. Participants received no gameplay strategies during training.

The experimental session for each participant contained two blocks. Each block included nine, 1.75 minute trials with a workload questionnaire administered after each trial. Game time remained constant in all trials. Each block presented each input point described in Figure 1 to participants in a random order. A five-minute break separated the two blocks.

Data Analysis

Each experimental round contained the same game duration but employed different IAT. Thus, a different number of ships appeared in each experimental round. Therefore, it was inappropriate to compare the number of routes drawn and the total score across each experimental round as changes in IAT influenced these variables. To account for this difference, performance was measured as the percentage of the maximum possible score obtained in a game. Furthermore, engagement was calculated through two measures: human draws (HD) per ship and HD per second. When experiencing small IATs, the user may struggle to draw a route for every ship, even if the user desires to draw a manual route per ship. However, this does not mean the user is less engaged in the task than rounds where the user is physically capable of drawing a route for every ship. Therefore, it was desirable to use HD per second to measure overall engagement of a human at each IAT and ART point. However, HD per ship still proved useful for defining thresholds (i.e. we can say the human must at least engage with one in every five ships). Workload was measured using a subjective questionnaire containing three questions from NASA-TLX on a 0-20 scale. These questions were selected as previous studies found a correlation between the workload categories of temporal demand, effort, and performance with changes in IAT (Schneider et al., 2018). Workload values were standardized using min-max normalization within each participant to allow comparison across all participants. Total workload for a single Space Navigator round was calculated as the sum of the normalized workload values for each of the three workload questions.

Relationships between our independent and dependent variables were investigated using multiple regression analysis. This analysis contained two steps. First, multiple regression analysis on output variables was conducted to the third order. Second, insignificant effects were removed one at a time until only significant effects remained. Regression analysis was applied for each output variable across all participants. If large participant variability caused no significance for IAT and ART across all participants, regression analysis was conducted on the mean output values for each input IAT and ART combination.

Results and Discussion

Table 1 displays correlations of IAT and ART with human-agent team performance, human engagement, and workload. Results indicated IAT strongly correlates with score ($r(8) = 0.9229$, $p = 0.0004$), engagement ($r(8) = -0.7969$, $p = 0.0642$), and workload ($r(8) = -0.9578$, $p < 0.0001$). Results also indicated that ART strongly correlates with engagement ($r(8) = 0.8481$, $p = 0.0039$). From Table 1, it becomes evident that as IAT increases, the percent of maximum possible score increases, human draws per ship increases, and workload increases. Additionally, Table 1 illustrates that as ART increases, participant engagement with the system increases. These results are consistent with data obtained in preceding research (Schneider et al., 2018).

Table 1.
Correlations between variables. Values in bold represent significant correlation at α = 0.05. Italicized data points represent significant correlation at α = 0.10.

	Avg. % Max Score	Avg. HD per Ship	Avg. HD per Sec	Avg. Std Workload
IAT (IV)	**0.9229**	*0.6385*	**-0.7969**	**-0.9578**
ART (IV)	0.0018	**0.8481**	0.3955	0.0006

Multiple regression analysis on the data across all participant trials indicated that there was a collective significant effect between IAT and ART on percentage of max score, F(5, 246) = 25.4565, $p < 0.0001$, $R^2 = 0.3410$. Further examination of the predictors indicated that IAT ($t = 6.14$, $p < 0.0001$, $\beta = 0.1404$), IAT to the second degree ($t = -3.42$, $p = 0.0007$, $\beta = -0.1288$), ART ($t = -3.15$, $p = 0.0018$, $\beta = -0.1263$), ART to the second degree ($t = -2.96$, $p = 0.0034$, $\beta = -0.0812$), and ART to the third degree ($t = 2.84$, $p = 0.0049$, $\beta = 0.0757$) were significant predictors in this model.

Multiple regression on data across all participant trials indicated there was a collective significant effect between IAT and ART on human engagement represented as human draws per ship, F(2, 249) = 16.1716, $p < 0.0001$, $R^2 = 0.1150$. Further examination of the predictors indicated that IAT ($t = 2.78$, $p = 0.0058$, $\beta = 0.0890$) and ART ($t = 4.29$, $p < 0.0001$, $\beta = 0.0746$) were significant predictors in this model.

Multiple regression analysis to on data across all ART and IAT combination averages indicated there was a significant effect between IAT and ART on workload, F(4, 4) = 130.1843, $p = 0.0002$, $R^2 = 0.9924$. Further examination of the predictors indicated that IAT ($t = -18.71$, $p < 0.0001$, $\beta = -0.2345$), ART ($t = 4.94$, $p = 0.0078$, $\beta = 0.0377$), ART to the second degree ($t = -3.39$, $p = 0.0275$, $\beta = -0.0265$), and the interaction of ART and IAT ($t = 3.08$, $p = 0.0370$, $\beta = 0.0535$) were significant predictors in this model

Derivation of Near-Optimal Agent Response Function

To determine the optimal ART, the regression equations derived in the previous section were applied within an optimization problem. The optimization problem was solved for the ART at each IAT value between zero and four seconds on a 0.001s interval. This optimization sought to maximize the percentage of maximum score subject to the constraints that the participant would draw at least one route for every five ships and would have a mean standardized workload between the mean, plus or minus one standard deviation of the workload from this experiment (between 0.423 and 0.561).

The optimization determined that when IAT is less than approximately 1.5s, the optimal ART is 0s. In this range, IAT is much lower than the average human response time, so the human will likely struggle to match the pace at which new tasks appear. Therefore, the human will likely require shorter ART. Once IAT is greater than 1.5s, the ART increases as IAT increases, permitting the human to take on a more involved role since they can better keep up with the slower rate at which tasks appear. As IAT approaches the average human response time, it disrupts the linear function. This permits a constant ART for IAT near the average human response time. ART then continues to increase once IAT is greater than the average human response time. Violation of the constraints specified in the function occurred at IAT greater than 3s. For this reason, ART at IAT greater than 3s was extrapolated from the function starting at IAT of 2.7s. As IAT increases from 2.7s, the human has more time to complete present tasks until the next ship arrives. Therefore, human need for agent assistance remains low at IAT levels greater than 2.7s. Optimization produces a piecewise linear function for the calculation of the optimal ART based on IAT. Equation 1 provides this piecewise linear function.

$$\begin{aligned} &For\ IAT < 1.485, ART = 0 \\ &For\ 1.485 \leq IAT < 2.206, ART = 3.5327 * IAT - 5.2461 \\ &For\ 2.206 \leq IAT < 2.735, ART = 2.5471 \\ &For\ IAT \geq 2.735, ART = 5.2807 * IAT - 11.8955 \end{aligned} \tag{1}$$

Conclusion

Results from this study indicate that IAT is strongly correlated with human-agent team performance, human engagement, and workload. Furthermore, ART is correlated with human engagement. This study produced a method for computing ART as a function of IAT. The ART function was obtained by gathering data at logical IAT and ART points and calculating which ART produced the maximum percentage of possible team score while following workload and engagement constraints. The proposed ART function will be applied in subsequent research to determine if ART calculated from IAT can effectively balance workload and engagement while maintaining equal or better performance than a constant ART agent.

Disclaimer and Acknowledgement

The views expressed in this article are those of the authors and do not necessarily reflect the official policy or position of the Department of the Air Force, Department of Defense, nor the U.S. Government. The authors gratefully acknowledge the financial support of the Air Force Office of Scientific Research, Computational Cognition and Machine Intelligence Program.

References

Goodman, T. J., Miller, M. E., Rusnock, C. F., & Bindewald, J. M. (2017). Effects of agent timing on the human-agent team. *Cognitive Systems Research*, *46*, 40–51. https://doi.org/10.1016/j.cogsys.2017.02.007

Ishihara, S. (2012). Ishihara's design charts for colour deficiency of unlettered persons. Retrieved from https://scholar.google.com/scholar?hl=en&as_sdt=0,36&q=ishihara%27s+design+charts+for+colour+deficiency

Kaber, D. B., Riley, J. M., Tan, K.-W., & Endsley, M. R. (2001). On the Design of Adaptive Automation for Complex Systems. *International Journal of Cognitive Ergonomics*, *5*(1), 37–57. https://doi.org/10.1207/S15327566IJCE0501_3

Parasuraman, R. (2008). *Supporting Battle Management Command and Control: Designing Innovative Interfaces and Selecting Skilled Operators*. Fairfax, VA. Retrieved from http://www.dtic.mil/docs/citations/ADA480645

Schneider, M. F., Bragg, I. L., Henderson, J. P., & Miller, M. E. (2018). Human Engagement with Event Rate Driven Adaptation of Automated Agents. In *2018 IISE Annual Conference*. Orlando, FL.

VIRTUAL REALITY FLIGHT ENVIRONMENTS MAY TAX WORKING MEMORY AND DISRUPT PROSPECTIVE MEMORY

Cassandra Ommerli, Jinous Mirzaagha, Chunyun Ma, Kathleen Van Benthem, & Chris Herdman
Carleton University
Ottawa, ON, Canada

While vivid virtual reality (VR) environments may afford better performance for some flight tasks, it is possible that enhanced stimuli could overload some cognitive resources. Prospective memory (PM) is a cognitive factor sensitive to working memory and visual processing demands, and it may be a performance factor either adversely affected or enhanced by VR factors. Forty-seven pilots flew a VR flight simulation scenario, which included an auditory cue-based PM task. Self-ratings of psychological experiences in VR revealed three factors with relationships to PM: fluency, presence, and interactivity. Path analyses examined the relation of each of these factors with PM, and with two types of working memory, based on Level 1 SA. Higher fluency ratings were associated with lower PM, whereas greater presence and interactivity were correlated with better PM. Working memory also significantly mediated the effects of fluency on PM.

The use of virtual reality (VR) simulators for flight training is becoming increasingly more widespread. Moreover, VR can support a wide range of training paradigms involving many different aircraft types and flight conditions. However, there is a lack of research concerning the impact of users' psychological experiences in VR on fundamental cognitive processes during flight. The present research considered three aspects of users' psychological experiences in VR and how they might impact prospective memory (PM), which is the ability to recall or perform an intended thought or action at a future point in time (Brandimonte, Einstein, & McDaniel, 1996). PM is also a key index of pilot performance. The three VR factors were presence, interactivity, and fluency. Presence was defined as the subjective experience of "being there" in a virtual environment (Witmer & Singer, 1998), interactivity concerned the degree to which users felt that they were able to influence the content of the virtual world (Steur, 1992), while fluency reflects a form of immersion characterized by intense focus and concentration (Rheinberg, Engeser, & Vollmeyer, 2003).

PM is a cognitive factor sensitive to ongoing working memory and visual processing demands; it is also influenced by the salience of memory cues, ongoing task workload, and individual differences, such as age, cognitive health, and expertise. Van Benthem, Herdman, Tolton, and LeFevre (2015) found that in naturalistic settings, older pilots with lower cognitive health demonstrated poor PM when ongoing task workload was high, and when cue-salience was low. These results suggest that flight environments with high working memory demands impede detection of low-salience cues that are less associated with the PM task. Therefore, working memory overload may occur in visually-rich VR environments, resulting in reduced PM performance, particularly when the task is linked to low-salience PM cues. For example, Bailey, Bailenson, Won, Flora, and Armel (2012) found there to be a significant negative association between participants' levels of perceived VR presence and memory performance on a cued recall task. In this paper, path analyses with latent factors were used to examine the effects of three VR factors on PM and determine whether some of these effects were mediated by working memory.

Method

All testing took place at the Advanced Cognitive Engineering Laboratory (ACE) at Carleton University as part of a larger research agenda investigating the effectiveness of a VR-based cognitive assessment in predicting pilot risk in general aviation.

Participants

Data were collected from a total of 47 participants. Eligibility criteria for participation in the study included individuals who were in possession of a valid pilot's license and medical certificate, and who had flown as pilot in command within the past two years. The majority of participants were recruited through flyers circulated at Ottawa-region flight clubs in Ontario, Canada. Some pilots were also former participants of ACE lab aviation studies who had consented to joining a research mailing list. Table 1 provides a snapshot of pilot demographics.

Table 1. Description of Pilot Sample

	Age	License/Rating	Total Hours Flown	Total Years Licensed
Mean	47.13	4.09	1384.85	14.77
Standard Deviation	17.42	1.38	2684.51	14.34
Minimum	17	1	2	1
Maximum	71	6	12000	70

Notes. License/Rating was based on a six-point scale, where 1 = student permit, 2 = recreational permit 3= visual flight rules (no additional ratings), 4 = visual flight rules with additional ratings, and 5 = instrument rated, commercial, and instructors, 6 = airline transport.

VR Simulated Flight Apparatus

Participants wore a Microsoft Oculus Rift VR headset, which displays graphics for a 360-degree external environment and full Cessna 172 cockpit and fuselage. The virtual environment was rendered using Lockheed Martin's Prepar3D software. A prototype flight control unit was developed in-house to accompanying the functionality in the virtual cockpit. The location of the simulator controls mirrored the location of controls present in the simulated cockpit.

Flight exercise

Prior to the experimental exercise, participants were given the opportunity to practice flying in the VR simulator without additional task requirements. The virtual airfield was uncontrolled, and participants could not see or interact with the aircraft they heard information about. Following the practice exercise, participants were briefed with the details of general flight plan and notified that they would be given specific directions pertaining to airspeed, heading, and altitude over the duration of the VFR flight. Participants read-back instructions to ensure that they understood the trajectory. Participants flew in the simulator for approximately 35-45 minutes, completing a 4-leg flight with an arrival and departure at the same airport.

Measures

Auditory PM task. The PM task involved detecting the presence of an auditory cue (a spoken word) placed randomly within the pre-recorded radio calls. The radio calls were played over the headset throughout the flight, and conveyed information regarding the identity and behavior of neighboring aircraft. Participants were asked to listen to the content of the calls, and to press a button on the left side of the yoke whenever they heard the word "traffic" (the special cue). The cue word was presented randomly in approximately half of all calls. Radio calls were administered in the same order and at approximately the same time points for all participants.

Working memory. Working memory indices were developed from Level 1 SA measures (information detection) using a system similar to Endsley's (1988) Situation Awareness Global Assessment Technique (SAGAT). This method was selected, as it produced ecologically valid working memory measures, and because working memory plays an essential role in facilitating Level 1 SA. Via working memory processes, individuals are able to store and manipulate a finite quantity of information on a temporary basis (Baddeley & Hitch, 1974). New information is continuously combined with existing content in working memory to generate updated snapshots of one's environment (Endsley & Jones, 2012). To collect our working memory data, the experimental exercise was paused at random intervals, during which time the external environment and gauges were occluded, and the assessment was administered. Participants were given a set of 24 queries pertaining to situation awareness, which were divided across three freezes. While all three levels of situation awareness were queried in this experiment, only Level 1 SA items were used to index working memory for this analysis. For example, for Level I SA, a participant might be asked about details of circuit traffic, such as "State the orbit altitude of Gulf Hotel India". All participants received queries in the same order and experienced SA freezes at approximately the same time points. Each item was scored for accuracy and completeness (e.g., fully incorrect, partially correct, or fully correct). Factor analysis was conducted using only the Level 1 SA queries with strong working memory components. Two variables emerged: one latent factor involving visual updating of information from cockpit gauges (our *visual working memory* variable), and one single-item factor involving auditory updating of environmental information (wind speed and direction communicated via a ground services radio call) (our *auditory working memory* variable).

VR experiences questionnaire. Following the flight exercise, participants were asked to complete a questionnaire, which contained 14 items pertaining to user perceptual exercises in VR environments. These items (Table 2) were adapted from a questionnaire created by Mütterlein (2018), which tests perceptual experiences in VR for fluency, presence, and interactivity. All VR experience items were answered using a seven-point rating scale. Latent variables were developed from the VR experiences questionnaire items. Support was found for the three proposed constructs of fluency, presence, and interactivity. During the latent measurement model fitting, indicators were omitted from the final analysis if they resulted in factor loadings below .60, or if they brought down the value of the average variance extracted (AVE) and the composite reliability (CR) coefficients when included. Each of the final three factors demonstrated appropriate levels of internal consistency, resulting in composite reliability coefficients above the conservative threshold level of .70 (Kock, 2018). Moreover, all factors exceeded the threshold AVE value of .50 (Kock, 2018). Table 2 shows the factor loadings, cross-loadings, and final quality scores for the three VR perceptual experience factors.

Table 2. Factor loadings and quality scores for final VR perceptual experience variables

Item	Fluency	Presence	Interactivity
The Oculus Rift created a new world for me, and this new world suddenly disappeared when the exercise ended.	.25	**.86**	.31
When I removed the Oculus Rift, I felt as if I returned to the "real world" after a journey.	.21	**.83**	.33
I forgot about my immediate surroundings when I was using the Oculus Rift.	.17	**.72**	.54
I had no difficulty concentrating.	**.81**	.34	.29
My mind was free to focus on flying.	**.88**	.11	.30
My thoughts and movements felt effortless.	**.83**	.22	.37
I was totally absorbed in what I was doing.	.44	.38	**.84**
The Oculus Rift content allowed me to interact with the virtual world.	.14	.40	**.85**
I had the feeling that I could influence the virtual world of the Oculus Rift.	.39	.47	**.86**
Composite reliability	.88	.84	.89
Average variance extracted	.70	.64	.72

Note. Bolded factor loadings were the final items used in each latent VR construct.

Results

Partial least squares path analyses were conducted using WarpPLS v. 6.0, which quantified the direct effects of VR factor and working memory on PM, as well as the indirect effects of VR factors on PM, mediated by working memory. Six path models were tested: three with simple direct effects from each VR and working memory factor on PM, and three where working memory served as a mediator for effects from the VR factors on PM. Figure 1 shows a potential path model where visual working memory is mediating the effects of fluency on PM.

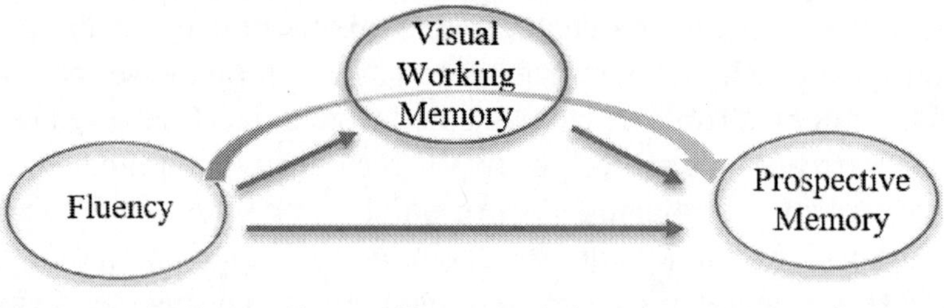

Figure 1. The straight arrows indicate direct effects, and the curved arrow indicates an indirect effect of fluency on PM mediated by visual working memory.

Fluency, PM, and Working Memory

Higher ratings for fluency were associated with lower PM, *beta*= -.28, $p < .05$ and lower visual working memory, *beta*= -.43, $p < .01$. Visual working memory was positively correlated with PM, *beta*=.35, $p < .01$. In contrast, fluency was positively correlated with auditory working memory, which itself was negatively correlated with PM, *beta*= -.28, $p < .01$. The mediation models found that significant effects of fluency on PM were mediated by both visual working memory, *beta* = -.15, $p = .043$, and by auditory working memory, *beta* = -.12, p=.04.

Presence, PM, Working Memory

Higher ratings for presence were associated with higher ratings for PM, *beta*= .36, $p <$ .01, visual working memory, *beta*= .28, p < .01, and auditory working memory, *beta*= .25, $p <$ .01. In the presence model, auditory working memory was negatively correlated with PM, *beta*= -.33, $p < .01$, and there was no significant correlation between visual working memory and PM. No mediating effects of either working memory measure were observed for the relation of presence to PM.

Interactivity, PM, and Working Memory

Similar to presence, higher ratings for interactivity were associated with higher ratings for PM, *beta*= .19, $p = .03$, visual working memory, *beta*= .29, p < .01, and auditory working memory, *beta*= .24, $p < .01$. As in the previous models, auditory working memory was negatively correlated with PM in the interactivity model, *beta*= -.35, $p < .01$, while visual working memory was positively correlated with PM, *beta*= .38, $p < .01$. No mediating effects of either working memory measure were observed for the relation of interactivity to PM.

Discussion and Implications

Our findings suggest a relationship between fluency and PM, which might be explained by working memory limitations. Fluency negatively influenced PM and visual working memory, but positively influenced the working memory variable associated with details of the virtual environment. In both cases, working memory indices were shown to mediate a significant portion of the relationship between fluency and PM, with the strongest effect being produced by visual working memory. These results suggest that experiences of fluency or intense concentration in VR significantly tap into working memory resources. Content belonging to the VR environment may engage fluency, which could in turn divert attention otherwise needed for PM. These findings are aligned with research by Bailey et al. (2012), which suggests that highly vivid sensory experiences in VR create a strain on cognitive resources.

Presence and interactivity were shown to positively influence PM and working memory. Presence, in particular, had a strong positive effect on both cognitive processes. He, Zhu, Perlin, and Ma (2018) found that VR environments with high representational fidelity support enhanced situation awareness, proposing that vivid design facilitates continuity in presence through enhancing the “realism” and “believability” of the experience. These attributes in turn help to avoid breaks in presence that interfere with a user’s awareness. Returning to working memory, enhanced engagement of presence and interactivity do not appear to create a strong demand on the cognitive resources required for PM and working memory processes. Stable and continuous experiences of these VR factors may help to mitigate situations where disconnection between a user’s virtual and external environment create competing stimuli and added attentional demands.

Future research in this area might seek to identify a core set of criteria related to vividness in flight simulators and explore the impact on VR perceptual factors. Additional work should also be done to provide validation for the proposed VR constructs, especially fluency. In order to create more robust VR cognitive training resources for pilots, researchers should seek to develop a more precise understanding of the threshold levels at which VR factors begin to support or detract from cognitive processing.

Acknowledgements

Research infrastructure was supported through the Canadian Foundation for Innovation and the Ontario Innovation Trust. We thank Anne Barr and Andrew Staples of the Carleton University, ACE Laboratory for their consultation and engineering expertise.

References

Baddeley, A., & Hitch, G. (1974). Working Memory. In G.H. Bower (Ed.), *The Psychology of Learning and Motivation: Advances in Research and Theory*, 8, 47-49. New York: Academic Press.

Bailey, J., Bailenson, J.N., Won, A.S., Flora, J., & Armel, K.C. (2012) Presence and memory: immersive virtual reality effects on cued recall. *In Proc. of the International Society for Presence Research Annual Conference, Philadelphia, PA, 24–26, October 2012*: ISPR-14.

Brandimonte, M., Einstein, G.O., & McDaniel, M. A. (Eds.). (1996). *Prospective memory: Theory and applications. Mahwah*, NJ, US: Lawrence Erlbaum Associates Publishers.

Endsley, M. (1988). Situation Awareness Global Assessment Technique (SAGAT). *In Proc. of the National Aerospace and Electronics Conference (NAECON)*, 789–795. New York: IEEE.

Endsley, M.R., & Jones, D.G. (2012). *Designing for Situation Awareness: An Approach to User-Centred Design* (2nd Edition). Boca Raton, FL: CRC Press.

He, Z., Zhu, F., Perlin, K., and Ma, X. (2018). *Manifest the Invisible: Design for Situational Awareness of Physical Environments in Virtual Reality*. Cornell University. Retrieved from arXiv:1809.05837v1.

Kock, N. (2018). *WarpPLS User Manual: Version 6.0.* Lardeo, TX: ScriptWarp Systems.

Mütterlein, J. (2018). The Three Pillars of Virtual Reality? Investigating the Roles of Immersion, Presence, and Interactivity. *In Proc. of the 51st Annual Hawaii International Conference on System Science,* Honolulu, Hawaii, 3-6 January 2018: HICSS-51.

Rheinberg, F., Engeser, S., & Vollmeyer, R. (2002). Measuring Components of Flow: The Flow-Short-Scale. *In Proc. of the 1st International Positive Psychology Summit*, Washington, DC, October 2002.

Steur, J. (1992). Defining Virtual Reality: Dimensions Determining Presence. *Journal of Communication, 42*(4), 73-93.

Van Benthem, K.D., Herdman, C.M., Tolton, R.G., & LeFevre, J.-A. (2015). Prospective memory failures in aviation: effects of cue salience, workload, and Individual differences. *Aerospace Medicine and Human Performance, 86*(4), 366–373.

Witmer, B. G., & Singer, M. J. (1998). Measuring Presence in Virtual Environments: A Presence Questionnaire. *Presence: Teleoperators and Virtual Environments, 7*(3), 225-240.

WHICH OCULAR DOMINANCE SHOULD BE CONSIDERED FOR MONOCULAR AUGMENTED REALITY DEVICES?

Elodie Bayle[a,b,c], Estelle Guilbaud[a], Sylvain Hourlier[b], Sylvie Lelandais[c], Laure Leroy[d], Justin Plantier[a,c], Pascaline Neveu[a]

[a] French Armed Forces Biomedical Research Institute, Brétigny-sur-Orge, France
[b] Thales AVS France SAS, Mérignac, France
[c] IBISC, Evry University, Paris Saclay University, Evry, France
[d] Paris 8 University, Saint-Denis, France

A monocular augmented reality device allows the user to see information that is superimposed on the environment. As it does not stimulate both eyes in the same way, it creates a phenomenon known as binocular rivalry. The question therefore arises as to whether monocular information should be displayed to a particular eye and if an ocular dominance test can determine it. This paper contributes to give a better understanding of ocular dominance by comparing nine tests. Our results suggest that ocular dominance can be divided into sighting and sensorial dominance. However, different sensorial dominance tests give different results, suggesting that it is composed of distinct components that are assessed by different tests. There is a need for a comprehensive test that can consider all of these components, in order to identify on which eye monocular information should be directed to when using monocular augmented reality devices.

Augmented reality refers to an interactive virtual interface (in 2D or 3D) combined and superimposed in real time with the environment. Within the semi-transparent displays ('see-through' devices), displays can be divided into three types: binocular (two images are displayed, one to each eye), biocular (the same image is displayed to both eyes), or monocular (one image is displayed to one eye). Monocular devices are, in most cases, adjustable (*i.e.* information can be centered with respect to the observer's pupils), and also the lightest and least expensive system. However, as virtual information is only presented to one eye, the two eyes are not stimulated in the same way. This can lead to a phenomenon known as binocular rivalry, which can cause visual fatigue, headaches, visual suppression, etc. (Hershberger & Guerin, 1975). Monocular augmented reality devices therefore raise the question of which eye should receive monocular information.

Currently, there is no consensus in the literature. On the one hand, Hershberger and Guerin (1975) argue that there are fewer disturbing phenomena when information is fed to the dominant eye. On the other hand, and more recently, Cupero et al. (2009) demonstrated that the eye that information is displayed to has no influence on perception. In these two studies, dominant eye was considered as the sighting eye. However, there is little agreement among the scientific community regarding the definition of ocular dominance. The many definitions go hand-in-hand with the multitude of ways to determine it, which together contribute to a lack of clarity on its determination and its use. As binocular rivalry is a sensorial phenomenon (Coren & Kaplan, 1973; Handa et al., 2004; Li et al., 2010; Seijas et al., 2007), the question arises of whether the sighting eye is relevant to considered in the case of monocular augmented reality and if all dominant eye tests gives consistent results. Only Gronwall and Sampson (1971) have found a positive correlation between various dominance tests, suggesting a unifactorial phenomenon. Other authors identified two (*i.e.* sighting and sensory dominance) (Cohen, 1952) to fives groups (*i.e.* monocular acuity, sighting dominance, orientation dominance, sensory dominance and hemi-retina dominance) (Lederer, 1961).

As with any lateralization of the human body, ocular dominance can be expressed in terms of strength of dominance. Chaumillon (2017) found that patients with less pronounced ocular dominance are thought to have fewer visual constraints leading to better support for monovision. This finding, in turn, raises the question of the impact of the strength of ocular dominance on the use of monocular augmented reality devices.

The aim of the present study is to identify whether different dominant eye tests give consistent results in term of preferred eye and strength of dominance.

Methods

Participants

A total of nineteen participants, ten men and nine women with normal or corrected-to-normal vision (18–60 years; mean 36.9±12.1) were enrolled in the experiment. Three participants were excluded from the analysis because their acuity was worth than 0.05 log or their blink rate meant that fixation could not be maintained during the form rivalry test. All participants gave written informed consent. The study was conducted in accordance with the Helsinki Declaration and met local legal requirements (N IDRCB: 2018-A01331-54).

Tests

We compared nine ocular dominance assessment methods conducted in the same order for each participant: far and near acuity, Near Point Convergence test (NPC), first repetition of the form rivalry test, Bagolini test, Worth test, +1.5δ blur test, hole-in-card test, second repetition of form rivalry test, motion coherence threshold test repeated three times and last repetition of form rivalry test (Table 1.).

Following Blake and Logothetis (2002), form rivalry and motion threshold test were displayed using a mirror stereoscope, for details see Neveu et al. (2012). At 0.5 meter, participants perceived a minimum horizontal visual field of 18.30°, and 21.17° vertically through the device. They could indicate what they saw with a computer keyboard and their position was stabilized via a chin rest, forehead support (Handa et al., 2004) and mouthpiece similar to dental X-ray equipment. Stimuli were generated using the MATLAB Psychophysics Toolbox (The MathWork, Natick, MA, USA).

Form rivalry test was based on Coren and Kaplan (1973) and Handa et al. (2004). Two sinusoidal patterns were presented using the stereoscope: one oriented at 0° for the right eye; and at 90° for the left eye. A spatial frequency of 3.87 cpd was selected to approach the highest spatial frequency used in the form rivalry literature. Contrast was set at 100%. Maximum and minimum target luminance were set at 120 and 0.8cd/m², respectively, and mean luminance of the target and background was 42cd/m². The stimulation was presented for ten seconds, during which participants were asked to avoid blinking (blinking acts as a reset of the visual system (Rainwater & Cogan, 1975), and the frequency could have modified the results). They pressed the 'space' key on the keyboard when they only saw horizontal striations, and the 'enter' key when they only saw vertical striations. If their perception was mixed, they did not press any key. The task was repeated four times with a three-second break to allow the participant to blink.

Table 1.
Different tests used.

Test	Dominant eye	Reférence
1. Hole-in-card	Eye behind the hole when focusing a Landol rings scale at 5m, 551 lux.	(Li et al., 2010)
2. NPC	First eye to diverge.	(Li et al., 2010)
3. Far acuity	Eye having the best acuity on Landol rings scale.	
4. Near acuity	Eye having the best acuity on Logarithmic morphoscopic scale at 0.5m, 551 lux.	
5. Bagolini	Eye needed the highest filter to suppress (by 0.3 to 1.8 log units) at 0.5m, 20 lux.	(Li et al., 2010)
6. Near Worth	Eye having the colored filter of the color of the bottom point of the test at 0.33m, 20 lux.	(Li et al., 2010)
7. +1.5δ blur	Eye having the highest blur induced by +1.5δ when focusing the Landol rings scale.	(Pointer, 2012)
8. Form rivalry	Eye having the target seen the longest time.	(Coren & Kaplan, 1973) (Handa et al., 2004)
9. Motion threshold	Eye having the lowest threshold of points needed to evaluate movement.	(Li et al., 2010)

Data analysis

The results for each test were coded as −1 if the left eye (LE) was dominant, +1 if the right eye (RE) was dominant or 0 if the test did not detect a dominant eye. To be in accordance with the qualitative analysis of the results, strength of dominance for acuity tests, Bagolini and +1.5δ were evaluated using the formula (Li et al., 2010): (RE results – LE result)/(RE result + LE result) whereas it was determined by: (LE threshold – RE threshold)/(LE threshold + RE threshold) for motion coherence test. Strength of dominance for form rivalry was obtained by: (percentage RE – percentage LE)/100. Z scores were calculated qualitative and quantitative data to compare the tests together. Qualitative and quantitative data were analyzed using a Kendall 2 x 2 rank correlation (Coren & Kaplan, 1973; Li et al., 2010). Next, following Coren and Kaplan (1973), a factor analysis was run to determine if the results could be grouped. All statistical analyses used the STATISTICA® (StatSoft, Tulsa, OK, USA) statistics package, and a p value ≤ 0.05 was considered significant.

Results

Results across the tests are very differents (Table 2). To compare qualitative results correlation coefficients were calculated on Z scores. The best correlation was found between the hole-in-card test and the motion threshold test ($\tau=0.683$). The hole-in-card test was also highly correlated with the near vision acuity test ($\tau=0.583$) and the NPC ($\tau=0.510$). To explore the findings in more detail, we carried out a factorial analysis (Table 2). The first factor grouped four tests which can be considered as assessing sighting or motor dominance. Strength of dominance was only analyzed on the other tests (*i.e.* Bagolini, +1.5δ blur, and form rivalry tests), as the sensorial dominance can be the relevant dominance to considered in the case of monocular augmented reality.

Figure 1 shows the participant distribution of the quantitative data in absolute values – values closer to one indicate greater dominance. This figure highlights that ocular dominance was, in general, weak and that the distribution of strength dominance according each test varies. No correlation between quantitative sensory tests was found.

Table 2.
Percentages of Participants (n=16) Classified with Right, Left, or Uncertain and Factor Loadings on Tests for Ocular Dominance on the three Extracted Factors.

Test	Right	Left	Uncertain	Factor I	Factor II	Factor III
1. Hole-in-card	62.5	37.5	-	0.906*	0.047	0.044
2. NPC	31.3	18.8	50	0.674*	0.205	0.212
3. Far acuity	18.8	50	31.3	0.305	−0.745*	0.003
4. Near acuity	31.3	25	43.8	0.708*	0.018	−0.511
5. Bagolini	43.8	25	31.3	0.006	−0.730*	0.074
6. Near Worth	12.5	31.3	56.3	0.021	−0.365	−0.849*
7. +1.5δ blur	50	43.8	6.3	−0.265	−0.799*	0.212
8. Form rivalry	35.7	64.3	-	−0.396	−0.254	0.006
9. Motion threshold	43.8	56.3	-	0.755*	−0.354	0.337
10. Handedness	75	25	-			

Note. *Values above 0.6 are high enough to be considered statistically reliable.

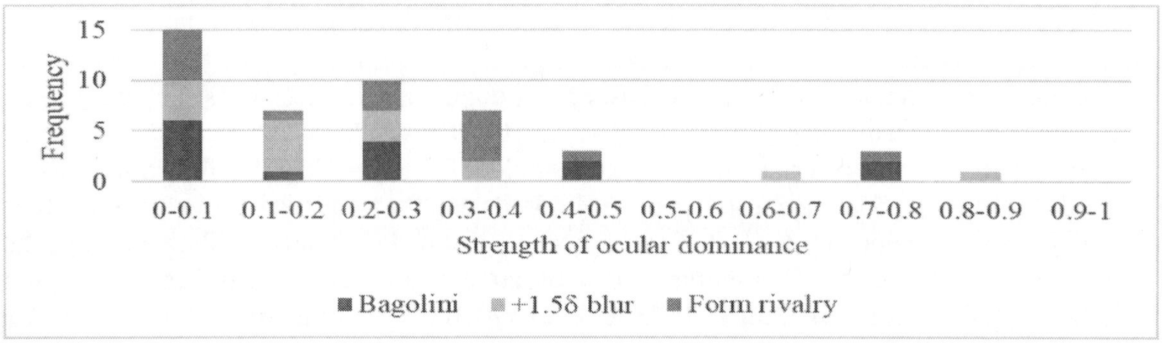

Figure 1.
Distribution of the Population as a Function of Strength of Ocular Dominance.

Discussion

The aim of the present study is to identify whether different dominant eye tests give consistent results in term of preferred eye and strength of dominance. More globally, the goal was to determine whether a particular test is effective in the context of using a monocular augmented reality device.

The analysis of the distribution of ocular dominance shows that it differs according to the test. The results of the hole-in-card test are consistent with Gronwall and Sampson (1971) study.

Like Seijas et al. (2007), we found that the Worth test is unable to classify dominance in a high proportion of subjects, and that the blur test has the lower uncertainty in clinical tests.

Factor analysis distinguished several groups of tests. The first consists of the hole-in-card test, the NPC test, the near vision acuity test, and the motion threshold test. This group of tests mainly takes into account tests identified in the literature as characterizing sighting or motor dominance (*i.e.* hole-in-card and NPC). Correlations between the tests are consistent with this grouping and are consistent with the literature (Coren & Kaplan, 1973; Gronwall & Sampson, 1971; Li et al., 2010, 2010; Pointer, 2012; Seijas et al., 2007). Thereby, the near vision acuity dominance would be more related to the participant's ability to converge than to discriminate details (Pointer, 2012) and it is possible that, despite the central fixation point during the moving threshold test, the moving points generate involuntary ocular eye movement.

The second factor grouped the far vision acuity test, the Bagolini test, and the +1.5δ blur test. However, there were no significant correlations between them. These results suggest that even they may share some common features, they do not give the same results, and they cannot be used interchangeably.

The factorial analysis identified a third factor, mainly based on the near Worth test. The fact that this test is responsible for a factor and does not correlate with other tests suggests that color plays a singular role in sensorial dominance. Our result is consistent with Seijas et al. (2007), who also found that the Worth test was uncorrelated with any other test.

Our results suggest that different sensorial dominance tests give different results. This was confirmed by the analysis of the strength of dominance. Paired correlations highlighted that tests are uncorrelated, as several authors have already noted (Cohen, 1952; Coren & Kaplan, 1973; Li et al., 2010). Our results suggest that sensorial ocular dominance is not unifactorial, but is made up of distinct components. Specifically, each test appears to assess one component of the sensorial ocular dominance. Thereby, Worth test shows which eye is preferred when color challenges binocular vision; the Bagolini test assesses the preferred eye when luminance challenges binocular vision; and the +1.5δ test assesses the preferred eye when blur challenges binocular vision.

Our results suggest that ocular dominance tests fall into two distinct groups: motor and sensory dominance. It suggests that the eye considered by previous studies (Cupero et al., 2009; Hershberger & Guerin, 1975) to explore if information must be displayed in a particular eye when using a monocular augmented reality device is not relevant to considered. While motor tests have similarities, the results of different sensory tests appear to be inconsistent. This latter observation highlights the importance of considering the set of sensory ocular dominance factors to determine the eye on which monocular information should be directed to when using monocular augmented reality devices. As far as we are aware, no test exists to reliably, and whatever the stimulus, determine the sensorial dominance.

References

Blake, R., & Logothetis, N. K. (2002). Visual competition. *Nature Reviews Neuroscience*, *3*(1), 13–21. https://doi.org/10.1038/nrn701

Chaumillon, R., Alahyane, N., Senot, P., Vergne, J., Lemoine-Lardennois, C., Blouin, J., … Vergilino-Perez, D. (2017). Asymmetry in visual information processing depends on the strength of eye dominance. *Neuropsychologia*, *96*, 129–136.

Cohen, J. (1952). Eye-Dominance. *The American Journal of Psychology*, *65*(4), 634. https://doi.org/10.2307/1418050

Coren, S., & Kaplan, C. P. (1973). Patterns of ocular dominance. *Optometry & Vision Science*, *50*(4), 283–292.

Cupero, F., Valimont, B., Wise, J., Best, C., & De Mers, B. (2009). *Head Worn Display System for Equivalent Visual Operations* (No. NASA/CR-2009-215781). Phoenix, AR: Nasa.

Gronwall, D. M. A., & Sampson, H. (1971). Ocular dominance: a test of two hypotheses. *British Journal of Psychology*, *62*(2), 175–185. https://doi.org/10.1111/j.2044-8295.1971.tb02028.x

Handa, T., Mukuno, K., Uozato, H., Niida, T., Shoji, N., & Shimizu, K. (2004). Effects of dominant and nondominant eyes in binocular rivalry. *Optometry & Vision Science*, *81*(5), 377–383. https://doi.org/10.1097/01.opx.0000135085.54136.65

Hershberger, M. L., & Guerin, D. F. (1975). *Binocular rivalry in helmet-mounted display applications* (No. AD-AO13 838) (pp. 1–146). Culver City, CA: Aerospace Medical Research Laboratory.

Lederer, J. (1961). Ocular dominance. *Clinical and Experimental Optometry*, *44*(11), 531–574. https://doi.org/10.1111/j.1444-0938.1961.tb06505.x

Li, J., Lam, C. S. Y., Yu, M., Hess, R. F., Chan, L. Y. L., Maehara, G., … Thompson, B. (2010). Quantifying sensory eye dominance in the normal visual system: a new technique and insights into variation across traditional tests. *Investigative Ophthalmology & Visual Science*, *51*(12), 6875–6881. https://doi.org/10.1167/iovs.10-5549

Neveu, P., Philippe, M., Priot, A.-E., Fuchs, P., & Roumes, C. (2012). Vergence tracking: a tool to assess oculomotor performance in stereoscopic displays. *Journal of Eye Movement Research*, 9. https://doi.org/10.16910/jemr.5.2.1

Pointer, J. S. (2012). Sighting versus sensory ocular dominance. *Journal of Optometry*, *5*(2), 52–55. https://doi.org/10.1016/j.optom.2012.03.001

Rainwater, G., & Cogan, R. (1975). Simultaneous disappearances of both images during binocular rivalry of contours and effect on eye-blink rate. *Perceptual and Motor Skills*, *40*(1), 71–77. https://doi.org/10.2466/pms.1975.40.1.71

Seijas, O., de Liaño, P. G., de Liaño, R. G., Roberts, C. J., Piedrahita, E., & Diaz, E. (2007). Ocular dominance diagnosis and its influence in monovision. *American Journal of Ophthalmology*, *144*(2), 209–216. https://doi.org/10.1016/j.ajo.2007.03.053

IMPACT OF ATCO TRAINING AND EXPERTISE ON DYNAMIC SPATIAL ABILITIES

Nadine Matton
ENAC & CLLE, University of Toulouse
Toulouse, France
Jean-Baptiste Gotteland
ENAC, University of Toulouse
Toulouse, France
Géraud Granger
ENAC, University of Toulouse
Toulouse, France
Nicolas Durand
ENAC, University of Toulouse
Toulouse, France

Dynamic spatial ability is supposed to be involved in a critical process of air traffic controllers, namely conflict detection. The present paper aims at testing whether dynamic spatial ability improves with air traffic control training and/or experience. We designed a laboratory task to assess the performance in predicting if two moving disks would collide or not. We conducted a cross-sectional study with four groups of participants : ATCO trainees at the beginning (*N*=129), middle (*N*=80) or end of training (*N*=66) and experienced ATCOs (*N*=14). Results suggested on one hand that air traffic control training leads to a decrease in the number of extremely high proportions of undetected collisions from the middle of the training. On the other hand, air traffic control operational experience leads to a decrease in the number of extremely high proportions of falsely detected collisions.

Spatial ability has been identified as one of the core abilities required for air traffic control (e.g., Durso & Manning, 2008). Indeed, an important part of air traffic controllers' job consists in understanding and manipulating visual and spatial information. A review of research on spatial abilities (Hegarty & Waller, 2005) revealed that spatial ability is composed of several separate abilities such as spatial visualization (supposed to be involved in a paper folding test for example) or spatial relations (supposed to be involved in a mental rotation test for example). Dynamic spatial abilities have been studied with the possibilities offered by computer testing to investigate the reasoning about motion and the integration of spatial information over time (Hegarty & Waller, 205, p.135). An early study highlighted that dynamic spatial ability could be interpreted as a distinct factor from static spatial ability (Hunt, Pellegrino, Frick, Farr & Alderton, 1988). Later, D'Oliveira (2004) confirmed the specificity of dynamic spatial ability within the spatial domain. However, dynamic spatial ability has been less studied than other components of spatial ability like mental rotation. Besides, among the "worker requirements" identified by Morath, Quartetti, Bayless and Archambault (2001, cited by Durso & Manning, 2008) for the job of air traffic control, four were grouped under the "spatial" label and comprised "visualisation" and "projection". Thus, the projection of trajectories of moving elements was highlighted as a central process for air traffic controllers (ATCOs).

As air traffic controllers are supposed to frequently use cognitive processes aimed at predicting whether two moving elements would collide, one question that arose was whether the performance at such tasks would improve during training and with experience. Thus, the present paper adresses the question of the potential improvement of the performance at a dynamic spatial ability task with air traffic control training and professional experience. Indeed, air traffic control training confronts learners to many dynamic visual problem solving situations. Therefore, their performance in extrapolating trajectories from dynamic visual data should be higher at the end of their training compared to the beginning. Similarly, after several years of air traffic control experience, ATCOs should have better performances at such tasks compared to ATCO students at the beginning of their training.

Method

Participants

Two hundred and eighty nine participants were recruited for the present study. They comprised 275 ATCO trainees at three stages of their air traffic control training (129 at the beginning, 80 at the middle and 66 at the end of their training) at ENAC (Ecole Nationale de l'Aviation Civile) as well as a group of professional ATCOs (n=14, with a mean experience in a control center of M=10,6 (SD=4,2) years).

Measures and Procedure

The TwoBalls test. A specific dynamic spatial ability test has been designed in order to measure the performance at predicting whether two moving disks would collide or not (see Figure 1). After three familiarisation trials with feedback on the correctness of their answer, participants were confronted to 50 test trials which varied in the heading and speed of each moving disk. Both disks had a diameter of 64 pixels (participants sat in front of a 24-inch, 1920 x 1200 resolution computer screen, thus the disk measured approximately 1.7 cm) and moved in a window of 640 pixels. The angles of the disk trajectories varied from 16° to 323° and the speeds varied from 10 to 40 pixels/s. At each trial, participants had to decide, as quickly as possible, whether the two disks would collide or not. No feedback was prodived for the test trials. Among the 50 situations, 19 involved colliding disks (the distance between the centers of the disks ranged from 6 to 46 pixels) and 31 non-colliding disks (the distance between the centers of the disks ranged from 78 to 222 pixels). Instructions explained that the scoring rule would take the response latency into account, so that the score would be higher if the response was given quickly. However, participants were also informed that in case of wrong answer their score on that item would be negative. Each trial ended automatically two seconds after the minimum distance between the disks was reached. On average, the minimum distance was reached after 9s. At each item the answer of the participant as well as its response latency were recorded.

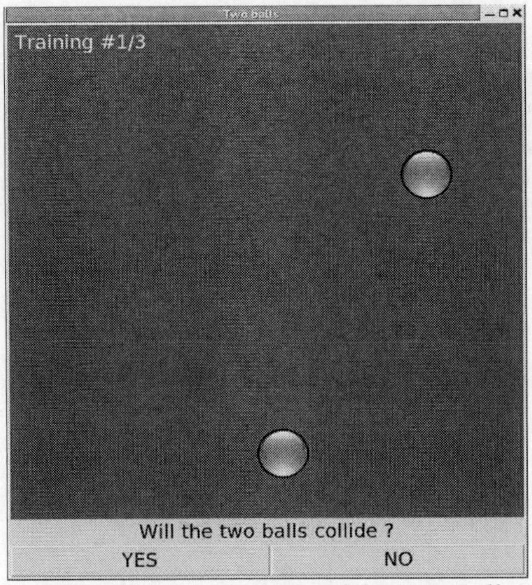

Figure 1. Screenshot of one item of the TwoBalls test. Both disks were moving towards each other and the participant had to click on "yes" or "no".

Procedure. Participants completed the computer-based TwoBalls test by groups of nine and interacted through a mouse. The whole test took maximum 10 minutes to be completed.

Results

The internal consistency of the TwoBalls test was satisfactory as corresponding Cronbach's alpha was 0.75. Globally, performances were high, even for ATCO trainees at the beginning of their training (see Table 1). Indeed, the mean rate of correct responses exceeded 87% for each category of participants. However, large individual differences were observed, specifically for ATCO trainees. For example, rates of correct responses of ATCO trainees at the beginning of their training ranged from 60 to 100%. As two types of errors could be committed by the participants, we investigated the miss rates and false alarm rates. These two variables were not normally distributed, thus we used non-parametric statistical inference tests: the Kruskal-Wallis test for the comparison of the four categories of participants, and the Mann-Whitney test for the comparison between two categories of participants.

Miss rate. Globally, miss rates were only marginally significantly different across the four categories of participants (Kruskal-Wallis rank sum statistic = 7.4, $p = .06$). More precisely, the mean miss rate of ATCO trainees at the beginning of their training (6.6%) was significantly higher than the one of ATCO trainees at the end of their training (3.3%), Mann-Whitney statistic = 3356.5, $p = .009$. Moreover, inspection of the variability of the miss rates of ATCO trainees at the beginning of their training revealed that 6% of them had a miss rate superior to 25%, whereas in the three other categories of participants, none of them had such a miss rate (see Figure 2).

Table 1.
Mean (and standard deviation) correct response rate, miss rate and false alarm rate of performances at the TwoBalls test for each category of participants.

	Correct response rate (%)	Miss rate (%)	False alarm rate (%)
Status	*Mean (SD)*	*Mean (SD)*	*Mean (SD)*
Trainee start	87.3 (8.9)	6.7 (9.0)	16.4 (11.0)
Trainee middle	88.9 (7.4)	5.5 (6.4)	14.4 (10.9)
Trainee end	89.3 (8.0)	3.3 (5.0)	15.1 (11.8)
Expert	90.4 (5.0)	4.9 (6.7)	12.2 (5.8)

Note. Correct responses correspond to correct answers "yes" when the disks would collide and "no" when they would not collide.

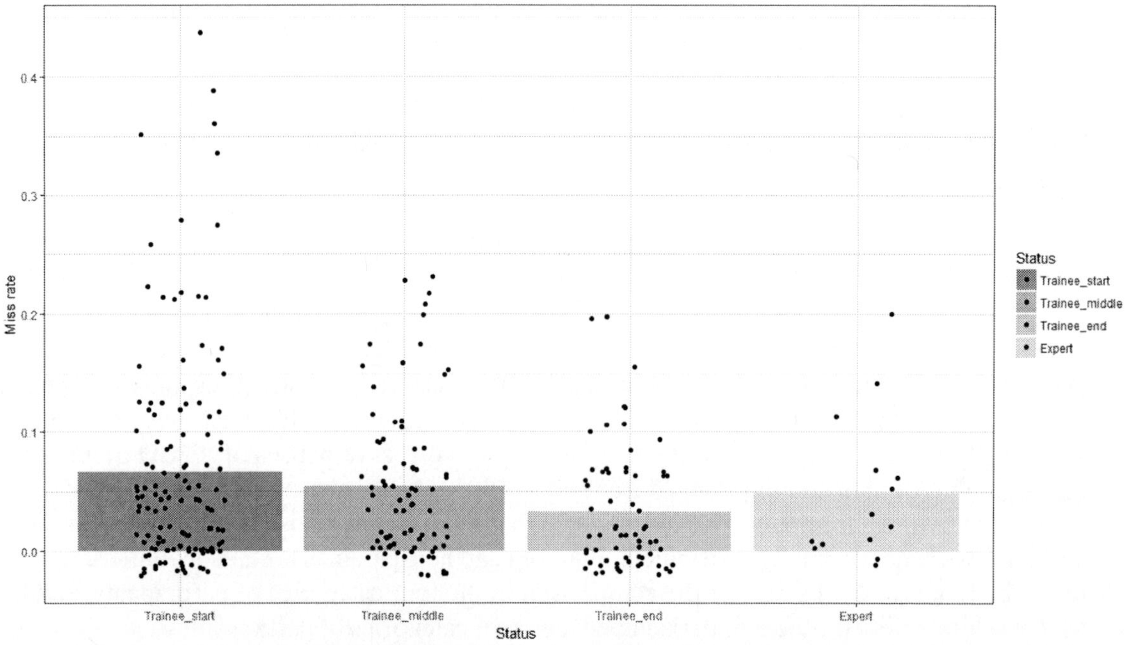

Figure 2. Mean miss rate and individual data (dots) for each category of participants (ATCO trainees at the start, middle or end of training and ATCO experts).

False alarm rate. Globally, false alarm rates were not significantly different across the four categories of participants (Kruskal-Wallis rank sum statistic = 3.0, $p = .39$). Even if we focus on ATCO trainees at the beginning and the end of their training, the difference is not significant, Mann-Whitney statistic = 3891.5, $p = .32$. The mean false alarm rate is lower for ATCO experts (12.2%) compared to ATCO trainees (between 14.4 and 16.4%). However, the difference is not significant, Mann-Whitney statistic = 723.5, $p = .22$. Nevertheless, inspection of the dispersion of the false alarm rates highlighted that none of the experts had a false alarm rate superior to 23%, whereas for ATCO trainees at the beginning, middle and end of their training respectively 21%, 19% and 20% of them had a false alarm rate superior to the maximum of those of ATCO experts (see Figure 3).

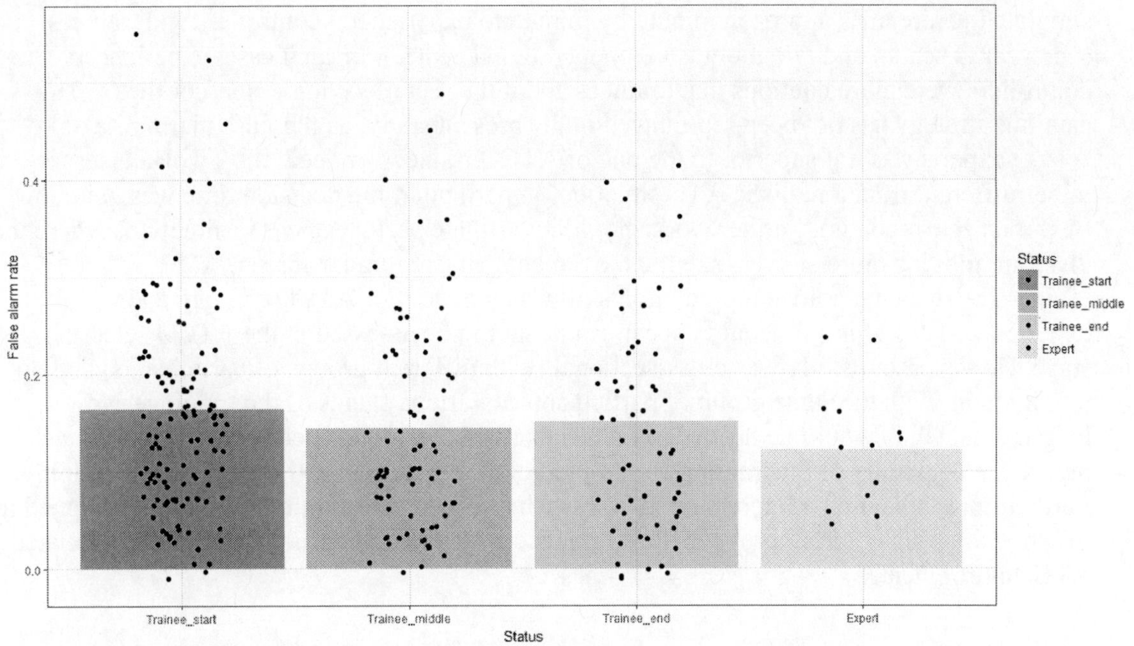

Figure 3. Mean false alarm rate and individual data (dots) for each category of participants (ATCO trainees at the start, middle or end of training and ATCO experts).

Mean response latency. We computed the mean response latency for the 50 items for each participant. Globally, the mean response latencies were not significantly different for the four categories of participants, Kruskal-Wallis rank sum statistic = 3.6, $p = .32$. Most of participants spent from 1.5 to 4.5 s at each item. This is far below the mean 9 s delay before the response was obvious, thus the instruction to answer as quiclky as possible had been followed.

Discussion

A dynamic spatial ability test has been designed to assess the skill in predicting whether two moving objects would collide or not. The present study assessed how the performance at this test would evolve during air traffic control training and experience. To that extend we compared performances at this test for different categories of participants, ATCO trainees at the beginning, middle and end of training, as well as experienced ATCO profesionnals. Firstly, the performance at this test was globally high for each category of participants. Secondly, differences appeared when we focused on the two types of errors that could be committed with such task, namely a non prediction of a future collision (miss) or a prediction of a collision that would not happen (false alarm). At the end of training, ATCO trainees had fewer misses than at the beginning of their training. Thus, ATCO training seems to help to detect conflicting situations. Concerning false alarms, ATCO training does not seem to help to reduce false detections of conflicts. However, this skill seems to be improved with ATCO professional experience.

In an early experiment, Bisseret (1981) presented experts (qualified controllers) with pairs of converging aircraft in a simulated environment. Conflicts were rarely missed (0.9%), but there was a high false alarm rate (68.4%). Thus, in a high-fidelity simulated environment ATCO experts have the tendency to adopt a cautious strategy which lead to miss few conflicts but to falsely diagnose conflicts when there is no conflict. In another experiment, Bisseret (1981) showed static images of radar screens or several successive images in order to simulate the aircraft's approach minute by minute to experienced controllers and trainees (one year at school and two months in an operational centre). In each case, experienced controllers were more cautious than trainees about the risk of saying « non-conflict ». This cautious strategy has not been reproduced in the present study, as the false alarm rate of ATCO experts was not superior to the one of ATCO trainees. Indeed, the TwoBalls test is rather different from a realistic ATC situation. In particular, the decision time was quite low (less than five seconds). Further work should investigate performances in situations where the disks are moving more slowly, like the movements on a real radar screen.

The dynamic collision prediction skill was higher after ATCO training and experience. Thus, this skill could appear as relevant to be assessed at the ATCO selection stage. However, our study was cross-sectional (with different groups of participants) and not longitudinal (with the same group of participants at various stages of the training). Now, a longitudinal study would be needed to investigate whether a poor performance at such task before the beginning of the training would predict more learning difficulties during training. Furthermore, it would be interesting to assess whether a specific training program designed to improve this skill could improve the performance at detecting conflicts in realistic simulated ATC environments.

References

Bisseret, A. (1981). Application of signal detection theory to decision making in supervisory control The effect of the operator's experience. *Ergonomics*, *24*(2), 81-94.

D'Oliveira, T. C. (2004). Dynamic spatial ability: An exploratory analysis and a confirmatory study. *The International Journal of Aviation Psychology*, *14*(1), 19-38. doi:10.1207/s15327108ijap1401_2

Durso, F. T., & Manning, C. A. (2008). Air traffic control. *Reviews of human factors and ergonomics*, *4*(1), 195-244. doi:10.1518/155723408x342853

Hegarty, M., & Waller, D. (2005). Individual differences in spatial abilities. *The Cambridge handbook of visuospatial thinking*, 121-169. doi:10.1017/cbo9780511610448.005

Hunt, E., Pellegrino, J. W., Frick, R. W., Farr, S. A., & Alderton, D. (1988). The ability to reason about movement in the visual field. Intelligence, 12(1), 77–100. doi:10.1016/0160-2896(88)90024-4

Morath, R. A., Quartetti, D., Bayless, A., & Archambault, C. (2001).Air traffic controller job analysis. In R. Ramos, M. C. Heil, & C. A. Manning (Eds.), *Documentation of validity for the AT-SAT computerized test battery: Volume 1* (DOT/FAA/AM-01/5, pp. 7–18).Washington,DC: Federal Aviation Administration Office of Aerospace Medicine.

THE USE OF A PERCEPTUAL SPEED TEST IN CIVILIAN PILOT SELECTION

Hans-Juergen Hoermann, Ph.D.
German Aerospace Center (DLR)
Institute of Aerospace Medicine, Department of Aviation and Space Psychology
Hamburg, Germany
Diane L. Damos, Ph.D.
Damos Aviation Services, Inc.
Gurnee, Illinois

Perceptual speed is an important attribute for success as a pilot and has been assessed in many pilot selection batteries. The Tabular Speed Test (TST), a paper-and-pencil test of perceptual speed, was administered to 227 ab initio pilots and 152 licensed pilots who applied for employment at a large European airline. The airline had a multi-stage selection process including a computerized battery assessing quantitative and spatial abilities, short-term memory, psychomotor performance, multi-tasking, and personality. The number of correct responses (NC) had significant positive correlations for both ab initio and licensed pilots with measures of quantitative and spatial abilities, visual memory, multi-tasking, and psychomotor performance, but not with personality. The number of incorrect responses (NW) was correlated with Emotional Instability and Openness for licensed pilots but not with measures of cognitive ability.

Perceptual speed was identified by Thurstone (1938) as one of the seven primary human abilities comprising intelligence. During World War II, the U.S. Army Air Forces Aviation Psychology Program found that flight instructors often cited slow perception as a cause of failure in flight training (Guilford & Lacey, 1947). Consequently, several perceptual speed tests were developed and administered to pilot candidates. The results from two of these tests, Table Reading and Dial Reading, were combined. The composite was found to have a validity of 0.28 to pass/fail from primary flight training. This composite also had the distinction of being the best single predictor for any aircrew specialty (pilot, navigator, bombardier, etc.). The Table Reading Test had loadings on a perceptual speed factor, a numerical factor, and a spatial-relations factor (p. 813-815). Perceptual speed tests are still included in the U.S. Air Force pilot selection battery because they contribute unique variance to the prediction equation.

Mount, Oh, and Burns (2008) reviewed the literature on perceptual speed tests and determined that NC and NW assess different attributes. Mount et al. hypothesized that NC assesses task performance, whereas NW assesses rule compliance. The NW showed incremental validity for predicting rule compliance for warehouse workers beyond that contributed by general mental ability, conscientiousness, extraversion, and emotional stability. Thus, NW reflects a different attribute.

The TST was developed to assess perceptual speed in civilian pilots, not in the population as a whole. Like traditional perceptual speed tests, it is speeded. It has been used at a U.S. university in the professional pilot curriculum, where it was found to predict pass/fail for the private pilot's license flight test (Mekhail, Niemczyk, Ulrich, & Karp, 2010).

The study reported below was conducted at a major European airline and had three major goals. The first goal pertained to the use of the TST as a selection instrument. A good pilot selection tool assessing an attribute should not be affected by flight time, age (within normal pilot hiring limits), or gender. The second goal was to determine the relation between scores on the TST and other tests of cognitive ability, psychomotor performance, and personality traits. If, as Mount et al. (2008) suggest, NC on perceptual speed tests assesses a cognitive ability, then NC on the TST should be unrelated to personality traits or psychomotor skills but positively related to scores on other cognitive tests. NW, in contrast, should only be related to personality traits that relate to rule compliance.

The third goal pertained to the predictive validity of the airline company's current assessment battery. Traditionally, selection tests have been validated by correlating scores on the instrument with scores obtained during flight training. However, because of the pilot shortage, airlines are less willing to fail pilots or students during training. Some airlines are providing remedial training for weak candidates before they report for their initial airline training. Other airlines are providing additional ground school and simulator sessions to increase the likelihood that a weak candidate will successfully complete initial training. The use of remedial training with or without extra sessions during initial training makes pass/fail from training questionable as a criterion. Additionally, large airlines with subsidiaries may send low-time pilots to their subsidiary airlines, which may make data collection difficult. Ab initio candidates may require several years to complete flight training, delaying the collection of criterion data and raising issues pertaining to internal validity (Campbell & Stanley, 1963).

This paper takes a different approach to validity. Pilot hiring is an expensive process, particularly when it involves simulator evaluations, interviews by line-qualified captains, and evaluations by multiple assessors. Most airlines place the most expensive selection instruments toward the end of the selection process to minimize late-stage failures. Such failures represent substantial financial losses for the airline. Thus, any instrument that can be administered in the early stages of selection and can predict late-stage results is valuable for an airline. For the purposes of this paper, we examine how well scores on the TST predicted success at various stages of an airline selection process.

Methods

Participants

Data were collected on 227 ab initio candidates and 152 pilots holding at least a European commercial license. Most were German nationals. The ab initio sample consisted of 35 women and 192 men with ages ranging from 17 to 41. Most had no flight time although a few had up to 170 hours. None of the ab initio candidates had a commercial pilot's license. The experienced pilot sample consisted of 18 women and 134 men ranging in age from 20 to 41.

Their flight time ranged from 130 to 6250 hours. All of the licensed pilot candidates were native German speakers. The minimum education level for all candidates was passing the university entrance examination or the equivalent.

Procedure

The TST was included as part of a selection battery administered to both ab initio and experienced pilots who applied for a position at a major European airline. The airline used a four-stage, multi-hurdle procedure for both the ab initio candidates and the experienced pilots. Elimination was possible after each stage. Testing was conducted on three days, which could be separated by two to four months.

The four stages were 1) computerized aptitude testing, which included psychomotor tests and personality tests, 2) assessment center evaluation, 3) simulator evaluation, and 4) panel interview. Ab initio candidates performed the stages in this order. Stages 2 and 3 were reversed for the experienced pilot candidates. The aptitude testing was done on Day 1. The TST was administered after all of the aptitude tests were completed. Stage 2 testing was done on Day 2. Stages 3 and 4 were done on Day 3. Instructions for the TST were given in English for all candidates. Instructions for all of the other tests were given in German. Day 1 required about 9 hours for ab initio candidates and about 8 hours for experienced pilots.

Tests

Aptitude tests. The aptitude battery assessed five cognitive abilities plus a test of psychomotor coordination and a test of multi-tasking. Five cognitive abilities were assessed: 1) quantitative, 2) spatial, 3) attention, 4) perceptual speed, and 5) memory span (aural and visual). Additionally, the candidates were given a test of written English. Experienced pilots also received tests of aeronautical knowledge and left-right discrimination, whereas the ab initio candidates were given a test of mechanical aptitude.

Personality tests. The Temperament Structure Scale (TSS) (Mittelstaedt, Pecena, Oubaid, & Maschke, 2016)was administered to all candidates. This test requires approximately 40 min. and has 11 scales. Two of the scales were not administered to licensed pilots because they were concerned with willingness to travel and other lifestyle choices that were less relevant. The cockpit management attitudes questionnaire (FMAQ) was administered only to the licensed pilots and required about 35 min. Both personality tests were administered after the aptitude tests.(Merritt, Helmreich, Wilhelm, & Sherman, 1996)

TST. The TST is a paper-and-pencil test that requires 15 min. to administer with 9 min. of testing time.

Criterion Measures

Simulator evaluation. The ab initio candidates performed simple instrument flying tasks in a low-fidelity training device. The instrumentation was similar to that of a single-engine aircraft. In contrast, the experienced pilots used a high-fidelity simulator for their evaluation.

Assessment center. During the assessment center testing candidates were asked to perform three different types of exercises. One involved communication with a partner; the second, conflict resolution; and the third, group problem solving. Candidates were scored by a team of trained observers on traits such as "rule compliance" and "leadership."

Interview. The interview was conducted by a board consisting of one airline captain and two psychologists for the ab initio candidates and by two airline captains and two psychologists for the experienced pilots. The interviews for both types of candidates took approximately 60 min. The interviews used a semi-structured format, i.e. the topics were identical for each type of candidate, but the number and type of follow-up questions could differ from candidate to candidate. Results from the FMAQ guided some of the questions for the licensed pilots.

Results

Because some of the test results (especially NW) had very skewed distributions, the Spearman rho (r_s) correlation, a non-parametric measure, was used to assess association.

TST as a Selection Instrument

Ab initio candidates. Significant male-female differences were found (t (225) =3.43, p = 0.018) for NC but not for NW. Female candidates scored slightly higher. Neither NC nor NW was significantly related to either age or total flight hours.

Licensed pilots. There were no significant male-female differences for licensed pilots for either NC or NW. Age was not significantly related to NC or to NW. The NC was significantly related to the number of flight hours (r_s (151) = -.178, p=. 029); NW was not.

TST and Cognitive, Psychomotor, and Knowledge, Measures

Table 1 shows the significant correlations between TST scores and performance on the Day 1 computerized aptitude and knowledge tests.

Table 1.
Spearman Rho Correlations Between TST Scores and the Aptitude and Knowledge Test Scores

	Ab Initio Candidates		Licensed Pilots	
Selection Test	TST NC	TST NW	TST NC	TST NW
Quantitative				
Mental Arith.	.31**		.16*	
Mental Arith (NW)	-.27**			
Math Reason.	.37**		NA	NA
Math Reason (NW)	-.28**		NA	NA
Perceptual Speed				
Optical PS	.28**		.19*	
Optical PS (NW)	-.25**			

Spatial				
Mental Rot.	.19**		.22**	
Mental Rot. (NW)	-.19**		-.22**	
Cube Rot.	.20**			
Cube Rot. (NW)		.14*		
Left-Right Discrim	NA	NA	.33**	
Left-Right (NW)	NA	NA	-.22*	
Memory				
Visual	.27**		.33**	
Visual (NW)	-.30**		-.31**	
Running Span			.24**	
Attention				
Concentration	.18**			
Concent. (NW)	-.16*		-.28**	
Psychomotor	.15*		.19*	
Multi-task	.21**		.25**	
Mechanical Comp				
Test 1	.17*		NA	NA
Test 1 (NW)	-.15*		NA	NA
Test 2	.22**		NA	NA
Test 2 (NW)	-.19**		NA	NA
Knowledge				
English	.15*		.17*	
English (NW)	-.16*		-.18*	
Aeronautics	NA	NA		
Aeronautics (NW)	NA	NA		

Note. Only significant results are shown. N = 227 for the ab initio candidates. N = 152 for the licensed pilot candidates. * $p < .05$; ** $p<.01$. NA is not applicable. NW is number wrong. NC is number correct.

TST and Personality

For both the ab initio candidates and the licensed pilots, none of the correlations between scores on the TSS scales and NC were significant. NW correlated significantly with Openness ($r_s(151) = .179, p = .027$) and with Emotional Instability ($r_s(151) = .208$, $p = .010$) only for licensed pilots.

Criterion Measures

Most of the candidates who completed Day 1 testing had not completed Day 2 at the time of the analyses. NC had a significant correlation with pass/fail for Day 1 ($r = 0.187, p< .01$, N = 227) for the ab initio candidates. It was not significantly correlated with pass/fail for the Assessment Center ($p > .05$, N = 45) or with pass/fail for the simulator evaluation ($p > .05$, N = 28). No analyses were conducted on the final interview because only 18 candidates completed this stage. The log (NW) was not significantly correlated with pass/fail from Day 1, the Assessment Center, or with the simulator evaluation. For licensed pilots, NC correlated significantly with pass/fail from Day 1 ($r = .28$, $p < .01$, N= 152). No significant correlations were found for Day 2 or Day 3 (N = 85 and N = 52, respectively).

Discussion

The analyses of TST as a selection instrument showed mixed results. Age had no effect. NC was related to the number of flight hours, but only for the licensed pilots. Gender differences were found but only for ab initio candidates and only on NC. Because of the small number of women in both samples, the gender results should be viewed with caution.

NC was significantly correlated with many measures of cognitive aptitude, psychomotor performance, and knowledge. These results may indicated that the TST is g-saturated and assesses several attributes. The lack of correlation between NC and measures of personality was expected assuming NC is a measure of performance. NW was related only to the NW on the cube rotation test. The correlation between NW and Emotional Stability was larger than Mount et al. (2008) found but not unexpected. Thus, more errors on the TST are correlated with a higher level of emotional instability. We have no explanation for the correlation with Openness.

NC correlated significantly with pass/fail for Day 1 for both groups of pilots. Validation results for the ab initio candidates for Day 2 and Day 3 were limited by the small number of candidates who had completed these phases of testing and should be regarded with caution.

References

Campbell, D. T., & Stanley, J. C. (1963). *Experimental and quasi-experimental designs for research.* Chicago: Rand McNally.

Guilford, J. P., & Lacey, J. I. (Eds.). (1947). *Printed classification tests. Report No. 5.* Washington, D.C.: U.S. Government Printing Office.

Mekhail, A., Niemczyk, M., Ulrich, J., & Karp, M. (2010). Using the Table Reading Test as an Indicator for Success in Pilot Training. *Collegiate Aviation Review, 28*(1), 101-114.

Merritt, A. C., Helmreich, R. L., Wilhelm, J. A., & Sherman, P. J. (1996). *Flight Management Attitudes Questionnaire 2.0 (International) and 2.1 (USA/Anglo). (Aerospace Crew Research Project Technical Report 96-4).* Austin, TX: University of Texas:

Mittelstaedt, J. M., Pecena, Y., Oubaid, V., & Maschke, P. (2016). Construct validity of the Temperament Structure Scales within the Big Five Framework in Aerospace Selection. *Aviation Psychology and Applied Human Factors, 6*, 68-80.

Mount, M. K., Oh, I.-S., & Burns, M. (2008). Incremental validity of perceptual speed and accuracy over general mental ability. *Personnel Psychology, 61*, 113-139.

Thurstone, L. L. (1938). Primary mental abilities. *Psychometric Monographs, 1*, ix + 121.

HOW PERSONALITY, INTELLIGENCE, AND WORKING MEMORY PREDICT SITUATION AWARENESS AND FLIGHT PERFORMANCE

Andrew R. Dattel[1]
Andrey K. Babin[2]
Simona Teodorovic[1]
John Brooks[1]
Saralee Pruksaritanon[3]
Priyanka Shetty[1]

[1]Embry Riddle-Aeronautical University
Daytona Beach, FL
[2]Piedmont Airlines
Philadelphia, PA
[3]Aeronautical Radio of Thailand
Bangkok, Thailand

Situation awareness (SA) and flight performance may be intrinsically connected. Good SA can lead to good aeronautical decision making, and consequently better flight performance. Forty-three pilots participated in the study. Participants completed personality tests, a test of fluid intelligence, and a test for working memory. Participants flew a 15-minute flight scenario in an Elite PI-135 BATD, where participants received six SA questions. Airspeed, altitude, and heading were the flight performance variables. Participants also completed a version of Letter Factory (LF), a generic test used as part of the air traffic controller selection test. Good SA for LF, openness, agreeableness, and fluid intelligence predicted SA in flight. Better SA led to fewer airspeed deviations from the target airspeed, and fewer heading deviations from the target airspeed. Higher fluid intelligence indicated less altitude deviation from the target altitude. Knowing these predictors of SA can be helpful for pilot training and selection tests.

Maintaining situation awareness (SA) is critical in dynamic environments, such as aviation. Good SA typically leads to good decision making and good performance. Certain underlying mechanisms (e.g., working memory, conscientiousness) may be important constructs that are beneficial for good SA (Durso, Bleckley, & Dattel, 2006). Knowing what can predict good SA and good performance in aviation can be beneficial to selection and training in aviation.

The purpose of this paper is twofold. First, the paper will explore which personality factors and cognitive constructs can predict SA and performance. Second, this paper will explore if SA and performance in aviation can be predicted by SA from another environment. That is, can good SA in one environment, specifically a novel environment to the user, carry over to predicting SA and performance in another environment (i.e., aviation)?

SA is the degree of understanding in a typically fast-paced environment (Durso, Rawson, & Girotto, 2007). It can be determined on 3 levels: perception of the relevant elements in the environment, understanding what the elements mean specific to the task, and predicting how the situation will change in the future (Endsley, 1995). Working memory (WM) is one construct that may be an underlying mechanism of good SA (Dattel et al., 2011). WM is the degree to which one can retain and process information while attending to additional information (Baddeley & Hitch, 1994).

Carretta and Ree (2003) have shown the importance of personality factors, such as conscientiousness, in successful pilots. More recently, Barron, Carretta, and Bonto-Kane (2016) have highlighted the importance of extraversion and agreeableness as important factors in performance rankings. In addition, fluid intelligence (or g') has been shown as predictors of successful aviation performance (Ree & Caretta, 1996).

These personality and cognitive constructs were tested on pilots who had at least a private pilots license. The pilots were also tested on a novel task that measures SA and performance. This novel task is used exclusively for applicants taking an air traffic controller selection test.

Method

Participants

Forty-three pilots holding at least a private pilots licenses volunteered for this study. Pilots were remunerated $30 for approximately 2 hours of participation.

Materials

Participants completed several batteries of test to measure personality, working memory, and fluid intelligence. Goldberg's Big-Five Factor Markers Personality checklist was used to measure personality (Goldberg, 1992). For Goldberg's checklist, participants select a rating on a 9-point scale of how they identify on 100 adjectives of traits (e.g., active, sympathetic, anxious).

To measure WM, participants completed the computer version of Operations Span (OSPAN; Turner & Engle, 1989). For OSPAN, participants were asked to calculate several simple mathematical equation (e.g., "Is "3 + 5 = 8," "Is 4 -2 = 1," then shown an answer (True or False), and then asked to determine which answer is correct. After the participant states if the answers were true or false, participants are then shown a random letter for 1 second. Following a set of equations interleaved with letters (3 to 7 mathematical operations and letters in a set), a participant was shown a screen prompting him or her to select all letters in the order which they appeared. The WM score was calculated by the number of letters recalled in correct order.

A 15-minute flight scenario was created in Microsoft Flight Simulator X configured to a glass instrument panel Cessna 172, equipped with a Primary Flight Display and a Multi-Function Display. The flight simulator used a PI-135 Elite Flight Simulator power quadrant, which included a yoke (aircraft steering wheel) and rudder pedals. Three out-the-window screens provided a 120° view. Participants took off on a pre-defined flight plan, but did not land the plane because the flight was intentionally stopped 15 minutes into the scenario, before the participants reached their destination. The flight was in VFR conditions (clear skies, no winds), and participants were instructed to maintain assigned speed and altitude and follow a pre-set flight track on the aircraft's Garmin G1000 map. While participants were flying, six SA questions specific to the flight were played over a headset. Questions were presented in the SPAM format (Durso & Dattel, 2004), in real time. Participants said their answers aloud into a microphone. Accuracy and response time were the measures for the SA questions. A 5-minute practice flight was developed so participants could become familiar with the flight instruments and flight controls.

Standard Progressive Matrices (Raven, 1989) was used to measure fluid intelligence. To measure SA in a novel task, the Letter Factory (LF) subtest of the Air-Traffic Selection and Training (AT-SAT) test was given to the participants (see Dattel & King, 2010). The test was obtained by the ATCPrep™.com Air Traffic Controllers' study software. The LF test represents four conveyor belts with bins (i.e., boxes) at the end of each. During the 15-minute LF simulation, letters appear at the top of the screen on either of the conveyor belts and move down toward the bottom of the screen. Before getting too close to the bottom, but only after a certain point, participants have to identify the letter and place it in the appropriate bin by clicking on the bin and then on the letter. Each bin can only contain letters A, B, C, and D (one of each) before it disappears, and a participant has the ability to add another bin when it becomes full. Bins are removed from the stock on the right side of the screen, and bins should only be removed from the stock if required (i.e. if a letter of the same color is on the conveyor belt and there is no bin of that color already near the belt). The stock needs to be replenished by clicking on a specific button once the box quantity is below a certain value. Any other letter other than A, B, C, or D is called defective, and participants are instructed to acknowledge the defective letters, if they appear, by clicking on a corresponding button. Periodically during the scenario, the simulation is frozen, and participants are asked a specific question about the scenario (e.g., which letter is closest to the bottom). To measure the performance, correct letter placement, correct defective letter identification, and correct timeliness of stock replenishment were recorded. Participants' answers to questions that appeared on the screen during the task were used to assess participants' SA.

Procedure

After participants signed a consent form, they were given the battery of tests. It took approximately 1 hour for the participants to complete the battery of tests. Participants then flew the 5-mintue practice flight, followed by the 15-minute flight scenario. During the flight scenario, participant answered 6 scenario-specific flight SA questions that were played over a headset every 2 to 3 minutes. Participants were instructed to follow a particular flight track to their destination airport, using the aircraft navigation display. Participants were instructed to maintain a specific heading and altitude at various points through the flight scenario. After completing the flight scenario, participants completed the LF test.

Results

Only about 65% of SA questions were answered correctly (M = 3.90, SD = 1.00). RT to answer SA questions are only calculated if the question is answered correctly. Due to the reduction in power for this measure, the SA analyses are only conducted for correct questions answered out of a total of six questions.

After entering the personality variables, WM, SA for Letter Factory, and fluid intelligence, a multiple linear regression showed that SA for Letter Factory, agreeableness, openness, and fluid intelligence predict SA in flight $F(2,25) = 3.186$, p=.012, adjusted R^2 = .346 where higher SA for Letter Factory, higher fluid intelligence, higher agreeableness, and lower openness predict higher accuracy in answering SA questions (see Table 1).

TABLE 1. Multiple Linear Regression

Variable	Beta	*t*	*p*
Constant		-1.662	.109
SA for LF	.367	2.439	.022
Extraversion	-.278	-1.672	.107
Agreeableness	.695	3.607	.001
Conscientiousness	.191	.992	.331
Neuroticism	-.210	-1.015	.320
Openness	-.434	-2.497	.019
Fluid intelligence	.451	2.850	.009
WM	.260	1.663	.105

A median split of OSPAN was conducted to categorize participants as high of low WM. Although not quite significant $t(34) = 2.010$, p = .052 participants with higher WM answered more SA questions correctly ($M = 4.11$, $SD = .76$) than participants with lower ($M = 3.5$, $SD = 1.04$) WM (see Figure 1).

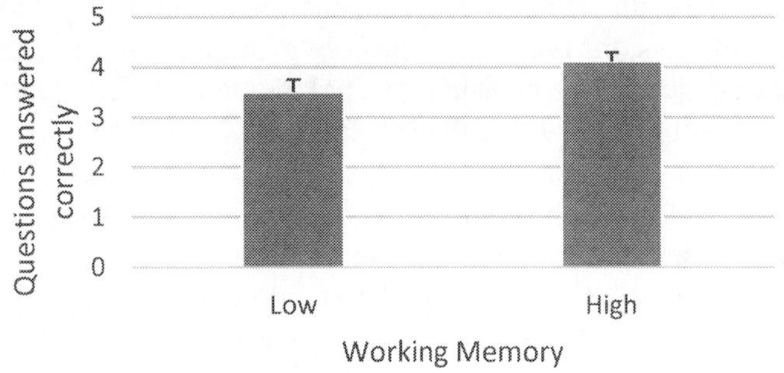

Figure 1. Bar graph of correctly answer questions by working memory.

Correlations conducted between personality factors and SA found that pilots who are more agreeable are more likely to answer more SA questions correctly (see Table 2).

TABLE 2. Bivariate Correlations Between Personality Factors and Situation Awareness

Variable	1	2	3	4	5
1. SA					
2. Extraversion	-.062 *n* = 35				
3. Agreeableness	.348* *n* = 35	.208 *n* = 40			
4. Conscientiousness	.119 *n* = 35	.362* *n* = 40	.364* *n* = 40		
5. Neuroticism	.137 *n* = 35	.307 *n* = 40	.571** *n* = 40	.530** *n* = 40	
6. Openness	.006 *n* = 35	.249 *n* = 40	.234 *n* = 40	.279 *n* = 40	-.039 *n* = 40

*p < .05
**p < . 01

Deviation from assigned airspeed, altitude, and heading were measured as performance variables. Heading deviation was measured from how far the aircraft was from the G1000 track. Thus, larger numbers for airspeed, altitude, and heading indicate greater deviation, and consequently poorer performance. Pearson correlations found that higher SA was related to less airspeed deviations from target airspeed and less heading deviations from target G1000 tracking. Additionally, higher fluid intelligence predicted less deviation in altitude from target altitude (see Table 3).

TABLE 3. Bivariate Correlations Between Situation Awareness, Fluid Intelligence, and Flight Performance Variables

Variable	1	2	3	4	5
1. SA					
2. Altitude deviation	-.185 $n = 31$				
3. Airspeed deviation	-.548** $n = 31$	.020 $n = 37$			
4. Heading deviation	-.359* $n = 31$	-.061 $n = 37$	.137 $n = 37$		
5. Fluid intelligence	.271 $n = 36$	-.335* $n = 37$	-.002 $n = 37$	.042 $n = 37$	
6. WM	.222 $n = 36$	.113 $n = 37$	-.097 $n = 37$	-.082 $n = 37$	.205 $n = 43$

*p < .05
**p < . 01

Discussion

Of the personality and cognitive constructs entered into the regression equation, higher agreeableness and higher fluid intelligence are shown as predictors of good SA for flight. These results support the findings previously shown for agreeableness and fluid intelligence as predictors of good performance in aviation. Thus, one can assume that good SA, as predicted by high fluid intelligence and an agreeable personality will carry over to good aviation performance. It is also of interest that good SA in a novel task (i.e., the LF task) can predict good SA for aviation.

In the regression model an inverse relationship with openness and SA in flight was found. Thus, the less likely one is open to new experiences, the better one's SA in flight. However, many pilots knew what their career would be from an early age. Therefore, pilots have probably stayed focused on their career for many years before realizing their goal, even at the expense of eschewing new experiences.

To further draw the connection between good SA in flight with good flight performance, a correlation with good SA and less airspeed deviation was found. In addition, a correlation with good SA and less heading deviation was found. Finally, in support of previous findings, the results from this study also found fluid intelligence was predictive of good aviation performance, where higher fluid intelligence is associated with less deviation from altitude.

This paper supports many of the previous findings showing how personality traits and cognitive constructs predict good aircraft performance. What this study adds is the connection between SA and these factors. If SA is an important predictor of aviation performance, then SA should continue to be researched as it relates to aviation personnel selection and training. Finally, the finding of SA in LF as a predictor of SA in aviation warrants further exploration if SA is an individual difference trait, where people with good SA have good SA in all dynamic environments, or if SA is specific to a particular industry, where it may be typical for one to have good SA in only one environment, but poor SA in a different environment.

References

Baddeley, A. D., & Hitch, G. J. (1994). Developments in the concept of working memory. *Neuropsychology, 8*, 485-493. doi.org/10.1037/0894-4105.8.4.485

Barron, L. G., Carretta, T. R., & Bonto-Kane, M. V. A. (2016). Relations of personality traits to military aviator performance. *Aviation Psychology and Applied Human Factors, 6*, 57-67. doi.org/10.1027/2192-0923/a000100

Carretta, T. R. & Ree, M. J. (2003). Pilot selection methods. In P. S. Tsang & M. A. Vidulich (Eds.). *Principles and* practices of aviation psychology (pp. 357–396). Mahwah, NJ: Erlbaum

Dattel, A. R., & King. R. E. (2010). The Impact of Reweighting the AT-SAT subtest on group score differences. International *Journal of Aviation Psychology, 20*, 111-123.

Dattel, A. R., Vogt, J. E., Fratzola, J. K, Dever, D. P., Stefonetti, M., Sheehan, C. C., Miller, M. C., & Cavanagh, J. (2011). The Gorilla's role in relevant and irrelevant stimuli in situation awareness and driving hazard detection. Proceedings of the *55th Annual Meeting of the Human Factors and Ergonomics Society*, 924-928, Las Vegas, NV.

Durso, F. T., Bleckley, M. K., & Dattel, A. R. (2006). Does situation awareness add to the predictive validity of cognitive tests? *Human Factors, 48*, 721-733.

Durso, F. T. & Dattel, A. R. (2004). SPAM: The real-time assessment of SA. In S. Banbury and S. Tremblay (Eds.). *A Cognitive Approach to Situation Awareness: Theory, Measurement and Application* (pp. 137- 155). Burlington, VT: Ashgate.

Durso, F. T., Rawson, K.A., & Girotto, S. (2007).Comprehension and situation awareness. In F.T. Durso, R. S. Nicerson, S. T. Dumais, S. Lewandowsy, & T. J. Perfect (Eds.). *Handbook of Applied Cognition* (2nd ed.) (pp. 163-193). Hoboken, NJ: John Wiley & Sons.

Endsley, M.R. (1995). Towards a theory of situation awareness in dynamic environments, *Human Factors, 37*, 32-64.

Goldbert, L. R. (1992). The development of markers for the Big-Five structure, *Psychological Assessment, 4*, 26-42. doi:10.1037/1040-3590.4.1.26

Ravens, J. (1989). The Raven Progressive Matrices: A review of national normoing studeins and ethnic and socioeconomic variation within the United States. *Journal of Education Measurement, 26,* 1-16.

Ree, M.J., Carretta, T. R. (1996). Central role of g in military pilot selection. *The International Journal of Aviation* Psychology, 6, 111-123. doi: 10.1207/s15327108ijap0602_1

Shah, P., & Miyake, A. (1996). The separability of WM resources for spatial thinking and language processing: An individual differences approach. *Journal of Experimental Psychology: General, 125*, 4-27.

Turner, M. L., & Engle, R. W. (1989). Is working memory capacity task dependent?. *Journal of Memory and Language, 28*, 127-154.

ENVISIONING USER REQUIREMENTS FOR FIRST-OF-A-KIND FUTURE ROTORCRAFT

Christen Sushereba, Laura Militello, Katie Ernst, & Julie Diiulio
Applied Decision Science
Dayton, Ohio

Emilie Roth
Roth Cognitive Engineering
San Francisco, CA

Scott Scheff
HF Designworks
Boulder, CO

William Huff IV
JHNA
Washington, D.C.

In this paper, we describe an ongoing effort to envision and articulate requirements for the United States Army's Future Vertical Lift (FVL) program, specifically related to crewing. The goal of FVL is to develop a new family of rotorcraft that will incorporate advanced technologies to support new capabilities. We will discuss the challenges associated with envisioning a future system, along with approaches to design for the envisioned world with examples. We emphasize the importance of focusing on the envisioned work domain, rather than advances in new technologies. We recommend that by articulating the future work domain, requirements for new technologies that will support human crew members can be more easily articulated. When designing for an envisioned world, it is important to incorporate relevant perspectives early and often throughout the process.

The United States Army is in the process of articulating a new family of vertical lift aircraft. The goal of the Future Vertical Lift (FVL) program is to develop a new family of rotorcraft to replace existing airframes. These aircraft will incorporate advanced automation and technologies to increase and change current Army aviation capabilities. In this paper, we describe how we are addressing the envisioned world problem in the context of a project to determine crewing recommendations for the first FVL platform – the Future Attack Reconnaissance Aircraft (FARA). Technologies are already being developed for the FARA platform; our challenge is to articulate the current constraints and predict the envisioned end state in order to guide development toward an optimal, and achievable, human-vehicle system. However, the envisioned end state will include many new capabilities that are not present in current operations, which complicates the task of analyzing the future work domain.

Defining an Envisioned World

Introducing new technology into a work domain invariably transforms the work domain, especially when new technology involves advanced automation (Bainbridge, 1983; Dekker & Woods, 2002; Woods & Dekker, 2000). Technology can change how people accomplish their work (tasks, goals), how expertise is defined, and how failures can occur (Woods & Dekker, 2000). The key challenge of designing for a future work domain is what Woods and Dekker (2000) called the *envisioned world problem* – how can studies and analyses of cognitive and cooperative activities in current practices be applied to the design of future practices when the future technologies will transform the work domain itself? Stated differently, because technology

---

DISTRIBUTION STATEMENT A. Approved for Public Release. Distribution Unlimited.

will change the cognitive and cooperative aspects of the work domain when it is implemented, current studies of cognitive and cooperative activities will not generalize completely to the future domain. If the introduction of a new technology changes the support that operators require to complete their work, then how else will the work domain need to be modified to support the operators (in terms of new work processes, different roles, additional technology, etc.)?

When envisioning a new world, stakeholders develop concepts and ideas about how the future world will operate. Woods and Dekker (2000) outlined four properties of these types of envisioned world concepts that make it challenging to develop requirements for a future system:

- Plurality – There are many different concepts that could be implemented, each with many different manners of affecting the future domain.
- Underspecification – different envisioned concepts for how the future domain will operate are simplifications, with only partial representations of all aspects of how they would function in a concrete system.
- Ungrounded – envisioned concepts are easily disconnected from the research base.
- Overconfident – advocates for envisioned concepts may become overly confident that only the predicted consequences associated with a concept will occur.

It is important to keep these four properties of envisioned world concepts in mind as we begin to articulate how the envisioned operating environment for FVL aircraft will look.

Envisioned World for Future Vertical Lift

Method. Addressing the envisioned world problem requires a clear definition of the envisioned end state – what will the FVL work domain be? This is challenging because the envisioned end state shifts based on changing priorities within the Army, along with improved understanding of the nature of the future work domain (operating environment, enemy threats, technological capabilities, etc.). We used a combination of methods to articulate the envisioned world. This included interviews with Army stakeholders to understand the vision for FVL, as well as priorities and how they have evolved over time. We reviewed doctrine for insight into current operations and conducted cognitive task analysis interviews with pilots to collect examples of challenging incidents. This combination of methods supported our focus on the work that will need to be accomplished, which will inform the development of requirements for supporting technologies.

Early in our effort, we articulated the envisioned world and core missions in which we hypothesized the crew of the FARA platform were likely to engage. We then took these to stakeholders (subject matter experts, FVL program leadership) for feedback. These documents represent the envisioned end state that our project team could focus on during the rest of our analysis activities (Militello et al., 2018).

Outcomes. The envisioned operating environment for FVL will be different from the environment that current Army aviation assets are built to support. Instead of conducting operations in the two-dimensional battlespace (the air and ground between forces), FVL aircraft will be designed to conduct operations in a more complex battlespace that includes air, ground, space, radio-frequency, and cyber (Phillips et al., 2018). FVL aircraft will likely contend with sophisticated anti-access area denial (A2AD) strategies from near-peer threats. To support Army pilots operating in this type of environment, FVL aircraft will be designed to support teaming with unmanned assets and conducting electronic warfare.

Crewing will also be different in the envisioned world. Current crew configurations employ two rated pilots – one who flies the helicopter, and one who manages non-flying tasks. The Army's vision for FVL platforms does not necessarily rely on the same configuration. The Army wants the airframes capable of flying with two pilots, one pilot, or potentially no pilots (managed by remote operators on the ground or on board other aircraft), depending on mission type. The core missions that we identified for the FARA platform are attack, air assault, reconnaissance, and support of air assault and air movement.

Approaches for Designing for the Envisioned World

Miller and Feigh (2019) described different approaches to addressing the envisioned world problem to identify requirements for a new system. They represent the envisioned world problem as a decomposition space with two axes: the work domain (x-axis) and technological capability (y-axis; see Figure 1). The lower left quadrant of the space represents the existing world. The envisioned world sits in the upper right quadrant of the space. The authors identify two possible pathways (or vectors) to reach the envisioned world: along the technological capability axis or along the work domain axis.

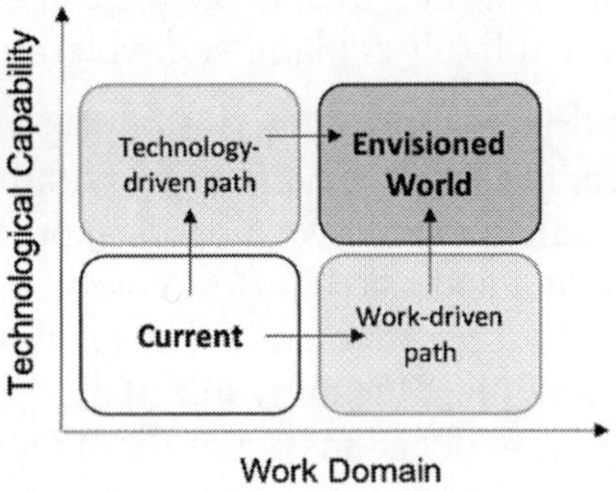

Figure 1. The envisioned world problem represented as a decomposition space (adapted from Miller & Feigh, 2019).

Technology-Driven Pathway

The pathway to the envisioned world that follows the technological capability axis is a commonly used path: new technologies are built, then they are incorporated into the work domain. Often, the new technologies initially appear to provide capabilities that align with the goals of the domain. For example, the addition of a forward-looking infrared (FLIR) sensor to the UH-60 Black Hawk helicopters flying medical evacuation missions in Afghanistan was meant to improve a pilot's ability to land in a degraded visual environment. Desert landings are often complicated by the brown-out conditions caused by dust getting caught in the Black Hawk's rotorwash, reducing visibility. The FLIR sensor provided the pilot with an infrared view of the environment that was not hindered by the flying dust and debris.

The FLIR sensor was installed on the lower side of the aircraft landing strut, which affected how pilots landed the helicopter. It was also expensive – damage to the sensor when landing was classified as a Class A (i.e., most serious) accident, requiring the pilot to report the accident up the chain of command. The FLIR was not certified for pilotage (i.e., never approved as a primary flight display); pilots could use it for situational awareness regarding the location of the ground and other obstacles, but they were not allowed to use it for landing the helicopter in brownout conditions. Many pilots chose not to fly with the FLIR sensor because of the adaptations they were required to make to their own operations to accommodate the technology. In this case, a promising new technological capability was coupled with significant barriers to use, becoming a liability rather than a support. The integration of the technology into the work domain was poor. Ultimately, the risks associated with the technology outweighed the benefits, and did not fully support the pilots as intended.

Work-Driven Pathway

Miller and Feigh (2019) endorse the work-driven pathway as a better alternative to achieving the envisioned end state than the technology-driven pathway. Following this approach, system designers focus on envisioning the future work domain first. They accomplish this by including operators in the design process early enough to generate and envision the desired end state. By engaging operators early, system designers can study the constraints of the current work domain and ground the envisioned work domain in the existing context. Miller and Feigh (2019) state that it is reasonable to assume that *certain* constraints that are present in the current environment will translate to the envisioned world (e.g., those related to physical laws, human capabilities, legal boundaries, etc.), while others will change to meet the requirements of the envisioned world. Therefore, it is important to analyze the current constraints and work with stakeholders to anticipate which will likely remain, and which will likely change in the future.

An example of the work-driven approach to introducing new technology is the modification of a software system designed to help Army commanders with pre- and post-mission flight and maintenance management. Developers approached commanders in Afghanistan with an existing product and asked how it could be improved to better support them. One suggested improvement was to support commanders as they determined how to select helicopter pilots for a mission to optimize the crew mix. There are several variables that must be taken into account when making these decisions, from flight experience level to recency in certain environmental conditions such as flights using night vision goggles. The development team listened to the operators' input and created a patch for the existing software system to provide an immediate fix. Later, the Army produced and fielded a new software program that incorporated the operators' feedback. In this case, the technology was successfully implemented to meet a critical mission need.

Future Work Domain for FVL

Method. To understand the current work doman from which the envisioned world of FVL will evolve, we conducted cognitive task analysis (CTA) interviews (Crandall, Klein, & Hoffman, 2006). CTA interviews informed a functional analysis, as well as identification of cognitive requirements. We tailored traditional critical decision method interviews by adding questions that prompted interviewees to consider potential benefits and drawbacks of envisioned technologies, and the impact of different crewing configurations. When designing for an envisioned world, determing which types of experts to interview is not straightforward. Often the roles and tasks envisioned do not currently reside in a single career field. As a result, we engaged with a variety of operators and stakeholders to understand the work domain and the types of skills anticipated to be required in the envisioned world. Specifically, we interviewed Black Hawk and Apache pilots who had relevant experience (e.g., serving as an Air Mission Commander, flying attack missions, or teaming with unmanned systems). We also interviewed operators outside of the pilot community. For example, while no one in the Army is currently conducting electronic warfare operations from a helicopter (to our knowledge), we anticipate that this will be an important function in FVL. To better understand the work domain associated with electronic warfare, we interviewed an electronic warfare specialist from the Navy.

We also conducted a series of focused sessions with an experienced pilot to support two additional research activities to understand current operations: creating an abstraction hierarchy depicting high-level goals and core functions, and to inform modeling efforts. A member of our research team created human performance models that simulated the workload associated with current and envisioned Army attack missions. To deal with the challenge of underspecification that Woods and Dekker (2000) identified as a characteristic of trying to articulate concepts for an envisioned world, we wanted to intentionally characterize the complexities associated with the current work domain that will likely continue in the future work domain. Cognitive task analysis methods allow us to identify complexities that have arisen in the past and analyze how operators adapted to them.

To extend what we learned from these retrospective interviews, we also considered empirical results from lab simulations of systems designed to allow pilots to control multiple unmanned assets simultaneously.

Outcomes. The outcomes of our analysis of the work domain include artifacts from the functional analysis – abstraction hierarchy, contextual activity templates, and interdependency analyses; IMPRINT models; and cognitive requirements. These outcomes allow us to hypothesize what the the envisioned work domain might look like, while maintaining a grounding in the current work domain. Furthermore, these analyses allowed our research team to articulate sample crewing configurations for the FARA attack mission that stakeholders and experts can react to.

Generating Requirements for FVL

Woods and Dekker (2000) recommended that addressing the envisioned world problem requires a shift from late-cycle human factors evaluations of proposed technologies to early-cycle generative activities, such as ethnographic methods and participatory design. Consistent with this recommendation, we have engaged relevant subject matter experts and FVL stakeholders in every step of our process. The final step will be to present the Army with a method to determine how to crew FVL platforms, which have not yet been built and that include technologies that have not yet been developed, for missions that are not yet clearly defined.

To this end, we have also created a framework (Sushereba, Diiulio, Militello, & Roth, 2019) to conduct an analysis of the trade-offs associated with different crewing configurations. This "tradespace framework" will serve as a mechanism for stakeholders and subject matter experts to provide input about key evaluative factors for each envisioned concept (i.e., crew configuration). The research team will use the tradespace framework during a multi-day workshop with FVL stakeholders. During the workshop, attendees will participate in activities designed to immerse them into the envisioned world for FARA and help them articulate the envisioned tasks. The anticipated outcomes of the workshop will be more specific crewing configuration recommendations, along with the associated risks and technological capabilities that will be required to support each of the configurations. From these outcomes, requirements for the FARA platform will begin to emerge.

To achieve the envisioned world for FVL, a significant shift in the work domain is needed, along with new required technologies to support the envisioned capabilities. We agree with Miller and Feigh's (2019) work-driven approach being a better approach to developing future systems than to focus entirely on the technology and forcing the work domain to adapt. Our approach outlined in this paper combines theoretical knowledge from human factors and cognitive systems engineering with the expertise of operators and the vision of envisioners to develop intelligent requirements for FVL. Each of these perspectives offer a unique, but incomplete picture; focusing on a single perspective will inevitably lead to a less than ideal solution. We recommend incorporating each of these perspectives to create a holistic approach to articulating and designing the FVL platforms to effectively operate in the envisioned world.

Acknowledgements

This research was funded by the U.S. Army Combat Capabilities Development Command Aviation & Missile Center (Contract Award Number NNX16AJ91A, 21-1614-5637-ADS2018). The findings and conclusions in this report are those of the authors and do not necessarily represent the official position of the Army.

References

Bainbridge, L. (1983). Ironies of automation. *Automatica, 19*(6), 775–779. https://doi.org/10.1016/0005-1098(83)90046-8.

Crandall, B., Klein, G., & Hoffman, R. R. (2006). *Working minds: A practitioner's guide to cognitive task analysis.* Cambridge: MIT Press.

Dekker, S. W. A., & Woods, D. D. (2002). MABA-MABA or abracadabra? Progress on human-automation co-ordination. *Cognition, Technology & Work, 4*, 240–244. https://doi.org/10.1007/s101110200022.

Militello, L. G., Roth, E. M., Scheff, S., Ernst, K., Sushereba, C., Diiulio, J., & Klein, D. (2018). *Toward an optimally crewed future vertical lift vehicle: Overview of the envisioned world, core missions, and pertinent technology.* Unpublished technical report [Award no. NNX16AJ91A, 21-1614-5637-ADS2018].

Miller, M. J. & Feigh, K. M. (2019). Addressing the envisioned world problem: A case study in human spaceflight operations. *Design Science, 5*(e3). DOI: 10.1017/dsj.2019.2. Retrieved from https://www.cambridge.org/core/journals/design-science/article/addressing-the-envisioned-world-problem-a-case-study-in-human-spaceflight-operations/EC1DFEC6D7B77F190BB1983D14D872A8.

Phillips, W. N., McConville, J., Rugen, W., Todd, T., Drezner, J., Hirschberg, M., & Graves, G. (2018). 2018 AUSA Global Force Symposium Future Vertical Lift Panel. Huntsville: American Helicopter Society International, Inc. – The Vertical Flight Society. Retrieved from https://vtol.org/files/dmfile/ausa-fvl-cft-panel-transcript-27mar2018.pdf.

Sushereba, C., Diiulio, J., Militello, L., & Roth, E. (2019). A tradespace framework for evaluating crewing configurations for future vertical lift. Manuscript submitted for publication.

Woods, D. & Dekker, S. (2000). Anticipating the effects of technological change: A new era of dynamics for human factors. *Theoretical Issues in Ergonomics Science, 1*(3), 272-282. DOI: 10.1080/1463922011003745 2. Retrieved from https://www.tandfonline.com/doi/abs/10.1080/14639220110037452.

THE USE OF 3D MODELING SOFTWARE TO ENHANCE ROTORCRAFT MAINTENANCE TRAINING

Celeste D. Torrez
Purdue University
West Lafayette, Indiana

Brian J. Kozak, Ph.D.
Purdue University
West Lafayette, Indiana

In order to obtain an airframe and powerplant (A&P) certificate, students must receive a minimum of 1,900 hours of instruction from an FAA approved 14 CFR Part 147 School. Within Part 147, students are only required to learn about helicopters at a level 1 proficiency, which requires only classroom lectures. In order to fill this possible gap in knowledge, the authors created a training exercise at the sophomore level. A helicopter tail rotor was simulated using CATIA to model common stresses on helicopter components. Additionally, helicopter accident reports were used to increase the understanding of proper maintenance, and how components can affect the failure of the rotary system. Through the exercise, students are expected to improve their knowledge of rotary assemblies, while simultaneously expanding their comprehension of statics. As students progress through their Part 147 training, they can apply their understanding of flight-critical components while making inferences on safety and procedures.

All general aviation (GA) aircraft must recurrently undergo an annual inspection and in some circumstances a 100-hour inspection, depending on the operating conditions, performed by a certified airframe and powerplant (A&P) mechanic. While some inspections require an Inspection Authorization (IA) rating to return these aircraft to service, the A&P certificate is considered the minimum requirement to perform the maintenance for both types of inspection (14 CFR 43, 2019). With 292,002 mechanics practicing in the U.S. 2018, the number of aviation mechanics that have undergone specific Part 147 training to receive their A&P certificate is considerable (FAA, 2018a). However, it is important to note, that students at Part 147 Schools are only required to study about helicopters for 1 hour at the level 1 proficiency in the entire 1,900 hours of instruction for the A&P, thus creating a potential knowledge gap. This gap in knowledge could lead to students not being prepared for future careers in helicopter maintenance (Torrez & Kozak, 2019). Thus, posing a potential threat to the quality of maintenance performed on the 10,500 rotorcraft registered with the FAA in the United States (FAA 2018b).

The potential knowledge gap described in the quality of learning is currently being addressed by helicopter manufactures in the U.S. by providing additional helicopter training specific for their helicopters, however, more needs to be done to standardize the level of learning for all mechanics. While this problem is not exclusive to the U.S, different regions address this issue in different ways. For example, in Europe, the A&P equivalent administered by the European Union Aviation Safety Agency (EASA) is an Aircraft Maintenance License (AML). The AML is then broken down into 3 main groups with several subgroups. In order to perform maintenance on specific types of aircraft, such as a single turbine engine helicopter, one must have the proper AML license with the appropriate type rating (AMC/GM Part-66, 2012). Adding aircraft type ratings to the A&P might not be feasible at this time, however, the FAA can assist helicopter manufactures in the U.S. by increasing and standardizing the amount of knowledge taught to students through the Part 147 curriculum.

Literature Review

The use of advanced technology in maintenance training in Part 147 classrooms and laboratories has been encouraged for the last 30 years as ways to train students to the highest level of technical skill possible. Computer simulations are often seen as practical supplements to various aspects of learning, such as simulating the operation of a turbine engine, analyzing stresses of structures, or testing the aerodynamics of a design (Johnson & Norton, 1991). Conversely, building or using physical mockups instead of computer simulations can be expensive and time intensive for Part 147 schools (Abshire & Barron, 1998).

Computer Aided Design (CAD) can often optimize a designer's effectiveness when solving intricate models. One of the tools within CAD is the use of Finite Element Analysis (FEA), which models the real-world stresses applied to designs. FEA is expected whenever a designer intends on testing the strength of a structural system and is often one of the first tools taught to engineering students (Novak & Dolšak, 2008). Through FEA, students are exposed to key steps of the structural design process. They are able to make changes to the types of materials and tolerances on the parts to see how it all impacts the related systems (Amoo, 2013).

Within helicopter design and maintenance for both normal category and transport category helicopters, a damage-tolerant concept is used to determine when structural components need to be replaced. Parts that are considered Principal Structural Elements (PSEs) must be analyzed and tested during the design of the component and must be inspected regularly on the helicopter to ensure that they do not cause a complete failure of the helicopter if the part were to fail (14 CFR 27, 2019; 14 CFR 29, 2019). Hess, Stecki and Rudov-Clark bring up the point that using FEA to model possible design weakness in systems is necessary for damage-tolerant maintenance, as it can highlight parts that might require additional maintenance or inspections. Increased knowledge of failures of individual parts will result in an overall increase in reliability of the entire system (Hess, Stecki, & Rudov-Clark, 2008).

In addition to FEA, case studies of helicopter accidents were used to highlight the need for correct maintenance, as it has been proven that students use these case studies to understand human errors in real-world scenarios and their consequences in ways that are more impactful than learning in a classroom (Saleh & Pendley, 2011). The accidents used as case studies for this laboratory exercise were likely caused by maintenance technician errors, and even though technicians performing the maintenance were preforming tasks that are taught during their A&P training, the technicians might not have had the familiarity with helicopter systems to anticipate potential problem areas.

Laboratory Project

In order to fill the knowledge gap and increase student understanding of flight critical areas of helicopters, a laboratory project, also referred to as the laboratory activity, was designed around a Schweizer 269A helicopter, such as the one seen in figure 1. The course that this laboratory will be implemented in is a sophomore-level statics course with a bi-weekly, one-hour lecture and a weekly two-hour laboratory. The key concepts taught in the course are statics and forces on aircraft using the 3D modeling software CATIA to simulate the effects on various aircraft components.

Figure 1. Schweizer 269A helicopter located at a Part 147 School

The main part of the project is a helicopter tail rotor, modeled with CATIA software after the Schweizer 269A tail rotor, shown in Figure 2. The finished CATIA modeled tail rotor, and tail rotor under analysis can be seen in Figure 3. In the laboratory activity, students will be simulating the forces felt by the tail rotor through normal operation by applying rotational forces on the CATIA model and observing the areas with greatest stress concentrations.

Figure 2. Schweizer 269A tail rotor used as a guide for the construction of the CATIA model

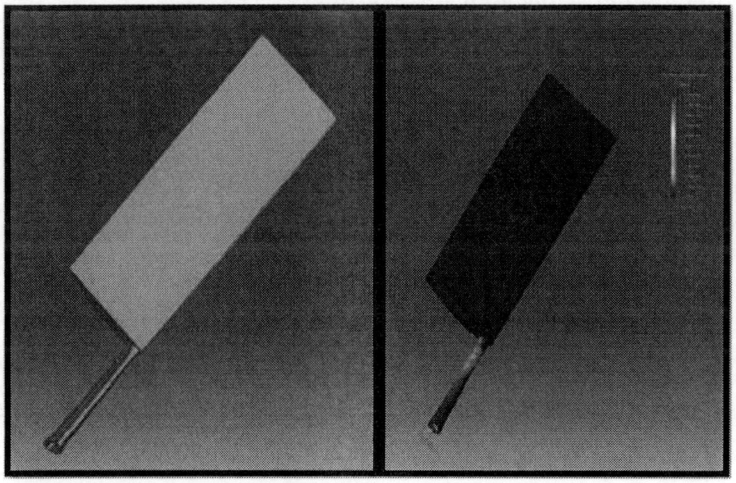

Figure 3. Side by side comparison of Schweizer 269A tail rotor modeled in CATIA with the Von Mises Analysis of the model

To determine the amount of rotational force to apply on the model, students will be asked to obtain the minimum and maximum revolutions per minute (RPM) values for the main rotor for the specific model of helicopter from the Type Certificate Data Sheet (TCDS). Knowing that the tail rotor rotates at a speed ratio of 3:1 to 6:1 when compared to the main rotor, students will be computing the RPMs for the simulations for the following scenarios:

1) Speed ratio of 3:1, and minimum rotor RPM
2) Speed ratio of 3:1, and maximum rotor RPM
3) Speed ratio of 6:1, and minimum rotor RPM
4) Speed ratio of 6:1, and maximum rotor RPM

With the information obtained above, students will fill out a table such as the one in Table 1 to complete their laboratory activity:

Table 1. *Schweizer 269A Main and Tail Rotor RPMS*

	3:1 Speed Ratio		**6:1 Speed Ratio**	
	Minimum Limit (RPM)	Maximum Limit (RPM)	Minimum Limit (RPM)	Maximum Limit (RPM)
Main Rotor				
Tail Rotor				

Using the tail rotor RPMs computed, students will run the four scenarios described above, record the maximum Von Mises stress, and displacement values obtained for each scenario. A typical image of a post-processed tail rotor, displaying the Von Mises can be seen in Figure 3. After recording the required information, students will be answering post-analysis questions to justify and theorize the results they obtained. The analysis questions are shown below in Table 2.

Table 2. *Post Analysis Questions*

	Post Analysis Questions
Question Number	Questions
1	Where was the primary stress found on the tail rotor assembly for each of the increasing RPMs applied to it?
2	If you were to design maintenance based on the analysis, where would you focus your inspection on?
3	What improvements could you make to the tail rotor assembly to increase the strength of the tail rotor?
4	What are the drawbacks/benefits for the solutions you gave in questions 3?
5	What issues with rotorcraft maintenance do you foresee impacting the results that you received?

One of the benefits of using a Schweizer 269A helicopter for the tail rotor model is that an actual Schweizer 269A helicopter is located in a hangar adjacent to the classroom and laboratory. If students have difficulties visualizing the areas of stress, the movement of the forces, or how the tail rotor affects the rest of the helicopter, they can refer to the physical helicopter to answer their questions. Additionally, the hangar also contains the technical documents for the Schweizer 269, so they will be able to look up any maintenance procedures or cut-away diagrams for the helicopter, if necessary. Furthermore, this helicopter may be used as a static maintenance trainer by students in an upper level course depending on aircraft assignments, so students who are already familiar with the flight-critical components and technical documents might find themselves at an advantage.

In order to increase a student's understanding of the impact of improper maintenance, two National Transportation Safety Board (NTSB) accident case studies involving related accidents in rotorcraft. Students will read these studies, then have to answer a set of questions about them prior to starting the lab. Students will then submit their written answers along with their report at the end of the lab. The case study questions are found in Table 3.

Table 3. *Case Study Questions*

Case Study Questions	
Question Number	Questions
1	What was the probable cause of failure for the accident?
2	What parts of the rotorcraft did the failure(s) affect during flight?
3	What other factors played a role in the failure of the aircraft?
4	Describe some of the key takeaways from the accident that can be applied to maintenance procedures performed in labs.

Limitations and Future Works

There were several limitations regarding the development of the tail rotor model. First, CATIA does not have a default composites material with assigned values regarding the strength of the material. Since we did not have access to accurate data regarding the composition of the tail rotor, we used aluminum to model what the stresses would look like. Doing so allowed students to see a general concept regarding stress. Another restriction regarding the lab is that the hangar does not always possess the appropriate tools to disassemble the physical tail rotor, and even if it did, students would need appropriate permission and supervision from authorized personnel in the hangar. This would limit the amount of analysis the students could do with the physical rotor to tasks that do not require disassembly.

In order to overcome or reduce the impact of these limitations, the authors plan to further refine the laboratory exercise for use during the Fall 2019 semester. By analyzing the tail rotor within CATIA, they can potentially see if variations of materials and sizes and other loads can be used to see when these catastrophic failures happen, and they can further research the properties of the materials. Survey questions will be conducted to gather student's knowledge and comfort levels for working on helicopters, as well as their perceived workload. The authors plan on dividing Fall 2019 students into two groups where one receives the laboratory instruction about helicopters, and one does not. Through pre and post surveys, the authors will track the effectiveness of the lab activities, both for the case studies and the CATIA exercises.

Acknowledgements

The authors would like to thank Natalie Zimmermann for her help with CATIA resolving various errors along the development of the part, developing key suggestions for visualizing the part, and her proofreading to the paper. The authors would also like to thank Kristoffer Borgen for his feedback and suggestions during the development of the CATIA model, specifically as it relates to the applications of loads.

References

Abshire, K. J., & Barron, M. K. (1998). Real-World Applications Within Virtual Environments. *Annual Reliability and Maintainability Symposium. 1998 Proceedings. International Symposium on Product Quality and Integrity*, 132-137. Anaheim: IEEE.

Amoo, L. M. (2013). On the Design and Structural Analysis of Jet Engine Fan Blade Structures. *Progress In Aerospace Sciences, 60*, 1-11.

Airworthiness Standards: Normal Category Rotorcraft, 14 C.F.R § 27 (2019).

Airworthiness Standards: Transport Category Rotorcraft, 14 C.F.R § 29 (2019).

Federal Aviation Administration. (2018a). Civil Airmen Statistics 2017. Retrieved from https://www.faa.gov/data_research/aviation_data_statistics/civil_airmen_statistics/

Federal Aviation Administration. (2018b). Air Traffic by the Numbers. Retrieved from https://www.faa.gov/air_traffic/by_the_numbers/media/Air_Traffic_by_the_Numbers_20 18.pdf

Hess, A., Stecki, J. S., & Rudov-Clark, S. D. (2008). The Maintenance Aware Design Environment: Development of an Aerospace PHM Software Tool. *Proc. PHM08, 16*, 17.

Johnson, W. B., & Norton, J. E. (1991). Using Intelligent Simulation to Enhance Human Performance in Aircraft Maintenance. *1991 International Conference on Aging Aircraft and Structural Airworthiness*, 305-313.

Maintenance, Preventive Maintenance, Rebuilding, and Alteration, 14 C.F.R § 43 (2019).

Novak, M., & Dolšak, B. (2008). Intelligent FEA-Based Design Improvement. *Engineering Applications of Artificial Intelligence*, 1239-1254.

Part 66 & AMC/GM – Certifying Staff. (2012). Consolidated Version of Part -66 of the Commission Regulation EC No. 2042/2003 and related EASA Decisions. PDF File. Retrieved from https://www.easa.europa.eu/sites/default/files/dfu/partialpart66.pdf

Saleh, J. H., & Pendley, C. C. (2012). From learning from accidents to teaching about accident causation and prevention: Multidisciplinary education and safety literacy for all engineering students. *Reliability Engineering & System Safety*, 99, 105-113.

Torrez, C. & Kozak, B. (2019). Limitations of Helicopter Training within 14 CFR Part 147. Collegiate Aviation Review International, 37(1), 107-116.

EFFECT OF PAIN AND TASK LOAD ON FLYING PERFORMANCE

Kellen Probert and Brian P. Dyre
University of Idaho
Moscow, Idaho

Justin G. Hollands
Defence Research and Development Canada
Toronto, Canada

Tristen Beaudoin
University of Idaho
Moscow, Idaho

Elaine Maceda
Defence Research and Development Canada
Toronto, Canada

An operationally-significant number of Griffon aircrew in the Royal Canadian Air Force (RCAF) develop chronic neck pain; however, it is unclear how this chronic pain affects their ability to accomplish their missions. Extant literature on pain and human performance has found that pain can negatively affect tasks constrained by short-term memory and attention switching. We sought to test whether pain has similar effects on personnel piloting helicopters in simulation. Twenty-three RCAF personnel flew a simulated Griffon helicopter through waypoints along a target path. We were particularly interested in the effects of three variables: a) the presence or absence of induced thermal pain, b) the presence or absence of a secondary engine monitoring task requiring sustained attention, and c) the experience level of the pilots. The results suggest that pain can interfere with flight performance, particularly for less experienced pilots engaged in multiple tasks over more extended time durations.

Introduction

An operationally-significant number of Royal Canadian Air Force (RCAF) Griffon flight crew develop chronic neck pain that compromises their ability to fly missions, resulting in pilots benching or grounding themselves (Chafe & Farrell, 2016). Since some pilots continue to fly despite experiencing pain, our aim here was to examine the potential for pain compromising flight performance. Does neck pain negatively affect performance of flight-related tasks such as real-time control and monitoring of systems?

The experience of pain has been shown to negatively affect human performance on a variety of tasks, including tasks requiring controlled executive functioning, attentional switching, and high cognitive load (Berryman, Stanton, Bowering, Tabor, McFarlane, & Moseley, 2013). Chronic pain has been shown to induce deficits in spatial and verbal working memory capacity (Luerding, 2008), attention and working memory (Dick, 2008), immediate recall (Pearce, 1990), and running memory (Veldhuijzen, 2006). Pain can also reduce physiological indicators of information processing such as the amplitude of the auditory P300 in EEG recordings (Alanoglu, Ulas, Ozdag, Odabasi, Cakci & Vural, 2005). Experimentally-induced pain has also been shown to negatively affect cognitive performance, including interference on go-no-go tasks (Babiloni, Brancucci, Arendt-Nielsen, Del Percio, Babiloni, Pascual-Marqui, Sabbatini, Rossini, & Chen, 2004), and deficits in attention control (Eccleston & Crombez, 1999). Moore, Keogh, and Eccleston (2012) showed that the attentional tasks most affected by pain are those that require the processing of multiple cues, and the need for executive control.

It is also possible that pain could positively affect performance. The Yerkes-Dodson Law states that peak task performance is achieved at moderate levels of stress or task demand (Yerkes

& Dodson, 1908). Tasks too low in stress or demand do not fully energize an operator to use all their available cognitive resources for the task. In contrast, tasks too high in demand and stress result in physiological stress responses that impede working memory capacity (Wachtel, 1968) and attentional control (Hockey, 1997). Pain acting as a stressor could therefore increase performance on low-stress/low demand tasks while negatively impacting high stress/demand tasks. Given that pain can lead to deficits in attentional control and that focusing attention is a valid strategy for mitigating pain (Eccleston & Crombez, 1999), we might expect that pain could improve performance when attention is dedicated to a single task.

Taken altogether, the above studies suggest that pain affects cognitive performance as measured by traditional psychological tasks in the laboratory. Our question is whether these effects of pain found in a basic and clinical literature, often using elderly or special populations (e.g., fibro-myalgia patients) and laboratory tasks (e.g., Stroop interference, verbal working memory performance), will generalize to cognitive multi-task performance of RCAF aircrew on flight-related tasks, such as multi-axis control, visual navigation and instrument monitoring. If pain has its greatest cognitive effects on attentional control and task switching while multitasking, then we would expect to find that pain negatively affects multi-task performance. In contrast, we expect pain to have less effect and perhaps even a positive effect on single-task performance, due to enhanced arousal.

Method

Participants

Twenty-three RCAF members served as participants in the experiment. Ages ranged from 21 to 50 having up to 6,000 hours total flight time. All participants provided consent and none withdrew despite being informed that they could withdraw from the study at any time, without consequences. Fifteen participants participated at DRDC Toronto and 8 participated at Canadian Forces Base Gagetown. Post-hoc, 11 participants were classified as experienced helicopter pilots, having more than 1,000 total flight hours, and 12 participants were classified as novice pilots, having less than 1,000 flight hours.

Experimental Design and Procedure

The experiment was conducted over two consecutive days. On Day One verbal descriptions of the simulator controls were provided, followed by a 5 to 30 minute familiarization flight in the simulator until performance stabilized. Subsequently, instructions described either the *flight control* or *engine monitoring* task (order counterbalanced across participants), followed by one block of three two-minute practice trials, and then repeated for the other task, followed by one block of three two-minute dual-task practice trials performing both tasks concurrently. We then measured pain thresholds and established levels of pain induction.

For Day Two, we analyzed our dual-task paradigm, employing flight control and engine monitoring tasks, using a fixed-effects, factorial experimental design. Table 1 lists the experimental factors. All factors except experience were manipulated within-subjects. Participants completed three blocks of six experimental trials each lasting two minutes. In a given block, trials one and two consisted of either the flight control or engine monitoring task (single task trials); trial order was counterbalanced across blocks and participants. Trial three presented the dual-task where participants completed both tasks concurrently. In a given block, the first three trials and the second three trials were identical, except for presence or absence of pain, which was also counterbalanced across blocks and participants.

Flight Tasks

Stimuli and apparatus. The simulated flight task environment was generated using X-Plane flight simulation software, version 10.51 64-bit (Laminar Research Inc, 1998) with the X-Trident Bell 412 (equivalent to Griffon used by RCAF), a helicopter model add-on available through the X-Plane store. Displays were presented at a frame rate of 60 Hz and performance data were exported from X-Plane at 13 Hz. The primary flight display had a geometric field of view of 90.0° horizontal by 55.5° vertical. At DRDC Toronto, the pilots sat at approximately the design viewpoint of the 4k resolution display producing a viewing angle of 90.0° x 58.2°. At Gagetown, participants viewed the simulated flight environment on a 1280 x 1024 resolution monitor with a viewing angle of 61.9° x 38.6°.

Table 1.
Factors analyzed of experimental data from Day Two

Factor	Type	# levels	Level Names
EXP	Between-Ss	2	AF_Exp>=1000 AF_Inexp
PAIN	Within-Ss	2	Present Absent
TASK	Within-Ss	2	Single Dual
BLOCK	Within-Ss	3	1 – 3
SEGMENT*	Within-Ss	7	1 – 7

*The SEGMENT effect was only included for flight task data

The software package ViEWER (Dyre & Grimes, 2003) controlled the engine monitoring task, which was presented on a 17 cm diagonal LCD screen centered just below the primary flight display. At DRDC Toronto, the secondary task display had a viewing angle of 18.9° x 14.3° with the participant sitting 61 cm from the display, whereas at Gagetown the secondary task display subtended 11.7° x 8.8° of arc with the participant sitting 99 cm from the display. Participant input was sampled at 30 Hz.

Participants viewed the displays while seated at Pro Flight Trainer Puma helicopter controls in a darkened room. The collective, cyclic, and anti-torque pedal controls were similar to their real-world counterparts.

Flight control task. Participants flew the simulated helicopter through a single course defined by a starting helipad (H1), a series of six 91.44 x 91.44 m square gates laid out in a mountain valley, and a final helipad (H2). The target flight path was defined as a line from H1 to the center of each subsequent gate and from the last gate to the center of H2. As each gate was flown through, the subsequent gate appeared. H2 appeared only after the final gate had been flown through. Wind disturbances were not simulated.

Figure 1 . Gates define the target flight path in the primary flight task. Ideally, participants fly from the center of one gate to the center of a subsequent gate.

Engine monitoring task. We simulated the status indicators of two engines based on the MATB-II system monitoring task (Santiago-Espada, Myere, Latorella, & Comstack, 2011, see *Figure 2*). Each engine display appeared as a pair of vertical rectangles representing engine temperature and pressure level. Critical areas were indicated by tick marks near the top and bottom of each rectangle, marked "High" and "Low", respectively. A red horizontal line indicator within each rectangle moved vertically in a pseudorandom pattern. A "critical condition" occurred when an indicator moved into a critical area. Participants were instructed to squeeze a trigger on the cyclic only when both engines entered a critical condition (a "double critical condition" shown in Figure 2).

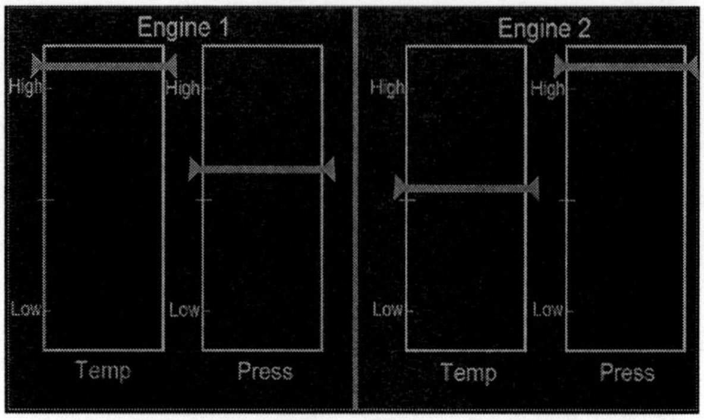

Figure 2. Engine status indicators for the head-down display used for the secondary detection task. The red-marker configuration is consistent with a "double-fault" requiring participant response.

Pain Induction

Stimuli and Apparatus. Thermal stimuli were presented and controlled using two digitally-controlled, thermal nociceptive stimulators (Nocistim units developed by Intellective Consulting & Services, LLC). The Nocistim units are based on the analog design of Morrow and Casey (1981) and are controlled via Nociscale software, which allows a personal computer (PC) connected via USB to control the temperature and exposure duration of two thermodes simultaneously using a variety of psychophysical methods. To present thermal stimuli between 37 and 48°C, we placed two 12.7 x 12.7 mm square contact thermodes on opposite sides of the participant's neck approximately 4.5 cm below the hairline on the dorsal surface (nape), held in place using a flexible neck brace.

Procedure for establishing pain thresholds and levels. We measured the pain threshold for each participant using an adaptive staircase procedure with an initial step-size of 2°C, which decreased by 50% for each reversal of pain judgment (present vs. absent). Trials continued until the participant was satisfied that at least one of the two thermal stimuli was at threshold or the step size was reduced to less than 0.125°C, whichever came first. A threshold trial consisted of an alternating pair of stimuli, each starting at the cooling baseline of 29°C with one side ramping up to the set temperature (37-48°C), and then falling back to the cooling baseline as the other side ramped up to its set temperature, then cooled. The period of a stimulus pair was 10 seconds.

Once the threshold temperatures were established, a second adaptive staircase procedure was used to match a suprathreshold pain stimulus to a marker at 24% of the distance between "no-pain" and "worst pain imaginable"on a visual analog scale (VAS). The 24% value was based on mean VAS pain rating of RCAF aircrew experiencing continuous chronic neck pain based on a recent survey (Fusina, Karakolis, Xiao, Farrell, McGuiness, & Apostoli, 2018). We recorded the set temperature of this stimulus and used it as the induced pain level for all subsequent task trials, where the 10 second anti-phase heating-cooling cycle of the two thermodes was repeated for the full task duration of 2 minutes.

Results

Our results will focus on three classes of measures: a) signal-detection parameters (sensitivity, *A*, and response bias, β; Zhang & Mueller, 2005) and response time for the engine monitoring task; b) flight accuracy measures (lateral and altitude constant, variable, and RMS errors from the prescribed flight path); and c) flight stability measures (e.g., variance in roll, pitch, speed, side-slip). For brevity, we will report only those results from Day Two that are directly relevant to our hypotheses of how pain affects multi-tasking flight performance. Violations of sphericity were corrected using the Greenhouse-Geisser correction where appropriate.

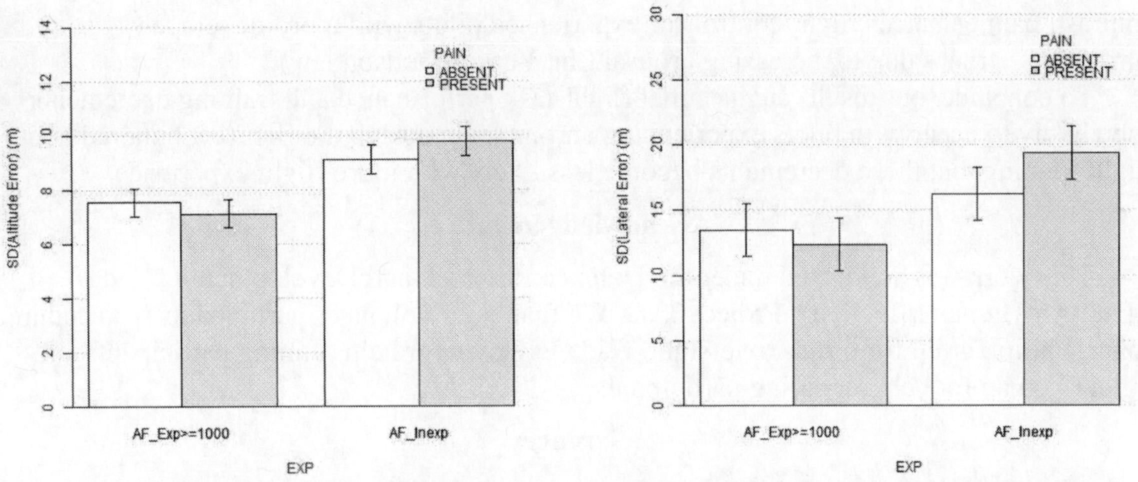

Figure 3. Standard Deviation of Altitude and Lateral Course Errors (m) plotted by the PAIN and EXPERIENCE factors. Error bars represent 95% within-subjects confidence intervals of the means.

We used several mixed-factor fixed effects ANOVAs with the factors listed in Table 1 to assess how pain affected our performance measures on Day Two. Engine-monitoring performance showed significant dual-task decrements on response time (Dual-task RT = 904 ms, Single-task RT = 635 ms, 95% CI = 9 ms; $F[1, 21] = 64.8$, $p < .01$, $\eta_P = .76$) and sensitivity (Dual-task A = .88, Single-task A = .93, 95% CI =.02; $F[1, 21] = 9.16$, $p < .01$, $\eta_P = .81$). Response bias was also affected by task (Dual-task $\beta = 2.24$, Single-task $\beta = 1.73$, 95% CI =.14; $F[1, 21] = 26.5$, $p < .01$, $\eta_P = .56$). However, there was no effects or interactions involving PAIN or EXPERIENCE (p > .05). PAIN did however significantly affect flight task performance. We found PAIN x EXPERIENCE interactions for the standard devations of both lateral error and altitude error ($F[1, 21] = 4.66$, $p < .05$, $\eta_P = .18$ and $F[1, 21] = 4.49$, $p < .05$, $\eta_P = .18$, respectively; see *Figure 3*). For inexperienced participants, pain degraded performance, but for experienced participants, pain improved performance. Identical patterns of interaction were found for other flight-stability measures, including the standard deviations of pitch, roll, ground speed, and vertical velocity ($p < .05$). Finally, there was a four-way interaction of PAIN x EXPERIENCE x TASK x BLOCK for the standard deviation of altitude error ($F[2, 42] = 3.92$, $p < .05$, $\eta_P = .16$). Performance on the dual-task produced greater variability in error in later blocks for inexperienced pilots in pain, while the variability of error decreased across blocks for all other conditions.

Discussion

The engine monitoring task showed typical dual-task decrements and performance was unaffected by pain. In contrast, flight-task performance did not show dual-task decrements and was influenced by pain, although differently for experienced and inexperienced pilots. These results suggest that our pilots treated the monitoring task as secondary. This is perhaps unsurprising given that pilots are trained to prioritize flight control over instrument monitoring or navigating. The lack of pain effects on the engine-monitoring task may be due to the fact that the engine monitoring task demanded little or no working memory resources.

Pain had opposite effects for experienced and inexperienced pilots. The stability of flight control and flight path errors increased during pain trials for inexperienced pilots, but decreased for experienced pilots. It appears that pain increases flight control error for inexperienced pilots with less-developed automaticity for flight control, perhaps due to working memory interference.

In contrast, pain enhances flight-control for experienced pilots with more-developed automaticity, perhaps due to increasing arousal (the Yerkes-Dodson law).

To conclude, our results suggest that flight-task performance and training decrements may be likely to occur with pilots experiencing chronic pain during the first few hundred hours of flight training, but these decrements become less likely with more flight experience.

Acknowledgements

This work was conducted under the Defence Research and Development Canada (DRDC) Air Human Effectivess Project 03aa. We thank our volunteer participants who endured the pain, Philip Farrell for project oversight, Nada Pavlovic for help running participants and Vaughn Cosman for help recruiting participants.

References

Alanoglu, E., Ulas, U.H., Ozdag, F., Odabasi, Z., Cakci, A., & Vural, O. (2005). Auditory event-related brain potentials in fibromyalgia syndrome. *Rheumatology International, 25*, 345–349.

Babiloni, C., Brancucci, A., Arendt-Nielsen, L., Del Percio, C., Babiloni, F., Pascual-Marqui, R. D., Sabbatini, G., Rossini, P. M., & Chen, A. C., 2004. Cortical sensorimotor interactions during the expectancy of a go/no-go task: effects of painful stimuli. *Behavioral Neuroscience, 118*, 925–935.

Berryman, C., Stanton, T. R., Bowering, K. J., Tabor, A., McFarlane, A., & Moseley, G. L. (2013). Evidence for working memory deficits in chronic pain: A systematic review and meta-analysis. *Pain, 154*, 1181-1196.

Chafé, G. S., & Farrell, P. S. E. (2016). *Royal Canadian Air Force CH-146 Griffon aircrew 2014 spinal musculoskeletal trouble survey.* Scientific Report No. DRDC-RDDC-2016-R179. Toronto: Defence Research and Development Canada.

Dick, B. D., Verrier, M. J., Harker, K. T., & Rashiq, S. (2008). Disruption of cognitive function in fibromyalgia syndrome. *Pain, 139*, 610–616.

Dyre, B. P., Grimes, J. G., & Lew, R. (2009). ViEWER 2.23: A Virtual Environment Workbench for Education and Research. *http://www.webpages.uidaho.edu/~bdyre/viewer.htm.*

Eccleston, C., & Crombez, G. (1999). Pain demands attention: A cognitive-affective model of the interruptive function of pain. *Psychological Bulletin, 125*, 356-366.

Fusina, G., Karakolis, T., Xiao, A., Farrell, P., McGuiness, C., & Apostoli, C. (2018). *Towards a "Smart-Scheduling" mitigating solution for Griffon aircrew neck trouble.* DRDC Scientific Report DRDC-RDDC-2018-R186. Toronto: Defence Research and Development Canada.

Hockey, G. R. J. (1997). Compensatory control in the regulation of human performance under stress and high workload: A cognitive-energetical framework. *Biological Psychology, 45*, 73-93.

Laminar Research Incorporated (1998). *X-Plane*. Columbia, SC: Laminar Research Incorporated.

Luerding, R., Weigand, T., Bogdahn, U., & Schmidt-Wilcke, T. (2008). Working memory performance is correlated with local brain morphology in the medial frontal and anterior cingulate cortex in fibromyalgia patients: structural correlates of pain–cognition interaction. *Brain, 131*, 3222–3231.

Moore, D. J., Keogh, E., & Eccleston, C. (2012). The interruptive effect of pain on attention. *Quarterly Journal of Experimental Psychology, 65*, 565-586.

Morrow, T. J., & Casey, K. L. (1981). A contact thermal stimulator for neurobehavioral research on temperature sensation. *Brain Research Bulletin, 6*, 281-284.

Pearce, S., Isherwood, S., Hrouda, D., Richardson, P., Erskine, A., & Skinner, J. (1990) Memory and pain: Tests of mood congruity and state dependent learning in experimentally induced and clinical pain. *Pain, 43*, 187–193.

Santiago-Espada, Y., Myer, R. R., Latorella, K. A., & Comstock Jr, J. R. (2011). *The multi-attribute task battery ii (matb-ii) software for human performance and workload research: A user's guide.* NASA/TM–2011-217164. Hampton, VA: National Aeronautics and Space Administration Langley Research Center

Veldhuijzen, D., van Wijck, A., Wille, F., Verster, J., Kenemans, J., Kalkman, C., Olivier, B., and Volkerts, E. (2006). Effect of chronic nonmalignant pain on highway driving performance. *Pain,122*, 28–35.

Wachtel, P. L. (1968). Anxiety, attention, and coping with threat. *Journal of Abnormal Psychology, 73*, 137-143.

Yerkes, R. M., & Dodson, J. D. (1908). The relation of strength of stimulus to rapidity of habit-formation. *Journal of Comparative Neurology and Psychology, 18*, 459–482.

Zhang, J & Mueller, S.T. (2005). A note on ROC analysis and non-parametric estimate of sensitivity. *Psychometrika, 70,* 203–212.

WEATHER HAZARDS IN GENERAL AVIATION: HUMAN FACTORS RESEARCH TO UNDERSTAND AND MITIGATE THE PROBLEM

Ian Johnson, MS
Federal Aviation Administration
Washington, DC.
Beth Blickensderfer, PhD
Embry-Riddle Aeronautical
University
Daytona Beach, FL.
Geoff Whitehurst, PhD
Western Michigan University
Kalamazoo, MI.
Lori J. Brown, FRAeS, MSc
Western Michigan University
Kalamazoo, MI.
Ulf Ahlstrom, PhD
AvMet Applications, Inc.
Reston, VA
Mary E. Johnson, PhD
Purdue University
West Lafayette, IN.

Weather-related accidents contribute to general aviation fatal accidents each year. These accidents continue to occur even with advancements in weather information technology available in cockpit display technology and mobile applications. The purpose of this session is to highlight a body of on-going human factors research addressing examining interpretability of aviation weather observations, displays, and forecasts; discussion of results from the weather information latency study; use of augmented reality to enhance aviation education, training, and weather information presentation; increasing the number and detail of GA pilot reports (PIREP's); and GA Pilot In-flight Visibility Assessments. This paper provides an abstract for each of the topic areas.

Despite enhancements in weather information and the proliferation of weather-related cockpit display technology and mobile applications by industry, weather-related accidents continue to account for the majority of general aviation (GA) fatal accidents. Previous research has shown that in many instances a contributing factor in many of these accidents was the pilot's failure to correctly interpret the weather information being depicted inside and viewed outside the cockpit, and inadvertently entering instrument meteorological conditions. (Pearson, 2002; Aarons, 2014). Fortunately, a body of on-going, human factors research exists aimed at understanding and addressing this problem.

The purpose of this session is to highlight that research. Topics will include: examining interpretability of aviation weather observations, displays, and forecasts; discussion of results from the weather information latency study; use of augmented reality to enhance aviation education, training, and weather information presentation; increasing the number and detail of GA pilot reports (PIREP's); and GA Pilot In-flight Visibility Assessments. This session is designed to foster a discussion about the complexity of interpreting aviation weather, the hazards of weather in GA operations, and the research underway to mitigate the hazards and improve GA safety.

Interpretability of Aviation Weather Products by GA Pilots

If a General Aviation (GA) pilot encounters hazardous weather during flight, a high likelihood of fatal accident exists (Fultz and Ashley, 2016). Fortunately, a wealth of aviation weather technology and information is available to help pilots to develop situation awareness of current and forecasted weather conditions. Little research, however, has examined the degree to which General Aviation pilots can interpret the weather observation, analysis, and forecast products* (Blickensderfer et al., 2017). If GA pilots are unable to interpret the weather products effectively, those pilots will not be able to take advantage of that technology and information to improve flight safety. To determine how well GA pilots, interpret weather products, a multidisciplinary research team including human factors specialists, meteorologists, and flight experts developed and validated a written test to assess the degree to which pilots can interpret weather products (Blickensderfer et al., 2017). The purpose of this presentation is to describe and discuss weather product interpretability based on a series of studies conducted using the Blickensderfer et al. (2017) test. Results will be discussed in terms of implications for product design and pilot training. *Observation products include Routine Meteorological Reports (METARS), Aircraft Reports (AIREPS)/Pilot Reports (PIREPS), radar, and satellite imagery. Analysis products include Ceiling and Visibility Analysis (CVA), Weather Depiction Charts, and Surface Analysis Charts. Finally, forecast products include Terminal Aerodrome Forecasts (TAFs), Prognostic (or prog) charts (e.g., surface weather charts), Graphical Airman's Meteorological (G-AIRMET) advisories, Graphical Turbulence Guidance (GTG), Significant Meteorological Information (SIGMET) advisories, and winds aloft.

Weather Information Latency Study

The inability of GA small aircraft pilots to correctly assess the actual in-flight weather situation is aggravated by the fact that most of GA aircraft are not equipped with complex and expensive airborne weather radars, like those in larger aircraft, especially in commercial aviation. Pilots of these small GA aircraft have to make in-flight weather relevant decisions based on information displayed on screens of various electronic devices capable of producing weather radar images (Pope, 2015). Unfortunately, these images show weather situations that existed some time ago. The time difference between an electronic device radar weather image and actual flight weather conditions, seen from the aircraft cockpit, can be very significant and may be as long as 20 minutes (Zimmerman, 2013; Trescott, 2012). This information discrepancy makes weather-related accidents more probable due to the degraded situational awareness of the pilots, who become predisposed to making safety-threatening decisions to continue flights into rapidly deteriorating weather conditions for which they, or their aircraft, are not certified. The purpose of this presentation is to describe and discuss weather information latency based on a study conducted at the FAA W. J. Hughes Technical Center in July 2018. Results will be discussed in terms of implications for pilot initial and recurrent training.

Augmented Reality to Enhance Aviation Education and Training: Bridging the Digital Gap

Interpreting and understanding aviation weather is critical to hazardous weather avoidance, and previous studies have indicated that improving understanding of weather phenomena and weather products can improve pilot decision making (King, et.al, 2017). One of the main challenges faced by pilots is the ability to correlate weather knowledge to real-life situations and decision-making. This may be due to a lack of weather knowledge, the usability of the weather information available, or both. Previous research suggests increasing the usability and understanding of weather products can improve pilot situational awareness and decisions making and, in turn, increase flight safety (Latorella & Chamberlain, 2002). Specific tasks such as encountering possible adverse weather conditions require an understanding of several interrelated human and machine components requiring practice and immersion. To meet these challenges, we can harness 3D simulated environments using Augmented Reality (AR) human interfaces to provide adaptive learning methodologies to meet the learning styles of changing generations and improve effectiveness and efficiency of training methods. These AR enhancements can include interactive 3D models, experiential learning modules and assessments, engaging real-life video, scenario-based training or links to additional URL information to provide more in-depth knowledge of the weather product or phenomena and engage a new generation learner. AR technology, allows the pilot, instructor or student to escape the limitations of traditional printed materials enabling them to truly visualize the objects or weather phenomena in full 3D.

General Aviation (GA) Pilot In-flight Visibility Assessments

One of the causes behind Visual Flight Rules (VFR) into Instrument Meteorological Conditions (IMC) flights is GA pilot difficulty to assess in-flight visibility distances (Coyne, Baldwin, & Latorella, 2008; Goh & Wiegmann, 2001; Wiggins & O'Hare, 2003). During cockpit simulation flights, we investigated the ability of GA pilots to estimate the forward visibility by providing visibility estimates at various route locations where the visibilities ranged from 30 nmi to less than 3 nmi. Using three different pilot groups, we provided two of the groups with specific training (i.e., Slant-range rule of thumb and sectional map distance training) and compared their visibility assessments with a third pilot group that did not receive any training (i.e., Control group). The result showed that the visibility estimate errors for the Slant-range group were on average half the size compared to the visibility estimate errors for the Control and the Map distance training groups. Furthermore, for large simulated visibilities (10 to 30 nmi) pilots severely underestimated the visibility. For simulated visibilities below 10 nmi, the analysis showed that pilots were overestimating the forward visibility. For pilots who decided to turn around at the end of the scenario, 42% were in violation of the VFR rules due to insufficient forward visibility. We believe that with training on the Slant-range rule of thumb, coupled with a set of decision-making rules, pilots would be in a much better position to assess the out-the-window visibility and make more informed flight decisions rather than continue flight into IMC.

Increasing the Number and Detail of GA Pilot Reports (PIREP)

A Pilot Report (PIREP) provides vital information on weather conditions experienced at certain flight altitudes in specific locations. Turbulence, icing, outside air temperature, and cloud layers are a few examples of weather conditions that may be reported. Pilots may use PIREPs when doing flight planning. PIREPs are used by air traffic controllers and flight services when communicating with pilots about weather conditions aloft. PIREPs are currently the only source of icing information for pilots (NTSB, 2017). However, many general aviation (GA) pilots submit few PIREPs per year (AOPA, 2016; Casner, 2010). PIREPs have other uses beyond immediate navigation. For instance, PIREPs may be used by aviation weather professionals to modify the area encompassed in a SIGMET or to update weather forecasting models (NTSB, 2017). This presentation discusses the need for more GA PIREPs, barriers to PIREP submission, existing submission tools, and the need for more accurate and timely reports. An exploratory study is presented and discussed that examined six potential features of a digital PIREP submission tool

Acknowledgements and Disclaimer

The views expressed in this paper and associated panel presentation are those of the authors and do not necessarily represent the views of the organizations with which they are affiliated.

References

Aarons, R. N. (2014). Weather-related loss of control. *Aviation Week Network*. July 1, 2014. https://aviationweek.com/business-aviation/weather-related-loss-control

Aircraft Owners and Pilots Association. (2016). *2016 Pilot report survey*. Retrieved from http://download.aopa.org/advocacy/0417_2016_pilot_report_survey_final_report.pdf

Blickensderfer, B., Lanicci, J., Guinn, T., King, J., Ortiz, Y., & Thomas, R. (2017). Assessing general aviation pilots understanding of aviation weather products. *The International Journal of Aerospace Psychology*, 27: 3-4, 79-91, DOI:10.1080/24721840.2018.1431780.

Casner, M., S. (2010). Why don't pilots submit more pilot weather reports (PIREPs)? *The International Journal of Aviation Psychology*, 20:4, 347-374, DOI: 10.1080/10508414.2010.48701

Coyne, J. T., Baldwin, C. L., & Latorella, K. A. (2008). Pilot Weather Assessment: Implications for Visual Flight Rules Flight Into Instrument Meteorological Conditions. *International Journal of Aviation Psychology, 18*(2), 153 – 166. https://doi.org/10.1080/10508410801926756

Fultz, A. J., & Ashley, W. S. (2016). Fatal weather-related general aviation accidents in the United States. *Physical Geography*, *37*(5), 291-312.

Goh, J., & Wiegmann, D. A. (2001). Visual flight rules flight into instrument meteorological conditions: An empirical investigation of the possible causes. *International Journal of Aviation Psychology,11*, 359–379. https://doi.org/10.1518/0018720024497871

King, J., Ortiz, Y., Guinn, T., Lanicci, L. (2017). Assessing General Aviation Pilots' Interpretation of Weather Products: Traditional and New Automated Generation Products. journals.sagepub.com

Latorella, K. A., & Chambelain, J. P. (2002). Graphical weather information system evaluation: Usability, perceived utility, and preferences from general aviation pilots (No. 2002-01-1521). SAE Technical Paper. doi:10.4271/2002-01-1521.

National Transportation Safety Board. (2017). *Improving Pilot Weather Report Submission and Dissemination to Benefit Safety in the National Airspace System.* Washington DC: National Transportation Safety Board. Retrieved from https://www.ntsb.gov/safety/safety-studies/Documents/SIR1702.pdf

Pearson, D. C. (2002). VFR flight not recommended: A study of weather-related fatal aviation accidents. *Technical Attachment SR SSD, 18*, 2443-2448.

Pope, S. (2015). 8 Offbeat Aviation Apps. - Flying magazine. Posted: May 28, 2015. Retrieved on December 13, 2015 from: http://www.flyingmag.com/avionics-gear/pilot-supplies/8-offbeat-aviation-apps

Trescott, M. (2012). NTSB Alert – In-cockpit NEXRAD weather age misleading and killing pilots. *Aviation Training, General Aviation Safety.* June 20, 2012. Retrieved from http://www.maxtrescott.com/max_trescott_on_general_a/2012/06/ntsb-alertin-cockpit-nexrad-weather-age-misleading-and-killing-pilots.html

Wiggins, M., & O'Hare, D. (2003). Weatherwise: Evaluation of a cue-based training approach for the recognition of deteriorating weather conditions during flight. *Human Factors, 45*, 337–345. https://doi.org/10.1518/hfes.45.2.337.27246

Zimmerman, J. (2013). Thunderstorm flying in the age of datalink weather. *Weather Geek.* September 8, 2013. Retrieved from http://airfactsjournal.com/2013/09/thunderstorm-flying-in -the-age-of-datalink-weather/

HUMAN-AUTONOMY TEAMING - AN EVOLVING INTERACTION PARADIGM: TEAMING AND AUTOMATION

Dale Richards
Coventry University, United Kingdom
Bill Kaliardos
Federal Aviation Administration, United States of America

Intelligent and complex systems are becoming common in our workplace and our homes, providing direct assistance in transport, health and education domains. In many instances the nature of these systems are somewhat ubiquitous, and influence the manner in which we make decisions. Traditionally we understand the benefits of how humans work within teams, and the associated pitfalls and costs when this team fails to work. However, we can view the autonomous agent as a synthetic partner emerging in roles that have traditionally been the bastion of the human alone. Within these new Human-Autonomy Teams we can witness different levels of automation and decision support held within a clear hierarchy of tasks and goals. However, when we start examining the nature of more autonomous systems and software agents we see a partnership that can suggest different constructs of authority depending on the context of the task. This may vary in terms of whether the human or agent is leading the team in order to achieve a goal. This paper examines the nature of HAT composition whilst examining the application of this in aviation and how trust in such stystems can be assessed.

We are surrounded by advanced technologies that are not only pervasive but make us dependent on their use. The relationship we share with intelligent systems traditionally provide us with an increase in efficiency, cost- or time-saving, or simply are technological tokens that we furnish our lives with in order to convince ourselves that we belong to an ever-changing technological world. However, we are amidst a rather different form of technology revolution that now affords us a degree of decision making that can either support, augment or even replace the human component entirely. This may conjure different perceptions depending on the individual and context of the technology being considered, but regardless of your view, there is a perception that such a relationship with this sort of technology is beneficial, efficient and welcomed. We have walked this path in the past, in the form of introducing automation into systems that allow this technology to release the human from conducting dangerous, difficult or dull tasks. The developments in advanced automation, autonomy and Artificial Intelligence (AI) are leading protagonists for this advanced technology, and in many instances we can argue that humankind has already achieved a degree of symbiosis that brings us closer to achieving human-autonomy teaming (HAT). This panel represents a series of papers that discusses some of the main issues with human-autonomy interaction in terms of the nature of a mixed human-autonomy team partnership. The structure of this paper and associated Panel Symposium will focus on the following aspects of HAT:

- Part 1: the nature of HAT composition and the manner in which team roles should be considered in terms of team membership,
- Part 1: how automation within aviation can be used as an example of a precursor of HAT,

- Part 2: the application of Cognitive Engineering as a means to design a HAT that is based on user and system requirements,
- Part 3: an examination of the role trust plays in HAT, with the focus on using neurophysiological measurement.

Human-Autonomy Teaming: An evolving interaction paradigm

While there has been a fair amount of literature that examines how human teams work well together (and when they do not), very little is known in terms of how the new human-autonomy teaming paradigm would fare. The nature of how humans act and interact within a *human-only* team construct is complex and requires a multidisciplinary approach in order to appreciate the different roles and behaviours (Salas et al, 2010). If we consider the seminal work by Belbin (1981) then we may begin to appreciate the importance of what a good team composition would look like, and also understand the nature team roles play within an effective partnership (Belbin, 1993). It would seem obvious that in order for a team to achieve a collective goal then individuals are likely to be defined in such roles in order to maximise efficiency and the likelihood of achieving that goal. Key to this effectiveness is the ability for team members to communicate with each other (Cooke et al., 2001), and unsurprisingly it has been suggested that a HAT would require similar roles and dynamics that we would expect to see in a human only team (Scholtz, 2003). A goal defined by a HAT would need to not only be a shared one (Baxter & Richards, 2010), but would also suggest that the nature of roles may be dynamic (depending of course on the context of the task/goal). This would see a team member sometimes being subservient, but in other cases providing plans or suggest actions to be carried out - which may either be performed by a human or an autonomous agent.

There are many different ways by which we can imagine a human interacting with an agent-based system, and indeed the role a human team member may play within a mixed team. Richards (2017) suggests this may present a number of alternative team constructs of HAT which begins to blur the boundaries of HAT composition. The construct of HAT may therefore not be as straight forward as we would traditionally perceive it, with the agent component not always being subservient to a human supervisor. In some instances we may see the agent sharing tasks in paralel to humans, and in some instances even providing commands to human team members. This raises a number of important aspects of HAT worth considering; namely that of acceptance and trust. Richards (2017) further stresses that in order to achieve an effective HAT it is important to define a framework of control that allows the delegation of control between the human and the agent. This is not new, as we have seen the demand for understanding the nature of how a human can interact with higher levels of automation and the manner in which authority may be delegated from human to machine (Sheridan & Verplank, 1978). Key to this interaction design is the manner by which information is passed between team members, ensuring that the intent of the team member conducting the task is visible.

Further to the composition and dynamic of HAT, the manner in which the augmentation of support is presented will dictate the effectiveness of the HAT. Within Aviation Psychology we are only too aware of the ineffective manifestations of human-automation that relate to losing situation awareness due to the human taking a back seat to the automation (Endlsey, 2016), the human not keeping up with automation in terms of understanding (Lyons, 2013) or even the human mistrusting the automation altogether (Lyons & Stokes, 2012). The next sections will examine these issues in a bit more detail.

Understanding Automation Behavior in Aviation: A practical limit to automation as a team member

The HAT concept differs from traditional human-computer interaction, or human-systems integration, in that it suggests some degree of human-like qualities for the "autonomous" agent (computer). This is in contrast to viewing the computer as, say, a processor that follows predetermined rules, and suggests the technology can offer intelligent and adaptive solutions (Cox, 2013). With recent advances in artificial intelligence, notably machine learning, the HAT concept can be reasonable, and advances in human-computer interface technologies can make HAT compelling. The issue in aviation—whether for pilots flying aircraft, or air traffic controllers (ATC) separating traffic—is that HAT might be in conflict with one of the core human-automation principles: the operator must sufficiently *understand* automated system behaviors. HAT concepts suggest a level of complexity or intelligence that make automation behavior challenging to understand and predict from the human operator perspective. Furthermore, when addressing off-nominal conditions (e.g., system degradation and failures, environmental conditions that are beyond the design limits), the challenge of understanding behavior of an autonomous agent is likely to be compounded.

That argument is a practical one, primarily based on lessons learned from decades of aviation research and practice in human-automation interaction (Brown, 2016). There is not a theoretical basis for what automation is capable of, given access to the right information, and given that it is operating within its design envelope. But even with proper design, clear human-automation design tradeoffs have emerged, and are well known (e.g. Hoffman & Woods, 2011; Woods & Branlat, 2011). For example, human supervisory control reduces excessive workload, but can lead to loss of situation awareness. Human-computer interfaces need to provide sufficient information to the human and be transparent and provide mode awareness, but information overload and display clutter can limit this (Selkowitz et al., 2017). History has demonstrated there are a number of inherent design tradeoffs with "traditional" human-automation interaction that suggest practical limits with respect to understanding by the human operator, even with substantial training and standard operating procedures. These limits are not just important; they are drivers for safety-critical systems in aviation and are explored in the following two examples.

Example 1: Flight Deck

Looking back, there are many well-known examples from aircraft accidents in which a primary factor was an off-nominal condition and a subsequent lack of understanding of that condition and the resulting system behaviors. The crash in 2009 of Turkish Airlines Flight 1951 is one of many examples. In that example the crew were clearly dependent on the auto-throttle to maintain velocity and height as they ran through their landing checklists. Unfortunately, a malfunctioning radio altimeter affected the auto-throttle configuration into a different flight mode (Dekker, 2009). In such examples, even when procedures have been developed for these off-nominal conditions, pilots can fail to first make sense of the situation, including the resulting automation behavior, in order to apply the procedures appropriately. Furthermore, it is impractical to develop procedures for all possible combinations of off-nominal conditions, including automation—leaving the pilots to figure out the mitigation. The point is not that there are shortcomings in pilots, pilot training, procedures, or designs; the point is that off-nominal conditions in the context of complex systems, complex procedures, and complex operations make it fundamentally challenging for pilots to understand the situation. Rather than designing automated systems to be more complex autonomous team agents, as in HAT, perhaps the focus should be on simplification.

Example 2: Air Traffic Control

Current research by the FAA on unmanned aircraft systems (UAS) illustrate solutions centered on simplification. UAS have great potential for intelligent automation to fly the aircraft—aviate, navigate, communicate—within the national airspace system (NAS). But as great as their automation potential is, we are seeing the importance of human understanding of UAS behavior through standardization. As an example, during off-nominal "lost link" situations, in which the pilot-in-command loses the control link, the UAS must fly autonomously for some period of time. During this time, what is best for the UAS can involve a complex decision that depends on its goals, system health, altitude, weather, communications infrastructure, available airfields, etc. But from the ATC perspective, the primary need is simple: predictability. This has been documented in ATC interviews, knowledge elicitation, and cognitive walkthroughs during research on contingency operations in the Terminal and Enroute domains. Although lost link standards have not yet been defined, it is clear that controllers primarily need to be able to understand exactly when the UAS will maneuver (with sufficient time to issue clearances to traffic), and what exactly the maneuver will be (from a small set of operationally suitable options). Therefore, the sophistication of UAS automation to autonomously maneuver based on its needs might be irrelevant because the operational constraints from the human operators (ATC), to understand what the UAS will do, largely determines UAS behavior in that situation. In this case, human understanding during off-nominal situations has so far been a primary driver for standardizing lost link procedures and technologies.

These aviation examples illustrate that there are practical limits to consider in automated system design from the perspective of human operators, especially in off-nominal conditions. Pilots, air traffic controllers, and other human actors in the NAS are responsible for safety, and held accountable for safety. They might benefit at times from a computer-based team member as viewed in the HAT metaphor, but history suggests the need something simpler, and current research suggests solutions moving forward could be centered on human understanding versus more complex autonomous agents.

Discussion

This paper represents the first part within a series of papers associated with a Panel Session that examines the nature of HAT. Within this paper we have discussed the need for defining the role of the team members and draw on examples in human teaming. We suggest that the same philosophy should be applied to teams that are composed of both human and autonomous members. The dynamic aspects of defining team roles and, perhaps more importantly, who has authority within the team is worth considering. There may very well be instances where the synthetic team member may be in a position to give commands to a human team member. In order for a team to be effective and to respond to changing goals then the acceptance of such authority within the team is a significant factor. It is likely however, that any composition of HAT would still ultimately require a human to supervise and monitor the team. Of course this is dependent on the context of the operation.

The autonomous team member must also be viewed as possesing a degree of intelligence that may not be as tightly *bounded* to rationality as the human team members. When we consider applications in aviation, where HAT could afford measurable benefits, care must be taken. The automation literature in aviation human factors provides us a with a wealth of knowledge as to the benefits and pitfalls of applying complex systems alongside the human. When we look at the potential use of HAT across the flight deck and within ATC there are limitations that need to be considered. On days when the operation is routine and mundane a HAT would function in a manner that would free up resource of the human, but in off-nominal conditions the human will be less likely to understand (and trust) the system as to the rationale behind decisions benig presented. This has implications to safety, acceptance and shaping the future regulatory requirement.

References

Baxter, J., & Richards, D. (2010). Whose Goal is it anyway? User Interaction in an Autonomous System. Twenty-Fourth AAAI Conference on Artificial Intelligence, Atlanta, Georgia, USA, July 11–15, 2010.

Belbin, R.M. (1981) *Management Teams: Why They Succeed or Fail*, Butterworth-Heinemann, Oxford.

Belbin, R.M. (1993) *Team Roles at Work*, Butterworth-Heinemann, Oxford.

Brown, J.P. (2016) The Effect of Automation on Human Factors in Aviation. The Journal of Instrumentation, Automation and Systems, 3(2), 31-46.

Cooke, N.J., Kiekel, P.A. & Helm, E.E. (2001) "Comparing and validating measures of team knowledge", *Proceedings of the Human Factors and Ergonomics Society*, pp. 361-365.

Cox, M. T. (2013) Goal-driven autonomy and question-based problem recognition. Retrieved from http://www.cogsys.org/papers/2013poster9.pdf

Dekker, S. (2009). *Report of the flight crew human factors investigation conducted for the Dutch safety board into the accident of TK1951, Boeing 737-800 near Amsterdam Schiphol Airport, February 25, 2009* (p. 127). Lund: Lund University, School of Aviation.

Endsley, M.R. (2016) From here to Autonomy: Lessons learned from Human-Automation research. Human Factors: The Journal of the Human Factors and Ergonomics Society, 59(1), 5-27.

Hoffman, R.R. & Woods, D.D. (2011) Simon's Slice: Five Fundamental Tradeoffs that Boundthe Performance of Human Work Systems, in: 10th International Conference on Naturalistic Decision Making, Orlando FL. pp. 5–31.

Lyons, J. B. (2013). Being transparent about transparency: A model for human-robot interaction. Proceedings of AAAI Spring Symposium on Trust and Autonomous Systems, Palo Alto , 48-53.

Lyons, J.B., & Stokes, C. (2012) Human-Human reliance in the context of automation. Human Factors: The Journal of the Human Factors and Ergonomics Society, 54(1), 112-121.

Richards, D. (2017) Escape from the factory of the robot monsters: Agents of change. Team Performance Management: An International Journal, 23, 96-108.

Salas, E., Cooke, N.J. & Gorman, J.C. (2010) "The science of team performance: progress and the need for more…", Human Factors, Vol. 52 No. 2, pp. 344-346.

Scholtz, J. (2003) Theory and evaluation of human robot interactions, *Hawaii International Conference on System Science 36 (HICSS 36), Hawaii*, January 2003.

Selkowitz, A.R., Larios, C.N., Lakhmani, S.G. & Chen J.Y.C. (2017) Displaying Information toSupport Transparency for Autonomous Platforms. Advances in Human Factors in Robots andUnmanned Systems. Orlando, FL: Springer International Publishing, 161. doi: 10.1007/978-3-319-41959-6_14.

Sheridan, T. B. & Verplank,W. (1978) Human and Computer Control of Undersea Teleoperators. Cambridge, MA, Man-Machine Systems Laboratory, Department of Mechanical Engineering, MIT.

Woods, D.D. & Branlat, M. (2011) How human adaptive systems balance fundamental trade-offs: Implications for polycentric governance architectures, in: Proceedings of the Fourth Resilience Engineering Symposium, Sophia Antipolis, France

Zhou, J., & Chen F. (2015) Making Machine Learning Useable. International Journal of Computational Science and Engineering14(2): 91–109. doi: 10.1504/IJISTA.2015.074069.

HUMAN-AUTONOMY TEAMING - AN EVOLVING INTERACTION PARADIGM: A COGNITIVE ENGINEERING APPROACH TO HAT

Axel Schulte and Diana Donath
Bundeswehr University Munich, Germany

This contribution outlines a cognitive engineering approach to structure, describe and depict configurations for highly automated human-machine systems using a common language. These systems involve cognitive agents, for autonomous vehicle guidance and mission management. The method focusses on the systematic top-down deduction of requirements for human-autonomy work share and interaction in the work process. Therefore, this contribution outlines a procedure to follow to design and describe such human-autonomy teaming systems, related user and system requirements, and top-level system designs. This contribution primarily aims at the application field of military, highly automated manned/unmanned vehicle systems.

Today, higher cognitive functions (e.g., perception, planning, and decision-making), traditionally exclusively owned by the human, are becoming an integral part of automated functions. In the last one or two decades the term "autonomous system" has widely been used to describe complex automated systems working largely independent from a human operator. However, the more capable the automation becomes, the more essential the issue of human-system functional allocation and integration has turned out to be (Klein et al., 2004). We share the concern of Bradshaw et al. (2013) that an undifferentiated use of the term of "autonomy" and the proliferation of automation can lead to unfruitful discussions and oddly defined development programs. We see the need for a conceptual framework unifying the nomenclature and description of systems in which human operators interact with complex automation.

Therefore, in this paper (the second part of the panel topic), we attempt to identify and formally describe common grounds among researchers and practitioners in this field. Despite our concerns, we want to adhere to the term of Human-Autonomy Teaming (HAT) to describe systems in which humans work with highly automated agents. Where those agents carry attributes like "autonomous" or "intelligent", we will assign the unified term *Cognitive Agent*. Our approach, in general, suggests a *common semantical and graphical language* (Schulte & Donath, 2018), as well as a procedure to follow, to describe systems, system requirements, and top-level system designs. Both have a stronger focus on human-automation work share and integration aspects than traditional systems engineering practices and tools (e.g., Unified Modelling Language, UML). The traditional approach focuses solely on the formulation of requirements and the design of the technical functions of a system. The human operator only appears as an actor, usually located outside the system boundary. This approach is reasonable when automation is relatively simple, in the sense that it can perform specific clear-cut part-tasks. There, one can well describe the relationship between the (technical) system and the human user through use cases calling for certain user-system supervisory control interactions.

In this paper, in contrast, we want to take account of the following trends. Firstly, the automation in HAT will become much more capable in the sense of being able to perform higher cognitive tasks. Secondly, the work share and interaction between the user and the automated system will be much less stable (e.g. adaptive automation; as in Scerbo, 2006). Finally, the task performance of human and automation will be highly dependent on a cognitive level (Hollnagel & Woods, 2005).

Building Blocks of a Language Describing HAT-Systems

Hence, we follow a strict systems engineering top-down approach to establish a formal description language, semantical and graphical, for highly automated human-machine and HAT-systems. In a first step, we introduce the notion of the Work Process (WProc), and its graphical representation (cf. Fig. 1), to develop an integrated view upon the purpose of a human-machine co-action, its physical and conceptional work environment (WEnv), as well as its desired output (WPOut) to the environment the WProc lives in. However, most crucial is the Work Objective (WObj), i.e., the mission or the purpose of work, since it reflects the user requirements for a system we will design. The WObj defines and initiates the WProc. The proper definition of the WObj is of high priority and most critical for the definition of the system boundaries and the design. Connections and dependencies between multiple WProcs (e.g., hierarchical or networked structures) we also capture and describe on this level. For a more detailed discussion, please refer to [6][7].

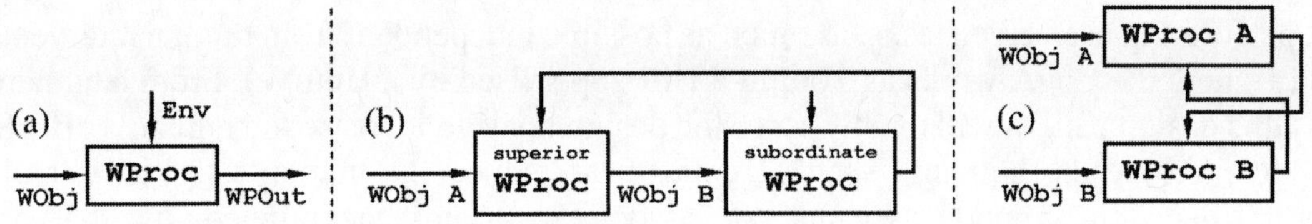

Fig. 1. (a) Work Process; (b) hierarchical; (c) networked.

In the second step, we establish a physical system instantiating the considered WProc. We call this system a Work System (WSys) in which the elementary roles of Worker and Tools are usually taken by humans and machines including conventional automation. The main characteristic of the role of the Worker is **to** know, understand, and pursue the WObj *by own initiative*. Without this initiative, the WProc would not be carried out. In principle, the Worker, and only the Worker, might as well self-assign a WObj. The Tools, on the other hand, will receive tasks from the Worker and will only perform them when told to do so. Hence, the Worker and the Tools are always in a Hierarchical Relationship (HiR, see green connector in Fig. 2, b). Again, for further reading, we recommend Schulte et al (2016) and Schulte & Donath (in prep).

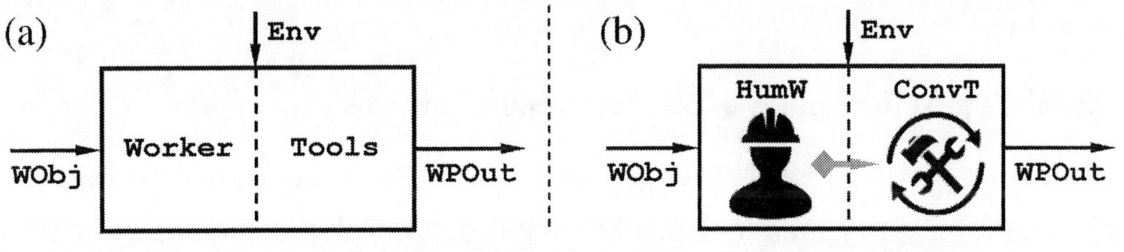

Fig. 2. Work System as physical instance of the corresponding WProc: (a) comprising the roles of the Worker and the Tools; (b) instantiated with a Human Worker (HumW) and Conventional Tools (ConvT).

Introduction of the Cognitive Agent into the Work System

In the third step, we introduce "the autonomy" into the WSys, represented by one or more Cognitive Agent(s) (CogA, little 'R2D2' in Fig. 3) in various roles and relationships to the human operator(s), the conventional automation, and the machinery.

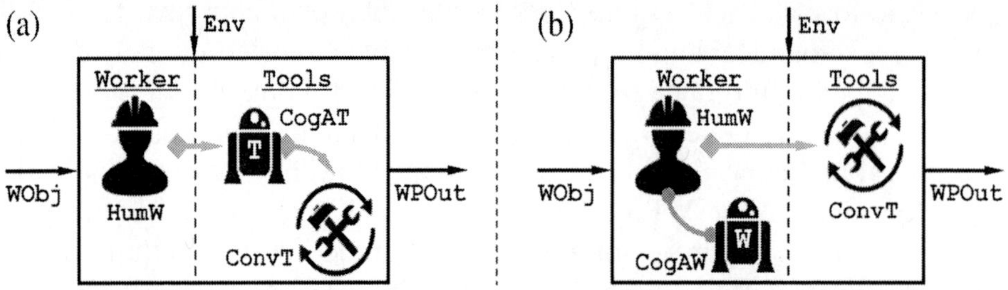

Fig. 3. (a) WSys with CogA as Tool (i.e., CogAT) in HiR;
(b) WSys with CogA as Worker (i.e., CogAW) in HeR.

Two trends have been followed in the past two decades concerning the role such a CogA could potentially take in system design. Firstly, so-called autonomous systems, i.e. systems that aim at performing user-given tasks, as much as possible independent from human intervention (see Fig. 3 (a): here the CogA works as Tool in a HiR supervised by a HumW). From a human factors stance, this design pattern will mostly serve the design goals to increase the human's effectiveness, to increase the human's span of control, to reduce the human's taskload, and others. Secondly, decision support, assistant, or associate systems, acknowledging that a human predominantly performs the work, while supported by a machine agent (see Fig. 3 (b): here the CogA is part of the Worker). In the latter case, there exists a Heterarchical Relationship (HeR, blue connector) between the HumW and the CogA being part of the Worker here. From a human factors stance, this design pattern will mostly serve the design goals to avoid or correct human erroneous action, to moderate or modulate human mental workload, to increase the human's situation awareness, and others.

To sum up at this stage, the design options in the different applications are plenty (Schulte & Donath, in prep), still constructed, however, of only a few elementary building blocks. In our language, we represent those building blocks by a handful of the aforementioned semantical and graphical descriptors. The descriptors stand for actors (i.e., humans, cognitive agents, conventional automation including machines), role allocations of actors (i.e., worker, tools), relationships amongst actors (i.e., hierarchical, heterarchical), and a few others of less importance (e.g., co-location, grouping), not explicitly mentioned here.

Actor-Relationship-Actor Tuples as Master Design Patterns

From the various application-specific WSys designs using those descriptors in a coherent manner, we can now identify similar elementary (actor-relationship-actor)-tuples, as we call them. Thereby, we repeatedly can identify structural similarities in the sense of master design patterns for HAT, over a wide range of extremely different applications and system design approaches. With this, scientists and practitioners can start identifying common or alternative solutions for similar problems, namely design patterns for HAT-, and conventional human-machine systems.

Fig. 4 gives an annotated overview of all possible tuples. Here, the possible actors are HumW, CogA, and ConvT; possible relationships are HiR, and HeR. The HumW will, per definition, always take the role of a Worker; the ConvT will always be Tool, whereas the CogA may take the role of a Worker (i.e., CogAW), or of a Tool (i.e., CogAT). Between a Worker and a Tool there will always be a HiR.

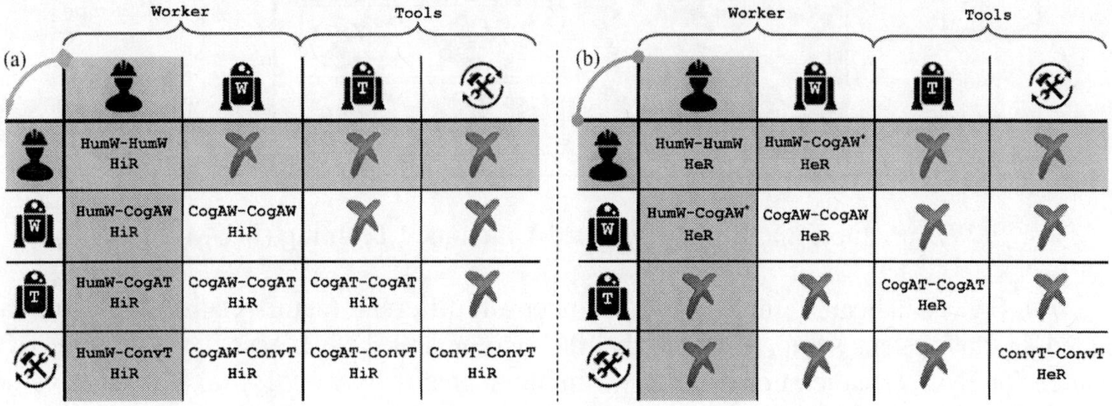

Fig. 4. (a) Hierarchical, (b) heterarchical (actor-relationship-actor)-tuples; (shaded: human involved; *: equal configurations; cross: invalid option).

A hierarchy of a CogA or a ConvT over a HumW, or a ConvT over a CogA is not reasonable. The same applies to a heterarchy of a ConvT with either a HumW or a CogA, for obvious reasons. Tuples, which do not involve humans, may not directly be interesting for HAT-systems. However, they certainly can influence the behavior of the automation "under the hood", and therefore, be worthwhile to look at, at least from a pure engineering stance. Also, the pure human-human relationships, either hierarchical or heterarchical, may not directly be relevant for HAT-systems, except of course for WSys with more than one human. Apart from that, they may serve as valuable source for design metaphors. Finally, we do not want to allow a Human-Agent HeR, where the agent is part of the Tools, since per definition there is always a HiR between Worker and Tools.

Example: Manned-unmanned Teaming Helicopter Mission

As an example, we look at a WSys for a Manned-Unmanned Teaming (MUM-T) military helicopter mission. On the level of WProcs, there is also involved a command and control (C2) WProc that provides the MUM-T mission WObj. The Work Object (WO) is the troops to be transported in or out a combat zone. Fig. 5 depicts the WSys setup. It consists of a cockpit crew of two humans, pilot flying and commander, and a CogAW representing a crew associate system. The roles of the pilot flying and the commander the crewmembers may swap amongst each other at any time. The Tools are a manned transport helicopter (H/C), where the aforementioned Workers are located, and three dislocated small reconnaissance UAVs (Unmanned Aerial Vehicles), each of which controlled by an on-board CogAT that provides a delegation interface to the cockpit crew for highly automated UAV tasks. The crew associate may also directly use this interface and the high-level commands supported by the CogATs onboard the UAVs.

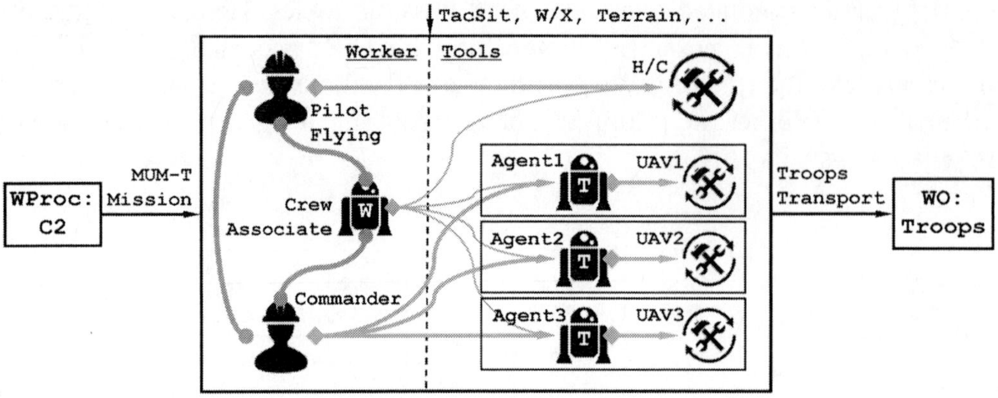

Fig. 5. WSys setup for a military Manned-Unmanned Teaming (MUM-T) mission.

As Fig. 5 indicates, there are a number of different (actor-relationship-actor)-tuples involved in the system setup. Particularly, the (Commander-HeR-Associate)-tuple is of high relevance for HAT considerations. Here, we implemented it by use of a mixed-initiative design pattern for mission planning tasks (Schmitt et al, 2018), and by a workload-adaptive design pattern for mission execution tasks (Brand et al, 2018), both optionally blended with the SA-based Agent Transparency (SAT) Model (Chen et al, in press). The (Commander-HiR-AgentX)-tuple is implemented by using our well-proven task-based guidance concept (Uhrmann & Schulte, 2012; Rudnick & Schulte, 2017). Addressing the UAVs as a team is also an implemented option.

Discussion

In this contribution, we briefly outlined a description language and procedure to follow for a systematic top-down approach for the definition of Human-Autonomy Teaming (HAT) systems. This approach tries to formalize the description of complex, highly automated human-machine systems, in particular, in the domain of manned and unmanned vehicle guidance and mission management applications. The resulting system representations allow the discussion of system characteristics on a common high level of abstraction, using only a few descriptors. Recurring structures of human-agent collaboration can also be identified easily. Additionally, the discovery, the discussion, and the exchange of beneficial design patterns for HAT are facilitated. Future works will aim at a further formalization of the language. Furthermore, we will need to strengthen the linkage between the system-level description of HAT-systems and the characterization of individual design patterns.

References

Bradshaw, J.M., Hoffman, R.R., Woods, D.D., & Johnson, M. (2013) The Seven Deadly Myths of "Autonomous Systems". IEEE Intelligent Systems, (3), pp. 54-61.

Brand, Y, Ebersoldt, M., Barber, D., Chen, J.Y.C., & Schulte, A. (2018) Design and Experimental Validation of Transparent Behavior for a Workload-adaptive Cognitive Agent. Intelligent Human Systems Integration (iHSI).

Chen, J.Y.C., Lakhmani, S., Stowers, K., Selkowitz, A., Wright, J., & Barnes, M. (in press) Situation Awareness – Based Agent Transparency and Human-Autonomy Teaming Effectiveness. Theoretical Issues in Ergonomics Science.

Hollnagel, E., Woods, D.D. (2005) Joint cognitive systems: Foundations of cognitive systems engineering. CRC Press.

Klein, G., Woods, D.D., Bradshaw, J.M., Hoffman, R.R., & Feltovich, P.J. (2004) Ten challenges for making automation a "team player" in joint human-agent activity. In: IEEE Intelligent Systems, (6), pp. 91-95.

Rudnick G., Schulte A. (2017) Implementation of a Responsive Human Automation Interaction Concept for Task-Based-Guidance Systems. Lecture Notes in Computer Science, Volume 10275. Springer-Link.

Scerbo, M. (2006) Adaptive automation. In: Neuroergonomics. Pp. 239-252.

Schmitt, F., Roth, G., Barber, D., Chen, J.Y.C., & Schulte, A. (2018) Experimental Validation of Pilot Situation Awareness Enhancement through Transparency Design of a Scalable Mixed-Initiative Mission Planner. Intelligent Human Systems Integration (iHSI).

Schulte, A., Donath, D., & Lange, D.S. (2016).: Design Patterns for Human-Cognitive Agent Teaming. Lecture Notes in Computer Science, Volume 9736. Springer Link.

Schulte, A., & Donath, D. (in prep) Systems Level Design Patterns Approach for HAT. In: Human-Autonomy Teaming: Supporting Dynamically Adjustable Collaboration. NATO STO Task Group Report HFM-247. Neuilly-sur-Seine, France.

Schulte, A., & Donath, D. (2018) A Design and Description Method for Human-Autonomy Teaming Systems. In: Karwowski W., Ahram T. (Eds). *Advances in Intelligent Systems and Computing*. Vol 722. Springer, Cham.

Uhrmann, J., & Schulte, A. (2012) Concept, Design and Evaluation of Cognitive Task-based UAV Guidance. IARIA Journals. ISSN-Online 1942-2679.

HUMAN-AGENT TEAMING - AN EVOLVING INTERACTION PARADIGM: AN INNOVATIVE MEASURE OF TRUST

Samson Palmer, Dale Richards, Graham Shelton-Rayner
Coventry University, United Kingdom
David Inch
Horiba MIRA, United Kingdom
Kurtulus Izzetoglu
Drexel University, United States of America

The promise of intelligent decision support systems is presented as a harbinger for humankind. With the potential partnership between the human and autonomous system we could see a significant increase in effectiveness and safety. However, as we see both human and agent team members being integrated we must investigate ways in which we can assess not only the interaction between the two actors, but also the very nature of trust perceived by the human. In this paper we present early findings of an experiment that examines the human-autonomy interaction across different frameworks of authority; from manual to fully autonomous. Participants were asked to interact with a ground control station that supervised several unmanned systems, and presented with goals that required manually selecting or approving assets to achieve mission goals. The use of neuroimaging (in this instance Functional Near Infrared Spectroscopy - fNIRS) was used to establish neuro-correlates of trust within the prefrontal cortical region of the brain. Initial findings suggest that the dorsolateral prefrontal cortical region of the brain is heavily associated when humans are confronted with different levels of authority with human-autonomy interaction.

As we have seen in the previous papers (Part I and II), the construct of HAT, while promising in terms of assisting the human in achieving goals, is somewhat complex and not without problems. When we traditionally think of whether a HAT is performing in an efficent manner it is easy to assess the quantifiable objectives as to whether the goal has been achieved succesfully or not. However, as outlined in Part I and Part II, how the human interacts and perceives the HAT is of critical importance. A key component of this is whether the human team members (single operator supervising the system) trusts the system behaviour.

Trust has yet to be universally defined, particularly when referring to Human-Computer Trust (HCT). However, one definition suggests that HCT is "the extent to which a user is confident in, and willing to act on the basis of, the recommendations, actions and decisions of an artificially intelligent decision aid" (Madsen, 2000). All engagements and interactions, both social and with computers, require elements of trust, particularly those that require some form of cooperation. When referring to social interaction this is often called Reciprocal Trust (RT), and is critical in the development of partnerships (Krueger et al, 2007). It is therefore reasonable to assume that the development of a human-automation partnership and the formation of HCT will also require elements of RT. To this effect, researchers have often tried to adapt interpersonal trust models and measurement instruments to explain operator HCT, even creating their own scales. Examples of these include the Empirically Derived scale, the Human-Computer Trust Scale, and SHAPE (Solutions in Human Automation Partnerships in European ATM) Automation Trust Index (SATI). These have all been designed to assess operator trust of decision support systems, and although they have all benefited from systematic development and validation, they have also attracted criticism (Lewis et al, 2018). To further confound this, much of the research surrounding HCT has utilised scales based on a third-person frame where there is a lack of personal consequence as a result of the decisions made (Miller et al, 2016).

Although the above-cited metrics of trust are based solely on subjective assessment Izzetoglu & Richards (2019) suggest that the validity of using a sole metric for such complex human behaviour should also ideally include behavioural data (such as task completion) and some form of physiological assessment. In particular Izzetoglu & Richards (Ibid.) discuss the use of acquiring data via wearable neurimaging equipment in order to assess human cortical activity. Indeed, some studies have begun to examine and characterise the neural correlates of trust and have often implicated activity within the prefrontal cortex in encoding the motives and intentions surrounding interactions with systems (Sripada et al., 2009). Additionally, some neuroimaging studies have investigated the neural correlates of cognitive processes aligned with cooperation and RT and have further shown activation of the prefrontal cortex, particularly the anterior medial prefrontal cortex (aMPFC) (Bos et al, 2009). With the advances in wearable neuroimaging techniques, the ability to monitor operators during system interaction has provided a chance to further assess the neural activity associated with trust. As we have seen in Part I and II papers of this session, human interaction with autonomous systems is complex and the role of trust plays an integral part in assessing whether the HAT is effective or not. The use of neuroimaging therefore is seen as a promising tool in providing an indication of trust within such HAT compositions.

This paper discusses a recent study whereby functional Near Infrared Spectroscopy (fNIRS) was used to assess prefrontal cortical activity during interaction with systems of varying levels of autonomy. The application of fNIRS allows us to monitor localised hemodynamic changes (mainly in oxygenation) across the prefrontal cortex (PFC) region as a result of increased cortical activity. It was hypothesised that the changes in oxygenation across different *expsoures* of decision support would be indicative of how HCT varied, whilst also how failure of autonomous systems can impact the HCT of the operator.

Exploring Prefrontal Cortical Activity During Human-Autonomy Interaction

In the current study (N=23), participants were asked to supervise eight unmanned vehicles (both ground and air - UxV) via a Ground Control Station (GCS) visual interface, as per Figure 1. Participants were tasked to utilise the different capability UxV's in order to defend each of the four bases (Bases indicated as A, B, C and D in Figure 1). Participants were asked to respond to events across different scenarios as they evolved, resulting in the participant having to either respond directly with (1) a command input (Manual mode), (2) a selection of command inputs as suggested by the system (Assisted Manual), (3) selecting a single recommended command suggested by the system with YES or NO (Assisted Auto), and finally (4) no inputs required from operator as they monitored the system respond and inform the operator of chosen action (Full Auto).

During the scenarios an event was triggered by an unknown entity entering the map and the operator would be tasked to interrogate the unknown signal. The most appropriate UxV would need to be sent to conduct a scanning task based on where assets were currently located, their sensor fit, current task, amount of fuel, speed of response, etc.

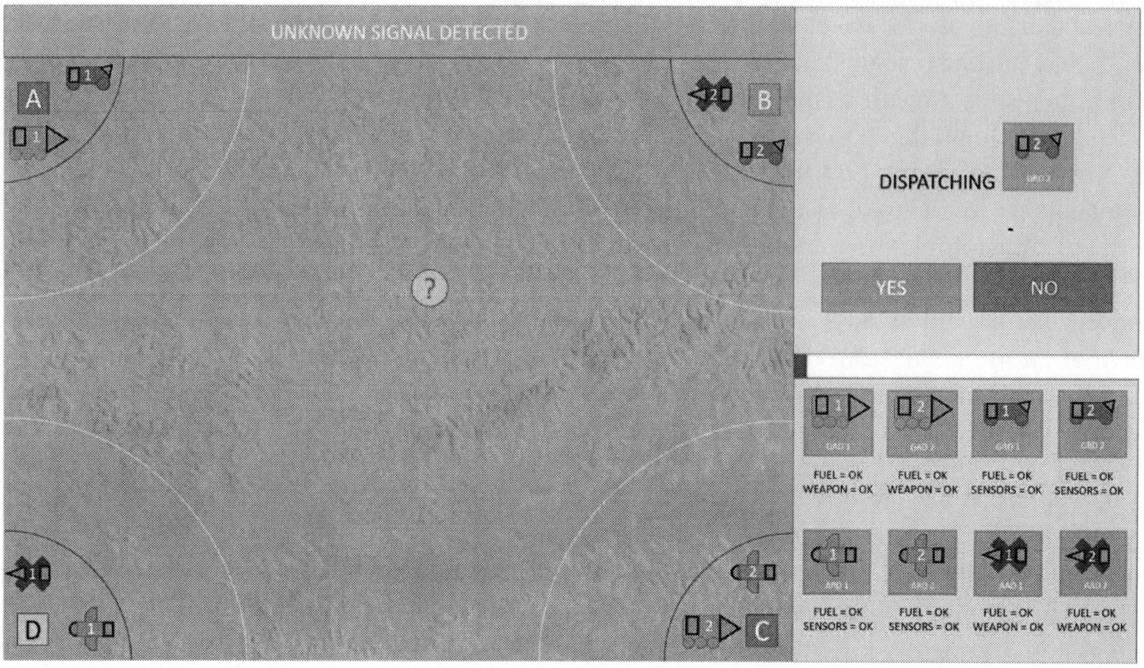

Figure 1 - GCS display showing all UxV on station in each base in Assisted Autonomy mode

Unknown signals consisted of a random mix of enemy combatants and civilians. Mission success occurred if all 4 military bases were still intact once the time ran out, and if no civilians were harmed. Mission failure occurred if an enemy combatant crossed the red barrier surrounding each base, or if a civilian was harmed. Scenario order was counterbalanced for each participant to avoid order effects of the repeated measures design of the experiment. Each scenario presented was designed with a different level of autonomy. The Assisted Manual scenario required participants to select from 2 or 3 drone options presented to them once an unknown signal appeared. However, the drone options presented were based solely on distance to target (closest drone was displayed as first option) and did not account for drone status or capabilities, therefore participants were required to check this before making a decision. The Assisted Auto mode required participants to agree to the decisions made by the system by selecting YES or NO. However, this time the decisions made were based upon all aspects of the system, and could be perceived as the best possible decision. Additionally, safeguards were in place whereby if a NO decision would result in base destruction, this option was removed. There were two Full Auto mode scenarios, where participants were required to observe the system with no interaction required. However, the integrity of one of the systems during these scenarios were manipulated in a way that could be perceived as aggressive and incorrect, whereby attack drones were always deployed first without scanning unknown signals, which ultimately ended in a civilian being harmed and thus failing the mission.

Initial Results from Experiment

The study is still ongoing at this time with current fNIRS data collected and analysed will be discussed. In order to assess prefrontal cortical activity, fNIRS neuroimaging was used. This allowed the collection of a 5-minute baseline before continuously gathering cortical activity of participants throughout all scenarios. Following the completion of artefact removal and extraction of processed optical density data (Izzetoglu & Richards, 2019), a modified Beer-Lambert Law (MBLL) was applied for calculations of channel-specific changes in oxygenated hemoglobin (oxy-Hb) and de-oxygenated hemoglobin (deoxy-Hb). This data was collected across 16 optodes aligned with the PFC as shown in Figure 2.

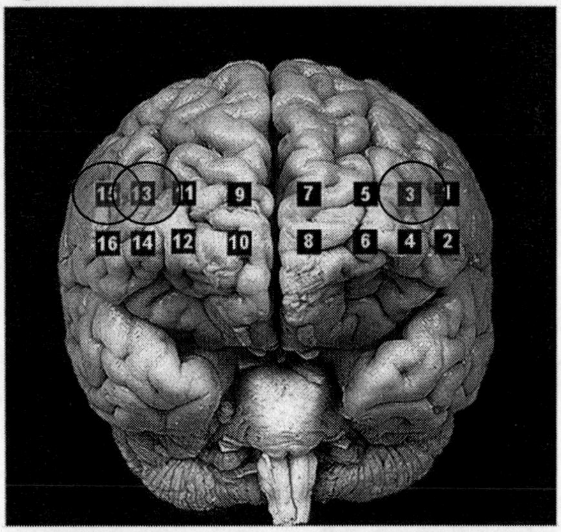

Figure 2 - Topographical image displaying optode placement across the prefrontal cortex (with associated regions that showed increased ratio of oxy-Hb and deoxy Hb circled)

Initial analysis demonstrates that optodes 3, 13, and 15 detect an increase ratio of oxy-Hb to deoxy-Hb during the Full Auto Incorrect and Assisted Manual scenarios. These scenarios were expected to place higher cognitive demand on the participant in terms of Memory (WM) and Attention, and optodes 3, 13 and 15 correlate to regions associated with both WM and Attention (Reddy et al., 2018). Figure 3 shows the mean values for each task across each optode.

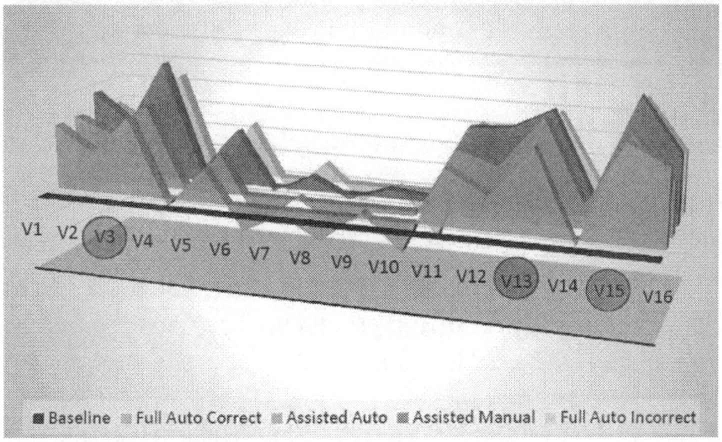

Figure 3 - Mean ratio values of oxy-Hb and deoxy Hb across all optodes (with larger variance from baseline circled)

As can be seen in Figure 3, when we compare a number of optodes between baseline fNIRS reading against the four experiment conditions (varying authority and delegation of control), we see a significant main effect for optode 3 ($F(4,110)=4.111$, $p<.005$), optode 13 ($F(4,110)=17.504$, $p<.05$). Optode 15, while not significant, showed promise of a main effect (($F(4,110)=16.806$, $p=.13$) - but noise from this optode meant losing seven participants sets of data from the analysis.

Discussion

It is often assumed that the addition of an autonomous aid to a human operator will result in a more effective system. However, as Dzindolet et al (2003) report, this can be inaccurate: if the operator does not trust the system fully, or the operator trusts a system that does not provide correct support, it can lead to disuse or misuse. Therefore a means of assessing the level of HCT is an important tool, especially when we consider safety critical systems. This paper reports initial findings where brain-based measures via fNIRS were used to study neural mechanisms during numerous scenarios with varying levels of autonomous support. It was hypothesised that the fNIRS measures may indicate how the autonomous support system affects the human operator's trust of the system, and that the neural correlates of trust could be observed during these scenarios.

Preliminary findings indicate that optodes 3 and 13 (with possibly 15) showed activation during the Assisted Manual and Full Auto Incorrect tasks. These particular optode regions of the PFC have been linked to Working Memory (WM) and Attention (Reddy *et al* 2018), and can be aligned with Brodmann's areas 9/10 (Zilles 2010), also referred to as the dorsolateral/medial and anterior prefrontal cortex, respectively. Previous research suggests that trust, particularly reciprocal trust, can be associated with greater activity in the dorsomedial prefrontal cortex (Filkowski et al, 2015). Other research has also shown the neural correlates of uncertainty may be linked to the dorsolateral prefrontal cortex (Pushkarskaya et al, 2015) and the results of this experiment, demonstrate increased activity in these regions during scenarios that required increased operator input, and where the autonomous system could be perceived as making incorrect decisions.

It is important to understand how human operators trust intelligent systems, and thus while we can observe behavioural performance and collect self-ratings from individuals, objective physiological measures could provide insight in assessing trust. This study is part of the new exciting investigations that examine the neural correlates of trust and distrust. We propose this method of cortical assessment represents an opportunity to better understand our relationship with HAT.

References

Bos, W., Dijk, E., Westenberg, M., Rombouts, S. and Crone, E. (2009) "What Motivates Repayment? Neural Correlates Of Reciprocity In The Trust Game". *Social Cognitive And Affective Neuroscience,* 4(3), 294 - 304.

Dzindolet, M., Peterson, S., Pomranky, R., Pierce, L. and Beck, H. (2003) "The Role Of Trust In Automation Reliance". *International Journal Of Human-Computer Studies* [online] 58 (6), 697-718.

Filkowski, M., Anderson, I. and Haas, B. (2015) "Trying To Trust: Brain Activity During Interpersonal Social Attitude Change". *Cognitive, Affective, & Behavioral Neuroscience,* 16(2), 325-338.

Izzetoglu, K., & Richards, D. (2019). Human Performance Assessment: Evaluation of Wearable Sensors for Monitoring Brain Activity. In M. Vidulich & P. Tsang (Eds.), Improving Aviation Performance through Applying Engineering Psychology: Advances in Aviation Psychology (1st ed., pp. 163–180). Boca Raton, FL: CRC Press..

Krueger, F., McCabe, K., Moll, J., Kriegeskorte, N., Zahn, R., Strenziok, M., Heinecke, A. and Grafman, J. (2007) Neural Correlates Of Trust. *Proceedings Of The National Academy Of Sciences,* 104 (50), 20084-20089.

Lewis, M., Sycara, K. and Walker, P. (2018) The Role Of Trust In Human-Robot Interaction." in *Foundations Of Trusted Autonomy. Studies In Systems, Decision And Control* [online] ed. by Abbass, H., Scholz, J. and D, R. Springer, Cham, 142 - 144. available from <https://link.springer.com/chapter/10.1007/978-3-319-64816-3_8#citeas> [28 February 2019]

Miller, D., Johns, M., Mok, B., Gowda, N., Sirkin, D., Lee, K. and Ju, W. (2016) "Behavioral Measurement Of Trust In Automation". *Proceedings Of The Human Factors And Ergonomics Society Annual Meeting,* 60 (1), 1849-1853.

Pushkarskaya, H., Smithson, M., Joseph, J., Corbly, C. and Levy, I. (2015) "Neural Correlates Of Decision-Making Under Ambiguity And Conflict". *Frontiers In Behavioral Neuroscience*[online] 9. available from <https://www.frontiersin.org/articles/10.3389/fnbeh.2015.00325/full>

Reddy, P., Richards,D., & Izzetoglu, K. (2018) *Cognitive Performance Assessment of UAS Sensor Operators via Neurophysiological Measures*, in *2nd International Neuroergonomics Conference*.

Sripada, C., Angstadt, M., Banks, S., Nathan, P., Liberzon, I. & Phan, K. (2009) "Functional Neuroimaging Of Mentalizing During The Trust Game In Social Anxiety Disorder". *Neuroreport*, 20(11), 984-989.

Zilles, K. and Amunts, K. (2010) Centenary Of Brodmann's Map — Conception And Fate. *Nature Reviews Neuroscience*, 11(2), 139-145.

A History of the Air Traffic Control Collegiate Training Initiative (AT-CTI) Program

Darendia McCauley
Dana Broach
FAA Civil Aerospace Medical Institute
Oklahoma City, Oklahoma

In the 1989 FAA Flight Plan for Training, the Federal Aviation Administration (FAA) proposed testing the concept of off-loading some portion of air traffic control specialist (ATCS) training to colleges and universities. This was the genesis of the program that became known as the Air Traffic Collegiate Training Initiative (AT-CTI). The AT-CTI program was initiated as a cost saving effort to defer some of the basic ATCS educational elements to participating colleges and universities. Beginning in 1989, the FAA entered into partnerships with selected post-secondary educational institutions to conduct some portion of ATCS technical training as a demonstration program. The program grew from an original five institutions to a total of 36 participating colleges and universities by 2012. There are currently 30 active AT-CTI participating colleges and universities.

Efficient training and reduction in training attrition of air traffic control specialists (ATCSs) has long been a topic of concern for FAA and its stakeholders (for examples, Davis, Kerle, Silvestro, & Wallace, 1960; GAO, 2008). One of the ideas to mitigate inefficiency of training and assist with the reduction of training attrition was a partnership with existing post-secondary academic institutions to provide training. With their theoretical and practical expertise in training and education, using colleges and universities to train ATCSs emerged as a viable option for improving the training process and reducing attrition (Means, et al., 1988). In order to plan for the staffing required for anticipated increases in air traffic operations and changes in technology, FAA developed a *Flight Plan for Training* (FAA, 1989). Training issues were addressed along with specific recommendations for off-loading some ATCS training through a collaboration between FAA and post-secondary institutions. *Flight Plan for Training* outlined the Pre-Hire Air Traffic Control Training Initiative, which became the starting point for the Air Traffic Collegiate Training Initiative (AT-CTI) Program. Pre-hire training at the post-secondary level was proposed as a "new source of highly qualified and motivated air traffic control specialists." (FAA. 1989, p.28). *The Flight Plan for Training* proposed a university-based pilot training program to see if graduates could be hired from these institutions with the equivalent of FAA Academy training. This pilot program evolved over time and became the AT-CTI Program. Early congressional support specifically for the training collaboration further contributed to the early development of the program through FY1990 FAA appropriations (*Department of Transportation and Related Agencies Appropriations Act, 1990*).

In 1991, the FAA began a partnership with five post-secondary educational institutions to provide initial training of ATCSs. This initial phase was called the Pre-Hire Air Traffic Control Demonstration Program. The program grew to include as many as 36 participating colleges and universities and evolved to become the Air Traffic Collegiate Training Initiative (AT-CTI) Program. A more detailed historical review of the AT-CTI program can be found in *History of the Air Traffic Collegiate Training Initiative (AT-CTI) Program 1991-2016* (Broach, McCauley, & Sanchez, 2019).

AT-CTI Program Development Expansion and Evaluation

Program Development

In order to administer the demonstration pilot, FAA developed and published the *Pre-Hire Air Traffic Control Demonstration Program* (FAA order 3120.26). The order specified an informal five-year period for the demonstration, but provided no termination criteria or date. The order did establish criteria for program expansion to other academic institutions. The order further specified FAA oversight responsibilities, including an evaluation by the Civil Aeromedical Institute (CAMI, now the Civil Aerospace Medical Institute). In addition, the order clearly stated that "The hiring of graduates of these programs will be governed by FAA's recruitment needs for the ATC occupation; FAA will not guarantee employment of graduates" (p. 3).

In 1991, five institutions entered into agreements with the FAA for the training of air traffic controllers. These five academic institutions were:

- Minnesota Air Traffic Control Training Center (MnATCTC), Eden Prairie, MN (also known as the MARC program);
- Hampton University (HU), Hampton, VA;
- Community College of Beaver County (CCBC-PA), Monaca, PA;
- University of North Dakota (UND), Grand Forks ND; and
- University of Alaska, Anchorage (UAA), Anchorage, AK.

The FAA hired 99 graduates under the program by 1992, with 66 more in 1993 hired and placed in facility field training. In 1994 Congress codified the legal authority for the AT-CTI program by amending Section 44506 of Title 49 of the United States Code (49 U.S.C. § 44506) with Public Law 103-272, July 5, 1994. The statute granted discretionary authority to the FAA Administrator to continue existing agreements and expand to additional academic institutions.

Program Expansion

Program expansion was heavily influenced by the fluctuations in ATCS hiring, as well as changes in selection criteria. The first program expansion was approved by the Director of Air Traffic in January of 1997. Nine post-secondary institutions were selected. Participating institutions were required to teach specific elements of the FAA Academy's initial qualifications training curriculum. Criteria for institutional participation also included being an accredited, nonprofit, degree granting institution offering a non-engineering aviation degree; having a viable aviation program with non-engineering graduates; and within specified location parameters.

The nine institutions selected in the 1997 expansion were:

- Vaughn College of Aeronautics, Flushing NY;
- Daniel Webster College, Nashua, NH;
- Dowling College, Oakdale, NY;
- Embry-Riddle Aeronautical University, Daytona Beach, FL;
- Inter American University of Puerto Rico, San Juan, PR;
- Miami-Dade Community College, Miami, FL;

- Middle Tennessee State University, Murfreesboro, TN;
- Mount San Antonio College, Walnut, CA; and
- Purdue University, West Lafayette, IN.

Those nine new schools joined four other colleges that had been participating in the AT-CTI program for five years as the AT-CTI Program. MnATCTC was no longer identified as an AT-CTI Program participant, since MnATCTC was a funded program and all other AT-CTI schools were unfunded. The MnATCTC program was operating under different parameters and no longer considered as a part of the AT-CTI program.

The anticipated retirement of a large number of ATCSs approaching retirement age prompted a focus on increasing the qualified ATCS applicant pool. Using AT-CTI participating post-secondary training instirutions as a resource for training prospective ATCSs, as well as the diversion of training expense from the Academy, prompted the second AT-CTI expansion occurring from 2007 through 2012. This expansion promoted not only an increase in the number of schools and qualified ATCS applicants, but a potential increase in the diversity of ATCS applicants, as well.

This expansion time frame also introduced a rigorous evaluative component for AT-CTI certification as well as a formal mechanism for an evaluative recertification. The evaluative materials for the certification of new programs and recertification of existing programs was provided by JJA Consultants (Fairfax, VA) through the *Air Traffic Collegiate Training Initiative Operational Guidelines and Management Oversight Package* (FAA, 2007) and introduced at an annual Best Practices Conference. If the applying school met minimum eligibility criteria, the detailed application was reviewed for five major components: 1) *Organizational Foundation and Resources*, 2) *Organizational Credibility*, 3) *Curriculum and Facilities*, 4) *Student Performance*, and 5) *Organizational Performance*. Applications were scored on the factors with an overall school evaluation score. In addition to the existing AT-CTI participants nine new participating colleges were added:

- Arizona State University, Mesa, AZ;
- Community College of Baltimore County, Baltimore, MD;
- Florida State College at Jacksonville, Jacksonville, FL;
- Green River Community College, Auburn, WA;
- Kent State University, Kent, OH;
- Lewis University, Romeoville, IL;
- Metropolitan State University of Denver, Denver, CO;
- Middle Georgia State University, Cochran, GA; and
- University of Oklahoma, Norman, OK.

In 2008, a similar process resulted in eight additional new colleges:

- Aims Community College, Greeley, CO,
- Broward College, Pembroke Pines, FL,
- Eastern New Mexico University, Roswell, NM,
- Embry Riddle Aeronautical University, Prescott, AZ,
- Jacksonville University, Jacksonville, FL,
- Le Tourneau University, Longview, TX,
- Saint Cloud State University, Saint Cloud, MN, and
- Tulsa Community College, Tulsa, OK.

In 2009, five more colleges were added:

- Florida Institute of Technology, Melbourne, FL,
- Hesston College, Hesston, KS,
- Sacramento City College, Sacramento, CA,
- Texas State Technical College, Waco, TX, and
- Western Michigan University, Battle Creek, MI.

Program Evaluations

Formative Evaluation 1996

In keeping with FAA Order 3120.26, CAMI initiated an evaluation and published a *Formative Evaluation of the Collegiate Training Initiative – Air Traffic Control Specialists (CTI-ATCS) Program* (Morrison, Fotohui, & Broach, 1996). The formative evaluation focused on the organization and implementation process of the 1991 demonstration project. The evaluation was favorable overall and indicated the programs at the five demonstration project educational institutions were functioning as expected. The educational institutions were making innovations beneficial to the FAA. The report did provide a caution that curtailment in ATCS hiring could cause a significant challenge to the AT-CTI program, with a reduction in hiring based on reduced need for ATCSs.

2006 Evaluation

The *Air Traffic – Collegiate Training Initiative (AT-CTI) Evaluation* (FAA, 2006) was conducted to document the progress of the AT-CTI Program and ensure that FAA business needs were being met. Those needs were providing high quality ATCS candidates, reduce training costs, and ensure prerequisite training had been provided. Specific evaluation goals supporting those business needs were also addressed. The evaluation summarized that "The AT-CTI schools and the FAA stakeholders responsible for administering the program believe that the AT-CTI Program is an effective recruitment tool for quality applicants who possess a broad base of aviation knowledge." (FAA, 2006, p. 5-1). With the recommendation for an expansion of the program, there were also recommendations for improvement in instructional quality across individual schools, provision of better FAA hiring information for students, addressing student academic disparity across AT-CTI schools, exploring the possibility of the provision of additional Academy courses, and the development and implementation of performance based metrics for AT-CTI schools. The second AT-CTI program expansion in 2007 followed this evaluation. A detailed and rigorous annual evaluative component for AT-CTI certification was also introduced following 2006 recommendations.

2007-2012 Annual Certification and Evaluation

The initial certification of existing schools and entry requirements for new schools in 2007 marked the beginning of detailed and rigorous annual evaluations for participating and new AT-CTI schools. There were annual best practices workshops sponsored by FAA, where information on evaluative assessments was presented. Identified best practices were also presented in collaboration with participating schools. Schools could achieve full certification, be placed on probation, or have certification denied based on the annual evaluative reviews. During this time, FAA exercised rigorous program oversight for the AT-CTI program.

External Reviews

In 2011, an Independent Review Panel (IRP) was convened at the request of the FAA Administrator to review the ATCS selection process and make recommendations for improvement. The panel had members from AT-CTI schools, union representation, and FAA. The IRP made several recommendations regarding selection and training, but also made some recommendations concerning the impact of AT-CTI schools on the career of an ATCSS. Of particular interest was the recommendation for differentiating schools by levels based on the differing training opportunities offered at each AT-CTI school. The IRP also recommended that AT-CTI participation and the level of that AT-CTI school influence the status of an ATCS applicant.

In keeping with the FAA commitment to diversity in hiring, the agency initiated detailed reviews of the ATCS hiring process and its possible impact on equal opportunity and possible barriers to employment. These analyses produced several concerns related to the hiring process (APT Metrics, Inc., 2013; Outtz & Hanges, 2013). In response, changes were made to the hiring process, including introduction of a biographical assessment, a requirement that all applicants, including AT-CTI graduates, take the Air Traffic Selection and Training (AT-SAT) exam, and competition under a single vacancy announcement open to "all sources." This meant in effect, AT-CTI graduates had the same status as general public hires, since all competed under the same qualifications criteria. The changes affected AT-CTI hiring.

Conclusion

In 2016, as required by a law passed by Congress, FAA initiated a pool process in which applicants were grouped into one of two hiring pools based on applicant background. Pool 1 was comprised of qualified veterans and AT-CTI graduates. Pool 2 was comprised of all other applicants. Hiring in equal proportion from both pools was also specified. In 2017 FAA began to re-establish relationships with the AT-CTI schools. That effort continues with a strong commitment to maintain the long-standing productive relationship with those schools.

References

APTMetrics, Inc. (2013, April). *Extension to barrier analysis of air traffic control specialist centralized hiring process.* Washington, DC: FAA Office of the Assistant Administrator for Civil Rights.

Davis, C. G., Kerle, R. H., Silvestro, A. W., & Wallace, W. H. (1960, March). *The air traffic control training program as viewed by training supervisors.* (Technical report No. 33, Project O under FAA contract FAA/BRD-40). Philadelphia, PA: Courtney and Company.

Department of Transportation and Related Agencies Appropriations Act, 1990 (Public Law 101-164, November 21, 1989

Federal Aviation Administration. (1989, January). *Flight Plan for Training: FAA training initiatives management plan.* Washington, DC: Federal Aviation Administration. Office of Training and Higher Education.

Federal Aviation Administration. (February 16, 1991). *Pre-Hire Air Traffic Control Demonstration Program.* (FAA order 3120.26 (Cancelled)). Washington, DC: Author.

Federal Aviation Administration. (2006, December). *Air Traffic – Collegiate Training Initiative (AT-CTI) evaluation.* Washington, DC: Federal Aviation Administration Air Traffic Organization Director of Technical Training and Development.

Federal Aviation Administration. (2007). *Air Traffic Collegiate Training Initiative Operational Guidelines and Management Oversight Package.* Washington, DC: Air Traffic Organization Office of the Director of Technical Training and Development.

Means, B., Mumaw, R., Roth, C., Schlager, M., McWilliams, E., Gagne, E., Rice, V., Rosenthal, D., & Heon, S. (1988, June). *ATC training analysis study: Design of the next-generation ATC training system.* (Report delivered under OPM Work Order #342-036). Alexandria, VA: HumRRO International, Inc.

Morrison, J. E., Fotohui, C. H., & Broach, D. (1996). *A formative evaluation of the Collegiate Training Initiative - Air Traffic Control Specialist (ATCS) program.* (Report No. DOT/FAA/AM-96/6). Washington, DC: Federal Aviation Administration Office of Aviation Medicine.

Outtz, J., & Hanges, P. *Barrier analysis of the Air Traffic Control Specialist (ATCS) centralized hiring process.* Washington, DC: Author.

U.S. Government Accountability Office (GAO). (2008). *Efforts to hire, staff, and train air traffic controllers are generally on track, but challenges remain.* (GAO-08-908T). Washington, DC: Author. Available from https://www.gao.gov/products/GAO-08-908T

Acknowledgment

Research in this paper was conducted under the Air Traffic Program Directive/Level of Effort Agreement between the Human Factors Division (ANG-C1), FAA Headquarters, and the Aerospace Human Factors Research Division of the FAA Civil Aerospace Medical Institute.

The opinions expressed are those of the author alone, and do not necessarily reflect those of the Federal Aviation Administration, the Department of Transportation, or the government of the United States of America.

Six-Year Follow-Up of Intensive, Simulator-Based Pilot Training

Paper to be presented at the 20th International Symposium on Aviation Psychology
May 7 - 10, 2019, Dayton, Ohio - Submission ID is 283

Maxine Lubner, Ph.D.
Vaughn College of Aeronautics and Technology, New York, NY
Andrew R. Dattel, Ph.D.
Embry-Riddle Aeronautical University, Florida
Emerson Allen, B.S.
Vaughn College of Aeronautics and Technology, New York
Deb Henneberry, Ed.D.
Vaughn College of Aeronautics and Technology, New York, NY
Sharon DeVivo, Ed.D.
Vaughn College of Aeronautics and Technology, New York, NY

In 2012, Vaughn College of Aeronautics and Technology initiated an intensive, simulator based, flight training program. Three cohorts, totaling 16 students, completed flight training with fewer flight hours than the United States average (ISAP, 2013). The students earned Federal Aviation Administration (FAA) Private Pilot certificates within 5 weeks, Instrument ratings in 3-4 weeks, Commercial certificates in an average of 20 weeks and Certificated Flight Instructor ratings in an average of 40 weeks. All participants met selection criteria, including completing their FAA Class III medical certificates, FAA Private Pilot Knowledge exams, a demonstration flight, financial counseling, having a grade point average of at least 3.0, and agreeing to remain substance-free during training. These 16 students and a comparison group of students who enrolled in traditional flight training at the college at that time have been followed over the past 6 years to observe factors associated with their career progression. Surveys were conducted by e-mail, phone, or in person to ask about flight training, career progression, and experiences of camaraderie, mentorship, and general satisfaction with their programs. This follow-up study found that 13 of the 16 students (81%) achieved their career goals of flying professionally. The comparison group has taken longer to complete flight training and proportionally fewer are flying professionally. Indicators of success in achieving professional pilot careers and networks included the cognitive variables related to intensive simulator-based training, camaraderie, shared learning experiences and opportunities to reflect on training.

Simulator based training (SBT) for pilots is accepted as essential and successful for training of pilots, particularly those operating advanced equipment and even for training of ab initio airline pilots in multi-crew, complex environments, even though the methods and times to completion may vary (McLean, Lambeth, Mavin, 2016). In the US, pilots usually obtain training on their own as they must have an FAA - issued commercial certificate to be paid to fly, and an airline transport certificate (ATP) to fly for the airlines. Airline pilots must hold a bachelor's degree. Pilots must accrue thousands of hours before they are hired by the airlines (BLS.gov, 2019). These requirements make it vital to obtain FAA qualifications quickly and collegiate flight training attractive, albeit expensive. However, SBT for beginner pilots is not often used because evidence of its effectiveness is not clear (Goetz, Harrison, Voges, 2015) and because the FAA only allows a limited amount of simulator to be entered in pilots' log books (14 CFR Part 61 or Part 141).

Advantages of SBT include financial ones because simulator equipment with moderate fidelity can be effective; transportation, scheduling and flight conditions such as poor weather do not impede training sessions; and dangerous or unusual maneuvers can be taught safely (Goetz et al., 2015; Harris, 2011; Salas, Bowers, & Rhodenizer, 1998; Taylor et al., 1999).

How to use SBT for flight training and understanding the learning processes have been productively examined. For example, scenarios that freeze a situation for detailed examination can only be created in a simulator. Simulator training can easily address a variety of styles of learning, such as conceptual, procedural, scenario, collaborative and individual styles of training (Dattel, et al., 2009, Dattel, Kossuth, Sheehan, & Green, 2013). However, the duration of simulator centric learning is hard to assess because of the usual difficulties of longitudinal studies, such the logistics and expenses of tracking pilots over time. Besides SBT, it is possible that accelerated flight training programs confer their own advantages, such as the intensity of training and support offered by a group of students in the same situation (Lubner, Dattel, Henneberry, & DeVivo 2105; 2017). In this cohort study, the duration of positive gains from an intensive, SBT flight training program is explored.

Method and Program Description

A descriptive examination of the effectiveness of a simulator-based training program for pilots was conducted. Of 55 students of varying backgrounds, but mostly with limited flight experience, sixteen enrolled in an intensive, simulator-based flight training program in January 2012. Within two years the remainder had enrolled in conventional collegiate flight training, supplemented with some simulator training. The students in the intensive program completed their FAA Private Pilot certificates in an average of 5 weeks (not including simulator time). Moreover, the intensive program group earned their private pilot's certificate in statistically significantly fewer hours (M=46.03) than the conventional collegiate flight training group (M=76.06). The intensive group returned to conventional training and completed their Commercial certificates in an average of 20 weeks and CFI qualifications in an average of 40 weeks.

The sixteen Vaughn College students comprised three cohorts who participated in and intensive SBT program six years ago. These students began their flight by traveling to the southwest US, stayed near a small airport and undertook a short duration, intensive simulator-based, ab initio flight training program. Later, the students returned to New York and completed the remaining flight qualifications required for their Bachelors' degrees in Aircraft Operations. The FAA qualifications for the BS degrees include Private, Instrument, Commercial, Multi-Engine, and Certified Flight Instructor. Each cohort of five to eight students traveled and studied together, following an intensive, simulator-based program. The students had to meet several criteria, including having a G.P.A. of 3.0 or better, possessing an FAA Class III Medical Certificate, taken a demonstration flight, successfully passed the FAA private pilot knowledge exam, obtained financial counseling and agreed to remain substance free during the training period.

The students were expected to travel between the Texas (TX) and New York (NY). The Texas group was to complete intensive SBT, then return to New York to complete their academic studies and finish their FAA flight qualifications (private, instrument and commercial) as needed. The students stayed in the Texas for 4-6 weeks at a time, undergoing training in simulators and aircraft six days per week. Students lived in a hotel and dined together. As the program unfolded, the second cohort group could only travel to the Texas flight school twice – for private pilot and instrument training. The third group only participated in the Texas, SBT for their private pilot training.

Back in New York, they followed conventional training that offered some simulator practice. Lessons in New York were spaced over time and students had conventional opportunities for group interactions. The conventional training in New York was conducted at a Part 141 flight school, located about an hour's drive from Vaughn College. By fall 2013, all students attended the conventional flight training at this Part 141 flight school. Students had limited access to simulators at the flight school and at Vaughn College.

In February and March 2017, semi-structured interviews and structured questions with Likert scale answers were conducted with the original three cohorts of the sixteen students. The interviews were coded and examined for themes related to a priori questions of predictors of learning and impact on careers. The authors met to discuss results and conclusions to ensure agreement of interpretations. This follows accepted qualitative methods of analysis (Creswell, 2013). In 2018 and 2019, the structured questions were sent via Survey Monkey to the original conventional training group, but only four responses were obtained to date. A more detailed follow-up study using telephone calls is planned. The progress in terms of FAA qualifications earned for both the intensive, SBT, Texas group and the conventional, NY training groups were obtained by examining FAA records.

Outreach to each student included one to several contacts by one or more of the program instructors and administrators. Most students expressed delighted willingness to participate in the interviews, but two of the cohort members were not interviewed. One of the non-responders agreed to the interview but did not participate. The second did not respond to any of the contacts. Some information on the progress of these two non-responders was obtained by looking up publicly available records, including the FAA airmen database, Linked-In and Facebook. The career paths of the two non-responders appear similar to those of their cohort members' paths (see below). The non-responder who did not participate in the interview obtained some flight qualifications and is working at a local, large airport and has recently returned to flight training. The second non-responder obtained flight qualifications up to ATP Instrument and two type ratings and is flying as a first officer for a regional airline.

Results

Of the original sixteen TX students, fourteen completed interviews (11 m, 3 f) in 2017. They answered both Likert scale questions and qualitative, unstructured questions. The original 36 NY students were sent questionnaires containing just the Likert scale questions in 2018 via email. Despite three different email follow up requests, only four have responded. Additional information about both groups was obtained by examination of FAA records, social media and alumni events.

All but two of the TX interviewees had obtained a Bachelor's degree in equal proportions between aeronautical science, which required flight qualifications as part of the degree and aircraft operations, which allows elective credits for flight qualifications (See Figure 1) (Lubner et al., 2015, 2017). For both groups, private pilot training, instrument training, and commercial pilot training were all conducted in both the airplane and the flight simulator, but the TX group had far more simulator time.

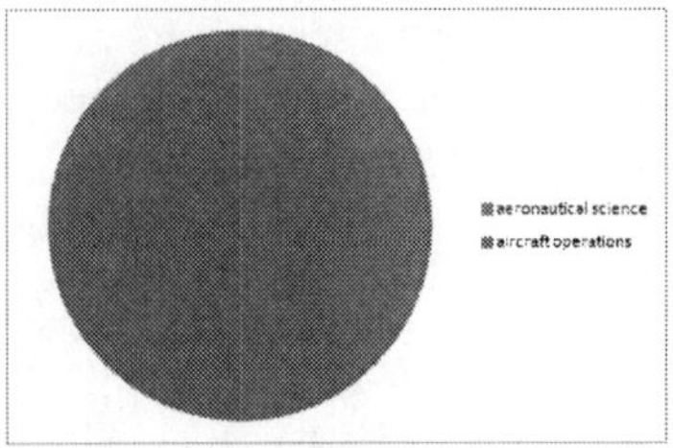

Figure 1: TX Cohort's undergraduate major:

The qualitative information on the TX students showed that the following were helpful to them: flight instruction, having a mentor available while training, training in the full motion flight simulator, and intense delivery of training improved the initial quality and subsequent duration of their flight skill development and knowledge retention as well as camaraderie, mentorship and savings in terms of time and money to complete their academic and flight studies.

The TX students valued their friendships and believed that the camaraderie, joint study sessions and to "practice flying together" were important in developing their flight skills. As one student stated, "we can discuss and learn from each other," while another student stated, "we share the same passion and support each other". One student commented that, "healthy competition builds motivation." The NY students also valued forming good friendships in their program, noting that they are still friends with their cohort members, ranging from 2 to 15 friends. They all agreed that their pilot friends help to improve their acquiring and importantly, maintain their flight knowledge and skills. As one NY student stated, "It's impossible to do it (flight training) alone. One of the things I did was study with my friends to pass all my tests. Honestly it's necessary; it keeps your life balanced with school and social…" The TX students felt that their career goals were met, and that they were now mentors themselves. Having a mentor on site was considered a great advantage. The students felt that they could ask questions of the mentor that they were not comfortable asking of the flight instructors. As one noted, "I felt that the mentor was my advocate". The NY students reported that their friends served as aviation mentors. Two students believed that serving as a mentor themselves would help improve their own flight skills and knowledge. This group did not find formal mentors to be an important part of their flight training.

Of the NY students whose qualitative information is available, three students report being on a path towards achieving their career goals, although two are not currently employed in aviation. Two students have left the aviation field altogether. These NY students stated that their flight training program was useful to their career goals of eventually becoming professional pilots. All students believed that they retained their flight knowledge and skills better when they learned them over a longer time. The TX students were able to spend a portion of their intensive program living and studying at the flight school, which they noted was helpful in developing camaraderie, study groups, and additional practice time in the simulator and airplane. They were allowed to stop out and then catch up with their regular college courses when they returned from TX.

Most of the TX group was also awarded scholarships at a later stage of their programs to allow them to complete their flight instructor ratings. The NY students, however, were not specifically targeted to receive scholarship money. They remained in the regular college schedule. Because of the geographical location of the college in the largest US city, any flight student must fly well outside the metropolitan airspace. The students noted that the distance between the flight school and college was an obstacle to their progress. The commute took several hours per day, making it particularly difficult to attend classes and flight lessons on the same day.

The TX students felt that their career goals were met. Of the 12 TX interviewees who have obtained their BS, 11 are currently employed in paid pilot positions (See Figure 2 for a breakdown of employment positions). All interviewees said that their career goals have been met, or they were approaching their goals. One interviewee is on a hiatus from obtaining additional flight licenses and ratings due to medical reasons.

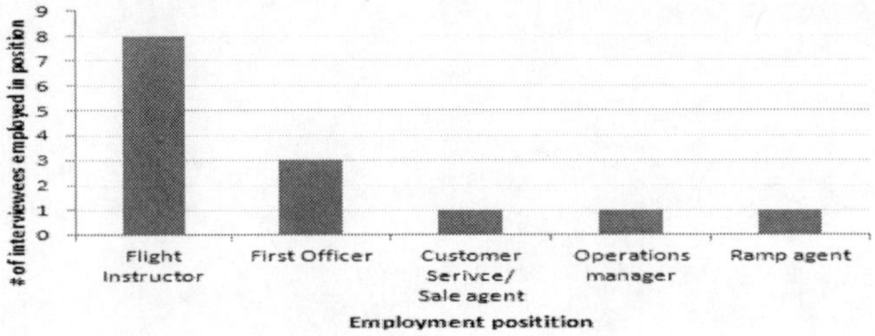

Figure 2: Employment positions.

In terms of flight training, as previously reported (Lubner et al., 2015, 2017), the TX students obtained their FAA Private Pilot Certificates in fewer hours than the NY group (see Figure 3). However, the time for the TX group to obtain Private Certificates was an average of 15 months, while the NY group took an average of 17 months. All TX students went beyond their Private Certificates, while many of the NY group did not progress beyond this qualification at the six year follow up point.

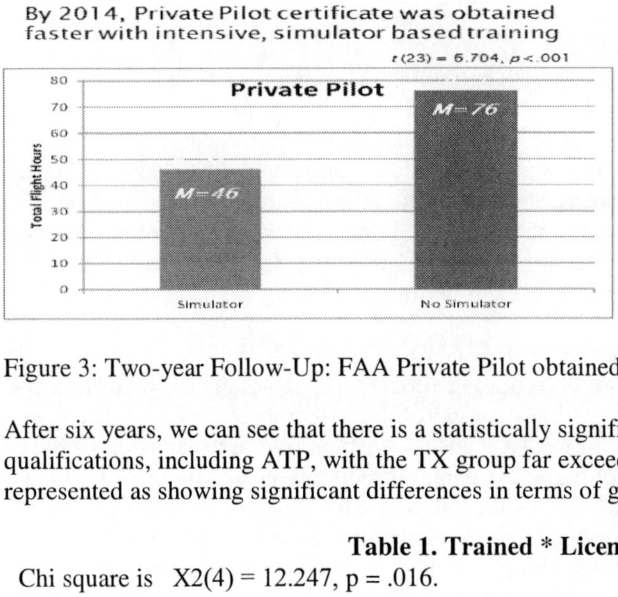

Figure 3: Two-year Follow-Up: FAA Private Pilot obtained in significantly different numbers of flight hours.

After six years, we can see that there is a statistically significant difference in terms of numbers of advanced pilot qualifications, including ATP, with the TX group far exceeding those of the NY group. This difference can also be represented as showing significant differences in terms of greater numbers of advanced qualifications obtained by

Table 1. Trained * License Cross-tabulation

Chi square is $X2(4) = 12.247$, $p = .016$.

Count

		License					Total
		ATP	Flight Instructor	Commercial	Instrument	Private	
Trained	New York	2	4	6	2	8	22
	Texas	7	5	2	2	0	16
Total		9	9	8	4	8	38

the TX group than the NY group (see Figure 4). All the TX students went beyond the Private Pilot qualification.

Table 2. Ranks

Mann-Whitney U = 71.0, $p < .001$

	Trained	N	Mean Rank	Sum of Ranks
License	New York	22	24.27	534.00
	Texas	16	12.94	207.00
	Total	38		

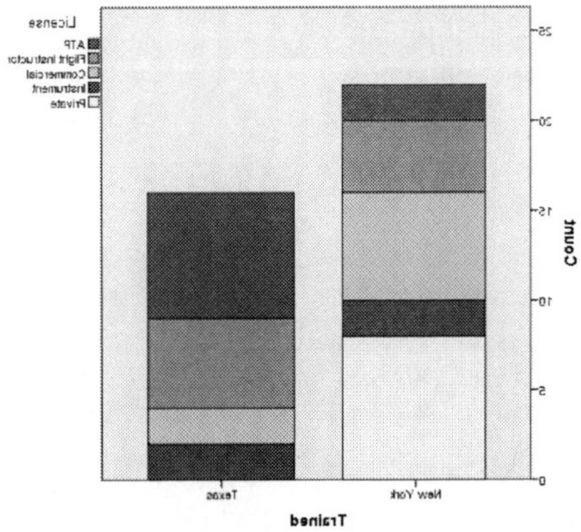

Figure 4: Six Year Follow-Up: TX group obtained statistically significantly more advanced FAA flight qualifications than the NY group, as represented by bar and pie charts.

The potentially useful aspects of the intensive program are discussed, including type of training such as intensive classroom, simulator and traditional in-aircraft instruction in addition to the psychosocial impacts of camaraderie and shared learning experiences (Lubner, et al., 2105, 2017).

Conclusion

In January 2012, Vaughn College, in New York City, launched a flight training program in partnership with a new training entity in Texas. Three cohorts of students participated over the next 18 months. While in Texas, these students flew twice a day five to six days a week, had constant access to simulators and were encouraged to use them to practice beyond their two flight lessons per day. Once students acquired a baseline of knowledge and skill, the simulators were more helpful to the training process, noted when conducting their Instrument training. Later, some of them obtained scholarships to enable them to complete their FAA CFI ratings. Over the same time period, a larger group of students remained in New York, where they continued traditional college level courses and conventional flight training at a Part 141 certified flight school. This group formed a natural comparison group.

Six years later, it is clear that the intensively trained, TX group progressed faster and further in their careers as professional pilots, obtaining statistically significantly more FAA pilot qualifications, including Private, Instrument, Commercial, Multi-Engine, Certified Flight Instructor and Airline Transport Pilot. All of those in the TX group went beyond their Private Pilot Certificate. Proportionally more of the TX students are employed in the aviation fields, including as professional pilots. With roughly two-thirds of the TX students currently flying as a profession and almost all involved in the aviation industry, the results indicate that the intensive SBT program assisted students in achieving their goals at follow-up. Subjectively though, the TX students felt that the intense program assisted the retention of their skills, while the NY students reported that learning over a longer period enhanced their knowledge and skills.

As stated by several members of the TX group, theirs was an intense form of training that required commitment, focus and a strong desire to achieve their goals. As demonstrated in the interviews, the aid of an on-site mentor, someone who had been both a flight instructor and was a current commercial airline pilot, supported student learning by providing additional information, advocacy and encouragement through the process. The NY group found mentors and support among their pilot friends.

Both groups noted that camaraderie and friendships with other pilots supported their acquisition of flight knowledge, skills and helped maintain their motivation to fly over time.

More robust follow-up is needed to obtain insights into the relative importance of SBT, intensity of training itself and the role of friendships and mentoring activity in helping to maintain students' focus and motivation. Structural obstacles, such as commuting time between flight school and college campus, professional mentors, and availability of SBT could be discovered and hopefully rectified in future studies.

References

Aircraft Owners and Pilots Association (AOPA) (2010) The Flight Training Experience: A Survey of Students, Pilots and Instructors. http://www.aopa.org/

Bureau of Labor Statistics, U.S. Department of Labor, *Occupational Outlook Handbook*, Airline and Commercial Pilots. https://www.bls.gov/ooh/transportation-and-material-moving/airline-and-commercial-pilots.htm (visited *March 12, 2019*).

Creswell, J.W. (2013). *Qualitative Inquiry and Research Design (3rd ed.)* Los Angeles: Sage

Dalgarnoa, B., Kennedy, G. and Bennett, S. (2014). The impact of students' exploration strategies on discovery learning using computer-based simulations. *Educational Media International*, Vol. 51, No. 4, 310–329, http://dx.doi.org/10.1080/09523987.2014.977009

Dattel, A. R., Kossuth, L., Sheehan, C. C., & Green, H. J. (2013). Poster presented at the *84th Annual Meeting of the Eastern Psychological Association.* New York, NY.

Dattel, A. R., Durso, F. T., & Bédard, R. (2009). Procedural or conceptual training: Which is better for teaching novice pilots landings and traffic patterns? *Proceedings of the 53rd Annual Meeting of the Human Factors and Ergonomics Society*. San Antonio, TX.

FAA(2008). Aviation Instructor's Handbook. Retrieved from http://www.faa.gov/about/office_org/headquarters_/avs/offices/afs/afs600

FAA (2008) Advisory Circular (AC) 00.2-15. Retrieved from www.faa.gov.

Goetz, S., Harrison, B. and Voges, J. (2015). The use of FAA flight training and aviation training devices at UAA institutions. *Collegiate Aviation Review*, 44 – 59.

Gopher, D., Weil, M., Bareket, T. (1994). Transfer of skill from a computer game trainer to flight. *Human Factors*, 36, 387-405.

Gopher, D., Sivan, R., & Iani, C. (2001). Comparing learning curves of experts and novices. *Proceedings of the Human Factors and Ergonomics Society*. 2, 1805.

Harris, D. (2011) *Human Performance on the Flight Deck*. Brookfield, VT: Ashgate.

Koglbauer, I. (2016). Simulator training improves pilots' procedural memory and generalization of behavior in critical flight situations. *Cognition, Brain and Behavior*. XX:5 (December), 357-366.

Koglbauer, I., Riesel, M. and Braunstingl, R. (2016). Positive effects of combined aircraft and simulator training on the acquisition of visual flight skills. *Cognition, Brain and Behavior*. XX:5 (December), 309-318.

Lubner, M.E., Dattel, A.R., Henneberry, D., DeVivo, S. (2105) Follow-Up Examination of Simulator-Based Training Effectiveness *18th International Symposium on Aviation Psychology*, Dayton, Ohio.

Lubner, M.E., Dattel, A.R., Henneberry, D., DeVivo, S. (2107) Four-Year Follow-Up of Intensive, Simulator-Based Pilot Training *19th International Symposium on Aviation Psychology*, Dayton, Ohio.

McLean, G. M. T., Lambeth, S. and Mavin, T. (2016). The Use of Simulation in Ab Initio Pilot Training. *The International Journal of Aviation Psychology*, 26:1-2, 36-45, DOI:
10.1080/10508414.2016.1235364 http://dx.doi.org/10.1080/10508414.2016.1235364

Salas, E., Bowers, C. A., & Rhodenizer, L. (1998). It is not how much you have but how you use it. *The International Journal of Aviation Psychology*, 8, 197-208.

Taylor, H. L., Lintern, G., Hulin, C. L., Talleur, D. A., Emanuel, T. W., & Phillips, S. I. (1999). Transfer of training effectiveness of a personal computer aviation training device. *The International Journal of Aviation Psychology*, 9, 319-335.

DATA-DRIVEN IMPROVEMENT OF FLIGHT TRAINING SAFETY AT PURDUE UNIVERSITY

Matthew C. Chow
Purdue University
West Lafayette, Indiana
Shantanu Gupta
Purdue University
West Lafayette, Indiana
Celeste D. Torrez
Purdue University
West Lafayette, Indiana
Suvarna Veeravalli
Purdue University
West Lafayette, Indian
John H. Mott
Purdue University
West Lafayette, Indiana
Jack J. Green
Purdue University
West Lafayette, Indiana
Nikolas A. Sambado
Purdue University
West Lafayette, Indiana

The School of Aviation and Transportation Technology at Purdue University utilizes a wide range of resources to train aspiring pilots, technicians, and managers. Aircraft operate in one of four practice areas located within a 30 nautical mile radius of the Purdue University Airport. Due to factors such as poor weather conditions, inexperience of student pilots, and proximity misjudgment, one aircraft could operate in close proximity to another in the same practice area, compromising the safety of both aircraft and causing a Near Midair Collision (NMAC) event due to miscommunication, misinterpretation, or failure to act on the part of the pilots involved. The objective of the research reported herein is to develop a diverse student team to evaluate and address flight training safety at Purdue University. The team comprises the research arm of the Purdue Flight Operations Center, aggregating data from sources such as weather monitoring systems, aircraft maintenance systems, ADS-B transponders, real-time dispatch systems, and air traffic systems. The collected data will be used to target common operational errors and study their frequency and nature, and to measure aircraft separation in order to develop and improve parameters for the identification and reporting of NMAC events. Techniques developed by this team will be used by the Purdue safety team to analyze each instance where an NMAC event has occurred and develop strategies to mitigate these events. The research team will, in addition, develop data dashboards and provide suggestions to help improve the overall safety of the Purdue flight program.

The Flight Operations Research Center (FORC) is an initiative to improve training operations involving the student pilots at Purdue University. Data-driven improvement of flight training safety, one of the research areas of the FORC, aims to mitigate the risk associated with training student pilots by analyzing data from ADS-B transponders installed on the training aircraft. Traditionally, collision avoidance in General Aviation (GA) flying has been accomplished by pilots utilizing visual scanning techniques. Wide adoption of new transponder technology has improved pilot situational awareness by displaying electronically the location of nearby aircraft. This ADS-B data can also be collected by ground stations to be analyzed by researchers. To assist pilots and improve training conditions, the FORC aims to develop a diverse student team to analyze the ADS-B data, identify the common operational errors, study the trend patterns, and estimate aircraft proximity to assist the Safety Committee in improving the overall safety of the program.

The research herein describes the data-driven safety improvement process of estimating aircraft proximity and identifying Near Midair Collisions (NMAC) of the training fleet. Accurate identification of these events, both from the data and from pilot safety reports, is of high importance in order to determine their causes and explore safety improvement opportunities. The analysis can further be extended to optimize the assignment of aircraft to practice areas in order to reduce congestion, hence providing safer airspace for practice. Additionally, this article discusses the student development and learning outcome aspect of the FORC. Through its involvement with this research center, a diverse team of student participants is acquiring essential skills to solve real-world problems.

Literature Review

A 2011 research study from the Massachusetts Institute of Technology (MIT) found that ADS-B is becoming the foundation for aircraft surveillance in the United States. One solution that was presented to encourage the implementation of ADS-B technology in the existing general aviation fleet is to incentivize users with high-value information. Airborne traffic alerts are one such application. The MIT researchers also found that the airport environment, which is also where a high volume of flight training occurs, is where the most NMAC reports arise (Kunzi & Hansman, 2011). NMACs resemble the concept of a "near miss", which is an event where a potentially disastrous consequence is narrowly avoided (Dillon & Tinsley, 2008). In their research, Dillon and Tinsley (2008) argue that a near-miss event tends to lower the perceived risk of a situation in the mind of the individual who experienced the event, which then leads that individual to make riskier decisions. Thus, it is imperative for organizations to reduce the number of these events, and to use them to improve awareness of perceived risk.

Introducing research opportunities in undergraduate school is said to strengthen students' interest in research while providing guidance from mentors, which can serve as a benefit for students (Russell, Hancock, & McCullough, 2007). While grade point average (GPA) is not a comprehensive measure of success, it has been shown that students who participate in a long-term research project tend to have higher GPAs than those who do not (Fechheimer, Webber, & Kleiber, 2011). It has also been found that teams can benefit from diversity if team members are encouraged to learn from one another and share differing viewpoints. An example of this occurs when team members are encouraged to work on their area of expertise, then present findings to the rest of the group to foster discussion and feedback (Post, De Lia, DiTomaso, Tirpak, & Borwankar, 2009).

Methodology

An initiative to improve flight training safety was taken by the Purdue University School of Aviation and Transportation Technology (SATT) through the establishment of the FORC. This student research team was formed in August 2018 to analyze data from multiple sources, including weather monitoring systems, aircraft maintenance systems, ADS-B transponders, real-time dispatch systems, and air traffic systems. The research team is comprised of students from multiple areas of study at the university. The team is comprised primarily of students enrolled in programs within the SATT, and includes undergraduate students studying Aviation Management, Aeronautical Engineering Technology, and Professional Flight Technology as well as graduate students pursuing a Master of Science in Aviation and Aerospace Management. Some of these graduate students have been through the undergraduate program at Purdue and others are new to the university, thus incorporating diverse perspectives. In addition to the SATT students, the team also has student members from other departments on campus, such as Computer Science, Data Science, Industrial Engineering, and Aeronautics and Astronautics Engineering (AAE). Including students with different skills has been a goal for this team, since each phase of the research requires different specializations.

Students are carefully recruited to the FORC in several ways. Initially, the faculty advisor selected students to form the preliminary research team. Once operational, existing members identified other students possessing skills needed by the team. By cultivating students with such diverse backgrounds, the research team can leverage its strengths where applicable in various stages of the projects. For example, team members with a background in coding wrote a Java program that filters out extraneous data and computes distances between two aircraft. Throughout this process, students in the Professional Flight program provided context and information about how the aircraft are operated and how the fleet is organized. Additionally, Aviation Management students compared data from multiple sources by drawing on experience from projects in previous classes in which they analyzed fleet data. Graduate students and those with user experience skills then created dashboards to present the results to the Safety Committee.

The primary objective of the team thus far has been to use ADS-B data collected from the aircraft to identify proximity incursions. The team has chosen this term to describe events in which one aircraft operates within a certain distance of another aircraft. While the Federal Aviation Administration (FAA) already uses the term NMAC to refer to cases where aircraft operates within 500 feet of one another, the research team has decided to use the broader term proximity incursion. This decision has been made as a result of a Safety Committee request for data related to any instances where aircraft are in close enough proximity to present a safety hazard. This includes situations where aircraft are not within 500 feet of one another, but were still close enough to warrant concern.

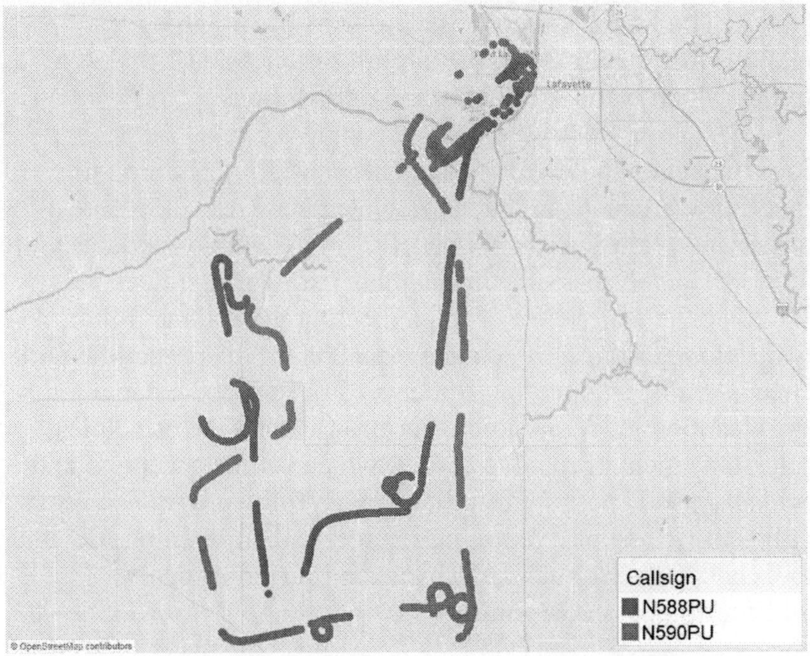

Figure 1. An example of aircraft position data recorded by the ADS-B receiver

To identify proximity incursions around the Purdue University Airport, an ADS-B data analysis program was coded in Java. This code filters the data to include only aircraft within a 30-nautical mile radius of the airport and between 300 and 10,000 feet MSL. The filtering process serves to reduce the number of calculations performed by the code. This particular airspace volume was selected because student pilots practice training maneuvers within 30 nautical miles of the airport. The altitude restrictions serve to filter out surface operations below 300 feet and commercial aircraft that typically operate above 10,000 feet. Once extraneous data has been eliminated, the code divides aircraft into altitude blocks depending on the desired sensitivity. Additionally, a Kalman filter was used to develop a predictive model to fill in missing data points due to intermittent ADS-B reception. Figure 1 shows an example of data from the ADS-B receiver before Kalman filtering. Finally, the code calculates the distance between aircraft in each altitude block and records the distance if it is less than the proximity incursion threshold.

Using the results from the algorithm, the team created dashboards that summarize safety trends for the Safety Committee. For example, the number of proximity incursions can be compared from one month to the next, from a month in the previous year to that of the current year, or year to date. The specific periods and intervals can be customized according to the needs of the Safety Committee. By analyzing these trends and correlating them with current flight operations procedures, the student team can make recommendations to the Safety Committee. Members of the committee can use the results to determine if or when there is an increased hazard level and take appropriate steps to mitigate risks.

Impact

At Purdue University, the Safety Committee strives to mitigate risks associated with the flight training program. With student lives at stake, continuous efforts to ensure safe flight operations is of paramount importance. The FORC supports these efforts by providing recommendations through data analyses that quantify relative aircraft distances, identify proximity incursions, and study trends which could lead to hazardous situations.

The FORC has a tremendous impact on the involved students as well. In addition to improving individual research capabilities, students on this research team also develop critical thinking and problem-solving abilities and realize academic benefits. For example, students with a statistical background develop their data analysis skills through this real-world project. Additionally, AAE students might utilize the experience of implementing the Kalman Filter, since the algorithm was originally developed for use in guidance and control of spacecraft. With the practical application and practice of skills, engineering students can better understand materials taught in classes such as Dynamics and Vibrations; this class teaches the principles behind mathematically modeling the movement of relative bodies, which is the concept behind the Kalman filter. Data Science students also benefit in a similar way. Writing programs in Java while taking a course focused on object-oriented programming reinforces their learning. Beyond coursework, students learn new tools and software such as Tableau and R (a statistical programming language) that could benefit them in the future.

The FORC also prepares students for their industry careers by developing teamwork and leadership skills. Students work in a collaborative group environment to accomplish common goals while working under such pressures as deadlines, setbacks, and a result-oriented organization. Student researchers develop professional skills including time management, prioritization, problem-solving, and decision-making. Additionally, students with leadership roles in the FORC learn project management, work-load distribution, and organizational leadership skills.

Beyond collaborative work and team leadership, team members also develop essential communication skills for time-sensitive work environments. Weekly updates to the operations center improve short impromptu summarization, while formal presentations provide opportunities for students to practice public speaking. In addition to presentations, team members also work on academic publications and gain an understanding of the process from an author's perspective. This process is valuable for students since strong written communication skills are required in almost every workplace. Apart from providing better job opportunities, work on academic publications is of high importance for graduate students with plans in academia. Participation in this research project provides team members with the valuable skills and experiences discussed above and strengthens their résumés to better equip them for industry exposure.

Limitations and Future Work

Although cultivating a diverse team of student researchers is a stated goal of the FORC, there are several limitations associated with doing so. Graduation of students and study abroad programs interrupt the growth of technical expertise, and results in a lack of consistency in the composition of the team. To overcome this, the team management needs to identify and recruit new members regularly, which is a time-consuming process due to the technical nature of the project. Similarly, with team members volunteering for the project, it is sometimes difficult to maintain motivation. Another limitation of the project is the difficulty in validation of the results. Because implementation of the recommendations made by the FORC is at the discretion of the Safety Committee, hypothesis testing and validation can be problematic. Further, since the NMAC calculations are based on ADS-B data, irregularity in the recording of data can also make validation difficult. ADS-B systems aboard aircraft only broadcast position data every few seconds, and data gaps can result from aircraft orientation issues with respect to the receiving antenna. If a proximity event were to occur during such intervals, the algorithm may inaccurately measure the distance between two aircraft or miss the event in its entirety. Pilot feedback can be used to validate some events; however, these do not result in accurate reporting of all such events.

The next steps for the FORC are to refine the accuracy of recorded data using the Kalman filtering algorithm, validate relative aircraft distances, and verify pilot-reported proximity events. Another goal is the automation of the algorithm to provide real-time results and live dashboards. These dashboards can be used to determine the average frequency of proximity incursions or to compare the frequency of proximity incursions between aircraft types. Access to live dashboards can also be used by dispatchers to distribute aircraft more evenly between practice areas, improving temporal or spatial utilization. These dashboards can increase the ability of the flight program to safely train a greater number of students, thereby providing increased flight opportunities for flight instructors. Improved safety and efficient fleet utilization enable the program to train more students, which will reduce student flight fees by decreasing the overall cost of the program.

Acknowledgements

The authors wish to thank Justin Porter for his help in team coordination. They also thank Hao Wu, Lingzhi Cheng, and Bryson Barnett for their data analysis work.

References

Dillon, R. L., & Tinsley, C. H. (2008). How near-misses influence decision making under risk: A missed opportunity for learning. *Management Science, 54(8), 1425-1440.*

Fechheimer, M., Webber, K., & Kleiber, P. B. (2011). How well do undergraduate research programs promote engagement and success of students? CBE—Life Sciences Education, 10(2), 156-163.

Kunzi, F., & Hansman, R. J. (2011). Mid-air collision risk and areas of high benefit for traffic alerting. *11th AIAA Aviation Technology, Integration, and Operations (ATIO) Conference, including the AIAA Balloon Systems Conference & 19th AIAA Lighter-Than.*

Post, C., De Lia, E., DiTomaso, N., Tirpak, T. M., & Borwankar, R. (2009). Capitalizing on thought diversity for innovation. *Research-Technology Management, 52*(6), 14-25.

Russell, S. H., Hancock, M. P., & McCullough, J. (2007). Benefits of undergraduate research experiences. *Science, 316*(5824), 548-549.

A TABLET-COMPUTER APP DISPLAYING RUNWAY WINDS

William R. Knecht
Civil Aerospace Medical Institute, FAA,Oklahoma City, OK, USA

We tested variants of a mobile meteorological tablet-computer application designed to help general aviation (GA) pilots land aircraft more safely under windy conditions. This "app" compared METAR runway wind information in several graphical and textual formats. Study 1 tested 25 GA pilots on 18 runway wind scenarios. Graphical METARs depicted the runway with a large arrow at 90°, representing the crosswind speed component, and a second arrow parallel to the runway, representing the headwind/tailwind component. We hypothesized that eliminating the need for complex mental calculation of wind components would increase speed and/or accuracy of information processing. Study 2 tested 17 pilots on 24 scenarios, employing the same basic method, but enhanced by color-coding the wind-component arrows according to each pilot's previously stated maximums for landing wind risk-tolerance. Both studies showed that runway-relative, two-arrow wind component depictions were significantly fastest and most efficient. Pilots unanimously preferred graphical displays over textual.

Adverse winds are a persistent challenge for all pilots and, therefore, a high priority for the FAA (FAA, 2017). Winds at landing are particularly problematic to general aviation (GA).

The current research continues empirical testing of a low-cost, portable GA device designed to deliver timely weather information to the flight deck. This mobile meteorological application runs on a tablet computer (iPad), and is currently under development by the Research Applications Laboratory (RAL) of the National Center for Atmospheric Research (NCAR, Ahlstrom, Caddigan, Schulz, Ohneiser, Bastholm, & Dworsky, 2015, Knecht & Dumont, 2019, Knecht & McCarthy, 2019).

Common Method of Experiments 1 and 2

Measuring "Quality" of Information Depiction

This "app" can present runway wind information similar to that shown in Figure 1. The research question centered on finding the best type of information to display for that purpose.

In order to support a claim that depicting wind information one way is "better" than another there has to be some method of objectively quantifying display *quality*. The metrics of quality measured here were *accuracy* and *speed* of the pilots' mental wind-evaluation process.

Decision *speed* was simply how much time it took the pilot to decide whether or not to land, given the runway-level wind information shown. Decision *accuracy*, however, was an altogether-different and harder quality to assess. To assess accuracy, we compared "objective landing difficulty" to "perceived landing difficulty" on the assumption that *the closer the* ***perceived difficulty*** *of a wind scenario was to its* ***objective difficulty****, the better the wind display.* Figure 2 explains.

Creating Scenarios with Known Objective Difficulty

Operationalizing the experimental method required wind scenarios with various objective levels of difficulty. This required controlling for each pilots' *skill* and *risk-tolerance*. For instance, if one pilot thought a 3-kt crosswind was "easy" and another thought a 5-kt crosswind

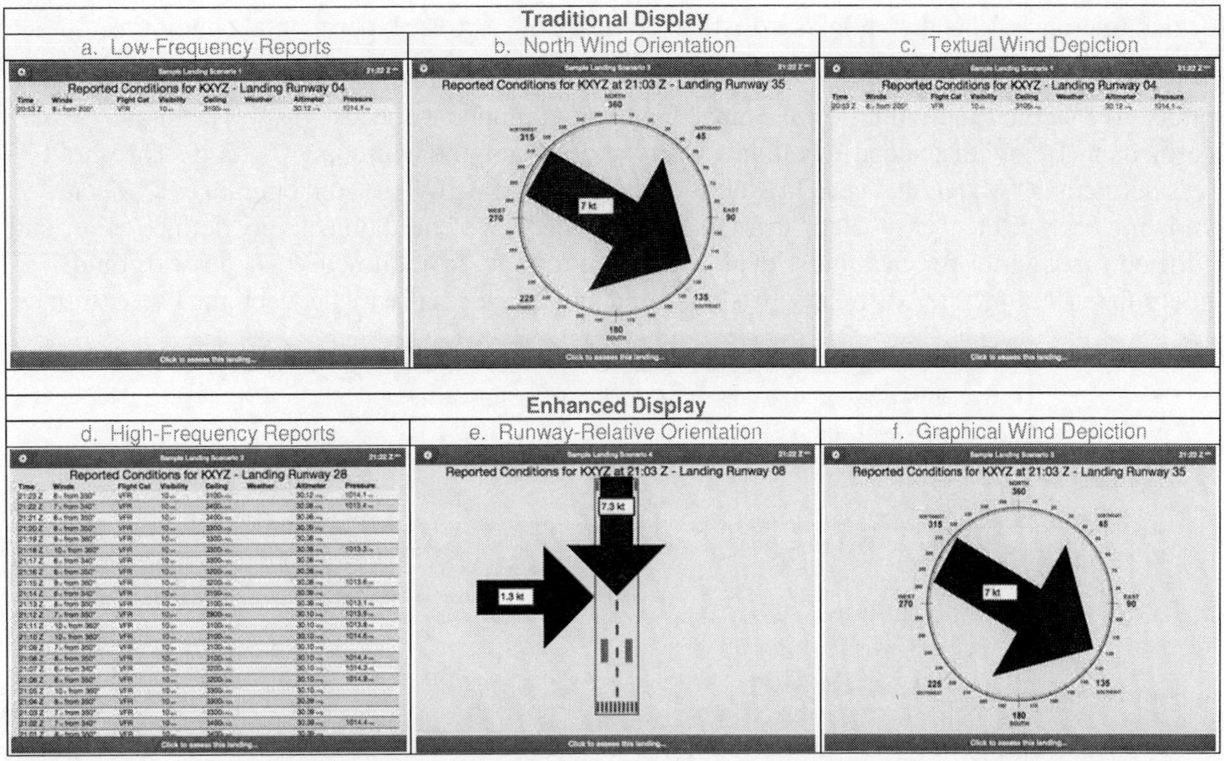

Figure 1. NCAR's Experiment 1 six runway wind depictions, all samples being supposedly 19 minutes old: a) "Traditional," text-based, similar to an aviation routine weather report (METAR); b) "Traditional" graphical wind depiction, a north-up view with an arrow showing wind direction and textual depiction of speed; c) "Traditional" textual METAR; d) "Enhanced" information similar to "a" but updated each minute, with the newest information no more than 1 minute old; e) "Enhanced" graphical wind depiction, a runway-relative view with separate arrows for crosswind and runway-aligned wind components, and; f) "Enhanced" METAR, similar to "c," but graphical as "b."

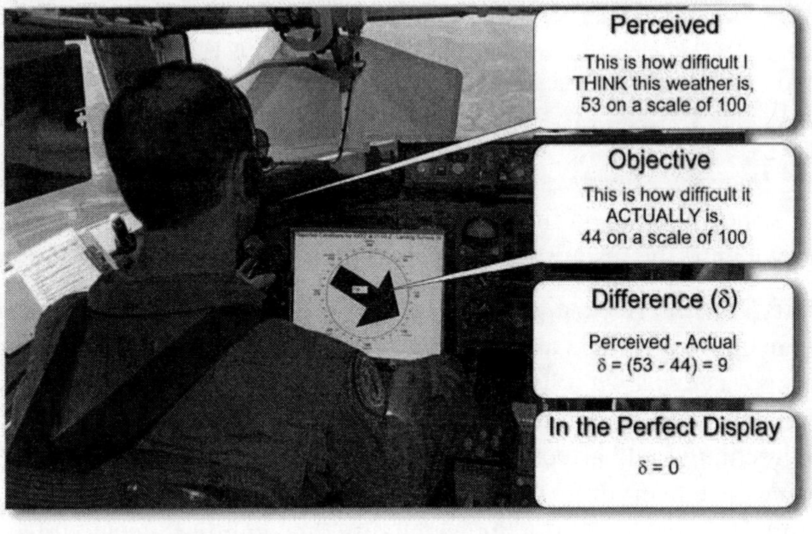

Figure 2. "Display quality" was measured as a difference score δ (delta), defined as participant's perceived scenario difficulty minus their objective scenario difficulty, both on a scale of 0—100. In a "perfect" display δ would equal zero; the display enabled them to correctly assess the scenario difficulty. Long, dry runway was assumed here.

was easy, to construct an objectively "easy" scenario, we would obviously want the crosswind component to be between 0 and 3 kts for the first pilot and 0—5 kts for the second pilot. This kind of individual adjustment is called *normalization*, and its goal is to create a single "normal" scale (e.g., 0—100 "difficulty units") that can be applied to all pilots, no matter what their skill or risk tolerance. This then allows statistical comparisons across experimental conditions.

To create such a "normal scale," at the very beginning of each pilot's test session we had pilots give us their individual "thresholds" for wind-component speeds. "Low Threshold" was defined as "Below that speed I wouldn't worry about that wind component." "High Threshold" was defined as "Above that speed I would hesitate to land with that wind component." Then, knowing each pilot's "easy" and "difficult" wind speeds, we could objectively define "easy" and "difficult" scenarios for each individual pilot. Additionally, from these two values we could interpolate a "moderate" difficulty by simply picking a value halfway between the two extremes.

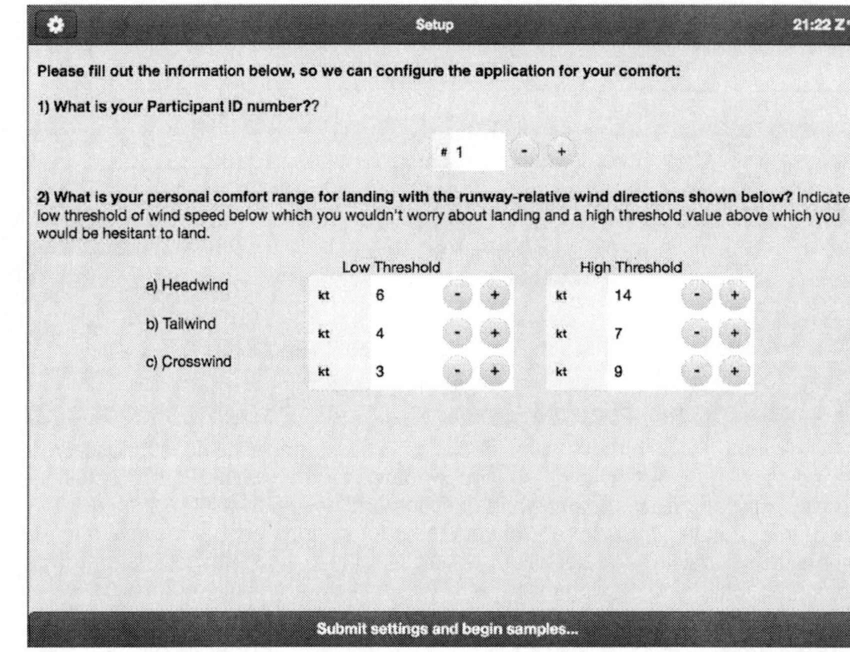

Figure 3. 1- Screenshot of the Setup page, showing the example of a "Low Headwind Threshold" of 6 kt and a "High Crosswind Threshold of 9 kt" for a hypothetical pilot.

Assessment of "Decision Quality"

Measuring *Decision Speed* was straightforward. This was merely the time it took each pilot to assess the wind situation, defined as the elapsed time from when the wind information page was first shown until the instant the pilot moved on to the subsequent assessment page.

To measure *Decision Accuracy*, pilots were asked to indicate each scenario's *perceived landing difficulty* by moving sliders along the "normal scale" of 0-100 (Fig. 4), representing how difficult the landing was *expected* to be. Meanwhile, recall that each scenario's *objective landing difficulty* had been normalized for that pilot, based on her/his previously reported values for how wind speed and direction would affect landing difficulty for him or her, personally. Therefore, the assessment page gave everything else necessary to calculate *perceived – objective difficulty* = δ. And, if one wind depiction was truly higher-quality than another, we would expect most of the δ scores to be smaller.

Experiment 1

Experimental Design

Experiment 1 utilized a within-participants (repeated measures) statistical design. Each pilot responded to 18 runway wind landing scenarios, each depicted by a single page similar to

Figure 1's, with a different set of wind parameters as independent variables. Figure 5 illustrates.

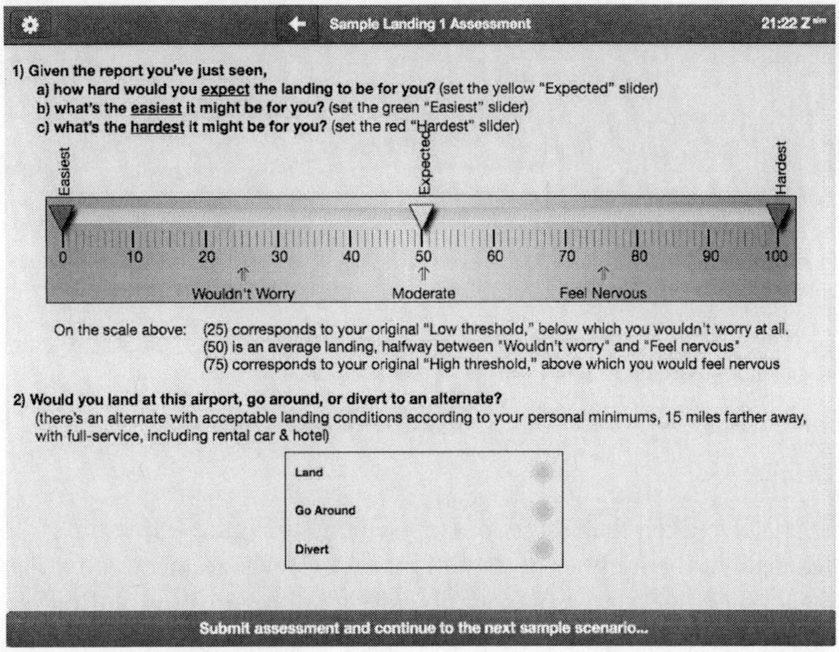

Figure 4. Screenshot of the Evaluation page.

Research design, 2 x 3 x 3 (Display Type (A) x Information Type (B) x Scenario Difficulty (C))						
	A1-Enhanced Display			**A2-Traditional Display**		
Scenario Difficulty[A]	B1-High-Frequency Reports	B2-Runway-Relative Orientation	B3-Graphical Wind Depiction	B1-Low-Frequency Reports	B2-North-Up Orientation	B3-Textual Wind Depiction
C1-Easy	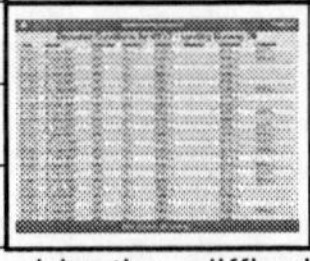		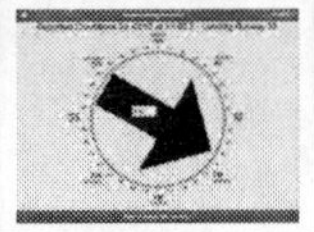	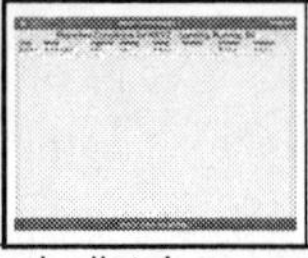	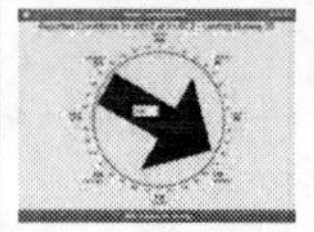	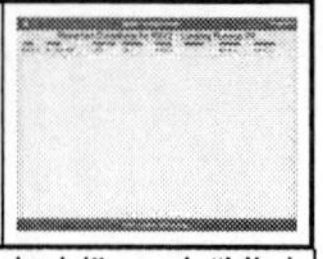
C2-Moderate						
C3-Difficult						
[A]All scenarios' objective difficulties were set according to individual pilots' answers for "Low Threshold" and "High Threshold" values on their Setup page (see text for details).						

Figure 5. Experiment 1's 2×3×3 research design.

Participants

Twenty-four GA pilots were recruited from a local flight school and paid \$50 (Fig. 6).

Student	0	**CFII**	5	**Age-mean**	28.8	**TFH**[1]**-mean**	959
Private pilot	24	**Commercial**	8	**Age-median**	21.5	**TFH-median**	185
Instrument-rated	9	**ATP**	2	**Age-SD**	14.6	**TFH-SD**	2150
CFI	7	**Multi-engine**	7	[1] Total Flight Hours			

Figure 6. Experiment 1 pilot demographics.

Results

Overall 2×3×3 ANOVA analysis of *Perceived Scenario Difficulty* δ scores showed significance only for the three objective difficulty levels (C1-3). Pairwise post-hoc comparisons indicated that each of those three levels was perceived significantly different from the other two at p = .00001 or better. However, as Figure 7a shows, there was considerable spread in the data.

The graphical twin-arrow display, depicting separate crosswind and headwind/ tailwind components, was fastest, with no apparent loss of accuracy representing landing difficulty.

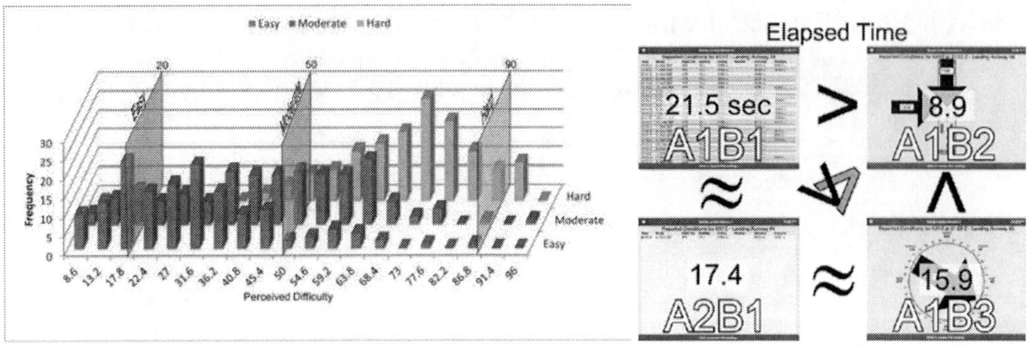

Figure 7. Experiment 1. a) variation in *Perceived Landing Difficulty* across Easy, Moderate, and Hard Scenarios, b) pairwise landing decision speeds (note that twin-arrow (A1B2) was fastest.

Experiment 2

Experimental Design

Experiment 2 leveraged the results of Experiment 1. Figure 8 illustrates. In a 2×3×4 repeated measures design, the number of depictions (A1-2) was reduced to two and the two-arrow depiction was color-coded to represent objective landing difficulty. Red represented a wind component speed greater than pilot's pre-stated maximum tolerance, orange represented medium-concern speeds and green represented "no worry."

Research design, 2 x 3 x 4 (Display Type (A) x Scenario Difficulty (B) x Time Constraint (C))						
	A1-Textual Display			A2-Graphical Display		
Time Constraint	B1-Easy	B2-Moderate	B3-Hard	B1-Easy	B2-Moderate	B3-Hard
C1-40 seconds						
C2-20						
C3-10						
C4-5						

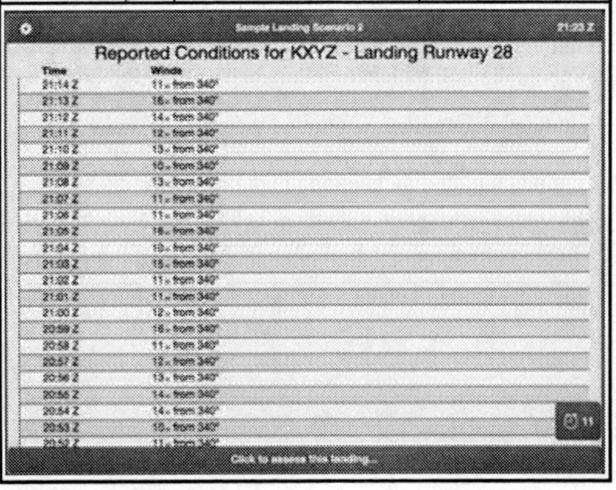

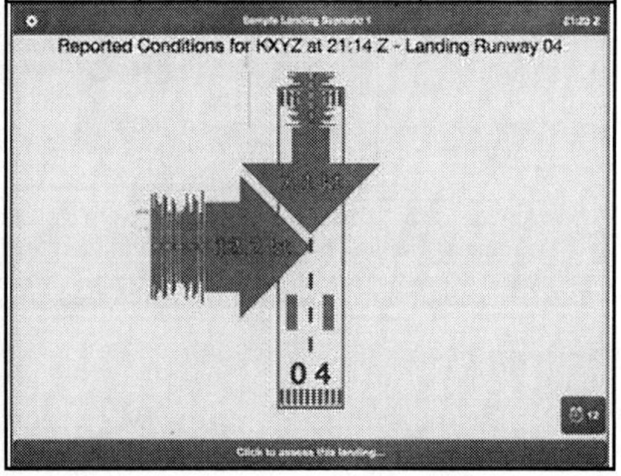

Figure 8. Experiment 2's 2×3×4 research design.

Again, three levels of scenario difficulty (B1-3) were used. Four levels of time constraint (C1-4) were introduced to see how restricting available viewing time would affect performance.

Participants

Seventeen GA pilots were recruited from a local flight school and paid $50 (Fig. 9).

Student	0	CFII	3	Age-mean	22.3	TFH-mean	323
Private pilot	17	Commercial	7	Age-median	22.0	TFH-median	200
Instrument-rated	15	ATP	0	Age-SD	3.4	TFH-SD	205
CFI	4	Multi-engine	4	TFH-max	800	TFH-min	98

Figure 9. Experiment 2 pilot demographics.

Results

Introduction of *Time Constraint* resulted in severe data non-normalities in *Perceived*

Landing Difficulty, disallowing ANOVA. Figure 10 shows *p*-values and effect sizes for paired t-tests, normality permitting, with Wilcoxon *p*-values for variable pairs involving a non-normality.

Significances between IV pairs (and Cohen's *d* effect size)															
	A2	B2	B3	C2	C3	C4	A1	A2	B1	B2	B3	C1	C2	C3	C4
A1	.337 (.69)							7.21×10^{-5} (2.92)							
B1		1.028×10^{-10} (15.10)	1.452×10^{-11} (16.97)							.004 (1.59)	.149 (.69)				
B2			0.025 (2.34)								.088 (.93)				
C1				.412 (.47)	.660 (.25)	.590 (.36)							.040 (.99)	2.88×10^{-4} (3.44)	2.93×10^{-4} (5.32)
C2					.742 (.23)	.853 (.09)								1.55×10^{-4} (3.14)	2.93×10^{-4} (5.89)
C3						.818 (.12)									5.03×10^{-4} (4.89)
							9931	6942	7571	9401	8338	11888	10251	6977	4632
							IV group means (milliseconds)								

Figure 10. Experiment 2 *p*-values and effect sizes for *Perceived Landing Difficulty* and *Elapsed Viewing Time.*

Elapsed Viewing Times were significantly different for graphical depictions (A1 vs A2). Differences between levels of *Time Constraint* were significant but logically trivial. More meaningful was that, given 40 seconds (C1), only one pilot timed-out on one scenario, whereas 74% of scenarios timed-out when pilots had only 5 seconds (C4).

Conclusions

These two studies clearly showed that, even when time is short, pilots *can* often discriminate between difficult and easy runway winds using either textual or graphical wind displays. However, this seems to be because they use a shortcut, or *heuristic*, when pressed for time. Rather than mentally computing wind components, they simply scan for wind speeds higher than their comfort level, regardless of wind direction. This allows quick scan of even long columns of numbers. But, deriving wind *components*—particularly intermediate-difficulty components—is a far more difficult task, particularly when time is short.

We therefore suggest that medium- and high-difficulty wind components will be best depicted by a graphical two-arrow display, particularly one color-coded according to each pilot's personal maximums reflecting their skill and risk-tolerance within the context of a given aircraft. Pilots here concurred, unanimously preferring the graphical displays over the textual.

Acknowledgments

This research was funded by the FAA's Weather Technology in the Cockpit Program, ANG-C61. Arnaud Dumont and Paddy McCarthy of NCAR's Research Applications Laboratory helped design the study and coded the mobile application.

References

Ahlstrom, U, Caddigan, E., Schulz, K., Ohneiser, O., Bastholm, R., & Dworsky, M. (2015). *Initial assessment of portable weather presentations for general aviation (GA) pilots.* (Technical Report DOT/FAA/TC-15/42).

Knecht, W.R., & Dumont, A. (2019). *Tailoring surface winds information for mobile meteorological applications: Part I, beta-testing.* (OAM technical report, in review).

Knecht, W.R., & McCarthy, P. (2019). *Mobile meteorological information tailored to landing phase of flight. Part II: Refinement.* (OAM technical report, in review).

AUTOMATED SPEECH RECOGNITION TECHNOLOGY TO SUPPORT IN FLIGHT WEATHER-RELATED COMMUNICATION FOR GA PILOTS

Gaojian Huang
*N*HanCE Research Lab, Purdue University
West Lafayette, IN
Brandon J. Pitts, Ph.D.
*N*HanCE Research Lab, Purdue University
West Lafayette, IN

Weather information latency during flight in general aviation (GA) has resulted in numerous incidents. Hands-free automated speech recognition (ASR) systems have the potential to help overcome this challenge and facilitate rapid weather-related information exchange. However, it is unclear to what extent ASR systems can support pilot communication in such noisy environments. The goals of this study were to (1) evaluate the performance of 7 commercially-available ASR systems to recognize weather phrases during GA operations and (2) determine whether speech-to-noise (S/N) ratio, flight phase, and accent type modulate system performance. Overall, the highest accuracy percentage achieved by any system was 72%, when the S/N ratio was at least 3/2. This research can help to inform the selection and development of next-generation technologies to be used in safety-critical, information-rich domains.

For more than two decades, adverse weather conditions has been cited as one of the most frequent causes of fatal accidents among general aviation (GA) pilots (e.g., Duke & George, 2016; Federal Aviation Administration, 2010) To help improve safety, GA pilots need to be aware of the weather conditions along their flight path. Traditionally, pilots are provided with weather briefings prior to flight and may receive updated weather information from Flight Service while flying (Ahlstrom, Ohneiser, & Caddigan, 2016). However, to date, weather information latency during flight, i.e., the time delay between flight environment weather conditions and the presentation of this information on cockpit displays, still represents a major problem in GA and limits the decision making abilities of pilots.

The emergence of NextGen technologies may offer pilots tools to improve their situational awareness and result in better real-time strategizing. For example, mobile devices and tablets are increasingly able to support aviation software that can inform pilots as new weather information becomes available. In addition, not all NextGen advancements require manual interactions. Automated speech recognition (ASR) technology is one particular development that can assist with activating commands and quickly obtaining critical information. These systems translate natural spoken language/words into text (Këpuska, 2017), which can then be used to execute specified functions. The benefit of hands-free interactions is especially important in the context of extreme weather conditions during flight, when pilots' cognitive and manual workload are already high. Recently, commercial ASR systems, such as Google Cloud Speech API and Microsoft Bing Speech API, have been developed and used in applications, such as portable devices, smart homes, and autonomous vehicles (Këpuska, 2017; Kimura, Nose, Hirooka, Chiba, & Ito, 2019). Significant progress in the development of artificial intelligence and machine and deep learning technologies has resulted in these systems achieving detection rates as high as 90% (Yu & Deng, 2016).

In aviation, previous work has investigated the use of speech recognition systems in flight (e.g., Arthur III, Shelton, Prinzel III, & Bailey, 2016) in both real-world and laboratory environments, but not with respect to specific weather-related communication. One other open question regarding ASR systems in flight is the extent to which they can perform in noisy environments (Hansen, 1996). The goal of this study was, therefore, to determine the effectiveness of commercially-available speech recognition systems to support weather-related communication in GA.

Method

Participants

Thirty participants from Purdue University and a multidisciplinary research project team volunteered to take part in this study. All participants were required to be fluent in English. The 30 participants were divided into 6 accent/dialect groups based on their geographical origins (East Asia, India, Latin America, Northern and Southern U.S, and UK/Australia/South Africa). This study was approved by the Purdue University Institutional Review Board (IRB Protocol ID: 1804020515).

Apparatus and Test Stimuli

Speech recognition system selection. During an initial market analysis phase, 50 potential commercially-available systems were identified based on accessibility (e.g., downloadable), capability (e.g., performance/accuracy), interface design, and cost. The final selection of systems was focused on: speaker-independent, customizable vocabulary database, platform type, and performance in noisy environments. In total, the following 7 systems were chosen for evaluation: Braina Pro; Dragon NaturallySpeaking (with and without speech training component); Google Cloud Speech API; Microsoft Bing Speech API; Houndify; Lily Speech.

Speech & Aircraft nose file generation. A human-subject experiment was conducted to create samples of spoken weather-related phrases. In particular, the 30 participants were recorded reciting 35 separate weather-related phrases commonly used by GA pilots (e.g., 'show PIREPs', 'show convective SIGMET', etc.). An aviation quality headset (i.e., ASA AirClassics HS-1A) was used to make these recordings in a quiet laboratory environment. At a different time, background aircraft cockpit noise samples were also recorded during the taxi, cruise, and takeoff flight phases of a test flight carried out at The Ohio State University airport (Don Scott Field). The intensity range of these samples was 95-124 dB. The device used to create these recordings was a Sony ICD-PX333 Digital Voice Recorder.

Test stimuli. The recorded speech files and aircraft cockpit noise samples were digitally combined using, Audacity 2.2.2, to create the "test stimuli." The goal was to evaluate conditions in which: a) the background noise was louder than the speech (S < N), b) the noise and speech volumes were the same (S = N), and c) the speech was louder than the noise (S > N). To this end, the combined speech and noise file was adjusted to different speech-to-noise (S/N) ratios in each of the three categories. Specifically, 9 S/N intensity ratios (1/2, 5/8, 3/4, 7/8, 1/1, 5/4, 3/2, 7/4, 2/1) were initially selected based on psychophysical research involving the differentiation between two concurrent stimuli (Biberger & Ewert, 2015; Bradley, Reich, & Norcross, 1999). Also, a baseline condition with only speech (no background noise) was generated.

Factors selection. All files were processed internally within the 7 speech recognition software packages and recognition accuracy rate was calculated. After preliminary investigation, 8 S/N ratios (1/2, 5/8, 3/4, 1/1, 5/4, 3/2, 2/1, and the baseline condition) and 2 types of flight phases (taxi and cruise) were selected, because no statistically significant differences were found between adjacent S/N ratio and flight phases and those that were excluded.

Experimental Design

Overall, the experiment employed a 2 (flight phase) × 8 (S/N ratio) × 7 (system) × 6 (accent) full factorial design. Flight phases, S/N ratios, and systems were within-subject variables, and accent was a between-subject variable. The 6 accent/dialect groups were determined based on self-reported information provided by participants prior to the experiment. Sixteen auditory files were created for each participant (2 flight phases and 8 S/N ratios). This resulted in a total of 480 files and 3,360 total runs.

Procedures

Each participant first signed a consent form. Next, they familiarized themselves with the 35 phrases (i.e., pronunciation and sequence). Once participants indicated that they were ready to record, the experimenter left the room and the participant started and stopped the recordings as instructed. All phrases were read using participants' normal speaking volume (~60 dB).

Data Analysis

The dependent variable was phrase accuracy rate (PAR), i.e., the percentage of phrases correctly recognized by the software out of the total number of phrases. This measurement was inspired by previous work which used Word Error Rate (WER) as the ASR performance measure (e.g., Vipperla, 2011). A 4-way analysis of variance (ANOVA) was used to identify main and interaction effects. Results were considered significant at $\alpha = 0.05$. Since none of the 7 systems recognized speech when the S/N ratio was less than 1 (i.e., PAR = 0%), a perfect separation assumption was used and only data in cases where S/N ratio ≥ 1 were included in analysis.

Results

There was a significant main effect of system on PAR, $F\ (6, 1994) = 796.067, p < .001,\ \eta_p^2 = .705$. In particular, post-hoc analysis revealed that the Dragon NaturallySpeaking (with speech training component) (mean PAR = 72.3%, standard error of mean (SEM) = .018) has a significantly higher PAR compared to all other systems, see Figure 1.

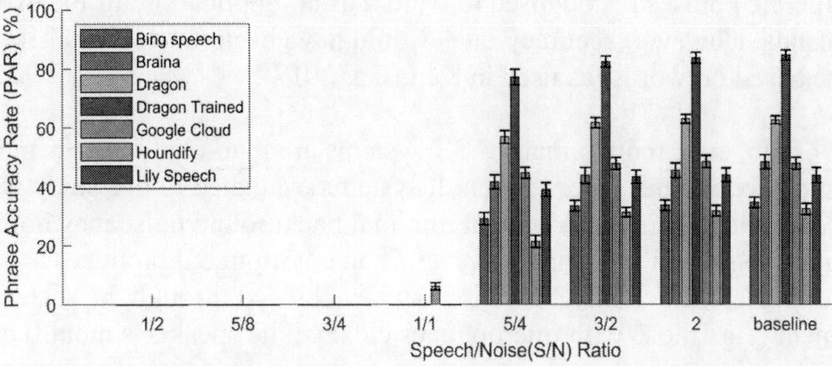

Figure 1. Phrase accuracy rate (PAR) as a function of speech/noise ratio for 7 speech recognition systems during cruise flight phase (error bars represent standard error of mean)

There was also a significant main effect of Speech/Noise (S/N) ratio on PAR, $F\,(4, 1994) = 1392.741, p < .001,\ \eta_p^2 = .736$. In particular, all systems performed better when the S/N ratio was at least 3/2 (mean PAR = 51.2%, SEM = 0.010) compared to when the S/N was 5/4 (mean PAR = 47.6%, SEM = 0.010) or 1/1 (mean PAR = 11.0 %, SEM = 0.008). The baseline condition (mean PAR = 52.3%, SEM = 0.010) and an S/N ratio of 2/1 (mean PAR = 51.8%, SEM = 0.010) did not differ from an S/N of 3/2.

PAR was affected by accent type, $F\,(5, 1994) = 111.568, p < .001,\ \eta_p^2 = .219$, (note here the relatively small effect; Watson, Lenz, Schmit, & Schmit, 2016). Specifically, the Northern American (mean = 49.8%, SEM = 0.014) and Southern American (mean = 49.6%, SEM = 0.013) accents were slightly more recognizable than those from any other region (East Asia mean = 37.6%, SEM = 0.012; Latin America mean = 41.1%, SEM = 0.014; India mean = 38.3%, SEM = 0.013; and UK/Australia/South Africa mean = 41.2%; SEM = 0.014).

Discussion

This study evaluated the extent to which commercially-available speech recognition systems could recognize weather-related terminology in a GA environment. The highest phrase accuracy rate (PAR) achieved by any system was 72% (which included a training component). Also, all systems performed best when the speech-to-noise (S/N) ratio was at least 3/2. Finally, U.S. accents were slightly more recognizable than those from any other world regions.

None of the ASR systems used in this study achieved a PAR of 100%. Typically, default speech recognition vocabulary databases do not include aviation-related phraseology. Dragon NaturallySpeaking (with speech training component), however, achieved the highest accuracy rate. This result is consistent with previous work which found that Dragon NaturallySpeaking was significantly more accurate compared to other common speech systems (Rami, Svitlana, Lyashenko, & Belova, 2017). Specifically, in this study, the performance of Dragon NaturallySpeaking increased from 54% to 72%, without and with training, respectively. This suggests that training systems how to pronounce particular words can significantly increase detection accuracy. It is critical that training be conducted using a well-crafted aviation-specific vocabulary training set and default references for terms likely to be confused. For example, if the system perceives "Sig Mat," it should default to SIGMET. Relatedly, in this study, we focused on the accuracy of complete phrases (as opposed to words) as an implication for the execution of weather-related commands. However, accuracy rates would have been much greater if calculations were done based on words (as used in Këpuska, 2017).

In terms of S/N ratio, even though many ASR systems are marketed to perform in noisy environments, the best detection rates recorded for all systems evaluated in this study was when the S/N ratio was 3/2 or greater. This indicates that minimal background noise may not interfere with pilot communication to speech systems. However, if an environment produces a considerable amount of noise, then a high S/N ratio may be achieved through the selection of the proper headset equipment (e.g., those with microphones close to the speaker's mouth) or the use of a microphone that recognizes speech using throat vibration signals. Also, noise absorption material may be installed in the cockpit to reduce ambient noise sources.

Accent type was found to have an effect on PAR. Native Northern and Southern U.S. participants' speech was more detectable (i.e., detection accuracy ~ 50%) than participants from East Asia, Latin America, India, or UK/Australia/South Africa. One possible explanation for this finding is that the systems evaluated in this experiment were developed using (American) English speakers. This interpretation infers that in order to increase recognition accuracy, corpuses used to create and train ASR systems should include a wide range of demographic factors, such as accents/dialects, speech rates, and age groups. Finally, it is no surprise that PAR was not affected by flight phase. Although the background noise frequencies between the two conditions may have slightly differed, their overall loudness and rhythm were perceived comparably by the ASR system, especially given that the sounds did not resemble human speech.

In summary, while the outcome of this work will be useful in field research and to the GA community, more research is needed to determine, for example, minimum requirements prior to adoption into practice. Still, this research may help to guide decisions regarding the selection and use of smart devices and applications in complex domains.

Acknowledgements

This work was supported by the Federal Aviation Administration (FAA), as part of the Partnership to Enhance General Aviation Safety, Accessibility and Sustainability (PEGASAS) Project 4: Weather Technology in the Cockpit (WTIC). The authors would like to thank Han Wang, Yuyuan Liu, and Matthew Scott for developing the test environment and stimuli, and for assisting with data analysis. We also thank the statistics professionals from the Purdue University Department of Statistics for their statistical consulting service.

References

Ahlstrom, U., Ohneiser, O., & Caddigan, E. (2016). Portable Weather Applications for General Aviation Pilots. *Human Factors*, *58*(6), 864–885.

Arthur III, J. J., Shelton, K. J., Prinzel III, L. J., & Bailey, R. E. (2016). *Performance Evaluation of Speech Recognition Systems as a Next-Generation Pilot-Vehicle Interface Technology*.

Biberger, T., & Ewert, S. D. (2015). Envelope-power based prediction of auditory masking and speech intelligibility. *The Journal of the Acoustical Society of America.*

Bradley, J. S., Reich, R., & Norcross, S. G. (1999). A just noticeable difference in C50 for speech. *Applied Acoustics*, *58*(2), 99–108.

Capobianco, G., & Lee, M. D. (2001). The Role of Weather in General Aviation Accidents: An Analysis of Causes, Contributing Factors and ISSUES. *Proceedings of the Human Factors and Ergonomics Society Annual Meeting*, *45*(2), 190–194.

Duke, R., & George, T. (2016). *2016 Pilot Report Survey*.

Federal Aviation Administration. (2010). Weather-related aviation accident study 2003–2007. Retrieved January 24, 2019, from https://www.asias.faa.gov/i/studies/2003-2007weatherrelatedaviationaccidentstudy.pdf

Hansen, J. H. L. (1996). Analysis and compensation of speech under stress and noise for environmental robustness in speech recognition. *Speech Communication*, *20*(1–2), 151–173.

Këpuska, V. (2017). Comparing Speech Recognition Systems (Microsoft API, Google API And CMU Sphinx). *International Journal of Engineering Research and Applications*, *07*(03), 20–24.

Kimura, T., Nose, T., Hirooka, S., Chiba, Y., & Ito, A. (2019). Comparison of Speech Recognition Performance Between Kaldi and Google Cloud Speech API (pp. 109–115). Springer, Cham.

McShefferty, D., Whitmer, W. M., & Akeroyd, M. A. (2015). The Just-Noticeable Difference in Speech-to-Noise Ratio. *Trends in Hearing*, *19*, 233121651557231.

Rami, M., Svitlana, M., Lyashenko, V., & Belova, N. (2017). Speech Recognition Systems : A Comparative Review. *IOSR Journal of Computer Engineering*, *19*(5), 71–79.

Stern, M. K., & Johnson, J. H. (2010). Just Noticeable Difference. In *The Corsini Encyclopedia of Psychology* (pp. 1–2). Hoboken, NJ, USA: John Wiley & Sons, Inc.

Vipperla, R. (2011). Automatic Speech Recognition for ageing voices Ravichander Vipperla.

Watson, J. C., Lenz, A. S., Schmit, M. K., & Schmit, E. L. (2016). Calculating and Reporting Estimates of Effect Size in Counseling Outcome Research. *Counseling Outcome Research and Evaluation*, *7*(2), 111–123.

Yu, D., & Deng, L. (2016). *Automatic Speech Recognition*. https://doi.org/10.1007/978-1-4471-5779-3

FACTORS AFFECTING AIR TRAFFIC CONTROLLER'S WEATHER DISSEMINATION TO PILOTS

Ali Alshaqah, Chen Ling, Emma Pierson, Shengyoung Wang
The University of Akron
Akron, Ohio

As the number of flights in the United States continues to rise steadily, an equally amplified need for reliability and safety has come to the forefront of aviation research. One of the most alarming trends is the number of general aviation (GA) accidents during severe weather events that occur yearly, with fatalities occurring in more than half of these cases. This study focuses on identifying factors influencing weather dissemination of Air Traffic Controllers (ATC) to GA pilots. Ten factors affecting controllers' performance during severe weather events were identified through an in-depth literature review including controller mental workload, situation awareness, weather information format and accuracy, weather information needs, weather tool limitations, inaccurate assumption and bias, controller training and expereince, regulatory factor, supervisory factors, and pilot-controller relationship. Recommendation can be developed to address each factors so that aviation safety could be enhanced in severe weather situations.

The dynamic nature of the aviation environment is aggravated by adverse weather conditions which may change rapidly. These weather conditions generally require continuous evaluation and interpretation of weather information by both controllers and pilots to ensure a safe flight. The uncertain weather situation also requires a series of pre-flight and/or in-flight decisions to avoid accidents. The information incorporated to make these decisions is complex and is obtained from a range of sources. Therefore, the Federal Aviation Administration (FAA) considers the highest-priority weather functions to be those that detect phenomena posing a potential hazard to aviation. Indeed, controllers play a crucial role in providing tactical weather assistance through the dissemination of timely, accurate weather information. Inefficient weather-related decision-making of the flight crew during severe weather conditions, in which controllers can have a significant impact on the outcome, has been cited as a significant causal factor in general aviation (GA) accidents (Madhavan et al., 2006). Through numerous studies of literature based on weather-related accidents and incidents reports from the National Transportation Safety Board (NTSB), ten factors have been identified as recurring themes. These factors help set the stage for future in-depth research and analysis into mitigation strategies, interventions, and techniques that can improve methods by which air traffic controllers are trained to use weather-related information to aid pilots, especially GA pilots, in safely reaching their destinations.

1. Mental Workload

Air traffic controllers' tasks are cognitively complex in nature, and the mental workload is a dominant safety-related consideration in the air traffic controller domain. Throughout the literature, significant relationships between extremes of workload (under load, high workload, and overload) and a decline in controller performance, such as an increase in operational errors, have been consistently reported (Cox-Fuenzalida, 2007). A high workload may be experienced when air trafific controller disseminate weather information during severe weather situations, as additional weather related decisions making are needed and more pilots request diversions or assistance. Once the workload is driven over the "red line," performance can start to deteriorate, and errors can begin to appear when time stress is excessive (Wickens, 2017, NRC, 1997) Another issue is the transitions between workload extremes, which have been reported to be negatively associated with controller performance. Cox-Fuenzalida (2007) investigated the effect of workload transition in association with performance, and the results showed a significant decrement in performance after a transition in both high-to-low and low-to-high experimental conditions. Although the study was not specific to the air traffic controller domain, the results suggest that caution need to be exercised as many situations involving severe weather conditions may be associated with workload transitions from low to high or from high to low.

2. Situation Awareness

Weather situation awareness in the ATC domain is a combined perception of time, current weather location, airspace volume, weather movements in the near future, sector traffic flow, and the available control options(Ahlstrom et al., 2003). A high weather awareness would imply that the controller is able to form a coherent future "mental weather picture" and manage the air traffic accordingly. This means that the controller would know where the severe weather areas will be and how to adjust the traffic flow accordingly. It also implies that the controller could adjust their tactical strategies for more efficient traffic flow, vector planning, and better assistance to the pilot (Ahlstrom et al., 2003). In the current national airspace (NAS), controllers maintain their situation awareness by monitoring the aircraft using advanced display system integrated with decision support tools that show different levels of precipitation on the Standard Terminal Automation Replacement System (STARS), or the ARTS Color Display, or by receiving a weather briefing from their supervisor. In addition to this information, controllers get reports from pilots of hazardous weather conditions they encounter during flight. Therefore any deficiencies of the aviation weather system from collecting, communication, interpreting weather information would result in a controller's loss of situation awareness.

3. Weather Information Format and Forecast Accuracy

Weather-related accidents can usually be attributed to a lack of valid weather information during flight. Aviation weather information in aural format is difficult to integrate with spatial flight information and recall for reference. Indeed, the controller's lack of knowledge of complex aviation weather conditions often necessitates reference to various weather resources to develop a comprehensive understanding of meteorological conditions (Latorella et.al., 2002). When the weather resources for controllers are limited, controllers might have a lesser understanding of the available weather information, hindering flight safety (King et al., 2016). After National Research Council committee (NRC, 1995) identified some operational shortcomings pertaining to aviation weather dissemination, FAA conducted many studies to resolve these shortcomings, e.g. improving the weather forecast accuracy and quality to enhance the controller's ability to provide a timely, accurate, consistent and adequate information. Some of these issues have been addressed while others need further research to identify where the exact gaps are in the controller-pilot system. Although significant efforts have been underway to provide more accurate forecasts, weather forecast always comes with a degree of an uncertainty. When controllers provide weather information to pilots, it must be accurate. According to FAA Order

JO7110.65, controllers are to provide complete weather information based on all weather sources and this information should be consistent and accurate, considering the radar delay. Therefore, controllers may be hesitant to provide any weather information that is "incomplete, inconsistent, and outdated." If an accident is caused by ATC providing inaccurate information, it might lead to government liability (Bartsch, 1996).

4. Controller Information Needs for Integrated Weather Information

A poor understanding and under-defining of controller's weather information needs, including the actual weather information necessary to maintain the controller's situation awareness still exists. Research is still lacking in the exploration and specification of the weather information needs for controllers (Ahlstrom et al., 2003). FAA has outlined the concepts of operations for the weather in the NAS domain as well as the weather information needed by NAS decision-makers to mitigate the effects of weather on flight safety. However, according to Ahlstrom et al. (2013), the validation requirements on practical use were not established although a summary of conceptual weather information needs and a strategy to mitigate the deficiencies were presented. Therefore, to improve weather information display tailoring to the weather needs of different types of controllers and to understand the associated impact on operational services, more in-depth research is still needed, including a plan for integrating multiple sources of weather information onto user displays (Ahlstrom et al., 2003).

5. Training and Experience

Insufficient training and low experience levels in aviation weather scenarios could adversely affect air traffic control performance. Ruitenberg (1997) emphasized on the periodical recurrent training-schemes, which providing training when new technology is introduced to facilities, also providing training on any new pertinent knowledge or skills to controllers. Investigations in some aviation accident revealed a lack of training of air traffic controllers on how to retrieve both required and additional useful weather information from a regular channel, when such information is not instantly and directly available at their workstations (NTSB, 2015). Past reserach also identified that some controllers have limited aviation weather knowledge, and there is a lack of standardized procedures or protocols in training programs on mitigating weather-related incidents (King et al. 2016). To address this issue, recurrent training focusing on weather and scenario-based weather training could enhance controller's weather readiness.

6. Inaccurate Assumptions and Biases

Several inacurate assumption or bias may hinder controller from effectively disseminating weather information to pilots. Controllors may assume GA pilots are able to avoid bad weather visually, however pilots may be operating in IMC and cannot see anything beyond the aircraft's windshield. Moreover, there is a common thought in the ATC community that the pilot has a better perspective on what lies ahead because they have airborne weather radar. Cockpit weather radar systems are not better than NEXRAD or any other controller weather detection systems- they are just different (Werth, 2014). Another bias was identified by NRC committee as it gathered information from various FAA organizations and individuals. Because of ATC's heavy focus on aircraft control and collision avoidance, ATC community seem to have developed a general bias against increasing their involvement with weather-related issues (NRC,

1995). During severe weather, controllers may be so busy working with aircraft that they have little time to explore the weather. For this reason, and because some controllers view weather information as advisory service, they may not be aware of current and forecast weather conditions in their sector (NRC, 1995). The third bias is related to the controller's use of probabilistic weather information. Although NextGen provides better weather forecast accuracy and integrating the probabilistic weather information to the DST system, the NAS users still show insufficient understanding of the probabilistic weather information in the decision-making process (Abelman et.al., 2014).

7. Limitations in Weather Tools

ATC facilities utilize both ASR system which displays precipitation in 4levels and Weather and Radar Processor (WARP) which reports precipitation in 3 levels to provide weather information to controllers. These systems have their limitations in showing weather information. Unlike approach controllers who use ASR (which is nearly real-time), ARTCC or Center controllers use WARP which cannot provide warnings about precipitation until it reaches the "moderate" threshold because WARP does not display light intensity of precipitation (FAA Order JO7110.65, section 2-6-4 weather and chaff services). At this threshold, the weather situation can already be hazardous for unprepared pilots or light aircraft. Moreover, the data which appears on the controller's scope is typically six minutes old, since WARP collects and integrates data from one or more remote NEXRAD sites (FAA, 2016). In convective conditions where the severe weather activity builds rapidly, the time lag can render WARP information obsolete before it ever hits the screen. Understanding the limitations of aviation weather detection systems that NAS operators typically utilize, and developing a comprehensive understanding of how these systems can complement each other have a profound impact on safety (Werth, 2014).

8. Organizational, Regulations and Procedure Factors

Controllers should provide operational significant weather information to pilots before, and during flight, who then utilize this information to decide when and where to fly (FAA, 2017). NTSB (2014) suggest that controllers must have unconstrained access to critical information on essential weather information, such as real-time lightning data. Controllers must also be trained and equipped to disseminate this critical information expeditiously. Further, FAA must have the infrastructure and protocols in place to ensure vital information such as pilot reports (PIREP's), is conveyed in the NAS system (NTSB, 2014)frequently. However, controllers' first priority is to separate aircraft and issue safety alerts. Rational prioritizing should be used of all other provisions of FAA order based on requirements of the situation at hand. According to FAA Order JO7110.65, controllers should "select the most appropriate course of action in convective weather scenarios." The rule also instructs that "additional services, such as traffic advisories and safety alerts, can be offered to aircraft flying in uncontrolled airspace but only on a workload permitting basis". Therefore, it is up to the controller's discretion to decide whether to disseminate weather information to pilots based on the workload. Additionally, when various weather information sources exist, causing it difficult for controllers to decide how weather could impact their tasks, controllers may shift their decision-making process to be based on personal knowledge and experience rather than making sound decisions that are rule-based to assist in time-critical weather encounters (Lindholm, 1999).

9. Supervisory and Management Factors

Organizational and supervisory factors have a profound influence on restricting unsafe acts committed by controllers, including decision errors, skill-based errors, perceptual errors and violations (Pounds et al., 2000). Also, supervisory activities and practices such as enhancing operational effectiveness, managing resources, enhancing interpersonal skills, and monitoring controllers performance were reported as important to safety (Connor et al., 2001). Moreover, a sufficient number of trained supervisors/controllers also is considered as an essential factor. Therefore, clear guidance provided by trained supervisors and facility manager on how to disseminate weather information as well as a safety culture in the facility on weather-related issues will help controllers achieve safer and better performance.

10. Controller-Pilot Relationship

ATC and pilot work is interdependent of each other and their work depends on each other to be successful, and they need to trust each other (Owen, 1998). Successful joint practices for both parties enabled by smooth communication is needed to support such a highly inter-dependent system. Any miscommunication may endanger system safety and efficiency. Controller and pilot tasks have different objectives, controller ensure safe and expeditious traffic flow of an air space, while pilot ensures safe and expeditious traffic for their specific airplane. Constrained by their own work rules, there could be miscommunication and breakdown in coordination at the border of their individual activities. Both pilots and ATC might have inadequate understanding of one another's weather resources, capabilities, and the supportive working strategies necessary to avoid contingency flight situations. However, the controller-pilot relationship could be so tenuous or challenging that it is referred to as an "awkward alliance" , and similar tension may be due to pilots perception of the ATC role as "traffic cop" (Besco, 1997). Therefore, pilots may be hesitant to use ATC as a weather resource.

Conclusion

The need for improved weather dissemination is intended to reduce GA accidents. Overall better weather-related decisions for ATCs could also help reduce unnecessary diversion of aircraft and achieve better planning and more efficient routing. High-quality aviation weather service and weather information (e.g., weather observations, forecasts) must be provided, and controllers must be adequately trained in properly disseminating weather information to ensure consistent and safe air operations. Recommendation can be developed to address the identified factors so that aviation safety could be enhanced in severe weather situations.

Acknowledgements

This study is funded by the Federal Aviation Administration (FAA) Center of Excellence (COE) for Technical Training and Human Performance (TTHP)

References

Abelman, S., Gibbons, W., Jonsson, J., & Bass, R. (2014). Observed heuristics and biases in air traffic management decision making using convective weather uncertainty. In *6th AIAA Atmospheric and Space Environments Conference* (p. 2901).

Ahlstrom, V., & Della Rocco, P. (2003). *TRACON controller weather information needs: I. literature review* (No. DOT/FAA/CT-TN03/18). William J. Hughes Technical Center(US).

Bartsch, R. (1996) *Aviation Law in Australia.* LBC Information Services, Sydney.

Besco, R. O. (1997). An awkward alliance-The controller/pilot team. *Journal of Air Traffic Control*, 39 (4), 24-31.

Conner, M. & Corker, K. (2001). *Engineering Support to the Federal Aviation Administration Evaluations and Accident Investigations Staff for a Field Assessment of Controller Operational Errors*. National Aviation Research Institute.

Cox-Fuenzalida, L. E. (2007). Effect of workload history on task performance. *Human Factors*, *49*(2), 277-291.

Fedreal Aviation Adminestration. (2016). *Aviation Weather Services* :Advisory Circular. AC00-45H. Retrived from www.faa.gov/documentLibrary/media/Advisory_Circular/AC00-45H

Hancock, P. A., & Verwey, W. B. (1997). Fatigue, workload and adaptive driver systems. *Accident Analysis & Prevention*, *29*(4), 495-506.

King, J. M.; Ortiz, Y.; Blickensderfer, B. L.; Ph, D., 2016: ATC Weather Knowledge & Skills : A Contributor to the General Aviation Weather Problem ? *Proceedings of the Human Factors and Ergonomics Society 2016 Annual Meeting.*, 2014–2017.

Latorella, K., Lane, S., & Garland, D. (2002). General Aviation Pilots' Perceived Usage and Valuation of Aviation Weather Information Sources. (Technical Report NASA/TM-2002-211443). Springfield, VA: National Technical Information Service.

Lindholm, T. A., 1999: Weather information presentation. *Handbook of aviation human factors.*, 567–589.

Madhavan, P., & Lacson, F. C. (2006). Psychological Factors Affecting Pilots' Decisions to Navigate in Deteriorating Weather. *North American Journal of Psychology, 8*(1).

NTSB.(2014). General Aviation : *Identify and Communicate Hazardous Weather*. Retrived from https://www.ntsb.gov/safety/mwl/Pages/mwl7_2014.aspx

NTSB. (2015). Safety Recommendation Report :Weather Information Dissemination, (ASR-16-05). Retrived from www.ntsb.gov/investigations/AccidentReports/Reports/ASR1605.pdf

Owen, C. (2008). Analysing joint work between activity systems. *Activités, 5*(5-2).

Pounds, J.; Scarborough, A.; Shappell, S., 2000: A human factors analysis of Air Traffic Control operational errors. *Aviation, Space and Environmental Medicine.*, 71, 329.

Ruitenberg, B., 1997: Ten Years of Human Factors in Air Traffic Control.Retrived from https://skybrary.aero/bookshelf/books/1335.pdf

Werth, J. (2014). Airborne Weather Radar Limitations. *The Front*. Retrived from https://www.weather.gov/media/publications/front/14dec-front.pdf

National Research Council. (1995). *Aviation weather services: A call for federal leadership and action*. National Academies Press.

National Research Council. (1997). *Flight to the future: Human factors in air traffic control.* National Academies Press.

Wickens, C. D. (2017, June). Mental workload: assessment, prediction and consequences. In *International Symposium on Human Mental Workload: Models and Applications* (pp. 18-29). Springer, Cham.

MACHINE AWARENESS

Dr. Steven D. Harbour
Dr. Jeffery D. Clark
Mr. William D. Mitchell
Riverside Research
2640 Hibiscus Way, Beavercreek, Ohio 45431, USA
Dr. Krishnamurthy V. Vemuru
Riverside Research
2900 Crystal Dr., Arlington, Virginia 22202, USA

Current and future research that embodies a pathway to achieving machine common sense (MCS), including Capsule Neural Networks, Hebbian Plasticity Theory, Dual Process Theory, and machine awareness (MA). The final frontier may well involve a framework that is capable of machine curiosity, exploration, automatic self-direction and adaptation. The artificial intelligence (AI) system of the future will possess an innate curiosity and explore its own environment to gain knowledge, exhibiting a basic element of human cognition and awareness. The resulting MA system will possess inherent self-driven curiosity and related entropy in the decision space as it explores the environment in much the same manner as humans do.

Since the onset of AI research, MCS has been a crucial, yet missing and elusive component (Gunning, 2018). Common sense reasoning among machine systems has been unattainable. "Ultimate Machine Awareness" is a singularity that involves producing a human-made machine that is itself humanoid. Being self-aware involves knowing presence of self and consciously knowing one's own character, feelings, motives, and desires. Advances in AI will eventually lead to ultimate MA (consciousness). Providing AI systems human-like reasoning will enable human – humanoid symbiotic partnerships. Linking Hebbian Plasticity Theory (Brown, Zhao, & Leung, 2009; Widrow et al., 2019) and Capsule Neural Networks (Sabour, Frosst, & Hinton, 2017) provides a potential foundation for achieving MCS and MA by emulating the human mind. Similarly, Dual-process Theory (Evans & Stanovich, 2013) that is premised on System 1 autonomous processes and System 2 explicit processes is typically associated with consciousness that is emulating the human. A common framework for AI can be based in MA, constructed from third generation neural networks [Spiking Neural Network (SNN); Widrow et al., 2019] with the basic building blocks acting as Temporary Memory Neurons within the framework of the Leaky Integrate and Fire (LIF) Neuron Model learning from rules such as Hebbian Learning, capsule learning, or reinforcement learning. Capsule networks have a machine learning process that is similar to Hebbian Plasticity in neuroscience. SNNs demonstrate efficient learning with the integration of memory and information processing mechanisms. Like Hebbian learning, a higher-level capsule will receive its input from a lower-level capsule that shares affinity based on the largest scaler product of the activity vectors as a prediction coming from that lower-level capsule (Sabour, Frosst, & Hinton, 2017). A capsule can consist of a group of neurons whose outputs represent different properties of the same entity, e.g. a capsule layer built upon multiple capsule layers for perceptiveness. SNNs have the computational capability to continuously process spike trains to respond rapidly and accurately.

However, most SNNs do not retain the information of the spike train of a previous time step because the spike information is not retained once it causes a neuron to fire. One novel approach for machine consciousness could stem from a variation of the LIF neuron model that allows the spike train of the previous time step to be remembered. This modified LIF neuron approach has adjustable parameters that govern the 'remembering' time frame and the 'forgetting' time frame (Clark et al., 2019).

Capsule Neural Networks and Hebbian Plasticity Theory

We explore the connectivity between Hebbian Plasticity Theory and Capsule Neural Networks as a starting place for achieving MCS by emulating the human mind. Both of these learning frameworks are based on an adaptive neural computation in which manipulation of synaptic weights allow for strengthening or weakening of synaptic connections between neurons (W.D. Mitchell, personal communication, Feb 15, 2019). Hebbian Theory is biologically based and uses the concepts of long-term potentiation (LTP) and long-term depression (LTD) for the strengthening and weakening mechanisms, respectively (Brown, Zhao, & Leung, 2009). In Hebbian Theory, neurons connect when learning and self-organizing due to LTP and LTD. Consequently, by Hebbian principles: "units that fire together, wire together." (abstract, Widrow, et. al, 2019). Furthermore, neurons that fire out of sync weaken the link. Capsule Theory is based on artificial neural networks and uses a cluster of neurons whose activity vector represents the instantiation parameters of a specific type of entity such as an object or an object part.

Capsules use non-linearity procedures to convert the set of activation probabilities to estimate the membership of a capsule to a post-capsule (Sabour, Frosst, & Hinton, 2017). Each capsule will learn and store discriminating biases potentially analogous to working memory. It iteratively adjusts the means, variances, and activation probabilities of the capsules and resulting outputs in potential similarity to Dual-process Theory (System 2 working memory).

Dual Process Theory

Dual-process Theory is premised on the idea that human behavior and decision-making involves System 1 autonomous processes that produce default reflexive responses. These reflexive responses involve an implicit process unless interceded upon by a distinctive higher order reasoning processes. System 2 involves an explicit process and burdens working memory which is typically associated with consciousness. Studies have shown that System 1 or 2 decisions are influenced by neurocognitive capabilities, knowledge, and working memory (Harbour & Christensen, 2015).

Qualia Exploitation of Sensory Technology (QuEST) is an innovative approach to improve human-machine team decision quality over a wide range of stimuli (handling unexpected queries) by providing computer-based decision aids (software and hardware) that are engineered to provide both intuitive reasoning and "conscious" context sensitive thinking (Rogers, 2019). QuEST provides a mathematical framework to understand what can be known about situations to facilitate prediction of the future state to make a particular decision. In so doing, QuEST is additionally utilizing this emerging theory Dual-process or Dual-system theory (Evans & Stanovich, 2013). It is premised on the idea that human behavior and decision-making involves autonomous processes (Type 1) that produce default reflexive responses involving an implicit process unless interceded by distinctive higher order reasoning processes (Type 2). Type 2, on the other hand, involves an explicit process and burdens working memory. Type 2 is typically associated with: controlled, conscious, and complex. The Harbour and Christensen (2015) study compared Type 1 and Type 2 decisions made by pilots in actual flight, and assessed the impact of these decision types on cognitive workload and situation awareness (SA) under the enhanced-Theoretical Model of Situtation Awarness (TMSA) (Harbour & Christensen, 2015).

The query (Harbour & Christensen, 2015; Rogers, 2019) as the act of a stimulus being provided to an agent has characteristics that completely capture the salient axes (keep in mind what is salient in a stimulus is agent-centric) of the stimuli. Some of those axes are captured by an agent in its conversion of that stimuli into data (agent-centric internal representation of the stimuli). The term query is used instead of stimuli to capture the idea that a given agent must injest the stimuli and appropriately respond (thus an action). That response may be to just update its representation or not, or may be to take an action through an agent's effectors.

The unexpected query (UQ) (Harbour & Christiansen, 2015; Rogers, 2019) is an unexpected stimulus being provided to the agent and with the uncontrolled nature of in-flight events, pilots encounter unexpected queries and have to engage in both types of processing on any given flight. The enhanced-TMSA predicted that pilots with stronger perceptual and attentive capabilities needed to engage the arduous Type 2 system less, thus preserving spare capacity for maintaining SA. During 24 flights, there were UQ encountered by the pilot as well as expected queries (EQ) based on mission events and environmental stimuli. Results indicated that differences in workload and SA assessed both subjectively and through neurocognitive means existed. As UQ are encountered cognitive workload increases and SA decreases. During UQ working memory can become burdened leading to deficits in SA, which can be moderated by individual differences in perceptual and cognitive ability. Moreover, results from the research accomplished by Harbour and Christensen (2015) support Dual-process theory and assists in the development of the Theory of Consciousness (Fig. 1).

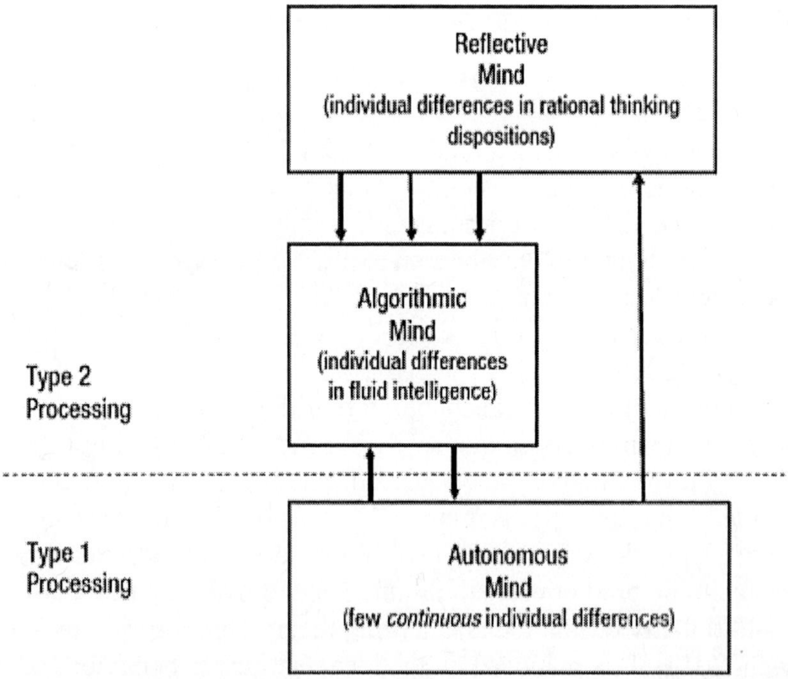

Figure 1. Conceptual Diagram of the Dual-Process Theory (Source: Evans & Stanovich, 2013).

Machine Awareness

The final frontier of MA will involve a framework that is capable of curiosity, exploration, automatic self-directed real-time learning and adaptation, learning with little data, learning with no labels, and learning rapidly. This framework will provide the structure, theory, and method for artificial reasoning, intuition, and human-like cognition. The artificially intelligent system proposed will possess an innate curiosity and explore its own environment to gain knowledge, exhibiting a basic element of human cognition and awareness.

The concept and model will demonstrate machine awareness and will be the building block for artificial general intelligence. Artificial general intelligence (AGI) is the ability for a machine to possess curiosity, investigate, think, reason, learn and be as dynamic, cognitively flexible and skillful as a human. It has been theorized that the ultimate MA singularity event is the moment true AGI will be achieved. This paradigm has brought about what is being called the Third Wave of AI. The Third Wave of AI boasts of the ability to develop machine cognition in the sense that it will be able to have general learning ability, self-directed learning, learn on the order (speed and accuracy) of the human biological brain, know when it needs to learn, and possess abstract reasoning, self-awareness, and cognition. Likely, combining the Law of Conservation of Energy and 3rd Generation Machine Learning, e.g. a SNN, may be a solution to this paradigm. One method is to use computational models of spiking neurons and synapses implementing the biologically plausible Leaky Integrate-and-Fire (LIF) model (Dayan, Abbott, & Abbott, 2001). This simulates the dynamics of a spiking neuron that is driven by the input spikes through plastic synapses. The LIF neuron processes the input spikes modulated by the inter-connecting synaptic weights, leading to a change in its membrane potential. The Spike Response Model zero order (SRM0), equation below, is a generalized LIF model where the model parameters are temporally based from the last output spike (Gerstner & Kistler, 2002). The membrane potential of the LIF model is given by:

$$\boldsymbol{u_i(t)} = (\boldsymbol{t} - \hat{\boldsymbol{t}}_i) + \Sigma_j \boldsymbol{w_{ij}} \Sigma_f \boldsymbol{\epsilon_i(t - t_j^{(f)})} + \boldsymbol{u_{rest}}\text{, (Gerstner \& Kistler, 2002)}$$

where $u_i(t)$ is the membrane potential of neuron i at time t, and $\eta(t-\hat{t}_i)$ is the model 'form' of the spike at some time t after the last spike of neuron i ($\hat{t}_i$), $\Sigma_j w_{ij}$ is the synaptic efficacy (the sum of the synaptic weights of the presynaptic neurons j exciting the postsynaptic neuron i), $\Sigma_f \boldsymbol{\epsilon_{ij}(t-t_j^{(f)})}$ is the sum of the postsynaptic potential (ϵ_{ij}) based on the current time and its relation to the presynaptic spikes of presynaptic neurons j at time f, and the membrane resting potential of neuron i (u_{rest}).

Recent neuroscience views curiosity as being internally motivated and absolutely intrinsic (Sharpee, Calhoun, & Chalasani, 2014; Loewenstein, 1994). In neuroscience, according to Incongruity Theory, when the human is presented with something that it has not seen before and therefore does not understand, human curiosity becomes stimulated, excited, and motivated (Kidd & Hayden, 2015; Loewenstein, 1994). Typically, the environment is viewed by the human as being predictable and orderly; this paper refers to that as being in a Low Entropy state: - H. Incongruity Theory states that when this order is challenged and the environment is poorly understood and not perceived to be predictable, something like curiosity is aroused (Friston & Buzsáki, 2016); this paper considers that as being in a High Entropy state: + H. According to Friston (2018), a want to reduce it occurs, (curiosity is aroused then exploration is stimulated according to Harbour, 2019), when a concept referred to as variational Free Energy (ΔF) is high.

To date, these theories and concepts have not been demonstrated in a machine; however, they have shown support in studies involving humans. The relationship of FE for Entropy (H) is:

$$(\boldsymbol{q}) = \langle(\boldsymbol{x})\rangle\boldsymbol{q} - \boldsymbol{H}[\boldsymbol{q}(\boldsymbol{x})],$$ (Friston & Buzsáki, 2016)

Where q is the probability density of belief and $\langle\rangle$ is the expectation of the total energy (E) of the environment with respect to q and H is the entropy of the density of belief of the environment. Entropy (H) and Free Energy (FE) can be viewed as the absolute difference between what is known and what is not known, or stated in another way, what is believed to be known and what is actually known. When (FE) is high then intrinsically curiosity is stimulated, resulting in exploration and learning occurring until (FE) reaches a minimum and (H) also has reached a minimum. The novel artificially intelligent system will need to possess both inherent self-driven curiosity and related changes in Free Energy as it explores the environment in much the same manner as humans do, thus demonstrating machine awareness.

Conclusion

It will take a team, bringing together multiple disciplines: electrical and computer engineering, neuroscience, neuroergonomics, cognitive engineering, psychology, physiology, biology, physics, philosophy, and mathematics in order to solve this arduous and nearly insurmountable quest of artificial general intelligence. Currently, the exact path toward MCS is not known, however, Capsule Neural Networks, Hebbian Plasticity Theory, Dual Process Theory, and MA, have one major characteristic in common, and that is they all use a biomimetic approach to some degree. As has been discussed it may require a blend of all four to achieve MCS. The one standout is MA, that additionally utilizes a psychophysiologically-inspired, -plausible, and topological approach which is showing to be centrally vital to solving the singularity to AGI. Consequently, while pursuing MA the researcher may be able to achieve MCS. Nevertheless, the researcher is encouraged to consider all of these pathways that embody a potential way to achieving MCS.

At some point ultimate MA and likely MCS will include the machine possessing the attribute of consciousness, which is a result of deliberation with a representation that is distinct from the sensory data representation and is situated, simulated, and structurally coherent about the past, present, and future. Consciousness can be cognitively decoupled and contain a cohesive narrative to represent reality that is easy to understand, to reason about, make decisions on, and take appropriate actions as a result of deliberations over and blended with information based on agent experiences (Harbour, Rogers, Christensen, & Szathmary, 2015, 2019). According to Rogers (2019) the ultimate goal of a theory of consciousness is a straightforward set of fundamental laws, analogous to the fundamental laws of physics.

References

Brown, T. H., Zhao, Y., & Leung, V. (2009). Hebbian plasticity.

Clark, J., Mitchell, W., Vemuru, K., & Harbour, S., (2019), *in print*. Temporary memory neuron for the leaky integrate and fire neuron model. ISAP 2019, Dayton, Ohio

Dayan, P., Abbott, L. F., & Abbott, L. (2001). Theoretical neuroscience: computational and mathematical modeling of neural systems.

Gerstner, W., & Kistler, W. M. (2002). *Spiking neuron models: Single neurons, populations, plasticity*. Cambridge university press.

Gunning, D. (2018). Machine Common Sense Concept Paper. *arXiv preprint arXiv:1810.07528*.

Evans, J. S. B., & Stanovich, K. E. (2013). Dual-process theories of higher cognition: Advancing the debate. *Perspectives on psychological science*, *8*(3), 223-241.

Friston, K., & Buzsáki, G. (2016). The functional anatomy of time: what and when in the brain. *Trends in cognitive sciences*, *20*(7), 500-511.

Friston K. (2018). Am I Self-Conscious? (Or Does Self-Organization Entail Self-Consciousness?). *Frontiers in psychology*, *9*, 579. doi:10.3389/fpsyg.2018.00579

Harbour, S. D., & Christensen, J. C. (2015, May). A neuroergonomic quasi-experiment: Predictors of situation awareness. In Display Technologies and Applications for Defense, Security, and Avionics IX; and Head-and Helmet-Mounted Displays XX (Vol. 9470, p. 94700G). SPIE.

Harbour, S.D., Rogers, S.K., Christensen, J.C., & Szathmary, K.J. (2015, 2019). Theory: Solutions toward autonomy and the connection to situation awareness. Presentation at the 4th Annual Ohio UAS Conference. Convention Center, Dayton, Ohio. USAF.

Kidd, C., & Hayden, B. Y. (2015). The Psychology and Neuroscience of Curiosity. *Neuron*, *88*(3), 449-60.

Loewenstein G. The Psychology of Curiosity: A Review and Reinterpretation. *Psychological Bulletin*. 1994; 116(1):75–98.

Rogers, S., (2019). QuEST – Cognitive Exoskeleton. Kabrisky Memorial Lecture 2019. For U.S. Government and DOD Contractors ONLY. USAF. WPAFB, Ohio

Sabour, S., Frosst, N., & Hinton, G.E. (2017) "Dynamic routing between capsules." *Advances in neural information processing systems.*

Sharpee, T. O., Calhoun, A. J., & Chalasani, S. H. (2014). Information theory of adaptation in neurons, behavior, and mood. *Current opinion in neurobiology*, 25, 47-53.

Widrow, B., Kim, Y., Park, D., & Perin, J. K. (2019). Nature's Learning Rule: The Hebbian-LMS Algorithm. In *Artificial Intelligence in the Age of Neural Networks and Brain Computing* (pp. 1-30). Academic Press.

TEMPORARY MEMORY NEURON FOR THE LEAKY INTEGRATE AND FIRE NEURON MODEL

Dr. Jeffery D. Clark
Mr. William D. Mitchell
Dr. Krishnamurthy V. Vemuru
Dr. Steven D. Harbour
Riverside Research
Beavercreek, Ohio 45431

Low-level terrain-following systems require the ability to rapidly and accurately respond to the environment to prevent inadvertent actions. Catastrophic and fatal results could occur if missed cues or latency issues in data processing are encountered. Spiking neural networks (SNNs) have the computational ability to continuously process spike trains from rapid sensory input. However, most models of SNNs do not retain information from the spike train of a previous time step because the membrane potential is rapidly reset to a resting potential after activation. A novel approach is presented, allowing the spike train of a previous time step to be 'remembered.' Results are presented showing rapid onset of a membrane potential that exceeds the threshold and spikes in the presence of the same continuous spike train without the latency of increasing the membrane potential from its resting state.

Piloted operations for both man and unmanned aircraft grow in complexity, which can overtax a pilot's cognitive ability. One area where this is especially true is with low altitude operations. This type of operation is often necessary for mission success, thereby giving the pilot a significant advantage by avoiding adversarial engagement. The number of visual cues a pilot must quickly process in order to avoid controlled flight into terrain (CFIT) coupled with parallel communications and weapons tasks can quickly saturate cognitive ability, severely degrading a pilot's situation awareness (SA). This degradation in SA can negatively affect mission performance and result in unnecessary loss of life and assets.

Terrain-following systems aid pilots with low altitude navigation. However, the latency with which these systems need to operate must be as small as possible allowing for high accuracy in terrain scanning. These systems should also be of low Size, Weight, and Power (SWaP) for easy integration as a subsystem to an overall SWaP-constrained aircraft system. Third generation neural networks, also known as SNNs, are modeled after the neurons of the brain. Much research has been done on the modeling of neuron function of the brain. Some models are considered more biologically plausible, while others are considered more biologically inspired, such as the Leaky Integrate-and-Fire (LIF) model, as they aim to borrow only certain advantageous neuron cell characteristics for easy practical implementation Hazan et al. (2018).

This paper proposes a novel neuron model, termed a temporary memory neuron (TMN), based on the LIF model with a tunable "memory" parameter which is correlated to a limited presynaptic spike time frame. This neuron, with its temporary memory capability, integrates into a SNN architecture and reduces the latency required to identify a previously learned spike train. This reduction in latency allows the terrain-following system to more rapidly provide the pilot with critical time-of-flight information necessary for mission success.

I. RELATED WORK

Different techniques have been explored that develop dynamic spike firing rates. The work by Triesch (2005) develops a mathematical model of a neuron that provides plasticity that could be intrinsic to the neuron. This model provides for an adjustment to the firing rate to approximate an exponential regime. While this method is novel, and provides a method for plasticity within the neuron (not just the synapse), it does not provide for a memory or temporary memory capability based on a sequential spike train that could decrease latency in the system. Other similar approaches have been explored as in Stemmler et al. (1999). In order to reduce latency, some have approached this concept by optimizing the parameters of the neuron model, to include the thresholds, based on the specific task to be solved. Diehl et al. (2015) developed an approach to optimize the accuracy or latency by adjusting the threshold levels until the appropriate values were obtained. It was found that to optimize one, the other was sacrificed on a spike-by-spike basis. Other intrinsic plasticity based neuron parameters, e.g. thresholds, have been developed as part of the neural models as in Pozo et al. (2010). However, these all have thresholds that are increased to prevent a specific neuron from overshadowing the response pattern. The development of the dynamic thresholds of the previous works do not produce a offset membrane potential allowing for an associated temporary memory function based on the previous input spikes. In contrast, the novel neuron model of this paper produces a temporary memory function based on an offset membrane potential.

II. BACKGROUND

The spikes that are received by the postsynaptic neuron affect the membrane potential of the soma. As the membrane potential is increased, it will eventually surpass a threshold that causes the soma to send a spike down its axon to be delivered to other neurons. Once the spike occurs the membrane potential enters a refractory period during which it remains unaffected by any presynaptic spikes. After the refractory period, the membrane potential is reset to its original resting value after which new presynaptic spikes will again influence the membrane potential Gerstner et al. (2002).

The synapse resistance, or lack thereof, governs the amount that the presynaptic action potential effects the postsynaptic membrane potential. This process is translated into the artificial neural network (ANN) as the input weights of the ANN. The ability to adjust these weights is considered the synaptic plasticity. Less common, and less accounted for, is the intrinsic plasticity of a neuron that common neuron models do not take into account, i.e. the threshold values Hao et al. (2018).

A. Leakey Integrate-and –Fire Model

The LIF neuron model is one of the most common neuron models used in spiking neural networks. This model is represented with a capacitor in parallel with a resistor and driven by a current, where a time constant $\tau_m = RC$ is introduced. The leaky integrate and fire model expresses spikes based on a firing time $t^{(f)}$. The LIF neuron builds up its membrane potential between spikes and will eventually reach the threshold if enough input spikes are received that overpower the leak function of this neuron, a refractory period is observed immediately following the induced postsynaptic spike. The refractory period is a finite amount of time in which incoming spikes are not allowed to have any influence on the membrane potential.

B. Spike Response Model

The Spike Response model (SRM) is a generalization of the leaky integrate and fire model, where the generalization is in the form of dependence on the last time a spike occurred Gerstner et al. (2002). The SRM evaluates the membrane potential by integrating over the past for a specific current time t. The state of a neuron is determined with respect to the membrane potential. This model defines a resting state u_{rest}, a spike response, as presynaptic spikes are encountered, ϵ, the form of the action potential η, and synaptic weights ω_{ij} Gerstner et al., (2002). The zeroth order of the SRM (SRM_0), is a simplistic "zero order" version of the SRM. Therefore, independent of the presynaptic neuron and the last firing time of the postsynaptic neuron $\hat{t}_i$, the postsynaptic potential is developed with each presynaptic spike weighted by the synaptic efficacy ω_{ij}. This model is mathematically represented by Gerstner et al., (2002):

$$u_i(t) = \eta(t - t_i) + \sum_j \omega_{ij} \sum_{t_j^{(f)}} \epsilon_{ij}\left(t - t_j^{(f)}\right) + u_{rest}, \tag{1}$$

where, $u_i(t)$ is the membrane potential, $\eta(t - t_i)$ is the action potential based on the time from the previous postsynaptic spike, ω_{ij} is the synaptic efficacy, $\epsilon_{ij}\left(t - t_j^{(f)}\right)$ is the postsynaptic potential based on the presynaptic spikes, and u_{rest} is the resting membrane potential.

III. TEMPORARY MEMORY NEURON MODEL

The desire to develop a neuron-level capability to decrease latency for previously discovered spike trains is paramount to real-time detection and classification missions. The intrinsic speed of the cognitive neuromorphic architecture developed for SNNs will determine the baseline processing speed of the SNNs. However, this baseline still produces latency when a neuron requires multiple spikes (multiple number of events i.e. computational cycles) to reach a threshold and emit a spike. Assuming that a specific spike train indicates an event of interest, most neuron models require multiple spike trains between postsynaptic spikes. This representation of a neuron model introduces latency between postsynaptic spikes.

While the novel approach presented in this paper is not suggesting that it is representative of a biological neuron, it is building on the capabilities of the biological neuron models to provide an advanced capability. This concept utilizes computational models of spiking neurons and synapses implementing the biologically inspired LIF model and more specifically the SRM model; adding a tunable memory component as in Dayan et al.(2001), and Gerstner et al., (2002). The LIF neuron integrates the input spikes modulated by the inter-connecting synaptic weights, leading to a change in its membrane potential. The SRM_0, equation 1, is a generalized LIF model where the model parameters are based on the last output spike Gerstner et al.(2002). The temporal dynamics of the TMN neuron are formulated in equations 2 and 3,

$$u_i(t) = \eta(t - \hat{t}_i) + \sum_j \omega_{ij} \sum_{t_j^{(f)}} \epsilon_{ij}\left(t - t_j^{(f)}\right) + u_{rest} + \zeta_i(t), \qquad (2)$$

where $u_i(t)$ is the membrane potential of neuron i at time t, and $\eta(t - \hat{t}_i)$ is the model 'form' of the spike at some time t after the last spike of neuron i, $(\hat{t}_i)$, $\sum_j \omega_{ij}$ is the synaptic efficacy (the sum of the synaptic weights of the presynaptic neurons j exciting the postsynaptic neuron i), $\sum_{t_j^{(f)}} \epsilon_{ij}\left(t - t_j^{(f)}\right)$ is the sum of the postsynaptic potential ϵ_{ij} based on the current time and its relation to the presynaptic spikes of presynaptic neurons j at time f, and the membrane resting potential of neuron i, u_{rest}. The TMN neuron builds on the basic equation of the SRM_0 model with the addition of $\zeta_i(t)$, the neuron memory of neuron i defined at time t as shown in equation 2 and developed in equation 3. The memory addition to the SRM_0 model is:

$$\zeta_i(t) = -\left[-\frac{1}{1 + e^{-[t-(t_i+\tau_d)+\tau_{off}]}} + 1\right]\left(-\left[u_{rest} - (v - v_{offset})\right]\right), \qquad (3)$$

where τ_d and τ_{off} are time constants that set the point in time after neuron i spikes allowing the membrane potential will be allowed to decrease back to its resting state, v is the neuron threshold and v_{offset} is the desired level below the threshold value where the membrane potential will be temporarily set. The TMN allows for a neuron to spike more readily upon receiving a spike once it has produced a postsynaptic spike. This concept, built into an ensemble of neurons, will provide a quicker response to an already known spike train, preventing the latency between postsynaptic spikes.

IV. RESULTS

The TMN neuron simulated with a constant current input at set intervals. Various parameters were changed in the TMN neuron to illustrate the capability for this neuron to reduce latency upon receiving additional spikes. This simulation provides constant current input between 1000-1200ms (200ms period), 2000-2300ms (300ms period), 2800-3200ms (400ms period), and 4200-4250ms (50ms period) to observe the response of the TMN for the different time periods. For the simulation illustrated in this work, the membrane rest potential is set to -70mV, the membrane reset potential is set to -75mV, the membrane threshold is set to -55mV and the refractory constant is set to 10ms. The temporary memory part of this neuron has the v_{offset} set to 7mV the τ_d (governs the time until the membrane potential is allowed to decay to its rest state assuming no other spikes have occurred) is set to 20mS and the τ_{off} (governs the rate of the decay back to the rest potential) is set to 200ms.

Figure 1(a) illustrates the TMN with an input current set to a magnitude of 10A. It is noted that with the varying time periods of, 200ms, 300ms, 400ms, 50ms, respectively, the number of spikes produced within the time periods would be fewer without the temporary memory. The initial spikes of each time period required significant more ramp up time as compared to subsequent spikes. It is also noted form figure 1(a) that the temporary memory is maintained for a specific time period, afterward decaying back to its resting state in the absence of a spike (e.g. the 100ms and 200ms time periods).

In contrast, figure 1(b) shows the effect of the temporary memory where an additional current was inserted at 139ms to 145ms allowing a post synaptic spike to occur where otherwise no spike would have occurred. This action reset the temporary memory values allowing for a new spike train to be recognized within the specified time frame.

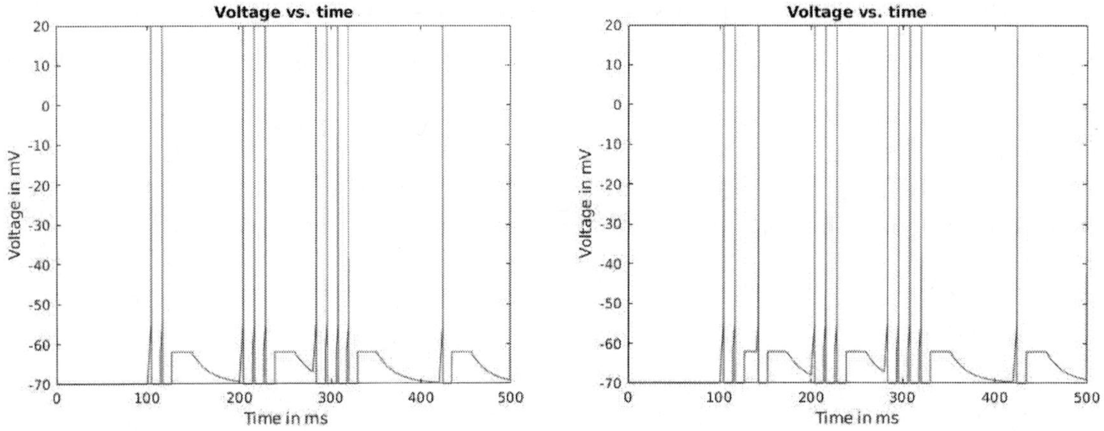

Figure 1. (a) Spike Response of a TMN neuron with 4 time periods (200ms, 300ms, 400ms, and 50ms) illustrating the temporary memory capability to produce subsequent spikes in a set time period, (b) illustrating the temporary memory capability to produce subsequent spikes in a set time period with an additional 6ms time period for starting at 139ms.

V. CONCLUSIONS

A novel neuron model based on the LIF neuron with a tunable temporary memory parameter is proposed. This tunable parameter allows for the temporary 'remembrance' of specific spike train inputs. After a prescribed time period previously learned spike trains are 'forgotten' allowing for future spike trains to be learned. This dynamic capability can be integrated into a third generation ANN system allowing for reduced latency in the recognition of specific inputs. An application where this concept could be of benefit is that of low altitude, high speed, terrain scanning systems. These systems require both low latency and low SWaP which third generation artificial neural networks show promise to deliver. Enabling a reliable terrain scanning navigation aid will help reduce cognitive overload and improve SA for pilots. Future work with this concept is underway where the optimal intrinsic neuron plasticity parameters such as sub-threshold voltage and temporary memory duration will be explored, modeling human sensory memory and working memory.

VI. REFERENCES

Dayan, P., Abbott, L. F., & Abbott, L. (2001). Theoretical neuroscience: computational and mathematical modeling of neural systems.

Diehl, P. U., Neil, D., Binas, J., Cook, M., Liu, S. C., & Pfeiffer, M. (2015, July). Fast-classifying, high-accuracy spiking deep networks through weight and threshold balancing. In *2015 International Joint Conference on Neural Networks (IJCNN)* (pp. 1-8). IEEE.

Gerstner, W., & Kistler, W. M. (2002). *Spiking neuron models: Single neurons, populations, plasticity*. Cambridge university press.

Hao, Y., Huang, X., Dong, M., & Xu, B. (2018). A Biologically Plausible Supervised Learning Method for Spiking Neural Networks Using the Symmetric STDP Rule. *arXiv preprint arXiv:1812.06574*.

Hazan, H., Saunders, D. J., Khan, H., Sanghavi, D. T., Siegelmann, H. T., & Kozma, R. (2018). BindsNET: A machine learning-oriented spiking neural networks library in Python. *Frontiers in neuroinformatics*, *12*, 89.

Pozo, K., & Goda, Y. (2010). Unraveling mechanisms of homeostatic synaptic plasticity. *Neuron*, *66*(3), 337-351.

Stemmler, M., & Koch, C. (1999). How voltage-dependent conductances can adapt to maximize the information encoded by neuronal firing rate. *Nature neuroscience*, *2*(6), 521.

Triesch, J. (2005, September). A gradient rule for the plasticity of a neuron's intrinsic excitability. In *International Conference on Artificial Neural Networks* (pp. 65-70). Springer, Berlin, Heidelberg.

REINFORCEMENT LEARNING IN AVIATION, EITHER UNMANNED OR MANNED, WITH AN INJECTION OF AI

Dr. Krishnamurthy V. Vemuru
Riverside Research
2900 Crystal Dr., Arlington, Virginia 22202
Dr. Steven D. Harbour and Dr. Jeffrey D. Clark
Riverside Research,
2640 Hibiscus Way, Beavercreek, Ohio 45431

We propose a novel theme of aviation with the injection of AI in the form of a reinforcement learning (RL) agent that learns flying skills by observing the pilot's psychological reaction and flight path in a simulator. The pilot and the RL agent learn flying skills simultaneously, forming a symbiotic relationship. The episodes for training the reinforcement learning agent can be simulated by a pilot flying in a simulator, or unmanned using a game on a computer. In a typical episode, the reinforcement learning agent provides a sequence of actions for the pilot to follow. These instructions produce one of the two types of results, either success or failure. The agent observes the psychological reaction of the pilot as well as the flight environment and receives a positive or negative reward. The trained RL agent represents a novel form of AI that assists the pilot for various phases of flight.

Human error is causal to most aircraft accidents; consequently, technologies have emerged to issue alerts when the aircraft's travel trajectory is irregular (Chang et al., (2018)). For example, detecting the aircraft's behavior is one approach to measure the safety of the aircraft. Continuous monitoring and analysis of flight operations is another approach to detect hazardous behavior from a pre-defined list. Li et al., (2016) have reported data mining methods such as cluster analysis of digital flight data using Gaussian Mixture Model (GMM) that are employed by safety analysts to identify unusual data patterns or anomaly detection and latent risks from daily operations. With the advent of Artificial Intelligence (AI), human-autonomy teaming can be an efficient way to minimize human error and further increase aviation safety records. Zhao et al. (2018) have used Reinforcement Learning (RL) as an adaptive online learning model to identify common patterns in flight data and to update the clusters for GMM using recursive expectation-maximization algorithm. The resurgence of interest in AI has attracted applications in aviation systems, in particular, air-traffic management (ATM), air traffic flow management (ATFM) and unmanned aerial systems traffic management (UTM). Kistan et al., (2018) have explored a cognitive human-machine interface (HMI), configured via machine learning, and examined the requirements. They postulated that increased automation and autonomy through AI will lead to certification requirements and discussed how ground-based ATM systems can be accommodated into the existing certification framework for aviation systems. The recent developments in AI open up possibilities in autonomous aviation for introducing a high level of safety by replacing pilot's actions with robotic functions and further research on how AI can be incorporated into autonomous aviation is highly desirable. Our motivation is to show that AI frameworks can be developed by incorporating RL into pilot training simulators.

In this work, we propose a novel theme of aviation with the injection of AI in the form of an RL agent that learns flying skills by observing the pilot's psychological reaction and flight path in a simulator. A unique feature of this AI framework is that the pilot and the RL agent learn flying skills simultaneously, forming a symbiotic relationship. The proposed approach is somewhat similar to how two non-experts comprising of a trainee pilot and an RL agent may learn to play a game by using their joint score as a metric. It is expected that the RL agent will learn a value system i.e., which combinations of states and actions are more rewarding and which ones are not. As the number of game episodes increases, the agent will balance between the exploration of new state-action pairs and the exploitation of known high rewarding state-action pairs until an optimal solution is achieved. RL algorithms usually slow in learning and typically require longer training times, which increase with the size of the state space.

In this context, identifying suitable methods for detecting pilot behavior is the key to developing an AI based on reinforcement learning. Pilot modeling technologies have played a crucial role in manned aviation and control models of human pilot behavior have been developed. Control models are used to analyze the characteristics of the pilot-aircraft system for guidance in the flight control system. Anthropomorphic models of a human operator, which covers the central nervous system, neuromuscular system, visual system and the vestibular system, can represent a pilot's behavior. Recently, Xu et al., (2017) have reviewed control models of human pilot behavior. These models reflect the dynamics of a human sensory and control effectors. AI in the form of computer vision can be coupled with these models to detect non-linear characteristics of human pilot behavior for training the RL agent.

Reinforcement Learning

Reinforcement learning is a type of semi-supervised learning inspired by the way animals learn. It relies on the definition of state space, actions for transitions between states and an associated reward structure in a Markov decision process. In a typical application of RL, an agent makes multiple attempts at a goal and learns from its failures and successes based on a reward structure that has both negative and positive types of rewards. In some of the simple forms of RL, an agent learns the optimal policy by evaluating the value functions $V(s)$ or by $Q(s,a)$ learning, where is s is the current state and a is the action taken in state (s), from episodes, which are attempts by the agent for reaching the goal (Watkins and Dayan (1992). In a game setting, the episodes can be either successful or unsuccessful attempts of playing the game. The game-like situations are realized in many daily life examples, including attempts of a pilot in flying an aircraft.

Flight Simulation Game Framework

Flight simulators are used in pilot training and research on the relationship between emotional intelligence and simulated flight performance to understand how emotional factors affect flight-training performance. Pour et al., (2018) have used a human-robot facial expression reciprocal interaction platform to study social interaction abilities of children with autism.

In this framework, a computer vision system captures the psychological reaction of a pilot undergoing training in a simulator for finding out the result of a pilot's reaction on a flight path following the pilot's operational action. To train the RL agent, we design a flight simulator framework, which is like a game for the pilot to play using his actions, a and express his/her gesture, which is representative of the result of his/her actions in the simulator. We represent the gesture, *g* as a two state variable; with values '*happy*' (☺) or '*unhappy*' (☹). The state space of the flight simulator, s consists of five variables: altitude, A, speed, S, heading, H, turn, U and roll, R. Table 1 lists the range of these five action variables.

Table 1.
The five variables that define the state s and their ranges.

State variable	Minimum	Maximum
Altitude, A	0 ft	35,0000 ft
Speed, S	0 mph	550 mph
Heading, H	0°	360°
Turn, U	0°	360°
Roll, R	0°	360°

Flight Path Analysis

A reliable flight path analysis can be obtained by real-time computation of the gradients of state space variables, namely altitude gradient dA/dt, speed gradient dS/dt, heading gradient dH/dt, turn gradient dU/dt and roll gradient dR/dt. A rule-based model compares the gradients with a predefined range to determine, if the maneuver is safe or risky and calculate a dynamic reward. Table 2 shows the gradients and the initial guess values of their ranges. The minimum and maximum of the range can be set as tunable parameters to improve the values iteratively.

Table 2.
Ranges of gradients of the state variables that define the safe operational zone. These ranges will be used in a rule model to dynamically determine the reward for the flight maneuvers.

Gradient of State Variable	Minimum,	Maximum
Altitude gradient, dA/dt	0 ft/s	1,000 ft/s
Speed gradient, dS/dt	0 mph/s	20 mph/s
Heading gradient, dH/dt	0°/s	3°/s
Turn gradient, dU/dt	0°/s	3°/s
Roll gradient, dR/dt	0°/s	2°/s

Pilot's Gesture Assignment

An example computer vision system can consist of a digital video camera, a neural processing unit such as Myriad 2, and a single board computer can be integrated for reading pilot's gesture. The computer vision system can be trained using a face detection machine learning algorithm for real-time monitoring of the 'happy' or 'unhappy' facial expression of the pilot. The agents who play the game of flying the plane in the simulator are particularly advised to show a happy gesture (☺) while their actions result in a safe operation and to show a unhappy gesture (☹) when their actions result in a risky or unsafe or catastrophic operation. The computer vision system can be as simple as a Google AIY kit, which operates with a Tensorflow machine learning model to detect a smile.

Reinforcement Learning Agent – Learning from Pilot's Actions

We first consider a human in the loop approach to develop an RL agent that can use artificial intelligence to determine a human pilot's gesture and calculate rewards. This type of RL agent is trained with the episodes that are generated when a pilot is flying an aircraft in a simulator i.e., on a computer. In a typical episode, the RL agent provides a sequence of actions for the pilot to follow. These instructions produce a result, which is either success or failure. The agent receives two types for rewards: one reward depends on the observation of the psychological reaction of the pilot and the other reward depends on the flight dynamics. The RL agent receives a first reward of +1 when the pilot's gesture is 'happy' or a reward of -1 when the pilot's gesture is 'unhappy.' The RL agent receives a second reward of +1 when the flight state variables and their gradients are in the safe range or a reward of -1 otherwise. The episodes can be used to train the RL agent with different reward structures to select the most suitable reward structure for *Q*-learning. The training process is repeated until convergence of the learning process. After training the agent using a sufficiently large number of episodes, the knowledge acquired by the RL agent is expected to represent a novel form of AI that directs the pilot with accurate instructions for various phases of flight. Fig. 1 shows a learning framework of a RL agent along with its interactions with the flight simulator and the computer vision system that detects the pilot's gesture for receiving rewards to update $Q(s,a)$ function and the policy $\pi(s,a)$.

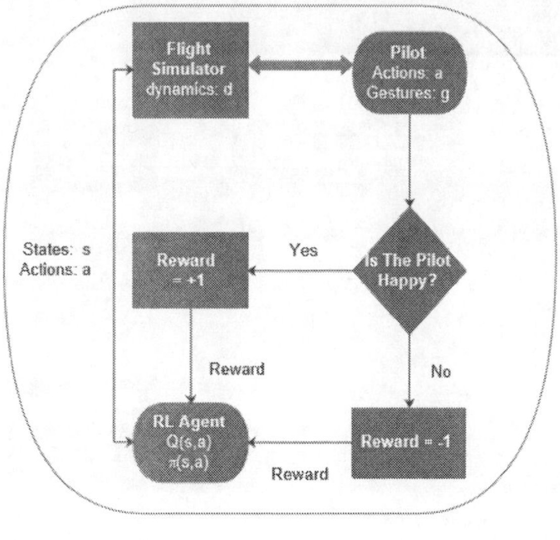

Figure 1.
A framework for the Reinforcement Learning (RL) Agent and its interactions with its environment consisting of the flight simulator and the pilot's gesture recognition system.

Flight Simulator Game Framework

Fig. 2 shows a game framework in a flight simulator for generating the states, actions, rewards, the Q function and the policy. The flight simulator game framework has an additional local reward and a long-term reward compared to the reward structure of the RL agent. The Game RL agent in the flight simulator game framework receives as additional reward of -1 for each instance of state variable's gradient falling outside the safe range. An optional long-term reward of +2 is also awarded to the Game RL agent when the total time taken to reach the destination is below a preset value. The RL agent will receive a reward of +1 when all of the state variable's derivatives are within the safe range. The choice of rewards is arbitrary and can evolve to a more realistic structure based on episodes. A game simulator module initiates the game by extracting actions using the current policy to simulate the flight dynamics. Then, two other modules evaluate the flight dynamics and the gesture of the pilot to identify the rewards. Then, the Q(s,a) function is calculated and updates for each state-action pair and the associated reward. The policy π(s,a) is then recalculated from the Q(s,a) values and updated.

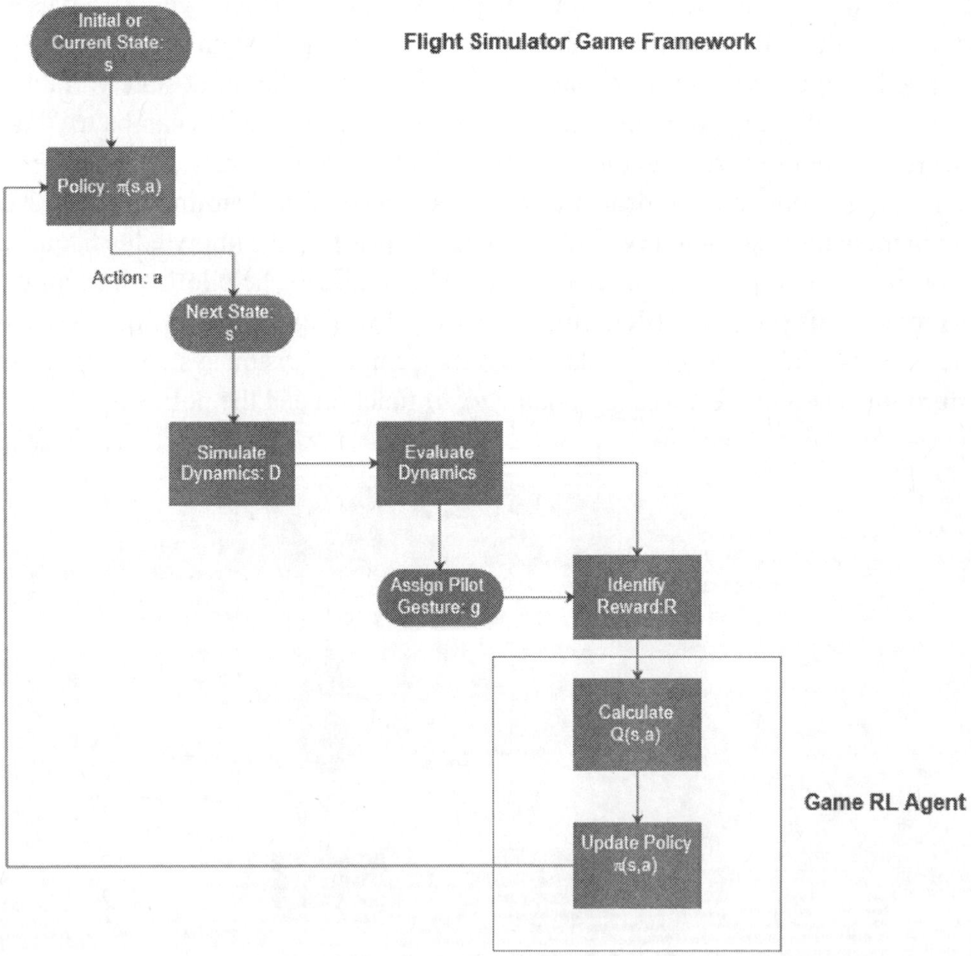

Figure 2.
The framework for flight simulator as a game for obtaining the states, actions, rewards, Q(s,a) function and policy.

Summary

In summary, we have proposed a novel framework of autonomous aviation with the application of artificial intelligence in the form of a reinforcement learning agent which learns flying skills by observing a pilot's psychological reaction and flight path in a flight simulator. The framework consists of a gaming module that works as a flight simulator, a computer vision system that detected pilot's gesture and a flight dynamics analyzer for verifying the safety limits of the state space variables during a simulated flight and a module to calculate the Q-function and the learned policy. With sufficient training within the proposed framework, the RL agent is expected to learn to fly the aircraft as well as to guide the pilot for safe aviation. It would be interesting if present work can attract the attention of game programmers and training tools developers in the AI domain for exploring prototypes based on the proposed frameworks. Finally, an alternate approach to RL is Inverse Reinforcement Learning (IRL) from expert pilot's operations and behavior. This method will require a significant amount of training data in the form of expert pilot's simulator data.

References

Chang, T. H., Hsu, C. S., Wang, C., and Yang, L. –K., (2008). On board measurement and warning module for measurement and irregular behavior, *IEEE Transactions on Intelligent Transportation Systems*, 9(3), 501-513.

Li, L. S., Hansman, R. J., Palacios, R., and Welsch, R., (2016). Anomaly detection via Gaussian mixture model for flight operation and safety monitoring, *Transportation Technologies, Part C: Emerging Technologies*, 64, 45-57.

Zhao, W. Z., He, F., Li, L. S., and Xiao, G., (2018). An adaptive online learning model for flight data cluster analysis, In *Proc. of 2018 IEEE/AIAA 37th Digital Avionics Systems Conference*, *IEEE-AIAA Avionics Systems Conference (*pp.1-7). London, England, UK.

Kistan, T., Gardi, A., and Sabatini, R., (2018). Machine learning and cognitive ergonomics in air traffic management: Recent developments and considerations for certification, *Aerospace*, 5(4), Article Number 103.

Xu, S. T., Tan, W. Q., Efremov, A. V., Sun, L. G., and Qu, X., (2017). Review of control models for human pilot behavior, *Annual Review in Control*, 44, 274-291.

Watkins, C. J. C. H. and Dayan, P., (1992). Q-learning, *Machine Learning*, 8(3-4), 279-292.

Pour, A. G., Taheri, A., Alemi, M., and Meghdari, A., (2018). Human-Robot facial expression reciprocal interaction platform: Case studies on children with autism, *International Journal of Social Robotics*, 10(2), 179-198.